WORLD SPATIAL METADATA STANDARDS:
Scientific and Technical Descriptions, and Full Descriptions with Crosstable

WORLD SPATIAL METADATA STANDARDS: Scientific and Technical Descriptions, and Full Descriptions with Crosstable

The ICA Commission of Spatial Data Standards

Editor

Prof. Harold Moellering
Commission Chair
Department of Geography, Ohio State University
Columbus, Ohio, U.S.A. 43210
FAX: +1 (614) 292-6213
E-mail: GEOHAL+@OSU.EDU

Associate Editor

Prof. Ir. Henri J.G.L. Aalders
OTB
Research Institute for Housing,
Urban and Mobility Studies
Delft University of Technology
Postbus 5030, 2600 GA Delft, The Netherlands
FAX: +31 (15) 278-2754
E-mail: H.J.G.L.Aalders@OTB.TUDelft.nl

Associate Editor

Mr Aaron Crane
V.P. Information Services, NAVTEQ
222 Merchandise Mart Plaza, Suite 900
Chicago, Illinois, U.S.A. 60654
FAX: +1 (312) 894-7050
E-mail: aaron.crane@navteq.com

ICA ACI

ELSEVIER

Published on behalf of the International Cartographic Association by Elsevier

Amsterdam – Boston – Heidelberg – London – New York – Oxford
Paris – San Diego – San Francisco – Singapore – Sydney – Tokyo

ELSEVIER B.V.
Radarweg 29
P.O. Box 211, 1000 AE
Amsterdam, The Netherlands

ELSEVIER Inc.
525 B Street
Suite 1900, San Diego
CA 92101-4495, USA

ELSEVIER Ltd.
The Boulevard
Langford Lane, Kidlington,
Oxford OX5 1GB, UK

ELSEVIER Ltd.
84 Theobalds Road
London WC1X 8RR
UK

First edition 2005

Library of Congress Cataloging in Publication Data
A catalog record is available from the Library of Congress.

British Library Cataloguing in Publication Data
A catalogue record is available from the British Library.

ISBN: 0-08-043949-7

⊗ The paper used in this publication meets the requirements of ANSI/NISO Z39.48-1992 (Permanence of Paper).
Printed in The Netherlands.

Dedication

This book is dedicated to my dearest wife Elizabeth who has endured the trials and tribulations of this ICA Metadata Standards book for longer than we have been married. As a writer extraordinaire, she has served as a reviewer of all of my writings in this book. Even so, it will still turn out to be a scientific book, and not a novel.

I also dedicate this book to my three loving grandchildren. Daniel, who is now eight years old, does not quite understand why his grandfather has been working on a book longer than he has been alive. Alexander, who is six, is still grasping to understand what spatial metadata is all about. And now little Claire, two years old, who will serve as more than a footnote to this book.

I also dedicate this book to all of the members of the ICA Spatial Data Standards Commission, who over a span of nine years have maintained their good humor, dedication and rich scientific insights to produce the best in content a work like this could offer. Henri J.G.L. Aalders, as Associate Editor has helped to bring together the Regional Summary authors and chapters, while Aaron Crane, and his assistant Chaz Ruhl, have spend many hours gathering and concisely editing the Part 4 chapter on the standards themselves. As an ICA scientific Commission, developing these scientific and technical metadata characteristics, field testing and reformulating them, sorting out the structure of the book, and then writing the requisite chapters is an exercise in true international scientific scholarship.

Harold Moellering,
Editor

Contents

Dedication v

Presidential Foreword xi

ISO/TC 211 Foreword xiii

GIS Foreword xv

Editors' Preface xvii

Part 1 – Introduction to Spatial Metadata Standards in the World 1

An Introduction to Metadata for Geographic Information 3
by Henri J.G.L. Aalders

Part 2 – Regional Summaries of Spatial Metadata Developments and Associated Activities 29
edited by Henri J.G.L. Aalders

European Efforts in the Field of Geographic Metadata and Related SDI Activities . 31
by Henri J.G.L. Aalders, François Salgé & Alexander I. Martynenko

North American Metadata Standards Developments in Canada and the United States of America 63
by Kian Fadaie, Julie Binder Maitra, Valerie E. Hume, Harold Moellering & Henri J.G.L. Aalders

Metadata Standards Development Activities in the Asia – Pacific Region 83
by Craig Macauley, *lead author, editor*, Du Dao-sheng, Kazuhiko Akeno & Tae-jung Min

Metadata Standards Development Activities in Central and South America and the Caribbean 103
by Tatiana Delgado-Fernández, Dora Inés Rey-Martinez & Martha Ivette Chaparro-Dominguez

Spatial Metadata in Africa and the Middle East 123
by Antony Cooper & Elizabeth J.O. Gavin

Global Spatial Metadata Activities in the ISO/TC211 Geographic Information Domain 141
by Olaf Østensen & David M. Danko

Part 3 – Scientific and Technical Characteristics for Assessing Metadata Standards for Geographic Datasets 167
edited by Harold Moellering & Aaron Crane

Introduction to the Scientific and Technical Characteristics for Assessing Metadata Standards for Geographic Datasets 169
by Harold Moellering & Aaron Crane

Scientific and Technical Characteristics of Metadata Standards for Geographical Datasets 179
edited by Harold Moellering & Aaron Crane

Part 4 – Scientific and Technical Assessments with Full Descriptions of the Spatial Metadata Standards 195
edited by Aaron Crane, Charles Ruhl & Harold Moellering

Introduction to the Scientific and Technical Assessments of Spatial Metadata Standards 197
by Harold Moellering

National Spatial Metadata Standards 199

Australia and New Zealand: ANZLIC Metadata Guidelines 201
by Craig Macauley

Canada: Directory Information Describing Digital Geo-referenced Data Sets; Information de répertoire décrivant les ensembles de données numériques à référence spatiale CAN/CGSB-171.3-95 219
by Kian Fadaie & Valerie E. Hume

China: 地理信息 元数据; Geographic Information – Metada GB/T 19710–2005 . . . 237
by Jiang Jing-tong & Liu Ruo-mei

Czech Republic: Standard ISVS pro strukturu a vymenny format metadat informacnich zdroju; Standard for Structure and Transfer Format of Metadata on Geo-data Sets 011/01.02 257
by Petr Rapant, Bronislava Horakova & Jan Ruzicka

Denmark: Infodatabase Om Geodata; National Danish GI Metadata Service 277
by Anders Neilsen & Peter Kjeld

Finland: JHS 137 Tietotuoteseloste; JHS 137A Tietotuoteseloste – Paikkatiedot JHS 137 Data Product Description; JHS 137A Data Product Description – Geographic Information JHS 137; JHS 137A 297
by Paula Ahonen-Rainio

Hungary: KIKERES Térinformatikai Profil; KIKERES Spatial Metadata Profile . 317
by Istvan Kadar & Tamas Prajczer

Israel: Israel Metadata Standard 2000 337
by Ammatzia Peled

Japan: 日本メタデータプロファイル; Nihon Metadeta Purofairu; Japanese Metadata Profile 355
by Kazuhiko Akeno, Shoji Okuyama & Shinji Takazawa

Korea: 국가지리정보체계(NGIS)의 자료이력서(메타데이터) 잠정표준; National Geographic Information System Metadata: Interim Standard 375
by Tae-jung Min & Heung-muk Cho

The Netherlands: NCGI Metadata 393
by Henri J.G.L. Aalders

Russian Federation: Государственный стандарт Российской Федерации ГОСТ Р 51353-99 “Геоинформационное картографирование. Метаданные электронных карт. Состав и содержание”; State Standard of the Russian Federation GOST R 51353-99 “Geoinformatic Mapping. Metadata of the Electronic Maps. Composition and Content”, November 11, 1999 411
by Alexander I. Martynenko

South Africa: Content Standard for Digital Geospatial Data 431
by Nic Scheepers & Antony Cooper

Spain: MIGRA (Mecanismo de Intercambio de Información Geográfica Relacional formado por Agregación); Aggregated Relational Geographic Information Interchange Mechanism UNE 148001 EX 451
by Sebastián Mas

United States of America: Content Standard for Digital Geospatial Metadata FGDC-STD-001-1998 469
by Richard A. Pearsall & Richard Hogan

International Spatial Metadata Standards 491

The Dublin Core Metadata Element Set ISSN:1041-5653 493
by Stuart L. Weibel

CEN/TC287: Geographic Information – Data Description – Metadata prENV 12657:1998 515
by François Salgé

ISO/TC211: Geographic Information – Metadata ISO 19115 535
by David M. Danko

Subject Matter Metadata Standards/Profiles 557

EuroGeographics: LaClef Core Metadata: Topography 559
by Marcus Wandinger & Claude Luzet

EU/JRC/SSSA: Recommendations on Metadata SSSA/ES/001 577
by Bernd Eckhardt

EuroGeoSurveys: GEIXS: Geological Exchange Information System – Description – Metadata 601
by Denis Bonnefoy

USA/NBII: Biological Data Profile of the Content Standard for Digital Geospatial Metadata, FGDC-STD-001.1-1999 619
by Susan Stitt

Part 5 – Crosstable of National and International Spatial Metadata Standards and Associated Characteristics 643
by compiled and edited by Harold Moellering

Introduction to the Crosstable of Spatial Metadata Standards and Their Scientific and Technical Characteristics 645
by Harold Moellering

Appendices 653
edited by Henri J.G.L. Alders

Annex A: Acronyms Used 655

Annex B: ICA Standards Commission Membership List 667

Annex C: Book Contributors 671

Subject Index 675

Presidential Foreword

During the last decades we have witnessed a tremendous growth in the building of geographical datasets. The transition from manual to digital systems for mapping, and for registration of real properties, population and natural resources resulted in a series of not compatible data. In 1980s work on standardization was initiated, first in the US and then country by country until the work became coordinated by ISO. Nowadays, it is obvious that a geospatial data infrastructure is needed, where data can be integrated both horizontally between themes and vertically from the local level to the entire world. The driving force for much of this development has been the more scientifically advanced countries and organizations throughout the world.

The initial situation in the 1970s was individual organizations operating insular spatial data systems with relatively little data sharing, primarily because of technological limits of not only hardware and software but also because of the insular mindset of most spatial data organizations. Advanced at that time were the Canadian Land Inventory for handling natural resource data and DIME (Dual Independent Mapping Encoding) for handling the US Population and Housing Census.

The first big breakthrough was the development of National and International spatial database transfer standards that began in the 1980s and continued through the decade into the 1990s. The resulting National and International spatial data transfer standards were carefully scrutinized and scientifically analyzed by the ICA Spatial Data Standards Commission that resulted in the authoritative book titled "Spatial Data Transfer Standards 2". This scientific work, led by Professor Harold Moellering, Chair of the Commission, examined all 22 National and International spatial data transfer standards in great detail with more than 220 tertiary characteristics. A major step forward was the large Crosstable that extracted the 67 most important characteristics about each transfer standard as rows across the 22 standards. This summary table has been used in many settings to learn about each standard and to compare them.

This current work "World Spatial Metadata Standards" has been designed by Prof. Moellering and the Commission members to examine all National and International spatial metadata standards existing in the world. It is organized in a structure similar to the transfer standards book, and serves as a companion piece to it.

Because of the complexity of the various national and international standards processes, this book has been nine years in the making. The first five years involved developing the characteristics that could be applied uniformly across each standard, although no standard possessed all of the characteristics. The next four years were involved in the Commission analyzing the standards against the characteristics, first in prototype tests, and then in a full analytical setting.

The Commission is made up of a group of world experts in the field who contribute their time and effort to the work of the Commission on a gratis basis. Special thanks are due to Professor Harold Moellering, Commission Chair and Editor, to Prof. Henri Aalders, Associate Editor who lead the work on the Part 2 Regional Summary chapters, and to Mr. Aaron Crane, Associate Editor who worked on the Part 4 Individual Assessment chapters.

This scientific work has been heavily peer reviewed throughout the development and execution process of the book. As such, the volume will serve as a comprehensive and definitive reference work on the scientific and technical characteristics of each major spatial metadata standard in the world.

In essence, spatial metadata is the conceptual grease that will lubricate the wheels of the spatial data infrastructure in various parts of the world. The scientific content of this volume will contribute greatly to the understanding of these various standards with a uniform set of characteristics that have been applied in an evenhanded and uniform way. Hence, this scientific work is a solid realization of the ICA to share scientific information on world spanning problems.

Dr. Bengt Rystedt
President of the International Cartographic Association 1999–2003

ISO/TC 211 Foreword

The ISO Technical Committee on Geographic Information/Geomatics was founded by the International Organization for Standardization in Geneva in April 1994. It has the daunting task to develop a set of world spatial data standards, that are, when possible, homologations of existing International or perhaps National Standards.

The ICA Spatial Data Standards Commission was in existence for four years before ISO/TC 211 was founded, and in some ways that provided an inspiration for the founding of ISO/TC 211. In fact, my own first experience with international standards was as a member of the Commission in 1990. Since the inaugural meeting of ISO/TC 211 in Oslo, 1994, the ICA Commission has been a Category A Scientific Liaison to ISO/TC 211. During that time this ICA Commission has contributed two scientific books to our effort. The first was "Spatial Database Transfer Standards: Current International Status", ISBN 185166677X, that was published by Elsevier Science in 1991, with Professor Harold Moellering as the Editor. It is a general overview of spatial data transfer activities in many parts of the world.

Their second scientific contribution to the ISO/TC 211 effort was their 1997 ICA book titled: "Spatial Database Transfer Standards 2", ISBN 0080424333, again with Prof. Moellering as the Editor. This work is a very deep examination of the various characteristics of spatial database transfer standards. It is comprised of 13 primary sets of characteristics, divided into 58 secondary sets of characteristics, and finally into about 220 tertiary characteristics. These characteristics are then applied across all 22 National and International spatial database transfer standards in the original language of the standard itself by an expert member of the ICA Commission who is a member of that standards body. These assessments are then translated into English to produce a uniform set of responses to the queries by the characteristics.

The result is a scientific book that has served as an authoritative reference to these transfer standards, and has been utilized on a number of occasions during the discussion of several of our TC211 Working Groups. For this reason alone this seminal work is recommended to scientific and technical personnel in the field of spatial data.

This new work titled "World Spatial Metadata Standards" has a similar structure and organization to the 1997 book on transfer standards. As such, this new book is designed and intended to serve as a companion volume to the existing 1997 book on transfer standards.

This new book utilizes a detailed set of assessment characteristics divided into 12 initial groups, 58 secondary classes and about 278 tertiary characteristics. Each National and International standard, plus a few application profiles (subsets) are examined in great detail. The result is a seminal and authoritative scientific work that is a wonderful reference to all of these standards. Each Part 4 chapter on the individual standard provides a wealth of detailed information about each standard that has been extracted by an expert member of the ICA

Commission in the original language of the standard itself. All the results are then published together in English.

Part 5 is a large Crosstable composed of the 70 most important spatial metadata characteristics crossed by the 22 National and International standards, including four application profiles. This large Crosstable is a valuable summary of each standard by column, with the 70 rows of characteristics that can be used to compare the various standards and profiles.

This ICA scientific book has taken several years to create because of the meticulous nature of the scientific and technical information it contains. However, it is an extremely valuable resource if one wants to learn more about a specific standard, or if one wants to compare each standard by a specific characteristic.

This work will serve as the preferred scientific reference for all those working in the area of spatial metadata standards at the National and International levels. It will also serve as a very valuable resource to all scientists who want to learn about these standards and what they contain, but cannot read them in their original languages.

As the Chair of ISO/TC 211, I highly recommend this book!

Olaf Østensen
Chair ISO/TC 211 Geographic information/Geomatics

GIS Foreword

Geographic Information Systems in the early years were essentially solitary systems that operated independently from other systems. By the early 1980s it was recognized that the data inputs were the largest cost to operate a major spatial information system. In fact, it was soon recognized that about 85 percent of the investment in a major Geographic Information System is in the databases it uses.

It also soon became clear to all that one major strategy to alleviate this problem and its attendant expense was to share spatial and geographic data between such systems. This required spatial data transfer mechanisms to transport the databases from system A to system B. In 1997 the ICA Spatial Data Standards Commission published an ICA book that exhaustively examined all of the National and International spatial database transfer standards in the world.

Sharing such spatial databases not only required some kind of data transfer mechanism, but also necessitated a way to understand something about the content of the database that one might access. This critical need in the spatial sciences resulted in the development of the concept of spatial metadata, which is, simply put, information about the data contained in the database in question.

Subsequently, it became clear that collecting spatial metadata about a specific spatial database was nice, but in order to share such databases effectively, it was necessary to employ effective, consistent and universal coding schemes of this information about spatial databases. Hence, the USA FGDC metadata standard was created in 1994. Similar scientific and technical work in other leading countries in Europe, Asia – Pacific and wider parts of the world resulted in a plethora of such National spatial metadata standards coming into existence in various parts of the world. The problem is that these various National standards define various spatial metadata elements in different and sometimes diverse ways.

If one truly wants to share or exchange databases, or just move limited pieces of data, on a systematic basis such as a computer network, or more contemporarily, the WWW, then a systematic and universal definition for the metadata elements referenced by such standards must be compatible. The spatial data community now realizes that spatial metadata is a critical element to building, exploring and running a spatial data network, or what is now called a Spatial Data Infrastructure. Frankly stated, spatial metadata is the essential ingredient that lubricates the wheels of the SDI. It is an indispensable element of any Spatial Data Infrastructure, national or international.

The world spatial data community came to see that a world spatial metadata standard was really needed to serve as a base on which the National metadata standards could be built and homologated. The world ISO/TC211 Committee on Geographic Information/Geomatics was founded in 1994, and recently completed work on the world metadata standard, ISO 19115.

This book "World Spatial Metadata Standards" developed and written by the ICA Spatial Data Standards Commission has produced an exacting examination of those National and International spatial metadata standards. Over a period of eight years the Commission members developed a meticulous set of scientific and technical characteristics, and then examined all of the 22 National and International spatial metadata standards using these scientific and technical characteristics. The resulting ICA book and large format summary Crosstable will be the seminal work for those interested in spatial metadata standards or comparing them in a systematic way.

Hence, this ICA book is useful to anyone endeavoring to learn about spatial metadata in a systematic way. It is also clear that this ICA book on spatial metadata standards, and the earlier ICA book on spatial database transfer standards make excellent companion volumes for those who are serious students of spatial data transfer mechanisms or spatial metadata. They make a fine companion set of books.

Jack Dangermond
President, Environmental Systems Research Institute

Editors' Preface

The current work represents nine years of work by the ICA Spatial Data Standards Commission ⟨current name⟩ across two ICA Terms of reference 1995–1999, and 1999–2003. Work began at the Commission's 1995 Costa Brava meetings, shortly before the Barcelona International Cartographic Conference. For the next three years the Commission proposed, discussed, and labored over developing a set of scientific and technical characteristics that would fully encompass the breadth and depth of a very wide range spatial metadata standard. Since the Commission meets only once each year, much of the work was done by correspondence, small team projects, and occasional informal minimeetings.

During the development of these characteristics, many rounds of peer reviews by Commission members took place, as well as empirical field testing of several of the mature standards against the then current set of metadata characteristics. This work continued at summer Commission meetings in Delft, Netherlands in 1996 and Uppsala, Sweden in 1997. As the scientific and technical solution of the final characteristics set was developed, the body of political support was built to support it. In the end the final set of characteristics achieved a unanimous vote. The result was a set of metadata characteristics with 12 basic levels, 58 secondary levels of characteristics, and more than 200 tertiary characteristics about each metadata standard or profile.

By the 1998 Commission meeting at Aix-en-Provence, France, the members developed a special subset of crucial metadata characteristics. That meeting resulted in about 70 characteristics being nominated for the Crosstable row variables, which were analyzed and tested during the intervening year. At the 1999 Strathmere Commission meeting, several small adjustments were made to the master characteristics list, and to the Crosstable list of row variables. At the end of this meeting the final lists of characteristics and Crosstable row variables were formally voted as the official Commission list for assessing all of the 22 National and International spatial metadata standards, and the four application profiles. A new innovation for the Crosstable is the addition of a set of obligation codes for the Part 2 substantive variables in the master Crosstable characteristics list. They are **O** obligatory, C conditional, or o optional. These are explained in detail in Part 5 of this book.

One must realize that no single spatial metadata standard or profile contains all of the characteristics found in the Commission master list of characteristics found in Part 3. Rather it is an ideal set of characteristics which serves as the Universe of metadata standards characteristics which all standards together possess in part. Since various metadata standards, and particularly the application profiles, may have rather different goals, it is easy to understand that the mix of scientific and technical characteristics found in any one standard or application profile can vary considerably.

The reader should be aware that this current ICA book on spatial metadata standards is purposely designed with the same fundamental structure and organization as our 1997 ICA book titled "Spatial Database Transfer Standards 2", ISBN 0080424333. These two books are designed to be used in tandem for those interested in conveying spatial data into some variety of Spatial Data Infrastructure, SDI. The spatial database transfer standards book looks at the problem from the perspective of the syntactic encodings and transfer problem. In this work the perspective is from the view of semantics and the fact that spatial metadata is the first step towards understanding the semantic meanings of the encodings contained in the spatial database at hand.

The book itself is structured into five fundamental Parts. Part 1 is the Introduction to the conceptual base for the following discussions in the book. Part 2 consists of six Regional Summary chapters that examine the general standards activities in the major continental areas of the world: Europe, North America, Asia-Pacific, Central and South America, Africa and the Middle East, and finally a nongeographical chapter on the ISO standards world. Antarctica is still pending. These chapters provide the reader with the regional standards milieu and organizations active in the area. This provides the context for the detailed discussions that follow. The initial chapter outlines were peer reviewed at the Commission Monaco 2000 meetings, while the fully written chapters were refereed in following Commission meetings.

Part 3 presents the detailed scientific and technical characteristics used for assessing the 22 National and International metadata standards and application profiles. There is also a glossary of definitions which are used in the discussions that ensue. Many definitions come form the World ISO/TC211 official definitions, but many come from our earlier ICA work and publications. Part 4 contains the 22 chapters on the individual spatial metadata standards and profiles. As such, the characteristics are posed as a questionnaire template to facilitate consistent replies which have consistent meaning across the spectrum of metadata standards assessed. Also included is a set of obligation codes to alert the reader if a characteristic is obligatory, conditional or optional. Each assessment chapter has been painstakingly prepared by a member of the Commission who is a member of the actual standards group that developed and prepared the metadata standard initially. This member has the ability to assess the standard in the *original language* in which the metadata standard was originally developed and written. After the initial chapter rendering, each chapter was peer reviewed by the full commission in a summer meeting and scrutinized as to the efficacy of the assessment. Many times Commission recommendations were made for requested improvements to the metadata chapter being examined. Through their collegial peer review process consistent assessments were achieved so that one can truly rely on the accuracy of the chapters. This scientific and technical assessment consistency makes authoritative comparability possible. It also makes it possible to summarize the information in a Crosstable.

Part 5 contains the Crosstable which utilized the 22 metadata standards and application profiles as columns, and 70 of the most significant scientific characteristics as row variables. The result is a Crosstable that summarizes all of the standards with the most important characteristics in a compact form of a wall poster that is both easy to read and follow the rows and columns. This metadata standards Crosstable has been purposely designed to be a companion piece to the 1997 Crosstable prepared as an analogous summary of the various spatial database transfer standards and their most important scientific and technical characteristics.

The tandem use of these two Crosstables will assist the reader interested in moving spatial databases in a Spatial Data Infrastructure milieu.

Finally, the Editors wish to thank the many Commission members actually involved in this massive undertaking. There are more than 50 authors from all over the world who wrote chapters for this book. Since they live in almost every time zone on earth, the challenge of coordination become even greater. Prof. Henri Aalders of Delft University took on the primary responsibility of coordinating the work of the Part 2 Regional Summary authors. Mr. Aaron Crane of Navteq Corporation was primarily responsible for coordinating the work of the Part 4 individual standards authors, and finely edited the chapters with the assistance of Mr. Chaz Ruhl of the same organization. The Part 3 characteristics prototype testing was carried out by Ms. Paula Ahonen of Finland, and Mr. David Danko of the United States.

Most of the members of the ICA Spatial Data Standards Commission are also members of their own National standards group(s) and many are also active participants in the ISO/TC211 world spatial data standards body. Since this Commission was originally founded in 1989 as an ICA Working Group, many of the members have been involved for many years. When ISO/TC211 was founded in November, 1994, most active Commission members became dual members of TC211 and the Commission. Many still are. In addition to their professional effort, we must also thank their families, friends, and loved ones for the large amounts of personal time they devoted to the work of this ICA Commission.

In closing, we acknowledge that developing the scientific and technical characteristics, peer reviewing and field testing them, developing about 30 chapters with 50 authors all over the world was not a simple task. Since the beginning of the research groundwork at the beginning of the book in 1995, the tempo of spatial data standards developments at both the National and International levels has accelerated considerably in the last nine years. Just like the old Greek philosopher who observed that one never steps into the same river twice, but now that river is moving faster than ever, and has some stiff rapids in it as well. We trust that the reader has developed a sense of the momentum that has developed in the spatial data standards world that will now have a fundamental impact on the larger world of spatial data users.

The reader can follow the work of the ICA Spatial Data Standards Commission and its new 2003–2007 task of modeling the Spatial Data infrastructure at the home page of: http://ncl.sbs.ohio-state.edu/ica.

Harold Moellering
Columbus, Ohio, USA
Editor

Henri J.G.L. Aalders
Delft, Netherlands
Associate Editor

Aaron Crane
Chicago, Illinois, USA
Associate Editor

PART 1

INTRODUCTION TO SPATIAL METADATA STANDARDS IN THE WORLD

Edited by

Henri J.G.L. Aalders

World Spatial Metadata Standards
H. Moellering, H.J.G.L. Aalders & A. Crane (Editors)

An Introduction to Metadata for Geographic Information

Henri J.G.L. Aalders

Delft University of Technology, Research Institute for Housing, Urban and Mobility Studies, The Netherlands
Katholieke Universiteit Leuven Faculty of Engineering, Belgium
e-mail: h.aalders@otb.tudelft.nl

1. Introduction

Traditionally, geographic information was produced and used by the geo-spatial community, such as experts in specialised fields as geography, cartography, geodesy, photogrammetry, remote sensing, bathymetry, hydrography, geology, soils, physical planning, architecture, etc.

The work of these experts resulted in the cartographic representation of the real world originally in maps and later as a digital computer model. Combinations of different geo-spatial datasets in various scales, projections and co-ordinate systems, content and appearance create a united computer model of the earth as a whole or in parts. This includes thousands of aggregates of digital geographic, topographic and thematic maps (such as ortho-photomaps, city maps, air- and space-photo images, (3D-) terrain models with gravimetric, photogrammetric and cartographic information) integrated under common idea and ordered by geodetic base, content, design, and reference information stored digital form.

To make the geographic information available to users, the geo-spatial community prepares and fulfils measures to develop and implement conceptual and methodological bases, normative and legislative regulations and dataset descriptive standards including formats for spatial data interchange. It also develops implementation methods, hard- and software tools and technologies for acquisition and the transfer of datasets to users [Moellering 1991], [Moellering 1994], [Moellering 1997], storage, analysis and processing digital cartographic data and the creation of analogue and digital representations of these datasets.

In daily life, the public is used to product labels to receive information about the products they acquire. These labels have several functions: they identify the product as well as promote and describe it. For instance in many food industries, legislation clearly states what information the labels need to contain and how it should be presented, while for technical equipment the product label summarises the content of a package and a guide describes its use. In addition, this information is widely controlled by international regulations. For geographic data, no such legislation or internationally controlled system exists. However, now that the geographic community enters into the age of global spatial data infrastructures to make these datasets available to non-specialised users, knowledge about widely distributed geographic datasets is essential to universally allow users to locate, evaluate, extract, and employ the data. As expert personnel changes and time passes, new workers may have little understanding of the datasets' content and consequently the information about the organisation's data will be lost and the data may loose its value. Therefore standardised descriptions of datasets are required following the rules set in metadata standards – both official and defacto.

Descriptions of geographic datasets have been in existence for some time now. In many cases these geographic dataset descriptions for different information communities on regional and national levels evolved in different ways and are incompatible. As the existing general dataset descriptions, that have become national or regional standards, provide hardly global interoperability, they are at least insufficient to support geographic information from different areas in the world. This incompatibility and insufficiency was the motivation for the development of a global standard for data descriptions: ISO 19115, Geographic Information – Metadata. The ISO (International Organisation for Standardisation) Metadata Standard was designed to:

- support geographic information;
- work with wider information technology standards and practices;
- serve the global community, in a multi-national, multi-language environment;

based on a foundation of national, regional, and special information community standards and experiences and a thorough requirements analysis, and implementation testing.

The ISO 19115 Metadata standard defines and standardises a comprehensive set of metadata elements and their characteristics, along with the schema necessary to fully, and extensively, document geographic data. The standard applies to all geographic data – it is applicable to datasets in series, datasets, individual geographic features, and their attributes. The standard defines the minimum set of metadata required to serve the wide range of metadata applications, as well as optional metadata elements to support a more extensive description of geographic data. It also provides a standardised manner for users to extent their metadata and still ensure interoperability allowing other users to comprehend and exploit this extended metadata when browsing the web for certain type of geographic data.

In the field of Information Technology (IT), there are two other prominent metadata initiatives: the Dublin Core Metadata Initiative (DCMI, Dublin Ohio, U.S.A., see http://www.xml.com/pub/2000/10/25/dublincore/) and the Institute for Electrical and Electronic Engineers (IEEE) Learning Object Metadata (LOM) Working Group, see http://grouper.ieee.org/p1484/doc/wg12/LOM-WD3.htm).

The Dublin Core metadata element set (or the basic interlocking brick) is intended to support cross-subject search and retrieval. It can be thought of as a simplistic or pidgin metadata language that helps the user navigate through disparate subjects, languages, and cultures. Adoption of the Dublin Core by governments, libraries, museums, archives, publishers, environmental science repositories, prints and e-print archives, to name a few, testifies to its success in this role. There are emerging applications in the commercial sector, as well, with health care organisations and financial industries using the Dublin Core as the basis for organising and exchanging information.

2. History of metadata

To many in the hard-copy map world, data description, metadata and marginalia are often considered to be the same. They help to describe the properties and characteristics of the dataset's content (in the hard-copy map world called the 'map').

The Greek geographers applied descriptions of map-objects as early as the beginning of Christian era. Many of the Middle Age geographers applied beautiful drawings to accompany their vision of the world in maps [Bertius 1630]. Often in the Middle Ages the symbolisation on maps were very religious and descriptions referred to religious descriptions but usually carried dates and author names.

From the early to mid 18th century, topographic maps and sea navigation charts often carried descriptions and explanations in order to enable the accurate reading of the map information, called marginalia. The marginalia captured dates, bounding co-ordinates, grids, scales, accuracy, author, etc.

The term **metadata** – or used in other forms as *meta data* [McIntosh 1968], *meta-data* [Homer 1978], Ziegler 1978], [Weber 197], [Schelling 1978] and *metadata* [DLC 1979] – first was used in Computer science/Information sciences literature in 1968. Since, it has been used in any of the three forms: firstly, metadata was very popular in the European community in the beginning of the seventies and later started to show up in USA literature (as metadata – all one word) in military documents.

The word 'metadata' stems from Greek and can be described as 'data about data'. The ISO/TC 211 defined metadata by data about the content, quality, condition and other characteristics of data. Although the ICA Commission on Spatial Data Standards adopts the ISO definition on metadata, originally it used a description as 'information given along with geographical information and which allows a better understanding of geographical data'. According to this definition, metadata information encompasses transfer format templates, counts of data items in the transfer, conceptual data model and catalogues, positioning references of geographical data, quality reports and the logical description of the transfer metafile.

The term metadata took hold in the USA in the geo-spatial data community in the early 1990's with the development of the concept of the National Spatial Data Infrastructure and the signing of Executive Order 12906 by President Clinton in April 1994 (see: http://www.fgdc.gov).[1]

3. Fundamentals of metadata

Following the concept of set theory, a geographic dataset can contain either several homogeneous geographic datasets or one or more occurrences of geographic entities or any set of attributes or occurrences of geographic entities and/or relationships. As such, one could also

[1] The Metadata Company trademarked the term 'Metadata' in 1986. The Metadata Company sells goods and services related to metadata. Since the FGDC does not sell any goods or services, the FGDC does not feel it is in violation of the trademark as determined by the Department of the Interior, Office of the Solicitor (see FGDC Official statement at http://www.fgdc.gov/metadata/meta_trademark.htm). Since the trademark issue involves goods and services, several companies that make metadata products have tended to avoid the term metadata and have opted to use the term meta-data or even different terms such as data properties and data cataloguing.

speak of levels of datasets. Therefore, data shall be given to describe the data, which occur in a dataset.

Metadata serve a wide range of applications in geographic information as, apart from data description also the organisation and maintenance of data. Complete metadata descriptions of the structure, contents and accuracy of all datasets is an important requirement for database design. Such descriptions also may provide protection for the misuse of the data in a dataset.

A model for a metadata includes the definition of mandatory or optional metadata and which metadata is mandatory under special conditions (conditional metadata). In addition, the minimal metadataset should be clarified. Metadata can be organised in the computer in many ways: as an incorporated part of data within the dataset, in a separate database or as a text file. One may choose the manner of storage of metadata according to a management strategy, budget or other institutional or technical factors.

The best time to define and collect metadata is while the data are being developed, i.e., when the information needed for metadata is known. Probably waiting until after data collection is finished results in less accurate information lacking the description for the surveyors of what ought to be collected in the database. Besides, searching information, when lacking metadata may become cumbersome and increase the cost. However, documenting existing data can be daunting. Details may be long forgotten and costs can be high. Though these concerns are valid in well-considered decisions, they should be taken allowing documenting metadata for existing databases.

Using and combining data from different sources (and maybe also applying processes to them) that are available at different places requires the GRID approach of distributed computing and resources to conduct new and innovative specific applications. This may result in a Virtual GIS (V-GIS) environment in which the Digital Universe resident in the new databases can be seamlessly explored across the entire spectrum carrying the characteristics and possibilities of each of the single databases and spatial processes. This 'digital era' has been a topic for international discussion in the past year by the G8 countries, United Nations, heads of Governments, the non-profit community and the international private sector.

3.1. *Hierarchy in standardisation*

The application of standards is different in different countries. In some countries, law or governmental (executive) order, regulates the application of standards, while in other countries the use is left in free will to potential applications between parties. Countries where standards are applied according to state regulations make distinction between:

- state standard adopted by the State Committee on Standardisation, a standardisation law or a governmental (executive) order;
- domain standards that have been adopted by some ministries (departments) or by a group of organisations in the same field;
- application standard used only within certain – private sector – enterprise(s).

An International Standard usually is the standard adopted by the ISO, while a Regional Standard is adopted by a regional international organisation, for example, interstate standards organisations such as the Comité Européen de Normalisation (CEN). Governments or national standardisation institutes joining the agreement on co-ordinated policy of standardisation and certification adopt such standards. A National Standard is the standard adopted by national standardisation authority of a certain country. Many standardisation processes result in a set of interrelated standards in a certain field (sometimes also called 'a complex of standards') that have the same purpose and establish co-ordinated requirements to the interrelated subjects of standardisation; e.g., the set of standards of ISO 191XX (see also the chapter on ISO standardisation in part 2 in this book).

Since metadata standards include the object definition, in many cases it will be impossible to create a definition for so many different types of users. For example, the definition of a road for a car driver is completely different from the definition for a road used by a road manager, while different road managers use specific definitions in their own application. Therefore, data definition (in the geographic community also indicated by ontology) should follow a standardised way for data definition but may occur at different levels:

1. international level;
2. national level;
3. domain level;
4. application level.

In this situation, the definitions at a lower level follow the definitions at the higher level and the transfer of the meaning of object definitions to be applied in different geographic datasets.

Standards may contain different aspects such as prescriptions for the transfer, ontology, quality, conceptual models, etc. A standard explaining technical terms is a kind of a normative document on certain production or service adopted by the developing organisation, as a rule, in co-operation with potential user organisation(s) and customer(s). Metadata standards follow this terminology.

All standards are based on generalised results obtained from scientific, technical and practical investigations, and they are intended to reach the optimal public benefit. In geographic information, the development of standards involves the methodology, theoretical basis (concepts, methods), technologies of creating and using digital datasets in GIS. In addition, metadata standards follow these general organisational, scientific, technical and application aspects.

3.2. *Metadata modularity and extensibility*

Another concept in metadata applications is modularity [Duval 2002]: metadata modularity is a fundamental principle where vastly diverse sources, styles and content management and approaches to resource description exist. Modularity in metadata allows designers to create new assemblies based on existing metadata schemas: in a modular metadata environment, data elements from different schemas as well as vocabularies can be combined in a syntactically and semantically interoperable way. This allows designers to benefit from re-usability while gathering exiting datasets of metadata and 'snap' them together into larges structures, even though the semantics do not appear as obvious to be combined as long as there exist a common syntactic foundation.

Metadata modularity includes also the possibility for extension in order to accommodate particular needs in a specific application. For example, the *identification* of the dataset will be described in any metadataset, while the *degree of cloud cover* will only appear in remote sensing data of aerial photography.

Furthermore, the metadata modularity requires sufficient refinement to make a more specific meaning of a metadata element, e.g.: *illustrator, author, composer or sculptor* are all particular types of the more general term *creator*, or *date of creation, date of modification, date of acceptance* all have a narrower meaning as the *date* attribute. In addition, the range of values for a given element is a refinement of the metadata element.

In addition, multilingualism becomes a very important aspect since the Web connects a diversity of linguistic and cultural aspects. The Web will fail to achieve its potential as a global information system, unless resources can be made available to users in their native language, in appropriate character sets and with metadata appropriate to resource management.

Here *internationalisation* and *localisation* become apparent, though they may be contradictory: while global resource discovery is best served by internationalisation (using common conventions of practice, languages and character sets throughout the world), the needs of any given community may be better served by supporting local conventions. Basic to the promotion of a global metadata architecture is to translate relevant specification and standard documents into a variety of languages. DMCI maintains a list of translations of its basic documents, as the European workshop on Learning technologies is maintaining translations of the LOM specifications.

At last, namespaces is a fundamental part of the infrastructure of the Web, and which are used in metadata queries through the Web. They are a part of the modular metadata systems. For example, the declaration of namespaces in the Web in a URL may be specified by a prefix:

- *dc* to indicate a search only in the Dublin Core metadata element structured sets, while
- *lom* recognises the IEEE-LOM metadata element set, searching only for data, listed in IEEE-LOM metadatasets.

3.3. *The Universe of Discourse (UoD)*

All human beings observe the reality with their senses and create a model of the reality in their mind. In creating digital geographic data, it is attempted to model and describe objectively the real world phenomena that have a relation to the earth' surface, using IT concepts in geographic and spatial computer analysis and graphic display. Any description of the reality is always an abstraction of the real world objects, always partial and always just one of the possible real world 'interpretations'. This are not an exact duplication of the real world: some things are approximated, others simplified, and some are ignored. There is no such dataset as perfect, complete and correct, or it would be the reality itself. To ensure that the dataset is not misused the assumptions and limitations affecting the dataset must be fully documented in the metadataset.

The Universe of Discourse (UoD) defines phenomena in the reality into an abstract form, by theme, position and time in order to make these objects intelligible to represent them in a database. The UoD applies an ontology – in GIS usually understood as 'the definition, classification and structure of objects considered for a specific database application' – as a limitative collection of unambiguously defined concepts [Uitermark 2001], following the Greek philosopher's Parmenides (500–450 BC) doctrine on ontology and the Indian Nyāya–Vaiśesika school of philosophy on the classification and structuring of objects. In this way, the UoD contains the limitative set of characteristics of selected objects in the real world and their descriptions. However, real world objects are modelled into an object-based database, in

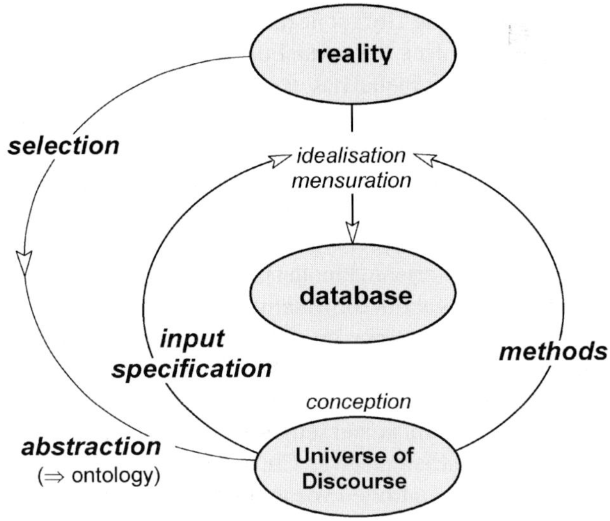

Fig. 1. From Universe of Discourse to Database. From reality the intended objects are selected and abstracted resulting in an object definition (database concept); with the input specification and the methods to be used (resulting in the surveying rules), the surveyed data are stored in the database.

which for each real world object a one-to-one representation. So, the process for defining the UoD is decomposed into two steps:

1. conceptualisation including the selection of what should be considered in the real world and the abstraction of the selected objects, resulting in a dataset content specification;
2. mensuration, specifying the measuring methods and equipment, the measurement requirements and the capturing and storage of the data itself.

Since the UoD (see Figure 1) includes any selection of objects from the reality as seen by the-potential-user, his (the customer's) satisfaction is determined by the perceived performance and expectation of the data for the user's application. This gives a product image, both emotional and cognitive factors and can be recognised in the process of the customer's satisfaction contributing to the customer's profitability of the data.

In America, often the term 'abstract view of the universe' is used for the concept of the Universe of Discourse. However, the 'abstract view of the universe' does not reflect the selection process, which is an integral part of the definition of the UoD. ISO/TR 9007, 1987 [ISO 1946] defines the UoD as 'all those objects or phenomena of interest that have been, are or ever might be'.

3.4. *Application of metadata*

No single metadata element set will accommodate the functional requirements of all applications, and as the Web dissolves boundary access, it becomes increasingly important to be able to cross discovery boundaries. This is served by:

- *conceptual models* – defining the UoD – needed to structure the representation of the semantics and syntax, and to replace the textual descriptions or lists of feature and attribute types, which are often found as metadatasets. Though lists of feature and attribute types give easily the impression of exact information, they often lead to misinterpretation when not applying the definitions of used terms. So, conceptual models limits to those parts and characteristics of the reality, which are relevant for the particular information system. A conceptual model names the object types in the UoD, the attributes of the object types and the relationships between the object types and the attribute domains (and the behaviour of the object types in an object-oriented model);
- *syntax and semantics*. Syntax is about form; semantics is about meaning. Agreements about both are necessary for sharing metadata; however, it is important to make a clear distinction between syntax and semantics. Where the syntax defines the manner of the transfer of data, the semantics will tell what is transferred. In addition, there is a third component for interoperability, that is beyond syntax and semantics: *content vocabularies*; this can be in an open and unconstrained natural language as Chinese, English, Japanese, Russian, Spanish, etc., but may also be a specific controlled vocabulary further narrowing down the scope and increase the precision of a description;
- *application profiles*. An application profile is a combination of metadata elements selected form one or ore metadata schemas, expressing the principles of modularity and extensibility, to tailor to the functional requirements of a particular application. Application profiles achieve this modularity through the following mechanism:
  - *cardinality*, referring to the constraints on the appearance of a metadata element (being mandatory, optional or conditional);
  - *value space restriction*, for specific applications differing from the general specification;
  - *relationship and dependency specifications*, by defining interrelations between data elements and their value spaces for a specific application;
  - *declaration of namespaces*, when supporting multiple namespaces.

One of the first steps in designing a database is the data analysis based on *user requirements*, based on how users perceive their application. The result is a set of external views representing exactly the data needed. External views represent each separate application of the information system; the conceptual schema is the combination of all external views of the information system. However, databases and their applications have life cycles and it is most probably that new ideas for data products appear later. Such new external views shall be derived to include in the conceptual model.

3.5. *Metadata element sets*

Metadata standards provide a mean to describe the content of a dataset at different levels of detail. The most common, and most vaguely defined metadata element is an *abstract* or an *overview*, providing a brief summary of the dataset's content, usually expressed in natural language without any structure. Although it gives the supplier the maximum freedom conveying the dataset content, its usefulness is fully depending on the evaluation by the data users. Because the natural language allows ambiguous and inconsistent expressions – despite of a defined grammar – it creates fuzziness and confusion requiring formal models for computer interpretation of metadata.

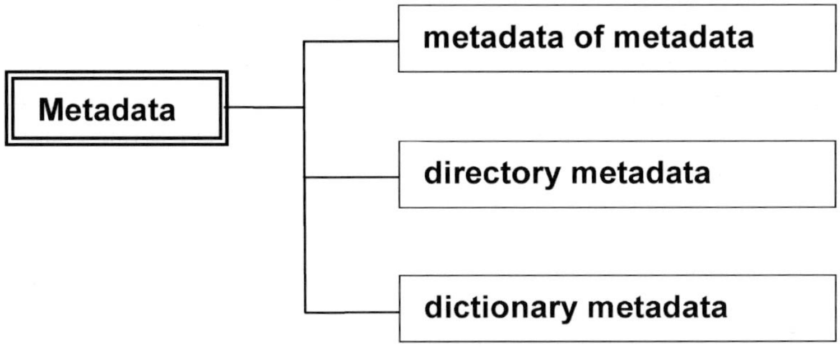

Fig. 2. Types of metadata.

More structured as free text is a list of objects types and their attributes, including hierarchies and object and attribute type definitions. A further approach is to provide a conceptual schema in a formal way. This leads to the necessity to list and define metadata elements. Metadata elements can be distinguished in three types, together forming the set of metadata elements (see Figure 2):

1. *content* of metadata (metadata of metadata). In order to understand the metadata of a dataset it is required to state the characteristics of the metadata such as the language and coding system of the metadata itself as well as the reference and coding systems that the metadata uses to define the thematic, geometrical and temporal extent;
2. *directory* metadata including the identification of the database, as name and origin as well as locating specific type of data as names and addresses of owner, distributor and manager and the description of the extents of the database in semantic, geometric and temporal sense, and quality and security measures, etc.;
3. *dictionary* metadata including the data definition, referring to semantic, geometric and temporal definitions of the data, data organisation by conceptual schemes for homogeneous datasets and quality information including the quality conceptual schema and the description of the parameters for the data as idealisation, accuracy, reliability, currency, completeness, consistency and currency.

3.5.1. *Metadata of metadata*

In order to understand the metadata of any dataset, the organisation of the metadata, the language used and the reference system for semantic, geometric and temporal extent should be given. (Metadata may refer to a homogeneous dataset series, a dataset or a set of occurrences of attributes of entities and/or relationships.) These characteristics can be given referring to other standards or by a description:

- *character set* used for encoding the data in the metadataset (using, e.g., ISO 8859-10);
- *language* for textual statements in the description of geographic datasets may follow a coding system as defined by ISO 639 or can be given by a text like 'Chinese (Hong Kong)', 'Chinese (Singapore)', 'Chinese (Taiwan)', 'English (Australian)', 'English (UK)', 'English (US)', 'French (Standard)', 'French (Belgium)', 'French (Canadian)', 'German

(Standard)', 'German (Swiss)', 'Portuguese (Standard)', 'Portuguese (Brazilian)', 'Russian (Moldova)', 'Russian (Russia)', etc.;
- *date description*, e.g.: by EN 28601, ISO 10303, part 41. The advantage of using ISO 10303 part 41 is that it also standardises the description of persons, organisations and addresses;
- *geometric reference system* applying geodetic standards as well for planimetric geometry, as for the vertical component;
- *quality element definitions* (Morrison, 1995); e.g., how to describe the original intended purpose of production by the producer, the usage prior to the present intended use, the lineage of the dataset describing the process history, the thematic, positional and temporal accuracy, the completeness and consistency of object types and the homogeneity of the given quality elements;
- *administrative data* about metadata, being:
  - *point of contact* in the organisation that created or manages the metadata;
  - *dates of creation*, last check and update future updates of the metadata;
  - *constraints and security* measures of the metadata.

The metadata language should be a mandatory element. The positional reference system whether it be geodetic (using co-ordinates) or non-geodetic (as addresses including postal codes, parcel identification, road networks identification, administrative subdivision of a country, etc.) are required when geometric data is given in the dataset. Here, conditional metadata may also exist: a dataset with geographic information may contain semantic geographic data only where no positional data is available (then no positional reference systems need to be given in the metadata).

3.5.2. *Directory metadata*

The elements of metadata to find the location of the dataset at least comprise:

- *identification* of the dataset by a code or a name, that defines the dataset uniquely and clearly amongst other datasets including the version number of the dataset when appropriate. Besides, other names can be given as alternative title(s). This may be the dataset's name in other languages or an abbreviation of the dataset name. Also the dataset overview, giving an overall description of the dataset including a summary (in text) and describing the content of the dataset can be given;
- *administrative* metadata to acquire a dataset information, regarding where and how the dataset is held as well as its procurement is given including:
  - *organisation*: name and abbreviations (if available) of the organisation, the address (postal address, visiting address, telephone number, telefax number and/or electronic-mail address, home page address on the web;
  - *organisation role*: as responsible authority for the dataset, producer organisation, etc.;
  - *point of contact*: in many cases it is evident that personal contacts lead to the best way of transferring data from provider to user. In such cases, direct contact between persons is necessary. Though personnel basis is an extreme aid for mutual use of data, the organisations should still appoint officials for this purpose. Here the name, function and addresses of the point of contact can be given;
  - *distribution* includes descriptions of restrictions on use, copyrights, units of distribution (e.g., per tile, per square kilometre, per administrative unit, etc.), pricing and discounts of

the data (types) per unit, data media on which the dataset can be recorded and retrieved or on-line accessed, delivering data formats, procurement, giving instructions for ordering the data and the delivery service and services for processing the data;

- *origin* of the data:
  - *producer's purpose* of the dataset, describing the original purpose by the producer for which the data in the dataset was captured. It may include the original intended application scale. Much of the digital geographic data is collected for direct data presentation on maps. The content definition of the dataset (selection and abstraction in the definition of the UoD) is usually based on what should and can be represented on those maps and is very dependent on the original purpose of the dataset;
  - *capturing method* and type of semantics describing the way of original collection and the type of data that can be found in the dataset, e.g., data from space, aerial data, landsurvey data, etc.;
  - *potential usage* give the provider's view on the potentials of the dataset, i.e., the possibilities of the data for different applications. In addition, the usage can be described to give the future users an idea for which applications the dataset was already used. Important in these cases is also to give an impression of the successes and failures that are experienced in the specific applications of the dataset. References to other relevant published documents or public available additional documentation may be provided;
  - *type of spatial schema* used in the dataset. This may be a standard or a user defined spatial schema, describing the main characteristics and components of the schema. On top, the definition of the spatial reference system should be given in the case of use of a spatial schema. The metadata may also describe geographical data that does not contain any spatial reference. In this case no spatial schema has to be defined in the metadata;
  - *examples* taken from the dataset and being representative for the whole dataset can be provided as samples, e.g., as a browse graphics either in raster or in vector format;
  - title and/or code and provider of *related datasets* of possible interest to a potential users can be enlisted for further information about the possible uses of the dataset;
- *dataset quality* elements. The dataset quality describes the difference between the actual dataset and the user's and producer's UoD. By describing the quality of the dataset, the user can determine whether the data has potential for the intended application before the transfer of the data. The dataset quality elements follow the designed quality conceptual model and comprises:
  - *source* describing the producers'-, owners'-, managers'- and providers' organisation as well as the point of contact within the organisation;
  - *homogeneity,* giving a description of how well the data in the dataset follow the overall uniformity of the data;
  - *usage* gives an overview of the applications for which the information in the dataset has been used previously and how well the data fitted in these applications. Also the *potential use*, is an important aspect in these which gives an indication of the possibilities of the data seen from the providers' perspective;
  - *lineage* giving a description of the origin of the information contents in the dataset and everything that has happened to the data since, until the moment of transfer;
  - *quality parameters* as positional accuracy, thematic accuracy, temporal accuracy, logical consistency, completeness and their values;

- *currency*, dating the information in the dataset as well as indicating how well the data is kept up to date;

- *extents*. Data about geographic objects are positioned in space and time. In order to judge for the user whether a dataset is suitable for the intended application the planar, vertical and temporal extent of the data in the dataset should be available. This can be done by one or more 'bounding range(s)' of the dataset giving the maximum and minimum co-ordinates appearing in the dataset or by one or more 'area boundary(ies)' delimiting the area(s) covered by the dataset. In addition, the currency of the extent, indicating the status, completeness and validity should be given. For the temporal extent an indication of the range (from start to expiring date) should be given (this may still be 'on-going', i.e., having no expiring date yet), but also a descriptive text may indicate such as 'medieval period', '20th century', 'annually update', 'continuously update', 'last update', etc.

3.5.3. *Dictionary metadata*

The dictionary data describe the semantics of the data in the dataset as well as the conceptual schema that has been used for data modelling. This allows search engines to not only access the data in the dataset but also interpret the semantics for determining the requested information and gives the user access both to the data and its semantics. Amongst others, dictionary metadata consists of the following elements:

- *data definition*. Objects may be defined in different ways in different datasets. To enable comparison between them, data describing the characteristics of objects are required. Data definition and classification, describing the differences between classes in order to distinguish between objects in different classes and to define the relationships between the classes do this. For this purpose, data is provided giving the definition of object types, attribute types and relationship types, wherever they exist in the dataset. Together with the application schema, all objects that are represented by data in the dataset should be defined and become understandable for the receiver of the data. Sufficient description of the data definition is given by:
  - *object type*, attribute type and relationship type name and definition should be given completely. If object types, attribute types or relationship types are indicated in the dataset by codes, also the coding should be part of the definition. Relationship types should also indicate the object types they relate (a 'from' and 'to' object type);
  - *classification*. The object, attribute and relationship types of a dataset can be described in a classification system, in which the object, attributes and relationship types may take part in a given hierarchical organisation. These hierarchies should be given in the form of a thesaurus according to the standard ISO 2788: 1988. In the case of using a thesaurus the name, version number or version date, the thesaurus administrator's point of contact and references to other publications about (use of) the thesaurus should be available to the user. If not all the thesaurus elements are used in the classification system, then those thesaurus elements that are used shall be listed in the classification part of the directory data within the metadataset. When a thesaurus is provided in the metadataset, besides the definition (if it does not belong to a 'de jure' standard) of each thesaurus element, also the code and synonym should be given; related, narrower and broader terms as well as pictures can accompany this;

– *spatial reference system.* The spatial reference system may be a direct, geodetic reference system based on co-ordinates, or a non-geodetic reference system, e.g., based on addresses or others. A dataset can contain objects referenced by a multiple of geodetic and/or non-geodetic spatial reference systems. Elements for the definition of the non-geodetic reference systems are: type (e.g., the system in which the references are given as country, county, municipality, etc.), title and owner and version date of the non-geodetic reference system. For geodetic reference systems the name of the reference system, the co-ordinate representation and its units (for both planimetry and height), the name of the datum, ellipsoid, map projection, etc.

4. Minimal metadataset

Minimally a dataset should have some metadata to stay understandable for any enquiry by users. It is noted that elements have descriptive text to convey a common semantic understanding of the elements. However, controlled vocabulary may be required to promote global understanding of the element values.

The development of the ISO 19115 Metadata standard has including a number of studies on different metadatasets available in different regions and organisations such as the ANZLIC Metadata: Core Metadata Elements (1995), Canadian Directory Information Describing Digital Geo-referenced Data Sets (1994), CEN ENv 12657, Standard for Geographic Information – Metadata (1996), FGDC Content Standard for Geo-spatial Metadata (1994), DIGEST (1994), IHO S 57 (1995). The system enables international users to locate datasets of interest, and contact the appropriate National Mapping Agency (NMA). Comparison to the from origin bibliographic Dublin Core Metadataset (that has much experience with metadata to structure the vast amount of publications for consulting), one can conclude for geographic dataset to the following minimal metadataset:

- *identification* of the dataset by a name given to the dataset by the original producer or publisher, being unique to distinguish the dataset from others. This might also be done by an unambiguous code for the dataset;
- *provider*, the organisation responsible for making the dataset available in its present form, such as a publishing house, topographic service, private company, municipality, etc.;
- *original producer*, the organisation responsible for the original capture of the data in the dataset;
- *reference system* for spatial reference, semantic definition by thesaurus and time definition;
- *extent*, describing the geographical, semantic and temporal coverage of the valid data in the dataset and whether or not spatial attributes are carried in the dataset. The geographical coverage may be a description by a set of quadrangles or geographically defined circumferencing boundaries. The semantic coverage may be a list of the type of objects by semantic attributes and the temporal coverage indicates the time period of the valid data in the dataset;
- *date*, indicating the date of validity of the data in the dataset. Many ways of writing dates are possible but if used they should be written in a clear and unambiguous manner;
- *language* of the metadata;

- *syntax* of the dataset transfer, to identify the software that will be required for reading the data;
- *quality*, describing the spatial, semantic and temporal quality parameters for the dataset. In the quality definition of larger datasets information should be available of the meta-quality indicating the quality indicators, e.g., for spatial quality a relative standard error should known as 'average', 'maximum', 'minimum', 'expected', 'required', etc.;
- *rights and management* indicating the copyrights, constraints on use, way of distribution (e.g., by tiles, by square kilometres, etc.).

5. Metadata query and distribution services

There are also many ways to transmit, communicate and present metadata. Metadata elements will be valued in different ways by different users or by one user for different tasks. The Internet and other technologies are causing rapid change in means to provide information although the need for metadata on physical means – including paper – will continue to exist.

The web has become an important universal information tool, embracing vast stores of information with many purposes, multiple disparate sources, and quite a few unpredictable users. There is a clear need to improve access to this mass of information and for the development of better search, retrieval, and organisational tools. Metadata is a fundamental part of the solution to these challenges. Browsing the web to find data resources three levels can be distinguished (see Figure 3):

- discovery metadata to find data resources (a front office application);
- inventory metadata to select the datasets, which will be acquired (also a front office application);
- data model access to enable the interface to load the geographical data into the application using the metadata that accompanies the dataset received from the data provider (a back office application).

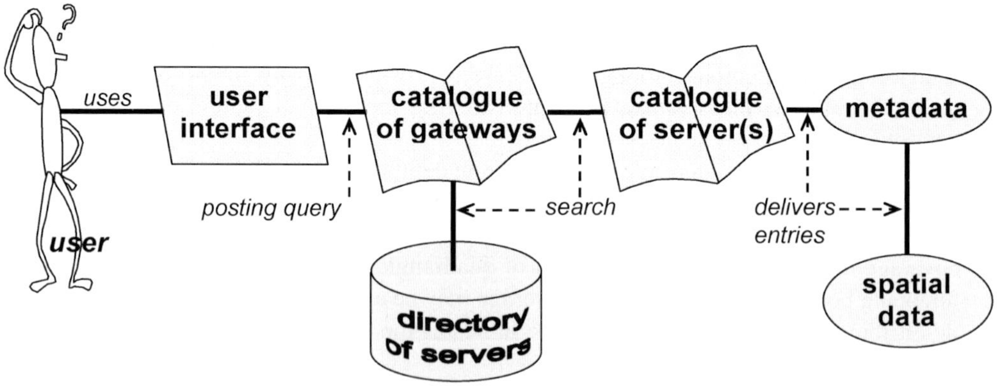

Fig. 3. Finding spatial metadata through a catalogue of gateways and servers to find the spatial data itself.

5.1. *Discovery Metadata – Metadata to find data resources*

Effective use of metadata in finding data on the web, requires three things:

1. set of commonly-understood terms to describe the content of information resources (semantics);
2. standard grammar for connecting those terms in meaningful metadata sentences (syntax);
3. framework that allows us to transfer and recombine those metadata sentences across different applications and subjects.

These three elements – standardised semantics, a definitive syntax, and a framework for transfer – provide architecture for resource description that can work across all subject areas on the web.

5.2. *Inventory Metadata – Descriptions of datasets that could be ordered*

In its simplest form, a central database is maintained, with data providers updating dataset descriptions (metadatasets) as necessary. Data providers are provided with a software update tool, which includes the current version of the central database. They can update information about existing datasets or add new datasets before returning the media to the central, when the new data are then transferred to the main central database. On the other hand, a Spatial Data Infrastructure (SDI, see Figure 4) may link the centre to all metadata providers [Williamson 2003]. Locally, the providers will maintain their datasets (both the metadata and the data itself). Users may consult the metadatasets for inventory by using the appropriate web browser,

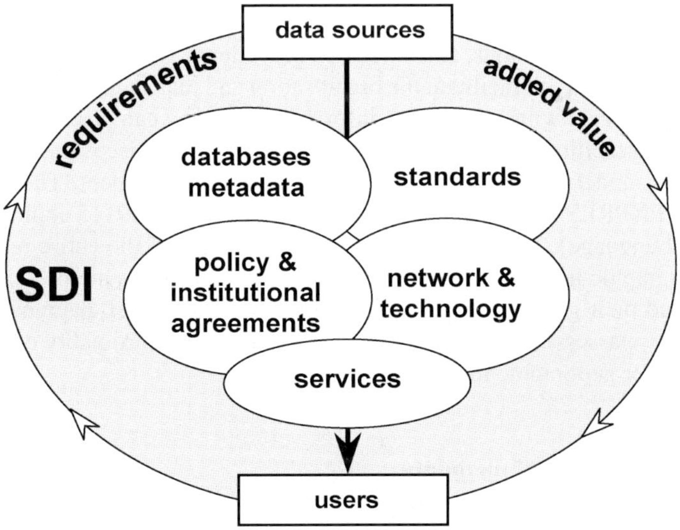

Fig. 4. Interactions between SDI components for the transfer of data to users: databases with metadata, standards, policies and agreements, network and technology and services. By using the data more experience is gained, allowing providers to improve the data quality.

protocols and standardised catalogue services to support technologically the interpretation of the (meta-)data.

Web pages are derived automatically from the database, and the information is made available free of charge to all Internet users. In a way, it can be described as an Internet gateway dedicated to Geographic Information. Regional and national examples of these are:

- MEGRIN Geographical Data Description Directory service in Europe, as well as the European Spatial Metadata Infrastructure (ESMI);
- Asia-Pacific Spatial Data Infrastructure (APSDI) as proposed by PC GIAP;
- Inter-American Geo-spatial Data Network (IGDN) in Latin America and the Caribbean;
- National Geospatial Data Clearinghouse in the U.S.A., following the EO 12906;
- Canadian Geospatial Data Infrastructure GEONet, developed by GeoConnections;
- Australian Spatial Data Infrastructure Distribution Network;
- the Japanese GSI clearinghouse gateway system with multi-byte character code sets using the GEO profile based on ISO 23950.

5.3. *Model metadata – Metadata for data transfer*

Most of the present transfer standards contain metadata on top of the geographical data that are transferred. Examples are the CEN ENv 12657: Geographic Information – Data description – Metadata, issued 1998 and the ISO 19115 Geographic information – Metadata issued in 2001. Apart from the used terminology in the standards, they also give the metadata elements and a schema, which, when properly implemented by a data producer, will enable users to locate, access, evaluate and employ geographic data. Users will be able to discover, select, and purchase geographic data; determine whether data in a holding will be of use to them, and be able to apply geographic data in the most efficient way by establishing a common set of metadata terminology, definitions, and extension procedures. Supplementary benefits of these standards for metadata are to facilitate the organisation and management of geographic data: it furnishes data producers with appropriate information so they can properly characterise their geographic data. It facilitates dataset cataloguing enabling data discovery, retrieval and reuse.

The standards models reflect their period in time for development. The ENv 12657 uses EXPRESS and EXPRESS-G as a modelling tool, while the ISO 19115 applies UML (Universal Modelling Language). These models are an integral part of the entire respective abstract models for geographic information. The metadata model diagrams are presented in computer readable text and their graphic representations in view. The model diagrams provide a good illustration of the classes and their attributes, and the type and cardinality of the relationships between classes for geographic metadata.

6. Experiences in the use of metadata

In the use of metadata some striking results were obtained; some examples given by [Ahonen-Raino 2001], [McCelland 2002] learn the following:

1. metadata are incomplete and contain errors. As users of data rely on descriptions given by suppliers, users with unexpected applications look for information in the metadata that is

not available because the UoD did not foresee in the apparent application. For the user it appears as if the metadata is incomplete or contains errors, while the supplier may have listed the metadata correctly according to his UoD;

2. specialised vocabularies complicate data import and understanding. Localisation of meta-datasets often implies specialised vocabularies, which are not understandable for many users outside the geographical area or outside the field of application. Suppliers do not often realise that users may use data for unexpected applications. The use of metadata for non-experts leads often to the Lego™method of building and children are not biased by the suppliers intention and may apply their play constructions in an unexpected water or desert type imagination. Non-experts users may apply the same philosophy which in their situation leads to inaccurate metadata descriptions;
3. in general standards are evolving together with the development of applications. Metadata specifications are not different. Therefore, conceptual models should follow these new applications.

In the past, many of the developments of metadata standards were focussed on the co-ordinate based geometry and raster geometry played a minor role, as is also correct for the ISO 19115 Metada standard. Therefore, in May 2003 ISO/TC 211 has decided to set up a project to develop ISO 19115-2 *Geographic information – Metadata – Part 2: Extensions for imagery and gridded data* . At the time of the press of this book the project was at WD stage (Working Group draft).

7. Regional and national efforts developing metadata

In some regions, specific developments have led to the use of metadata standards. Some of these were national oriented, while other were developed by some regional countries or regional bodies and accepted by its members. Part 2 of this book will describe these in detail.

7.1. *Europe*

The first attempt for the development of standards in the field of Geographic Information took place in Europe by the Technical Commission 287 on Geographic Information of the European Committee on Standardisation (CEN/TC 287) from 19991 until 1999. Its basic objective was to enable geographic information to be accessed by different users, applications and systems, and from different locations. This requires a standard way of defining and describing this information, a standard method for structuring and encoding it, and a standard way of accessing, transferring and updating via geographic information processing and communication functions, independent of any particular computer system. Suppliers and users of data and developers of GIS's and GI-applications may use the CEN family of standards to enable databases and applications that are different in structure form and content to interconnect and inter-operate.

The CEN ENv 12657, Geographic Information – Data description – Metadata is widely accepted in Europe and many countries use the standard for the creation of metadata in developing a National Spatial Data Infrastructure. Also pan-European projects as the Geographical Data Description Directory (GDDD), Geo-Scientific Electronic Information Exchange System (GEIXS), Added Value Information Dissemination for hydrographic Datasets (AVID),

European Spatial Metadata Infrastructure (ESMI), etc. apply the ENv 12657 for the dissemination of data (see the European chapter in part 2 of this book).

Especially in Europe the multi-lingual situation emerged in projects dedicated towards dealing with this issue like the Electronic Trade for Geographic Information (GISEDI), Methods for Access to Data and Metadata in Europe (Madame) and CLEF (Cross-Language Evaluation Forum); the last one attempting to develop information retrieval system that focuses on non-English approaches including Dutch, Finnish, French, German, Italian and Spanish. The Open Archive Forum supports projects and national initiatives using open archiving approach as an interoperability framework implementing Open Archives Initiative (OAI) metadata protocol in Europe.

Many countries in Europe have implemented the CEN ENv 12657 metadata standard; with the approval of the ISO 19115 Metadata standard, they have to decide to transform the metadata information into this new standard and introduce this into the geo-spatial data infrastructures.

Because of the developments of standards for Geographic Information within ISO/TC 211, CEN/TC 287 has ceased its activities in 1999. Facing the regulations of CEN, making ENv's mandatory for CEN-members to use, CEN/TC 287 has been revived in November 2003 to harmonise the existing ENv's with the new ISO 191XX suite of standards in the field of Geographic Information. Meanwhile CEN ENv's are abandoned and the acceptance of the ISO ISO 191XX standards within CEN is under a voting process. Due to this voting the following ISO standards have been accepted as European standards:

- EN-ISO 19101: 2002 Geographic Information – Reference Model;
- EN-ISO 19105: 2000 Geographic Information – Conformance and Testing;
- EN-ISO 19107 Geographic Information – Spatial Schema;
- EN-ISO 19108: 2002 Geographic Information – Temporal Schema;
- EN-ISO 19111: 2003 Geographic Information – Spatial Reference by Coordinates;
- EN-ISO 19112: 2003 Geographic Information – Spatial Reference by Geographic Identifiers;
- EN-ISO 19113: 2002 Geographic Information – Quality Principles;
- EN-ISO 19114: 2003 Geographic Information – Quality evaluation procedures;
- EN-ISO 19115: 2003 Geographic Information – Metadata;

while the following are still in process:

- prEN-ISO 19116 Geographic Information – Positioning services;
- prEN-ISO 19125-1 Geographic Information – Simple feature access, Part 1: Common architecture;
- prEN-ISO 19125-2 Geographic Information – Simple feature access, Part 2: SQL;
- prEN-ISO-TR 19120 Geographic Information – Functional standards;
- prEN-ISO TR 19121 Geographic Information – Imagery and gridded data.

Based on environmental issues in the EC FP-6 programme, the INSPIRE project was initiated in 2001–2003. The project aims to enable communication between users in a meaningful manner, despite the existing differences in language, context and content of spatial datasets in Europe. The project has two phases: a design phase (2004–2007) and an implementation phase (2007–2015) (see also chapter 1 of Part 2 in this book). INSPIRE intends to apply the ISO 191XX suite of standards.

7.2. *North America*

Between Canada and the U.S.A. a joint committee provided a forum to develop consensual positions, which then can be ratified by the main groups of users and providers. This process is being followed in both countries and has resulted in an identical Canadian and U.S.A. national metadata standard, becoming the North American profile of ISO 19115. In addition, there have been informal discussions between the United States and Canada on a common North American strategy to adopt and implement other ISO/TC 211 standards. Now, the United States has developed a Memorandum of Understanding (MoU) with Canada that will outline procedures for ANSI/INCITS L1 and SCC to develop common profiles of ISO/TC 211 standards.

Both, Canada and the U.S.A. are member of and actively participating in the Permanent Committee on Spatial Data Infrastructure for the Americas, which also contributes to the development of the Global Spatial Data Infrastructure.

7.2.1. *United States of America*

The Federal Geographic data Committee (FGDC), founded in 1990, has the task to develop geo-spatial data standards that enable the sharing of spatial data among producers and users and support the National Spatial Data Infrastructure (NSDI). FGDC working groups and sub-committees develop standards in the field of data transfer, metadata, cadastral data, ortho-imagery, and positional accuracy in consultation and co-operation with state, local, tribal, private academic and international communities through a structured open consensus. The FGDC sub-committees and working groups establish and implement strategic guidance and specific actions that support improved collection sharing, dissemination and use of geo-spatial data, and by doing so, contributing to the development of the NSDI.

The Metadata working group is established to promote an awareness of the metadata dimension to geo-spatial data, among FGDC members and geo-spatial data users. Presently most work of the working group deals with the harmonisation of the FGDC metadata standard with the ISO standard 19115 for Geographic Information – Metadata on order to adopt the ISO standard as an American National Standard. Proposals have been discussed for changes in the FGDC metadata standard, which may not be contained in the SDTS transfer set. Work is done or underway for profiles of the FGDC metadata standard for biological data, shoreline data, remote sensing data.

The U.S.A. is also member of Pan American Institute of Geography and History (PAIGH) and provides technical assistance, conducts training at research centres, distributes publications and organises technical meetings in the areas of cartography, geography, history and geophysics.

As consequence of the Executive Order 12906 of April 1994, to establish a National Spatial Data Infrastructure, also state, local and tribal governments are involved in the implementations of NSDI. In addition, in the U.S.A. three cross-sector stakeholders play also an important role in the development of NSDI and the use of metadata:

1. OpenGIS® Consortium, an international non-profit member consortium developing interface specifications for geo-spatial services;
2. I-team Geo-spatial Information Initiative, an interdependent partnership of federal, state, local and tribal authorities, academia and the private sector for interoperability implementations;

3. GeoData Alliance, a non-profit organisation promoting the creation and effective use of geo-spatial data.

The Geo data Alliance and the I-teams Initiatives will bring all geo-spatial actors in the U.S.A. together to improve the national co-operation in the field of GIS as part of the Geo-spatial One Stop electronic government.

The FGDC Metadata Content Standard for Digital Geospatial Metadata, STD-001-1998 Version 2.0, is well distributed and applied throughout the America's and the Caribbean by workshops to disseminate the information about metadata and clearinghouse implementation.

7.2.2. *Canada*

Since 1997, the Canadian GeoConnections develops the Canadian Geo-spatial Data Infrastructure (CGDI) co-ordinating numerous Canadian databases with geographic information and make them accessible through a common window on the Internet in order to promote partnership among federal, provincial and territorial governments, the private sector and the academic community. Geographic or geo-spatial information is seen as a key driver of the transition to a knowledge-based economy in Canada. In this case, everyone will benefit from having access to (geo-spatial) data and information.

The CGDI has seven GeoConnections programmes:

1. through cost-shared development, GeoExpress-Access provides the private sector, Canadians and government agencies *access* to government geo-information via a common Internet system;
2. national geo-information *framework* (foundation) with diverse data from various organisations, upon which other databases will be integrated or constructed;
3. *federal-provincial/territorial partnership* program to facilitate data sharing, and guide the vertical harmonisation of information sharing and access to geo-information holdings. The GeoConnections Secretariat co-ordinates federal/provincial/territorial ideas, programs and project activities with private initiatives;
4. *industry-partnership* (GeoInnovations) to accelerate the development of promising and advanced geo-information technologies by investments in the private sector to develop new tools and innovative technologies;
5. strengthen Canadian *communities* capacity of to effectively plan and manage their economic, environmental and social development using geo-information and services through selected pilot projects;
6. *National Atlas of Canada* to provide national perspectives on Canada's physical, environmental, economic, social and culture issues for students at all levels and the public in general. Examples of issues include climate change, industrial development and population;
7. *geomatics skills-matching network* fostering knowledge-based industry growth, with an Internet-based information clearinghouse to disseminate cross-jurisdictional information and facilitate employment.

7.3. *Asia-Pacific*

The Permanent Committee on GIS Infrastructure for Asia and the Pacific (PC GIAP) has proposed the Asia-Pacific Spatial Data Infrastructure (APSDI), being a network of databases, located throughout the Asia-Pacific region, that together provide the fundamental data needed to achieve the region's economic, social, human resources development and environment objectives. APSDI Data Node is a network of nodes established to distribute, manage, and maintain fundamental data in each country connected to APSDI. According to the project plan, Australia, China, Japan and Iran will act as demonstration data nodes in the first stage. In each country, a sub-node of the APSDI is planned and they will be linked with each other.

7.3.1. *Pacific*

The co-ordinating Spatial Information Council for Australia and New Zealand (ANZLIC), originally established in January 1986 as the Australian Land Information Council, aims to co-ordinate a widening range of activities and objectives to make land and geographic information readily available to different users. In particular, ANZLIC promotes the availability of land and geographic information for land resource management, environmental monitoring and economic development for both the public and private sector. A key concept of the Council is that each delegate represents a co-ordinating structure within his or her own jurisdiction and that close links are maintained with other co-ordinating bodies such as Intergovernmental Committee on Surveying and Mapping, the Public Sector Mapping Agencies, Registrars-General and Valuers-General.

Version 1 of the ANZLIC Metadata Guidelines was published in 1996 followed by an associated paper 'Recommended Guidelines for the Transfer of ANZLIC Metadata Core Elements'. These reference documents have been well received by the geo-spatial industry. However, with use, some inconsistencies and deficiencies have been identified, and some additional metadata elements have been defined. Consequently, in November 1999 the ANZLIC Metadata Working Group initiated a minor review of the ANZLIC guidelines. The review was seen as an interim measure pending the development of the international standard 'ISO 19115, Geographic information – Metadata'. A further revision of the ANZLIC guidelines is likely to be undertaken once the international standard becomes stable.

7.3.2. *Asia*

The Asian region is geographically and culturally diverse. Because of logistical reasons not many common project in the field of geo-spatial metadata have developed; most developments are country wise undertaken.

However, in the Asian region there are two organisations encouraging metadata initiatives, which are the Permanent Committee on GIS Infrastructure for Asia and the Pacific (PC-GIAP) and the Technical Committee 211 on Geographic Information/Geomatics of the International Organisation for Standardisation (ISO/TC 211). PC-GIAP, has 55 country members extending an region from Armenia, Iran, Azerbaijan and Kazakhstan in the north-west to New Zealand, French Polynesia, American Samoa and Cook Islands in the South East, and has the focus on SDI development in the region, including the Pacific area. The metadata work is part of the Date Node Project that started in 2000.

Although it was not in the scope of ISO/TC 211, developing ISO geo-spatial standards in the commission has brought together many experts and so created a forum in the Asian region for information exchange about metadata projects stimulating national developments based on ISO 19115.

7.4. *Central and South America and the Caribbean*

The region is very active in the developments of metadata as part of SDI developments. Four regional organisations stimulate these developments: PCIDEA, PROCIG, PAIGH and DIGSA.

In developing SDI's, the Permanent Committee on Spatial Data Infrastructure for the Americas (PC IDEA) is taking advantage from the lessons learned elsewhere in the world. Still, it has to face the differences in culture and politics, resulting in specific conditions for each country in the region. Therefore, two sub-regional committees are founded: for the Caribbean and for Central America. In addition, two surveys have been carried out identifying the key actors in the region. Although many countries in the region apply the FGDC metadata standard, the ISO19115 metadata standards is considered more and more.

The Central American Development Project for GIS (PROCIG) – the project ended June 2001 – promoted the integration of statistical and census data with other geographic information for public dissemination. Additionally, PROCIG supported national SDI activities, implementing metadatasets and clearinghouse nodes as well as training and consultancy to strengthen the national knowledge about SDI.

Pan American Institute of Geography and History (PAIGH) is an old institution in the Americas, dating back from 1928. Nevertheless, it encouraged the Inter-American Geo-spatial Data Network for the developments of a regional SDI for the Americas by USGS/Eros Data Center and USAID. Besides, PAIGH also co-financed a project called Atlas of the Americas to develop tools for sustainable decision-making. As part of this project in Argentina, Costa Rica, El Salvador, Mexico and Peru a SDI with metadata of existing datasets in the SDI was created.

The Directory of Geographic Institutes from South America, Spain and Portugal (DIGSA) is a policy-makers summit of national mapping institutions in South America, Spain and Portugal. The summits eases information exchange between members including information about SDI and metadata, although the focus is on licensing, copyrights, training, data production, updating policy and product quality.

Many of the application scenarios in the region are oriented towards the Environmental Decision-Making and Disaster Management due to natural catastrophes regularly returning in the area, as the hurricane Mitch in 1999.

7.5. *Africa and the Middle East*

Africa and the Middle East are entering a period that will show a rapid growth in capturing and publishing spatial metadata as a part in the developments of spatial data infra structures, although at present many of the NMA's stay with the traditional production of paper-based maps. However, the initial efforts to develop SDI, coinciding with the completion of the ISO 19115 Metadata standard, have resulted in an ISO equivalent development and working groups have been established for the standardisation of metadata and for creating a sound geodetic network for Africa.

The United Nations Environmental Programme (UNEP) has, in collaboration with the Regional Remote Sensing unit of the Southern African Development Community in Harare (SADC), drafted a strategy to develop the State of Environment metadataset for South Africa and will soon develop the same for West Africa.

8. Summary and conclusions

The applications of GIS has been globalised in the past and its cross-boundary use is demonstrated by applications as:

- environmental, forestry, marine and cultural protection analysis, as well as disaster regulations programmes, etc., applying GIS analyses in the future Virtual GIS environment using distributed databases and software programmes;
- decision support systems for policymakers in land, and socio-economic planning;
- location based services using GIS for route finding, building constructions, precision agriculture, etc.;
- and many more.

In order to make use of the many spatial datasets in such situations, the geographic information scene has realised to make well-equipped standard metadatasets enabling access to these datasets to the public. Even for the experts in the field, it was recognised that modular standards need to be developed for data dissemination and use. This has lead to different developments of metadata standards in several regions, as:

- in the USA, the FGDC standard Version 2.0 is developed in 1998 and applied in many American countries;
- in Europe, by the CEN/TC 278 designed the ENv 12657 Metadata standard in 1999, which is wide spread applied for geographic data discovery in national and regional SDI's and many project funded by the EU;
- in the Pacific region, ANZLIC published in 1996, a set of core metadata elements contained in ANZLIC's Metadata Guidelines, applied in Australia and New Zealand.

In the 1990's, the need for global standardisation of metadatasets for data discovery and transfer became apparent, which has lead to the development of the IS 19115 – Metadata standard by ISO/TC 211 – Geographic Information/Geomatics. Many countries in the world have stated they will harmonise the locally used metadata standard with the IS 19115.

As can been read in this book many countries are progressing in setting up the metadatasets. In Western Europe, North America, Australia, New Zealand the development of SDI's is well underway, and many metadatasets are already available. In Southern America and Asia, many countries have learned form the lessons from other regions and are creating SDI's and metadatasets. In Africa, the awareness of the need to describe datasets is still evolving.

Acknowledgements

We would like to acknowledge Olaf Østensen, Alexander Martynenko, Harold Moellering, Richard Pearsall and François Salgé for their contributions to this introduction.

We are also grateful to the Department of Geodesy and the Research Institute of Housing, Urban and Mobility Studies of the Delft University of Technology, as well as the Department of Geomatics of the Melbourne University to allow time to make these contributions to this book possible.

References

[Ahonen-Raino 2001]
Description of the content of geographic datasets
P. Ahonen-Raino (National Land Survey, Finland), in: J.T. Bjørke, H. Tveite (eds), Proceedings of the 8th Scandinavian Research Conference on Geographical Information Science, 25–27, June 2001, Ås, Norway

[Bertius 1630]
Geografie der Oudheid
P. Bertius, in: Antiquity Geography, Flanders, 1630

[DLC 1979]
Technical (Unclassified) Report UG478.A88, 1979
Defense Technical Information Center, Defense Logistics Center, U.S.A.

[Duval 2002]
Metadata principles and practicalities
E. Duval (KU Leuven, Belgium), W. Hodgins (Autodesk), S. Sutton (Univeristy of Washington), S.L.Weibel (DCMI), D-Llib Magazine April 2002, Vol. 8, no. 4, ISBN 1082-9873.

[Homer 1978]
A mathematical model of the flow of data in a management information system
E.D. Homer, 11th American Meeting of the Institute of Management Sciences, 1978.

[McCelland 2002]
Challenges for service providers when importing metadata in digital libraries
M. McCelland, D. McArthur, S. Giersch (Colleges Eduprise), G. Geisler (University of North Carolina), D-Llib Magazine April 2002, vol. 8 (4), ISBN 1082-9873.

[McIntosh 1968]
Information Processing 68
S. McIntosh, D. Griffel, Proceedings of International Federation for Information Processing (IFIP) Congress, August 1968.

[Moellering 1991]
Spatial database transfer standards: current international status
H. Moellering (ed.), Published on behalf of the International Cartographic Association by Elsevier Applied Sciences, London 1991. ISBN 185166677X.

[Moellering 1994]
Technical characteristics for assessing standards and for the transfer of spatial data and brief international descriptions
H. Moellering, C. Clement (eds), International Cartographic Association Standards Commission for the transfer of spatial data, Columbus, Ohio, U.S.A., 1994, 104 pp.

[Moellering 1997]
Spatial database transfer standards 2: characteristics for assessing standards and full description of the national and international standards in the world
H. Moellering (ed.), Richard L. Hogan associate editor. Published on behalf of the International Cartographic Association by Elsevier Applied Sciences, Oxford, UK and Tarrytown, N.Y. 1997. ISBN 0 08 042433 3.

[Morrison 1995]
Elements of Spatial Data Quality
J. Morrison, S. Guptill (ed.), International Cartographic Association, November 1995.

[Schelling 1978]
The use of IBM's data dictionary
G. Schelling, Proceedings of the Conference on Data dictionary Systems, London, November 1978

[Weber 1978]
Data Base theory and practice
H. Weber, A.I. Wasserman, Conference in Berlin, 13–15 September 1978.

[Williamson 2003]
Spatial DataInfrastructures – From Concept to Reality
I.P. Williamson, A. Rajabifard, M.-E.F. Feeney (eds), Taylor and Francis 2003. ISBN 0-415-30265-X.

[Uitermark 2001]
Ontology-Based Geographic Data Set Integration
H. Uitermark, PrintPartners Ipskamp Enschede, The Netherlands, ISBN 90-365-1617-X, 2001.

[Ziegler 1978] *Distribution: A New Impetus Toward Understanding Data*
K. Ziegler (International Business Machines, Inc.), August 1978.

Organisation	For on-line contact see URL:
Dublin Core Metadata Initiative	http://www.xml.com/pub/2000/10/25/dublincore/
INSPIRE	http://www.ec-gis.org/inspire
Institute for Electrical and Electronic Engineers, Learning Object Metadata Working Group	http://grouper.ieee.org/p1484/doc/wg12/LOM-WD3.htm
U.S.A. National Spatial Data Infrastructure	http://www.fgdc.gov
FGDC Official statement, Department of the Interior, Office of the Solicitor	http://www.fgdc.gov/metadata/meta_trademark.htm
International Organisation for Standardisation	http://www.iso.ch
ISO/TC 211 Geographic information/Geomatics	http://www.isotc 211.org/

URL's mentioned in this chapter were accessed on May, 2005, unless otherwise stated.

PART 2

REGIONAL SUMMARIES
OF SPATIAL METADATA DEVELOPMENTS
AND ASSOCIATED ACTIVITIES

Edited by

Henri J.G.L. Aalders

World Spatial Metadata Standards
H. Moellering, H.J.G.L. Aalders & A. Crane (Editors)

European Efforts in the Field of Geographic Metadata and Related SDI Activities

Henri J.G.L. Aalders[a], François Salgé[b], Alexander I. Martynenko[c]

[a]*Delft University of Technology, Research Institute for Housing, Urban and Mobility Studies, The Netherlands*
Katholieke Universiteit Leuven Faculty of Engineering, Belgium
e-mail: h.aalders@otb.tudelft.nl
[b]*CNIG, Conseil National de l'Information Géographique, France*
e-mail: francois.salge@cnig.gouv.fr
[c]*Institute for Informatics, Russian Academy of Sciences, Russian Federation, Russia*
e-mail: a.martynenko@mail.ru

1. Introduction

Geographic Information is often perceived in Europe as a sign of sovereignty and standardisation in that field was first seen as a national concern. In 1989, The AM/FM European Division organised in Montreux a workshop dealing with transfer standards. At the time, the challenge was national and discipline specific. National solutions such as the National Transfer Format (NTF) from the United Kingdom or discipline-oriented solutions such as DIGEST (on-line see: http://www.digest.org; DIGEST is an acronym for Digital Geographic Information Exchange Standard; the Digital Geographic Information Working Group – DGIWG – of NATO countries maintains this standard) were compared. In 1991, it was felt necessary to set up a specific, Technical Committee at the European Committee for Standards (CEN, Comité Européen de Normalisation) level: CEN/TC 287 – Geographic Information (since September 2004 see http://cen.iso.ch/livelink/livelink.exe). The technical committee finished its work in 1999 resulting in a set of ENv (European experimental standards) in the field of Geographic Information:

- ENv 12009: 1997, Geographic Information – Reference Model;
- ENv 12160: 1997, Geographic Information – Data description – Spatial schema;
- ENv 12656: 1998, Geographic Information – Data description – Quality;
- ENv 12657: 1998, Geographic Information – Data description – Metadata;
- ENv 12658: 1998, Geographic Information – Data description – Transfer;
- ENv 12661: 1998, Geographic Information – Referencing – Geographic identifiers;
- ENv 12762: 1998, Geographic Information – Referencing – Position;
- ENv 13376: 1999, Geographic Information – Data description – Rules for application schema; as well as some CEN reports:

- CR 12660: 1998, Geographic Information – Processing – Query and update: spatial aspects;
- CR 13425: 1998, Geographic Information – Fundamentals – Overview;
- CR 13435: 1998, Geographic Information – Vocabulary;
- CR 13568: 1999, Geographic Information – Conceptual schema language.

The CEN developed standards will be kept as experimental standards until an evaluation is done in relation to the appropriate DIS (draft international standard) or IS (international standard) of ISO/TC 211 (International Organisation for Standardisation/Technical Committee 211 on Geographic Information/Geomatics). This evaluation has occurred in November 2003 and the revived CEN/TC 287 (secretariat by NEN, the Netherlands Normalisation Institute) abandoned these ENv's ISO191XX standards are accepted in November 2004 by CEN members as EN-ISO standard.

CERCO (Comité européen des Responsables de la Cartographie Officielle) is the forum where the heads of the official National Mapping Agencies of Europe meet. It has been involved in co-operative activities of exchanging information on mutual problems and in collaborative initiatives towards a better integration of the products of its members. CERCO depended on periodic assemblies and working groups, from its start in 1980. In 1991, CERCO created its Permanent Technical Group to conceptualise a Multi-purpose European Ground Related Information Network (MEGRIN). Then, 18 CERCO's National Mapping Agencies (NMA) decided in 1993 to create the MEGRIN Group, which further developed into the GIE MEGRIN (Economic Interest Group), later with 19 members. As of first January 2001, CERCO and MEGRIN decided to fuse into EuroGeographics composed of more than 30 European NMA's. Metadata and quality is a central activity.

In 1990, the DG XIII (Directorate General XIII, in charge of Telecommunications, Information Market and Exploitation of Research) of the Commission of the European Communities organised a workshop in Brighton, U.K., on the activities that the Commission could undertake in the field of Geographic Information. (DG XIII is now renamed to DG Information Society with slightly different remits.) At the meeting, there was consensus on the possible benefits that could be gained by the Geographic Information economical sector from the creation of a European Umbrella organisation. A team of four prominent experts worked between 1991 and 1993 to investigate the feasibility, desirability and practical details of creating such an organisation. In 1993, EUROGI, the European Umbrella Organisation for Geographic Information, was set up. Amongst others, the promotion of standardisation and use of spatial data infra structures using metadata standards are their activities.

Realising present-day economic development, modern countries demand accurate and detailed geographic information to maximise the value of promising new geographic technologies such as Global Positioning Systems (GPS) and Geographic Information Systems (GIS). To realise all the potential benefits of geographic technologies at the European level, EUROGI stimulates the harmonisation of required geographical data by European co-operation, complementing the same efforts at the national level, trying to improve:

- transfer and integration of geographic information;
- sharing of experience gained by the numerous national and international geographic technologies research initiatives;

- contribute to the reduction of the costs of geographic information and geographic technologies and to their more widespread use.

The mission of EUROGI is to maximise the effective use of geographic information for the benefit of the citizen, good governance and commerce in Europe and to represent the views of the geographic information community. EUROGI achieves this by promoting, stimulating, encouraging and supporting the development and use of geographic information and technology. EUROGI tries to achieve this mission by the following objectives to:

- raise awareness of the value of GI and its associated technologies;
- encourage the greater use of geographic information in Europe;
- work towards the development of strong national GI associations in all European countries;
- facilitate the development of a European Spatial Data Infrastructure (ESDI);
- represent European interests in the Global Spatial Data Infrastructure (GSDI).

The European Commission has long been involved in standardisation, as one of its major projects CORINE was developed in the field of the environment. EUROSTAT, was leading the GISCO (GIS for the Commission) project (one of the outcome of the CORINE project) and is also using the SABE database (Seamless Administrative Boundaries in Europe), produced by MEGRIN and updated now by EuroGeographics for many European countries.

In the last decade, the DG XIII of the EU has set up framework programmes for the stimulation of the use of digital techniques and was executing the fourth and the fifth framework programmes for Research and Development (FP-4 and FP-5) till 2000. The European Union stimulated the use of digital information by the INFO2000 programme, which was supported by EUROGI for the field of Geographic Information.

The central theme of INFO2000 is the development of a European information content industry, capable of competing on a global scale and able to satisfy the needs of Europe's enterprises and citizens for information content, leading to economic growth, competitiveness and employment and to individual professional, social and cultural development. The programme aims to achieve this through four main Action Lines:

- stimulating demand and raising awareness;
- exploiting Europe's public sector information;
- triggering European multimedia potential;
- Support Actions.

The INFO2000 programme had a four-year work programme from 1996 until 1999. Several projects of the INFO2000 programme were dealing with metadata.

In the European context, a specific issue is the use of many (natural) languages used in the metadata services in the different European countries. Some EU funded projects are especially dealing with this problem.

Based on the experiences of several projects carried out under the 4th and 5th Framework Programme, the EC Directorate General Joint Research Council developed the INSPIRE project for Environmental issues in Europe. The background of INSPIRE (Infrastructure for Spatial Information in Europe) (see http://www.jrc.cec.eu.int) is the complexity of environmental problems in Europe with many distributed datasets, requiring an integral approach. Geo-information becomes an important factor for this European policy in the 6th Framework

Program of the EC (FP-6). INSPIRE aims at well-functioning national geo-information infrastructures, based on common standards. There are 17 projects defined within INSPIRE as an EC project by the DG JRC, containing technical standards and protocols, organisational aspects, information policy and management and access to geo-information both for policy makers and the citizens. From Technological point of view INSPIRE relies on the following principles:

- data collection and management at one site;
- seamless access from different sources;
- application at different scales;
- access for all policy makers under acceptable conditions;
- easy to find by good metadata;
- easy to understand and interpretation.

In the next parts the efforts and results on the developments of standards and the use of metadata in Europe will be discussed in more detail both at the European level (paragraph 2 of this chapter), as well as at the national level (paragraph 3 of this chapter).

2. European projects funded by the EU

The European geographical information market is still very nationally focused, and this means that users requiring pan-European datasets face a number of obstacles in acquiring the information they require, including:

- lack of awareness of available datasets, their quality and means to obtain them;
- datasets from different organisations are often available with different licence terms (which may even be contradictory);
- data itself is unlikely to be consistent across borders.

2.1. *ETeMII*

One of the EU funded projects from the fourth and the fifth framework programmes (FP-4 and FP-5) is ETeMII (European Territorial Management Information Infrastructure), aiming at organising a network of excellence, bringing together most of the stakeholders of the Territorial Management Information market, coming from research, industry and public sector. Territorial management means any management activity, related to the territory; it covers a wide scope of activities, including agriculture, transport, utility management, land planning, environment, fisheries, geo-marketing, etc. Particular attention will be given to user participation within this project, to ensure that all tasks are based on users needs. Awareness activities are an important component of the project.

Such a network will build consensus on the most important technical issues that are the foundation of ETeMII:

- reference data;
- data access policy;
- interoperability, standards implementation, including metadata;
- research and development challenges;

- integration of space tools: positioning systems, Earth Observation and telecommunications;
- active participation into global initiatives: ISO, GSDI, OGC, etc.

To move a step towards the creation of a European Information Infrastructure three themes of ETeMII are promoted: to be able to advertise best practice and promote contribution to and use of GI infrastructure issues.

2.2. *GDDD and La Clef*

As a commercial organisation MEGRIN represents, and is funded by, a group of 19 NMA's. It aims are to bring a European dimension into NMA national activities and to meet the increasing demand for pan-European data by improving international users access to national datasets (see Figure 1).

It does this by providing information about the digital data available now in 23 countries (see Figure 2), and is creating harmonised pan-European datasets. Although MEGRIN comprises only 19 full members, its everyday partners are all European NMA's, i.e., 30 CERCO members.

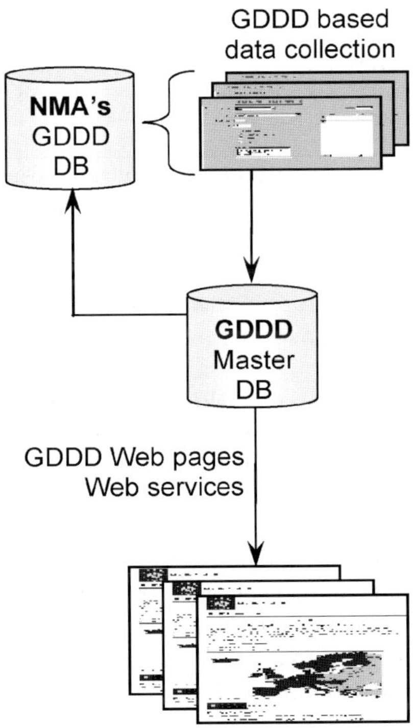

Fig. 1. As an example: Geographical Data Description Directory (GDDD), the central database at Multi-purpose European Ground-Related Information Network (MEGRIN), which is updated with data collected from National Mapping Agencies (NMA's) and from which static web pages are derived.

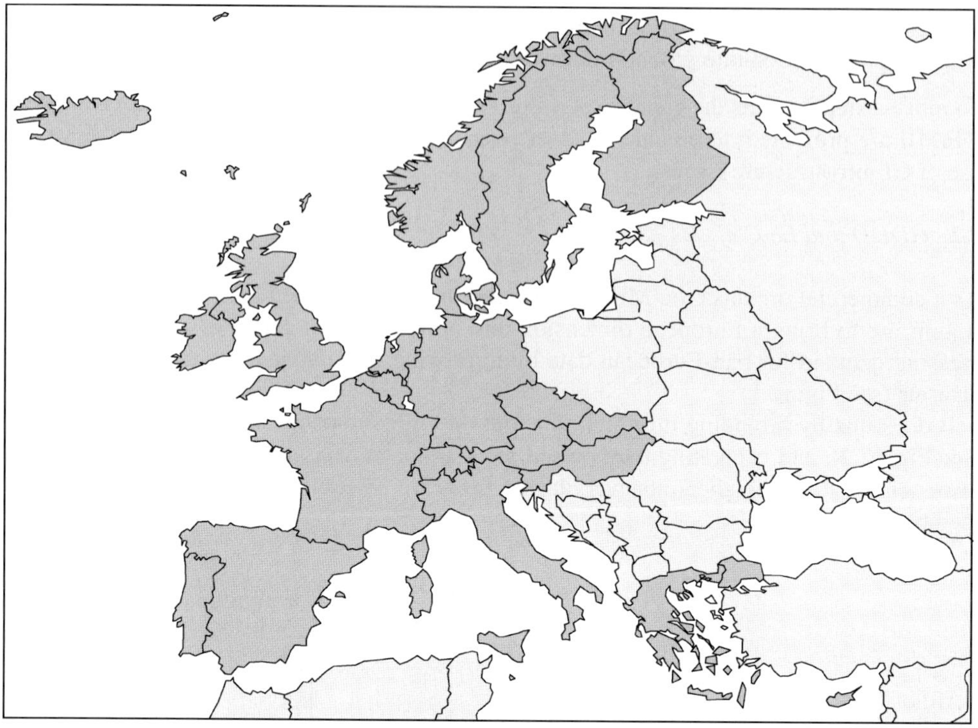

Fig. 2. The Geographical Data Descriptive Directory (GDDD) coverage: GDDD provides metadata covering the countries with dark colour.

Recognising the growing pan-European market demand, MEGRIN has been established to focus on two areas of activity: providing metadata and creating new harmonised digital datasets.

As of first January 2001, CERCO and MERGIN decided to fuse into EuroGeographics composed of more than 30 European NMA's. Metadata is a central activity.

2.2.1. *GDDD*

MEGRIN GDDD metadata service (Geographical Data Description Directory) provides information about 250 digital datasets available from the NMA of 23 countries of Europe. The GDDD was also the first pilot implementation of the pre European metadata standard prENv 12657 of CEN/TC 287. Its current structure has existed since its introduction in November 1994. In 1996, it became more accessible as the widespread use of World Wide Web (WWW) browser came more commonly into use.

Metadata information in the GDDD falls into the following categories:

- overview: short abstract, including contact address (organisation, web-site address and person);
- commercial information: containing some commercial details of coverage, copyright, format, price and other conditions related to the use of the datasets;

- technical information: describing the technical specifications of data sources, features and content, updates, data accuracy and other data quality parameters;
- descriptions of the provider are also held in organisation details.

2.2.2. *LaClef: developing GDDD to a future One-Stop geographic information shop*
Users require data which is ever easier to access, and would ideally like to use a 'one stop shop' to view, purchase, and be supplied with a wide range geographical data, including topographic data, remotely sensed imagery, geological and demographic data. So, LaClef must be a four-sided solution covering:

- semantic issues;
- distributed architecture;
- a wide range of services related to the metadata offered on LaClef;
- e-commerce facilities.

It is likely that such shops will be created incrementally. For the development of this metadata service the next steps should be:

1. distributed metadata systems will initially link the various (existing) national databases of metadata;
2. databases of metadata will then develop to enable on-line data sales.

Stage 1: Distributed systems

Although metadata systems such as the GDDD have large numbers of users, they tend to be nationally based and are often specialist restricted to services for specialist in a specific discipline (and, maybe a specific provider, such as topographic data from the government). In their current form, they do not allow users to access the geographic information directly, but are directories, which enable users to discover what exists. Accesses to the data itself need to be developed to enable an as wide audience as possible to reach geographic information. In the European context, a distributed metadata system will face many challenges to overcome the integration of various services, as well as different languages, application fields and standards. This section aims to provide an overview of the issues that will be important in developing distributed systems. All metadata systems need to provide information, which is easily accessible, reliable, and up-to-date. The following will have significant impacts on the success of distributed systems:

- semantics (terminology);
- (meta) data model;
- future data transfer formats.

Within LaClef, these topics will be investigated by connecting, e.g., GEIXS, the metadata service of EuroGeoSurveys (European National Geological Surveys, on-line see: http://www.eurogeosurveys.org), to the system. GEIXS is under development; a prototype is already accessible in the web. These topics were also addressed in the INFO2000 Project ESMI, co-funded by the European Commission DGXIII, where MEGRIN, a partner in the Consortium, will bring its practical experience in running a 'pre-distributed-metadata system'.

LaClef will use the XML standard to enable easy metadata transfer from the data provider to the central database, or alternatively offer a linked service to another metadatabase. The

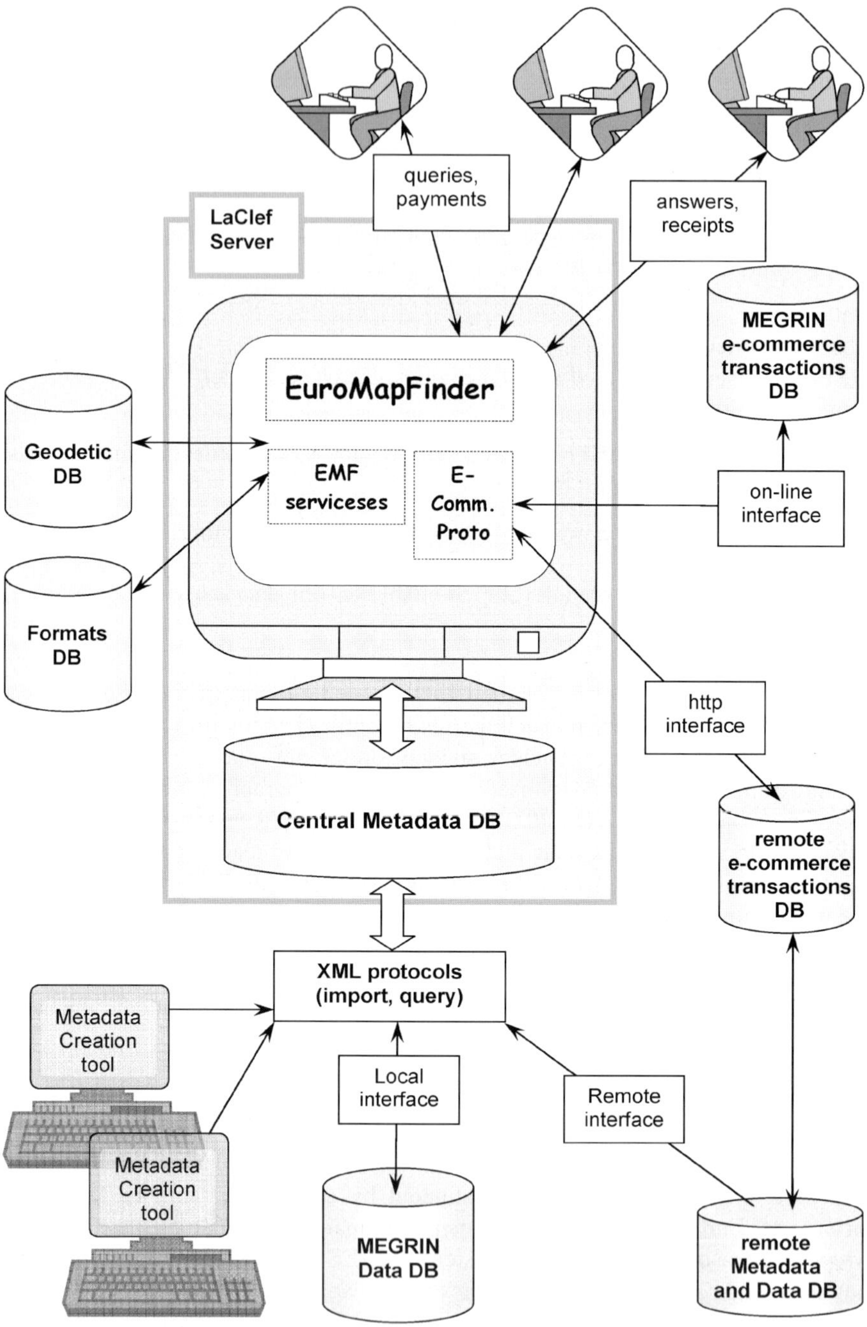

Fig. 3. Planned structure of new Multi-purpose European Ground-Related Information Network (MEGRIN) metadata service as it will be realised within the LaClef project.

data producer will be able to extract XML-formatted metadata from his own metadata in his own local metadatabase. The XML-formatted metadata will be sent to the LaClef metadata service where they will be imported with an import tool. From LaClef, they are available by dynamic HTML-pages to the users.

Stage 2: On-line data sales

Although the immediate aim is to connect metadata systems, there is growing demand for on-line access to data. To be truly useful, metadata services will need to be more closely integrated with the data purchasing process. Ultimately, users are interested in accessing actual geographic data. *Metadata is the means to achieve the end, not an end in itself.*

On-line supply is currently hindered by various factors; including the wide range of geographical data transfer formats. If on-line supply is to be successful, the interoperability between different systems needs to be improved. Various investigations are underway to include the formats of OpenGIS® and OGDI (OpenGIS® Data Interchange) consortia, standardisation bodies, while software vendors are developing products, which can change formats on the fly without the user needing to know. There are also the issues of data security, secure financial transactions and a range of marketing topics. For example, the existence of a network of agents and distributors acting for an NMA may make it only possible to introduce direct on-line sales slowly, due to existing long term commercial contracts which must be honoured.

2.3. *GEIXS*

This Geo-Scientific Electronic Information Exchange System (GEIXS, on-line see: http://geixs.brgm.fr/) will help in all European languages to find out, whom to approach for information on minerals, oil- and gas, groundwater, geology, natural disasters and geo-techniques anywhere in Europe. GEIXS is the European Geological Data Catalogue. It gives a dataset description through:

- geographic coverage of the data;
- key words from lexicons;
- free text.

Its aim is to provide a single point access to geological metadata because geology crosses borders. It also aims at reducing language difficulties. It is useful for:

- land use planners;
- minerals industries;
- civil engineering;
- waste disposal monitoring;
- energy sectors;
- water industries;
- environmental sectors;
- insurance and banking;
- pollution control;
- government infra-structural strategy;
- coastal flooding studies;
- global change;

- health studies (e.g. radon);
- research in schools, colleges and universities.

Geographic search allows the selection of geographic data using both geographic and thematic criteria. It is also possible to display the geographic data coverage. Thematic search allows to search data using more extensive thematic on top of the geographic search. GEIXS is a EuroGeoSurveys product and the European Commission has supported its development.

2.4. *AVID*

A wealth of information about the sea is stored in the databases of public hydrographic offices. AVID (Added Value Information Dissemination for hydrographic datasets, on-line see: http://www.ec-gis.org/avid.htm) will develop a prototype on-line service to provide access to this information for both general and specialised users. Data on bathymetry (depth measurement), coastal topography, sedimentology, waves, currents, tides, landmarks, buoys and beacons, lights and sea-limits will be made available.

AVID aims to demonstrate the effectiveness of European information service based on hydrographic data. An essential objective is to add value for potential users by providing for the integration of different sources of hydrographic data.

Additional objectives of the project are to:

- study user requirements (with special focus on fishery and coastal management);
- define products which will supply added-value information to end users;
- develop a prototype service for information delivery, including catalogues of available data;
- promote the use of the service and assess its performance.

A pilot, web-based system will be demonstrated in two application fields, which will be chosen for their applicability to the world of business. The service will include an on-line inventory of hydrographic data sources for at least three European countries. In addition, the AVID team will study and publish written reports on legal and copyright issues, pricing policy and charging systems, which will make the exploitation of public sector information easier to achieve. Workshops and web-based information will be targeted at potential users, in order to raise awareness.

2.5. *CLEAR*

When an infrastructure development or a business project crosses national boundaries, there can be particular problems in assembling all the geographic (and environmental protection) information needed. It is in the interests of both providers and potential users of this data that it is easily accessible, up-to-date and comprehensive. The CLEAR project (spatial data CLEARing house, on-line see: http://www.ec-gis.org/clear.htm) focuses on the Saar-Lor-Lux region, which includes the German Länder of Rheinland–Pfalz and Saarland, the Grand Duchy of Luxembourg, the French region of Lorraine and the Belgian Province Luxembourg, encompassing an area with some eight million inhabitants. The project team will develop a system giving information about available geographical data (a metadata system) including ownership, price and technical quality held in the participating countries. It will also provide a means to access the data itself.

The key objective of CLEAR is to develop a central, bi-lingual French/German information system for geographical data in the region (a metadata service) as well as a functional delivery system to provide direct access to information resources held by the public sector. The creation of a separate vehicle to market the spatial data will be considered. The project will produce:

- catalogue of digital spatial data for the region (the metadata system);
- definition of user requirements based on research through questionnaires and workshops;
- proposals for operational standards for the transfer, conversion and integration of data across national borders;
- proposals for a legal, organisational and political framework for a spatial data clearing-house for the region;
- bi-lingual information and delivery system for the region based on CD-ROM or Internet technology.

The CLEAR project will link users of digital geographic information: businesses, government departments, research and educational institutions as well as private individuals.

2.6. *ESMI*

The use of geographic information in all parts of European society is growing. Private business, government, research and educational institutions as well as private individuals are increasingly using geographic information as a key component in their activities. Increased international co-operation and competition means that there is a need to determine what geographic information is available in other countries or organisations and how to obtain it. The proposed European Spatial Metadata Infrastructure (ESMI, on-line see: http://www.ec-gis.org/esmi.htm) is an initiative to establish a framework for the distribution of geographic information by creating a universal metadata service. It is set up by several European public and private organisations (including CNIG in Portugal, Geodan in the Netherlands, University of Valencia in Spain, MEGRIN in Paris, France and Ordnance Survey of the United Kingdom).

There are already a number of existing metadata services in Europe and elsewhere. However, these systems are specific, nationally oriented and rely on central servers which may be separate from the actual geographic data. They do not communicate, so there is some duplication, which gives rise to inconsistency. ESMI has the objective to link these existing and future metadata systems. ESMI will create a European spatial metadata infrastructure by providing mechanisms to link spatial data users with metadata services using Internet. However, by allowing metadata services to communicate between themselves, a single access to the service by World Wide Web will allow access to all existing, connected metadata servers and provide multimedia tools to explore, search and compile relevant information. Links to data provision services will be made available where possible. ESMI intends to bring together existing technologies to add value to previous work. ESMI will develop interfaces for the user (to browse and query metadata) and the provider (to make metadata available and to maintain those metadata) as well as the system to connect the two. In addition, ESMI will simplify access and maintenance of metadata taking account of different systems, languages, etc. The participants who have wide experience and skills in these fields will bring all these components and functions together. Obviously, the implementation of ESMI can only be successful if national and international data providers are involved.

2.7. *GEOSERVE*

Europe has a wealth of high quality digital geographic information. The variety of data formats, reference systems, projections and quality standards reflect the wealth of Europe's history; but it is an obstacle to cross-border transfer and integrated use of geographic data. The access mechanisms are poorly developed. The GeoServe (Geographic Data Access Services, on-line see: http://www.ec-gis.org/geoserve.htm) project develops a brokering system that allows the user to identify the geographic data from a data catalogue of many providers in Europe and to order it in formats required by the users' applications. Additionally geographic services can be ordered. The system is based on distributed metadatabases that implement the CEN/TC 287 standard. Communication is done on Internet and Intranets. For casual users and simple Web clients, information kiosks are provided, interfacing a variety of geographic viewers, as well as GIS supports for professional users. Providing access to geographic information and geographic services is an important step on the road towards interoperability of GIS and geographic applications. GeoServe is a Research and Development project supported by the EC DGXIII Telematics for Administrations Programme. Siemens Nixdorf leads an international consortium of industry and data user administrations.

The project GeoServe aims to develop a network of Geographic Data Access Services for European Administrations and Data Providers. The Services shall support the identification, selection, and timely access to geographic data for data users on a European scale. For the provider of geographic data, the system will enable the marketing of geographic data internationally to a large user group.

The objectives of the project GeoServe are to:

- harmonise the requirements of European data providers and data users;
- develop a concept for a European geographic data access services network;
- guarantee that the user needs and user participation drive the project;
- build demonstrators for the network service nodes and for the data provider client, the geographic data user client and information-desk client;
- assure openness and flexibility of the system by interfacing with GIS's, geographic data formats, geographic viewers and application areas.

The project is driven by user demands. A consortium of leading IT industrial partners, service organisations, geographic data provider administrations and data user administrations shall develop the concepts and services and validate and test them with pilot applications. Pilot sites will demonstrate regional and international transfer of geographic data. The GeoServe system will provide Geo-Data service nodes, Geo-clients, and added-value functions on Wide Area Networks. The demonstrators in Finland, Greece, Germany, and Italy will link various international and regional data providers and users. They will support different geo-clients for data users in administrations based on PC/Windows and citizen information kiosks.

2.8. *GISEDI*

The GISEDI Europe project (Electronic Trade for Geographic Information, on-line see: http://www.ec-gis.org/gisedi.htm) aims to develop a European commercial and technical network infrastructure which will both facilitate and accelerate the transfer and trading of geographic information at the local, regional, national and international level. This requires the development of web-based GI query, view and retrieval functionality, integrated with secure transactional procedures (EDI and e-commerce solutions), which will be tried in four European countries. The principal output of the project will be the development of a Book of Specifications, which will provide the technical basis for development of GISEDI systems in other European and worldwide locations.

GISEDI is a project involving eight organisations from seven countries in Europe. These are the European Umbrella for Geographical Information EUROGI, Cara Broadbent & Jegher and URBA 2000 in France, Walter Research Centre from the United Kingdom, Instituto Engenharia de Sistemas e Computadores in Portugal, Ususimaa Regional Council from Finland, EDI Hellas SA from Greece, Indra SSI SA from Spain. Its main purpose is to provide a model for a trading infrastructure to help users to locate and purchase geographic information and suppliers to find a larger market for their data. For this open and flexible electronic market, the concept of a broker has also arisen.

The GISEDI Europe system will be demonstrated in several countries (e.g.: Greece and Finland). The users of the GI market are divided into three broad types:

- GI data suppliers (enterprises or institutional organisations);
- GI data users (either businesses or individuals);
- GI data brokers (enterprises or institutional organisations).

The GI data broker is a new entity, emerging with the development of GI electronic interchange systems. The GI data supplier and GI data user are relatively well-defined entities, although single organisations may act as both a user and supplier. The GI data broker, as a facilitator of data transfer between suppliers and users, is not well defined; indeed, it is an objective of the GISEDI project to define this role more closely. The survey of users and suppliers produces profiles that are real since these are based on organisations that have well-defined activities and characteristics. The survey of brokers is based mainly on organisations that have a perceived future role in this activity and hence the profile varies with these perceptions.

2.9. *MADAME*

The MADAME project (Methods for Access to Data and Metadata in Europe, on-line see: http://www.ec-gis.org/madame.htm) will identify solutions and best practices for making public sector data available across Europe. Focusing on geo-statistical and cadastral information in particular, it will evaluate current services, providing access to data and metadata from the perspective of current and potential users. The overall objective is to move from services that are producer-oriented to user-oriented services. An important strength of this project is that it will examine data provision at (and across) three levels: European, national and local. This approach will be critical to developing transferable guidelines of best business practice. It will inform public sector agencies in their approach to making their datasets more accessible to others. The objectives of the project are to:

- evaluate the current strengths and limitations of metadata service provision at European, national and local level. The project will consider the issues of copyright, data documentation, data access and pricing, data security, confidentiality and liability;
- compare the relevant institutional and legal frameworks in participating countries;

- identify the data needs of key users in public, private and academic sectors;
- develop a compendium of best business practice;
- increase the quality of support services based on best business practice;
- raise awareness of the results achieved and practical steps needed for further development.

The MADAME project has delivered:

- compendium of best business practice for access to data and metadata in Europe;
- comparative evaluation of data access policies and data infrastructures in the four participating countries at European, national and local levels;
- two manuals addressing respectively organisational and institutional issues and legal and economic issues affecting the increased exploitation of the public sector;
- user needs study in evaluating current service provision for data and metadata in each participating country;
- web-based data and metadata reference site for transfer of methodologies and practice in data documentation.

As follow-up of MADAME, the findings of several of the previous projects were proposed to develop:

- increasing awareness of the value of information as an organisational asset and metadata as the key to unlocking it;
- identification of specific user needs, and offer pragmatic solutions to the development of sustainable information policies;
- fostering interoperability through the adoption of international standards, in liaison with ISO and CEN, and national associations for geographic information.

This is particularly aimed at local government across Europe, focusing on geographic information as an important part of public sector information. It will be executed through a series of eight workshops and focus groups to gather user needs and disseminate findings, and guide activities with International standardisation bodies, the European Commission's Joint Research Centre, and OGC.

2.10. *INSPIRE*

The background of INSPIRE (Infrastructure for Spatial Information in Europe) (see http://www.jrc.cec.eu.int) is the complexity of environmental problems in Europe with many distributed datasets, requiring an integral approach. Geo-information becomes an important factor for the European policy in the 6th Framework Program of the EC (FP-6). INSPIRE aims at well-functioning national geo-information infrastructures, based on common standards. There are 17 projects defined within INSPIRE as a EC project by the DG JRC, containing technical standards and protocols, organisational aspects, information policy and management and access to geo-information both for policy makers and the citizens.

From technological point of view INSPIRE relies on the following principles:

- data collection and management at one site;
- seamless access from different sources;
- application at different scales;

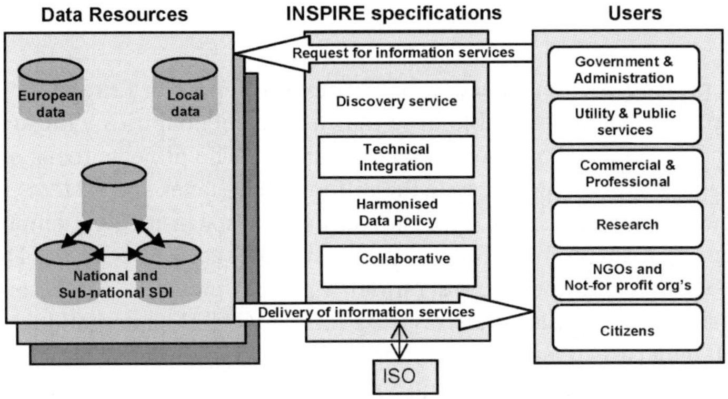

Fig. 4. INSPIRE information flow.

- access for all policy makers under acceptable conditions;
- easy to find by good metadata;
- easy to understand and interpretation.

The first expert groups to establish INSPIRE were founded in 2001, resulting in a position paper in 2002. This proposal was accepted by the EC in 2003, sending it to the European Parliament. The kernel of INSPIRE is the ability of two or more autonomous authorities with different systems, applications or data to communicate with each other in a meaningful manner, despite the differences in language, context or content. This should be achieved by the formulation of an ESDI (European Spatial Data Infrastructure), linked to national SDI's, which are the responsibility of the national EU members states, including metadata, reference data, thematic data and derived products. Also technical specifications, an EU portal and procurement guidelines will be created.

Because many large scale IT projects have proved to fail, the strategy in INSPIRE is to formulate many small manageable projects and learn from the successes and failure of these projects. The project is made in two phases:

- 2003–2007 – pre-implementation phase, creating and central information counter, demonstrating the interoperability, testing the process of intra-community consensus building, develop GI standards and broaden the activity field with extensions to e-Europe and e-Government;
- 2007–2015 – implementation phase.

To achieve the goals of the first phase an action plan has been enacted:

- create an INSPIRE metadata profile for discovery through a study and workshop by EUROSTAT and Joint Research Council;
- Revive CEN/TC 287 and CEN Workshop Agreements;
- Establish relationship with the e-Europe Steering Group;
- EuroSpec second workshop;
- FP-6 Integrated Project with applications based on INSPIRE principles.

2.11. *Summary*

Many different projects are developed or are being developing in Europe to enable the easy access of spatial data that are available in so many different countries with so many different types of data and systems and using different languages. The projects focus on such aspects, i.e., to make spatial data available despite the different languages and systems used within the Europe countries. The conclusions from the various European projects emphasise the need for providing mechanism to map ISO/TC 211 metadata standard with the Dublin Core for discovery purposes. In addition, the migration of present metadata databases created under CEN/TC 287 standards towards a system using the ISO/TC 211 international standard when available is well underway. Another main aim of the EU in funding these projects is to bring the data to the civilians and commercialise the data. However, these objectives are not GI specific but linked to general IT programmes. It is then up the GI sector to itself to propose projects connected to GI. This means that all these projects do not serve any European GI policy but are initiatives benefiting from funding opportunities.

Nonetheless, GI actors failed to obtain the political support from senior European Union decision makers although a position paper (called GI2000) almost reached the ‘communication of the European commission’ status. The new strategy of the GI sector is to channel the European GI infrastructure concept through applications area such as Environment: a new EC initiative (starting September 2001) is on its way to define a European Environmental GI infrastructure. Similarly, the GINIE project (Geographic Information Network In Europe), funded by the European Commission, will contribute to the development of the European GI infrastructure, in implementing EUROGI strategy paper.

3. National developments in the field of SDI in Europe

Apart from activities and developments by co-operating countries in Europe as described above, in many European countries also National Geographic Information Infrastructures (NGII) are set up to increase the interoperability between producers and users of geographic information. In this section, these developments are described for those countries contributing to this book. These countries were Republic of Croatia, Czech Republic, Germany, Ireland, the Netherlands, Poland, Romania, Spain, Turkey, and the Nordic countries, i.e., Denmark, Finland, Iceland, Norway and Sweden.

3.1. *Croatia*

The major national mapping organisations in the Republic of Croatia are:

- State Institute for Standardisation and Measuring (Drzavni Zavod za Normizaciju i Mjeriteljstvo);
- State Geodetic Administration (Drzavna Geodetska Uprava);
- Civil Engineering and Environment Protection Administration of the Ministry of Defence of the Republic of Croatia (Uprava za Graditeljstvo i zastitu okolisa Ministarstva obrane Republike Hrvatske);
- Croatian Hydrographic Institute (Hrvatski Hidrografski Institut).

The international spatial data organisations in which the Republic of Croatia actively participates are:

- Comité Européen des Responsables de la Cartographie Officielle – CERCO and the Multipurpose European Ground Related Information Network – MEGRIN; the national representative is the State Geodetic Administration;
- International Hydrographic Organisation (IHO) and International Maritime Organisation (IMO), where the national representative is the Croatian Hydrographic Institute.

Croatian metadata activities are at the very beginning. In 1994, the State Geodetic Administration founded six technical committees with the aim to propose priority to standards for the Croatian geodetic and spatial system. In 1998, the State Geodetic Administration together with the State Institute for Standardisation and Measuring changed the structure of the six technical committees into only one committee, having the same purpose but established formally according to the ISO/TC 211 and CEN/TC 287.

The Croatian maritime spatial system attempts to take over the metadata standard of IHO (S-57) and DGIWIG (DIGEST) as well as of other international organisations such as ISO/TC 211, SQL/MM of the OpenGIS® Consortium. The plans for metadata in Croatia are to learn about the spatial metadata from other European countries and international organisations and to design the national standards that will confirm the CEN and ISO directions.

3.2. *Czech Republic*

In the Czech Republic, the following organisations provide geographic information:

- Land Survey Office with the following datasets:
  - ZABAGED/1: containing vectorised features of the basic topographic maps 1 : 10 000, finished for about 80% of the Czech Republic;
  - ZABAGED/2: geo-referenced images of the basic topographic maps 1 : 10 000;
  - Geo-referenced images of base topographic maps 1 : 50 000;
- Military Topographic Institute with:
  - DMU200 containing vectorised features of (military) topographic maps 1 : 200 000;
  - DMU25 with vectorised features of (military) topographic maps 1 : 25 000;
  - DMR1 (DEM with 1 km grid);
  - DMR2 (DEM with 100 m grid);
  - archive of aerial images;
- Forest Management Institute, having:
  - Geo-referenced images of basic forestry maps 1 : 5000;
  - archive of aerial images of forests;
- Czech Geological Survey completed vectorised features of geological maps 1 : 50 000;
- many local GIS providers and local governments hold geographic data.

The application of metadata by providers is on a voluntary basis by using the central service of the Czech Association for Geographic Information (CAGI), where each provider can insert data record in system through the World Wide Web. CAGI Meta information system is an implementation of the CEN prEN 12657. Contemporary query services of CAGI Meta information system work on:

- text search in record name or abstract within selected system entity (i.e., dataset, person, project, event, organisation, software, etc.);
- selection of administrative unit in map;
- text search in dataset name and abstract within selected dataset class (i.e., topography, plans, DEM, aerial image, etc.).

3.3. *Denmark (including Greenland and the Faeroe Islands)*

In Denmark, since April 1997, the National Survey and Cadastre (KMS-DK) has developed a metadata service for spatial information on the web (on-line see http://www.geodata-info.dk some of the data descriptions are in English). The service aims at the Geo-ICT sector itself as well as professional GIS-users. Its purpose is to give an overview of the most important spatial datasets and data-collections within the private as well as the public sector. Hence, the datasets described are both public datasets made for administrative purposes and commercial products provided by private companies and both, direct and indirect referenced datasets. The service is free of charge for the users as well as for the data providers, and it is up to the data providers themselves to decide if they want to participate.

The Danish metadata-service is built as an implementation of the CEN GI-metadata standard (CEN ENv 12657); the majority of the CEN standard elements are supported, except the very detailed, descriptive elements in the 'data description' part of the standard. Data is stored in a relational database (at present MS-Access) and all the web pages are static HTML-documents, produced by an RDB-to-HTML application (developed by KMS-DK). Future developments encompass XML transfer procedures between heterogeneous systems, and change over from CEN to the ISO range of geographic standards.

In Greenland, the metadata-service is based on the same concept as the Danish one. A similar service for the Faeroe Islands has been considered.

3.4. *Finland*

The National Land Survey (NLS) has been responsible in the past decade for the maintenance of the National Geographic Dataset Directory (on-line see: http://www.nls.fi/ptk/aineistot). Finland has no national metadata standard. For the public administration, Finland has a recommendation describing data products called 'JHS 137 Data Product Description' and an additional part, specific for GI, called 'JHS 137A Data Product Description – Geographic Information'. Presently the Geographic Dataset Directory contains descriptions of over 300 geographic datasets of approximately 20 national organisations and more than 40 local and regional authorities. The dataset description in the National Geographic Dataset Directory consists of HTML-files, which are automatically generated from the ASCII source files.

Currently the Finnish Environment Institute (FEI) is developing a metadata system specifically for the environmental sector. The development work is carried out in collaboration with the Statistics of Finland and NLS. The FEI metadata-service will be able to communicate with the Statistics Finland's metadata service and national geographic dataset directory services in the future, possibly using XML.

Major Finish players in the GI field	Web site
Finland Post Ltd.	http://www.posti.fi/
Finnish Environment Institute	http://www.vyh.fi/syke/syke.html
Finnish Forest Research Institute	http://www.metla.fi/
Finnish Game and Fisheries Research Institute	http://www.rktl.fi/
Finnish Geodetic Institute	http://www.fgi.fi/
Finnish Maritime Administration	http://www.fma.fi/
Finnish Meteorological Institute	http://www.fmi.fi/
Finnish National Road Administration	http://www.tieh.fi/
Finnish Population Register Centre	http://www.vaestorekisterikeskus.fi/
Forest and Park Service	http://www.metsa.fi/
Geological Survey of Finland	http://www.gsf.fi/
Ministry of Justice	http://www.om.fi/
National Land Survey of Finland	http://www.nls.fi/
Statistics Finland	http://www.stat.fi/

3.5. *France*

Key players in the provision of geographic data are the National Statistical Institute (INSEE) in charge of gathering statistics at the commune level, the National Tax Office (DGI), responsible for the cadastre, the National Geographic Institute (IGN) for topographic mapping and the military hydrographical service (SHOM) carrying out nautical charting and several local governments.

The DGI launched in the early 90's a Digital Cadastral Plan (PCI) project to digitise its paper-based holdings through a series of partnerships with local authorities and utilities. This programme has been developed for the main urban areas and agglomerations but also in some departments such as Vendée or Haute-Savoie. Given the size of the task, over 100 million land parcels on 560 000 sheets, recently a decision has been taken to expedite the process by moving to raster images of most of the cadastral sheets with seed points for each land parcels.

The IGN has traditionally been responsible for the topographic maps at 1 : 25 000 scale or less. As an increasing number of digital products have become available, the government commissioned a thorough review of the IGN operations in April 1999. The result of this, the 'Lengagne' report of September 30, 1999 (on-line see: http://www.cnig.fr), recommended to focus on the completion and maintenance of national digital coverage for key datasets (i.e., Référentiel Géographique à grande Échelle, RGE, considered as the large-scale dataset for France) in partnership with the relevant organisations.

February 19, 2001, the French government issued an inter-ministerial decision following the instruction of the Lengagne report. Among the nine items one can find the assignment to IGN of the task of producing, integrating and distributing by 2007 the RGE composed datasets, containing four components. These are: imagery (i.e., ortho-photo colour, 50 cm pixel, 1m accuracy in rural area up to 12.5 pixel, 25 cm accuracy in dense urban area), topography (in its wider sense with a 1 m accuracy), land parcels and addresses. There is also a clear commitment to set strong relationships between DGI and IGN for the joint production of seamless version of the raster cadastral maps over the country and also about the integration of the vector cadastral map where agreements have been signed by DGI and local authorities or facility management companies. Beyond its role of integrating data from diverse sources including local government, IGN has the remits to facilitate the dissemination of RGE and the development of value adding services using the RGE data. The last decision is the signature of a contract between IGN and the government setting objectives for IGN with the appropriate governmental grant. The consequence of the agreement is that the RGE will be made available to users at a cost including the marginal cost of reproduction and a licence fee. This is to maximise the number of users and minimise the costs for the user, yet enabling a long term sustainable funding of the RGE creation and maintenance.

At present, the situation in relation to a GI infrastructure appears as follows:

- core datasets: the RGE is the answer to core dataset requirements at 1 m accuracy. Socio-economic and topographic data with 10 m and 100 m accuracy are in place including raster seamless datasets of the 1 : 25 000 topographic maps of France. Large urban areas and few departments are already covered by core datasets. These core datasets will be made available to populate the RGE concepts. Simplified street surveys with few cm accuracy are now considered to be in the RGE concept in dense urban areas. Main current issues are the setting up of agreements between IGN in its role of integrator for the RGE and contributors to the RGE with data they have or will fund to meet their own requirements;
- co-ordination: GI policies are co-ordinated through the National Council for Geographic Information (CNIG), which is an inter-ministerial council established in 1985 with representatives from all major government departments and GI-related agencies, and AFIGÉO, the French association for GI which includes representatives from the user community and the private sector. The February 19, 2001 inter-ministerial decision clearly states the role of CNIG to voice the user requirements for core datasets including the RGE. A significant co-ordinating role in matters of information policy is however played directly by the government through the office of the Prime Minister and the Ministry of Planning. An action Plan for the Information Society (PAGSI) was presented in January 1998. In conjunction, the French government funded the creation of information systems for departmental offices of the ministries (Système d'information territorial) in which there are plans to provide the 10 m accuracy core datasets quoted above;
- ENv 12657 Geographic Information – Metadata, the adopted CEN standard for metadata has been published by AFNOR as an experimental standard in August 1999 asking for comments by May 2002. Regardless the MEGRIN or the GEIXS activities (that are also European, although hosted in France) most of the activities in France related to metadata are co-ordinated by CNIG.

Under the CNIG umbrella, several Metadata activities took place, as:

- publishing of an information sheet on metadata for the governmental departments and local governments (on-line see http://www.cnig.fr/commun/proserv/ficheamo/fserie3/fiches/23catal.htm);
- development of a cataloguing tool based on the CEN experimental standard available on-line with CERTU (Centre d'Études sur les Réseaux, les Transports, l'Urbanisme et les constructions publiques, on-line see http://www.certu.fr/sitcert/geomat/minisi01/pg_home.htm);
- updating a catalogue of GI sources available in France (on-line see http://www.cnig.fr/commun/proserv/source/source.html);

- raising awareness and assistance to local governments to set up 'Region wide' metadata servers (Vendée, Provence Alpes Cotes d'Azur, Picardie, etc.);
- setting up of a metadata service related to the coastal zone within the frame of one of CNIG working parties.

3.6. *Germany*

In Germany, collecting and disseminating Geographic Information is mainly a task of the 16 Länder (states) and not of the Federal State. On behalf of the Länder the Federal Institute of Cartography and Geodesy (Bundesamt für Kartographie und Geodäsie, BKG) is maintaining a Metadata Information Server for all available topographical maps, datasets, and images. It is based on the CEN Metadata Standard and an Oracle database. The emerging ISO Standard 19115 Geographic Information – Metadata will be used in the future. By use of Internet, a possibility was created to allow all responsible institutions to update their parts of the metadata information and to have a central server for Germany for customers by accessing the German Metadata Server (on-line see http://www.atkis.de).

3.7. *Iceland*

The National Mapping Agency of Iceland has adopted the Danish metadata concept. The service is in Icelandic.

Iceland has a very ambitious national GIS co-ordination body, called 'LÍSA'. It is a venue of co-operation in the field of geographical database development and has become a forum for discussion and development of GIS in Iceland. In addition, the system stores and uses spatial information. LÍSA's main task is to promote co-operation between partners dealing with geographical information systems and encourage joint use of datasets by promoting:

- creation of joint rules of data transfer and communication;
- technical co-ordination;
- creation of standards for geo-coding;
- definition of copyright and related rights of digital data;
- availability of data and public access to it;
- Iceland's interests in international co-operation;
- education, courses and general promotion of GIS;
- forum for exchange of ideas and discussion between members;
- information dissemination to members and others about LÍSA's activities by reporting on LÍSA's work in conferences and meetings, issuing a monthly newsletter 'LÍSUFRÉTTIR' and creating an information centre at LÍSA's secretariat.

LÍSA's activities is distributed over working groups for:

1. inventory of the current status of geographical datasets in Iceland;
2. terminology, i.e., translation of words and concepts on GIS;
3. communication and standardisation, to get information about how GIS-partners transfer data today, what problems they have in communicating and make suggestions for a better data transfer;

4. technical aspects in GIS, which is founded by the IT sector board of the Icelandic Council for Standardisation in co-operation with LÍSA, to structure future care for Iceland's interests in international work within the field of GIS, especially standardisation.

3.8. *Ireland*

The national spatial data organisations in the Republic of Ireland are:

- Ordnance Survey Ireland (OSI, on-line see: http://www.megrin.org) for topographic mapping and aerial imagery;
- Geological Survey Ireland (GSI) for geological data;
- Environmental Protection Agency (EPA) for environmental data.

The Republic of Ireland actively participates in the regional (European) spatial data organisation MEGRIN GIE. In addition, other regional participation exists in:

- GDDD/LaClef: MEGRIN metadata service;
- EEA WebCDS: European Environmental Agency on-line metadata service (on-line see http://www.mu.niedersachsen.de/system/cds/);
- GEIXS: metadata service run by EuroGeoSurveys, the representative body of European Geological organisations;
- ESMI: European Spatial Metadata Infrastructure, distributed metadata service;
- European Committee for Standardisation (CEN/ISSS) Multimedia Metadata Initiative (MMI) Dublin Core DC Workshop.

Currently, metadata 'discovery' is available at a national level. The Geospatial Information Directory (GeoID), a pilot metadatabase for Ireland represents the most complete information available regarding public and private sector data. Besides, Ordnance Survey Ireland has an in-house metadata service based on CEN ENv 12657. Ordnance Survey Ireland datasets are currently described within the MEGRIN GDDD service at European level, which in turn is available through the ESMI service. As part of MEGRIN, Ordnance Survey Ireland is actively involved in the development of the GDDD service within the LaClef project and plans to implement the LaClef Fundamental Metadata, a GI core metadata description.

3.9. *Netherlands*

The main actors dealing with geographic information in the Netherlands are:

- Topographic Service in the Netherlands (TDN) providing datasets for the topographic maps at scales 1 : 10 000, 1 : 25 000, 1 : 50 000, 1 : 100 000, 1 : 250 000 and 1 : 1 000 000. Since 2004 the Netherlands Topographic Service and The Netherlands Cadastral Office has been merged into one NMA;
- Dutch Cadastre maintaining the national cadastral dataset, being a continuous topological dataset for the whole country of around 50 Gbyte with about 250 M vectors;
- Survey Department of the ministry of Traffic and Water Control, providing mainly road databases and height information in the Netherlands;
- National Co-operative Foundation for the Large Scale Base Map of the Netherlands (Landelijk SamenwerkingsVerband, Grootschalige Basiskaart van Nederland, LSV-GBKN) in

which co-operates the National Cadastre of the Netherlands, Society of Municipalities, Utility Company Organisations and Union of Water Control Boards to maintain the large scale topographic base map of the Netherlands;
- Alterra, the institute collecting and disseminating agricultural, forestry and soils information;
- Council for Geographic Information (Stichting Overlegorgaan voor Vastgoedinformatie) Ravi (on-line see http://www.ravi.nl) in which governmental bodies co-operate as ministries, provinces and municipalities but also utilities, cadastre, research institute NITG-TNO, and the Chamber of Commerce. In addition (Dutch representatives of), GI producers, GI software houses, consultant bureaux in the field of GI, private land-surveying and land-developing companies represent the private sector in the Ravi. The aim of the Ravi is to promote the use and development of the Geographic Information Infrastructure, GI Standardisation and GI policy development;
- Netherlands Normalisation Institute (NEN) for development of (also GI) standards for Dutch users;
- private members of Ravi as listed above.

In 1996, on initiative of the Ravi project group on standardisation, the first phase of a national GI metadatabase was developed by Geodan, called Idéfix, to make the many digital geographic databases that existed in the country available to all users. Idéfix provided only a metadata service on the Internet for the participating bodies. The system used the CEN prEN 12657 (the predecessor of the ENv 12657) as a basis (and by doing so, the Netherlands was the first country applying this standard), together with specific Dutch requirements. Idéfix was initiated by a development fund from the Ministry of Economic Affairs and was based implemented in an MS-Access database using 15 pre-defined keywords and free text searches as well as range searching.

In 1997, the system was named NCGI and started to operate privately. Since then, an internal investigation was performed on the operation of the system, resulting in a redesign in 2000 of the NCGI-site (on-line see http://www.ncgi.nl). The concept of the new design consists of:

- application of a central search engine (capacity 1800 searches/hour) by visiting circularly all connected providers, available through the Internet providing the same search possibilities as in Idefix;
- providing a central list of participating bodies that is updated continuously by the system;
- de-central metadata systems with appropriate servers. The geographic data(bases) are not available on the server: they reside on the participating institute's server. This server can be linked to the NCGI-site (on the providers' request) to download data directly;
- use of a ANSI Z39.50-1995 (ISO 10163-1995) protocol and metadata standard ENv 12657 (with specific changes as to the obligation to use certain elements). Also the system will use a catalogue service and apply data integration by web mapping as specified by OpenGIS®;
- de-central data protection by using CORBA servers and the allowance by the provider for certain groups of users to log on its system.

In 2004 a project was started to revise the national standard NEN 3610 "Terreinmodel Vastgoed" into the "Basismodel Geo-informatie", based on ISO 19109 – Geographic information – Rules for Application Schemas. The revised NEN 3610 contains a feature catalogue, a model structure and transfer model for the transport and feature definition of geographic data within the Netherlands, based on XML/GML. The commenting process for this standard run till July 1, 2005 (when this book was in press).

3.10. *Norway*

Within the framework of the National Geographic Information Centre (NGIS), 'Statens Kartverk' (Norwegian Mapping Agency), Norway is setting up a system called 'Geovekst' (English: Geo-Growth), aiming at the co-operation between national and local government (primarily at county level, but open to municipality level). In this way, it is ensured that national and international standards are implemented and that procedures for co-operation on all levels are optimised.

A prototype of a national spatial data catalogue (in Norwegian) is presented on the Internet. This catalogue, in its present status, is primarily aimed at in-house products of Norwegian Mapping Agency, but also 'Geovekst' projects will be shown.

One of NGIS outspoken goals is to implement the ISO range of geographic standards. Norway has a leading role in the development of these standards, and NGIS expects to let the expertise gained internationally give the Norwegian geographic society a great leap forward.

3.11. *Poland*

The Surveyor General of Poland is responsible for the creation and management spatial data but several other institutes contribute to the scientific aspects and standardisation in Poland. Therefore the Polish organisations dealing with spatial data are:

- Head Office for Geodesy and Cartography, managing all aspects of spatial data administration for the whole country (e.g., cadastral data, basic map data for civil use, maintenance of the reference system, etc.);
- Polish Society for Spatial Data dealing with scientific and practical aspects of geographic information management and use;
- Committee of Geodesy of the Polish Academy of Sciences, Section of Geo-informatics studying geographic information methodology;
- Polish Standardisation Committee, Commission for Standardisation No. 255 'Geodesy in architecture', Sub-commission on Geographic Information, elaborating of Polish Standards, promoting of European and International Standards (this committee also represents Poland in CEN/TC 287 and ISO/TC 211).

Now, no 'de jure' (so-called 'norms') metadata standards exist in Poland. However, there are several technical regulations issued by the Head Office for Geodesy and Cartography operating as 'de facto' standards. The most important amongst them is the regulation K-1, which can be recognised as a metadata standard for the content of the basic map of Poland. It comprises the following main issues:

- definition, goals and functions of the basic map;
- projection, system of co-ordinates and structure (partitioning in sections) of the basic map;
- specification, scales and content of the basic map;
- catalogue of objects and their symbols.

It is planned to adopt the CEN standard prEN-ISO 19115 Geographic Information – Metadata in the near future (together with other CEN standards).

3.12. *Romania*

Spring 1997, on the initiative taken by a group of members of the Romanian Academy of Sciences, the 'Forum for Information Society' was created. The role of the Forum is to highlight theoretical and practical aspects of Information Society, to help to harmonise different views and initiatives, to inspire efforts and push them up for and with actors capable to determine the building of a knowledge-based society in Romania. Having in mind the creation of an Information Society in Romania, in 1997 the Romanian Government adopted the Government Decision on the National Information Strategy (NIS), approved in 1998. The NIS included an Action Program regarding the development and large-scale use of information technologies in Romania to aim at:

- national information infrastructure in the short run (till 2000) for public administration and develop a national information communication technology (ICT) industry;
- extend the applications in the medium and long-run (by 2005).

To achieve this, the National Agency for Communications and Information (NACI) is founded under Romanian Government and works as a professional body for central public administration and has the mission to assure the elaboration, implementation, monitoring and evaluation of the politics in the field of communication and information.

Organisational framework in the GI-field

The National Office of Cadastre, Geodesy and Cartography (NOCGC) is a public institution to organise, manage and co-ordinate the cartography, geodesy, photogrammetry, remote sensing and cadastral activities at national level and following the legal directives in these domains. The office is responsible for the development of standards, transfer mechanisms, in the field of Geodesy and Cartography and controls the work of the County Offices of Cadastre, Geodesy and Cartography (COCGC) and the Institute of Geodesy, Photogrammetry, Cartography and Cadastre (IGPCC) in Bucharest, financed from state budget.

3.13. *Russia*

In the nineties, the Russian Information Society is created. Each member of this Society will be able to obtain complete and actual information, including geographic information, at any point of geographical space, by means of Internet using electronic libraries of content, including electronic map libraries. Completeness, methods and form of geographic information, temporal factors of the information supply can be considered as the main criteria for evaluating quality of the electronic geographic information resources. In the process of creating the electronic geographic information resources, standardisation in the area of geographic information technologies plays the main role. The Institute for Information Problems (of the Russian Academy of Sciences) and 29 Defence Research Institute are elaborating theoretical fundamentals of the Global Geographic information Mapping (GCM).

The theoretical basis of developing the National Standard for digital and electronic maps in Russia rests upon the State System for Standardisation of Russian Federation that had been put into operation since 1993 by the State Standard of Russian Federation GOST R 1.0-92. This Standard determines purposes and tasks of the standardisation process, main principles of standardisation in Russia, categories of normative documents and types of standards, main theses on the international co-operation in the area of standardisation, application of standards and technical terms, and state supervision of the observance of standards. The regulations of the State System for Standardisation of the Russian Federation are obligatory to be guided by for all enterprises, unions, joint-stock ventures, etc. (irrespectively of their forms of property and subjection), private business, technical committees on standardisation, ministries (departments) and other authorities of the Russian Federation and scientific societies.

Procedure for developing standards

In Russia, the development of state standards is being carried out by Technical Committees (TC) in accordance with assignments of state standardisation plans of Russian Federation, TC working plans and contracts on the development of standards. The development of standards is guided by the legislation of Russia, State System for Standardisation of Russian Federation and other normative documents. It also takes into account the documents of international and regional standardisation organisations, results of scientific and practical researches, patent investigations and other information about the achievements of domestic and foreign science and techniques. Then, the TC sets the deadline for all work and determines a sub-committee in which the standard is to be developed, or a working group for developing the standard. The development of standards for geographic information in mapping is being carried out by TC 22 – Information Technologies including the Sub-committee 051 'Geo-information Technologies' (chaired by A.I. Martynenko). This Sub-committee deals with standardisation in the field of GI, based on methods of acquisition, storage, analysis, displaying, processing and retrieving spatial data for the digital mapping.

International scientific co-operation

International co-operation in the area of standardisation is being carried out together with international and regional standardisation organisations, as well as on the base of bi- and multi-lateral agreements with correspondent foreign authorities. The edicts of the President of Russian Federation and decisions of the Government of Russia, international commissions on economic and scientific co-operation and Russia's participation in the activities of regional and international standardisation organisations are regulating such co-operation. The international co-operation between Russia and international standardisation organisations includes collaboration in the development of international and regional standards and implementation of these standards in business and interrelations with partner countries. The co-operation in the area of standardisation, both bi- and multi-lateral, includes the joint development of standards, fulfilment of joint scientific researches, information interchange, mutual consultations, education, etc.

The 29 Defence Research Institutes and the Central Research Institute of Geodesy, Aerial Survey and Cartography carry out the work on standardisation in the field of digital mapping in Russian Federation. Main efforts are directed towards the development of branch standards.

The work on standardisation in the area of terms and definitions of electronic and digital maps (including metadata) are also carried out in co-operation with cartographic boards of Ukraine and Belarus.

Description of GOST R 51353-99

'Geoinformatic Mapping – Metadata of Electronic Maps: Composition and Content'

TC 22 also includes committees on terminology, telecommunication and information transfer, programming languages and system program interfaces, computer graphics and image processing, encoding audio, image, multimedia and hypermedia information, automatic identification, methods and tools for data acquisition. Within the Sub-committee on Standards for Geographic Information Technologies, the State Standard of Russian Federation GOST R 51353-99 'Geoinformatic Mapping – Metadata of Electronic Maps: Composition and Content' was developed. In November 1999, the Russian Federation State Committee on Standardisation and Metrology adopted this standard. The areas of application of GOST R 51353-99 are information supply of state authorities, communication systems and tools, business, transportation, navigation, ecological monitoring, state and private cartographic enterprises. The need in development of GOST R 51353-99 grew from the necessity of:

- acquisition of metadata about electronic (digital) maps and information used for creation and storage of electronic (digital) maps;
- effective application of metadata by customers for the acquisition, storage, analysing, processing and transfer of geographic information;
- increasing quality of electronic (digital) maps;
- information compatibility of control, navigation systems and GIS;
- transfer of digital map data and metadata both inside Russia and world-wide.

GOST R 51353-99 establishes requirements for the composition and content of common metadata as well as metadata of geodetic, gravimetric, photogrammetric and cartographic information used for creation, updating and application of digital maps. Electronic (digital) maps are one of multiple forms of representing spatial data. The statements in this Standard are liable to application by all authorities and organisations located in Russian Federation, independently of their form of government, that are engaged in acquisition, systematisation, analysis, processing and transfer of spatial data, creation and application of electronic (digital) maps, organisation of metadatabases and electronic (digital) cartographic databases. This Standard refers to the following State Standards:

- GOST 34.003-90 Automated Systems. Terms and Definitions;
- GOST 22268-76 Geodesy. Terms and Definitions;
- GOST 24284-80 Gravitational and Magnetic Exploring. Terms and Definitions;
- GOST 23935-79 Air Photo Cameras and Photo Survey. Terms and Definitions;
- GOST 21002-75 Photo-topography. Terms and Definitions;
- GOST 21667-76 Cartography. Terms and Definitions;
- GOST 28441-90 Digital Mapping. Terms and Definitions;
- GOST R 50828-95 Geoinformatic Mapping. Spatial Data, Digital and Electronic Maps. Common Requirements;
- GOST R 52055-2003 Geoinformatic Mapping – Spatial Models of Terrain. General Requirements.

Metadata of digital maps is data describing contents, spatial extents, quality (accuracy, complexity, consistency and actuality) and other characteristics. This Standard operates with the following terms and definitions: Common terms – according to GOST 34.003, GOST 22268, GOST 24284, GOST 23935, GOST 21002, GOST 21667, GOST 28441 and GOST R-50828.

GOST R 51353-99 provides cartographic representation of the real world and creation of Digital Earth that is based on strict geodetic base by means of using millions air- and space high-resolution images, electronic maps of various themes and scale, and textual reference information. GOST R 51353-99 allows synthesising the digital (electronic) image of the earth as a spatial and temporal representation of our planet. This representation is based on mathematical and semantic modeling of spatial data changing dynamically and is realized as the aggregate of distributed electronic cartographic libraries (Electronic Maps Libraries) united in whole system by means of telecommunication technologies.

3.14. *Spain*

The national spatial data organisations operating in Spain are:

- National Geographic Institute (Instituto Geográfico Nacional);
- National Centre for Geographic Information (Centro Nacional de Información Geográfica);
- General Directorate for Cadastre (Dirección General del Catastro);
- Geographic Centre of the Ministry of Defence (Centro Geográfico del Ejército, Ministerio de Defensa).

Apart there are also several regional spatial data organisations as:

- Cartographic Institute of Cataluña (Instituto Cartografía de Cataluña) in Barcelona;
- Grafcan (on the Canary Islands);
- Tracasa (Comunidad Foral de Navarra);
- Municipality of Madrid (Comunidad de Madrid);
- Agencia de Medio Ambiente de Andalucía.

All organisations work in the field of GI production at different scales and for different purposes and applications, so a small overlap exists among all jobs developed by them.

AENOR, the Spanish organisation for standardisation, is working on the development of a national standard, which includes data definition and metadata structuring. At this level, most of the organisations are represented in AENOR. Besides, there are some minor approaches to metadata, through local projects such as the DIGA National Geographic Institute project and the MERCATOR project (a more academic approach).

3.15. *Sweden*

Sweden has for a number of years held geographic metadatabases on at least two levels:

- national level being served by Nationella Databaskatalogen operated by Lanmäteriet (NMA);
- county level, some of the regional GIS partnerships has build own catalogue services.

The latest development is a new national service based on a central CEN ENv 12657, based on an Oracle metadatabase and distributed access; the data collection modules are installed at the data producer's sites (on-line see http://www.megi.lm.se/).

3.16. *Turkey*

The General Command of Mapping (GCM) is the principal mapping organisation in Turkey for the production and distribution of topographic maps required for both national defence and development. The technical departments of GCM are:

- Geodesy, responsible for the establishment, maintenance and improvement of the National Control Networks at horizontal and vertical level but also for Gravity, Magnetic and GPS control;
- Photogrammetry for the photogrammetric dataset production with map-scales ranging from 1 : 1000 till 1 : 1 000 000;
- Cartography for map (re-)production and archiving;
- Information Systems and Support dealing with database creation for topographic data, digital elevation models, contour and raster maps;
- Flight Command Group for aerial photography;
- Survey Engineering School (for education in surveying and mapping).

In 1998, a modernisation program was initiated to create a national topographic databases at the scales 1 : 25 000 and 1 : 250 000 by automated cartography.

Other organisations dealing with Geographic Information are:

- General Directorate of Land Registry and Cadastre, dealing with large scale cadastral mapping by geodetic and photogrammetric methods;
- General Directorate of Highways and General Directorate of Forestry for specific map types;
- Private Sector with 4 to 5 big companies for digital photogrammetry, especially to produce large-scale digital topographical and ortho-photomaps and DEM's and to design information systems for municipalities.

On behalf of Turkey's society in the field of GI, CGM has the intention to participate in the OEEPE project on 'Topographical Map Production with High Resolution Satellite Images' (since 2004, OEEPE has been renamed into EuroSDR, European Spatial Data Research). CGM will also participate in other EuroSDR projects such as 'Laser Profiling', 'GPS-INS Integration' and 'Interferometric SAR' to improve our digital production capacity.

4. Summary and conclusions

In summary, there are many issues to be dealt with when developing a European metadata service. Generally spoken they are of four levels:

- data model standardisation;
- semantic standardisation;
- languages;
- search queries.

4.1. *Data model standardisation*

It is likely that implementing databases of metadata according to standards will provide information that is more reliable to users and help the data providers by:

- reducing the duplication of effort which currently exists in different databases;
- increasing access to their data descriptions, so increasing their possible sales (distributed systems need to agree on a common standard transfer format for the on-line transfer of information);
- ensuring that suppliers are able to store the same information (whether they choose to do so is another matter), possibly as a common core set of metadata.

Unfortunately, there are many metadata standards to choose from! Internationally, ISO/TC 211 is developing the metadata standard 19115 (which has been replaced in 2003 abandoned CEN/TC 287 ENv 12657) and national standards in Europe. Finally, activities such as the Dublin Core or the ETeMII recommendations may also provide useful input, as they propose a very simple and limited set of metadata, easy to comply with, and easily open to several sectors of data, while CEN and ISO are GI focused.

4.2. *Semantic standardisation*

The use of standard data structures will not necessarily ensure that the metadata is of consistent quality, completeness and accuracy. Indeed, the quality of the metadata may be more important than the data structure, since this is what users will see and use. Without reliable information, users are unlikely to use the service. All data providers and system users must have the same understanding of terms. Without this, the search results, which are presented to users, will be meaningless and/or incorrect. This means that all data providers must either attach standard keywords to their descriptions, or a thesaurus mechanism needs to be created to provide a common view of individual implementations. The task of harmonisation is likely to be complex, as there is considerable scope for differences to occur (for instance, between disciplines, within the same discipline in different countries, and even between different organisations in the same country).

The need for semantic harmonisation is widespread: for example, there is not yet a standard for location references, to combine with the geographic coded data and geographic information. Moreover, where standards do exist, they are not always implemented!

Europe's experience indicates that a reliable metadata service depends on the quality of the data descriptions. This is particularly significant in a multi-national environment. The initial data collection exercise raised problems because different organisations had used different keywords to describe datasets based on the same data model. This disagreement in choice of keywords was significant for complex themes, e.g., 'topography' and 'land cover', while straightforward keywords as 'road network' and 'railway network' were generally interpreted consistently. This clearly illustrated the importance of having the same understanding of terminology.

4.3. *Languages*

To propose a real solution for semantics consistency, the services aiming at covering all Europe, cannot be satisfied with only one language. Besides, other critical aspects indicate the inadequacy of a uniquely English language service:

- it may be expected that a fair number of technical experts are sufficiently fluent in English, but that will not necessarily be the case for all potential users of GI. As it is the wish of both the European Commission and European organisations to make GI, particularly public sector in GI, more available to a larger range of users through the INSPIRE project, it is desirable for the service to develop a friendly interface available in a number of national languages;
- for the best semantic comparability between countries using different languages, it is essential that multilingual thesauri and keywords are developed and tested. Above being tools for ensuring cross-border (and cross-language) metadata consistency, it will allow the easy creation of multilingual interfaces, and the automatic on-line extraction and translation of nationally based metadata are investigated.

4.4. *Search queries*

Ideally, the user who is looking for data would like to locate the data he is interested in, by pointing at the area covered by the project he is working on. Solving the main semantic and language issues should allow this possibility. A text based query such as 'FIND objects = roads' IN area = 'BENELUX' ' would be close to requests expressed in natural language. There are a number of ways to develop such a relatively user-friendly search query. Knowledge about regions or geographical locations can be stored by the way of geographical keywords. Different relatively common solutions for look-up tables have to be considered, when on the pan-European level, each having its strengths and weaknesses according to specific approaches:

1. geographical co-ordinates, based on a unique geodetic system, such as EUREF;
2. administrative units;
3. postal codes;
4. addresses.

Acknowledgement

We would like to thank the many scientists and commission members in Europe who provided us with the information on metadata activities in their countries and regions, which we have used to compile this overview (see list of contributors below).

We would also like to acknowledge the support of the Delft University of Technology, both the Department of Geodesy and the Research Institute of Housing, Urban and Mobility Studies as well as the Melbourne University, Department of Geomatics, which made this work possible.

Contributors:

Croatia	Dr. Miljenko Lapaine, University of Zagreb, Zagreb
Czech Republic	Prof. Milan Konecny, Masaryk University, Brno, mailto: konecny@dior.ics.muni.cz
Denmark	Mr. Anders Nielsen, National Survey and Cadastre, Copenhagen, mailto: asn@kms.dk
Finland	Mrs. Paula Ahonen, Helsinki University of Technology, mailto: paula.ahonen@hut.fi
France	Mr. François Salgé, GNIG, Paris, mailto: francois.salge@cnig.gouv.fr
Germany	Mr. Ulrich Düren, Landvermessungsamt Nordrhein-Westfalen, Bonn, mailto: dueren@lverma.nrw.de

Iceland	Mr. Thorbjörg Kjartansdóttir, Director, LÍSA, Reykjavik, mailto: lisa@aknet.is
Netherlands	Prof. Ir. Henri J.G.L. Aalders, Delft University of Technology, Institute of Housing, Urban and Mobility Studies, mailto: h.aalders@otb.tudelft.nl;
Norway	Mr. Olaf Østensen, The Norwegian Mapping Authority, Hønefoss, mailto: olaf.ostensen@statkart.no
Poland	Prof. Wojciech Pachelski, Space research Centre, Polish Academy of Science, Warsaw, mailto: wp@cbk.waw.pl
Romania	Dr. Angela Ionita, National Institute for Research and Development in Informatics – ICI, Bucharest, mailto: aionita@td1.ici.ro
Russian	Federation Prof. Alexander I. Martynenko, Moscow State University, Geodesy and Cartography, Moscow, mailto: a_martynenko@mail.ru
Spain	Mr. Sebastián Más Mayoral, Centro National de Información Geografica, Madrid, mailto: smas@cnig.ign.es
Sweden	Mr. Bengt Rystedt, National Land Survey, Gävle, mailto: bengt.rystedt@lm.se
Turkey	Director General Mapping Service, Ministry of National Defence, Harita Genel Komutanligi, Ankara

References

Organisation	For on-line contact see:
AVID	http://www.ec-gis.org/avid.htm
CEN/TC 287	http://cen.iso.ch/livelink/livelink.exe (latest accessed on October 20, 2004)
CERTU	http://www.certu.fr/sitcert/geomat/minisi01/pg_home.htm
CLEAR	http://www.ec-gis.org/clear.htm
CNIG	http://www.cnig.fr
CNIG GI sources	http://cnig.les-argonautes.fr/
CNIG metadata	http://cnig.les-argonautes.fr/
DIGEST	http://www.digest.org
ESMI	http://www.ec-gis.org/esmi.htm
EuroGeoSurveys	http://www.eurogeosurveys.org
European Environmental Agency	http://www.mu.niedersachsen.de/system/cds
Finland Post Ltd.	http://www.posti.fi/
Finnish Environment Institute	http://www.vyh.fi/syke/syke.html
Finnish Geodetic Institute	http://www.fgi.fi/
Finnish Forest Research Institute	http://www.metla.fi/
Finnish Game and Fisheries Research Institute	http://www.rktl.fi/
Finnish Maritime Administration	http://www.fma.fi/
Finnish Meteorological Institute	http://www.fmi.fi/
Finnish Ministry of Justice	http://www.om.fi/
Finnish National Geographic Dataset Directory	http://www.nls.fi/ptk/aineistot
Finnish National Road Administration	http://www.tieh.fi/
Finnish Population Register Centre	http://www.vaestorekisterikeskus.fi/
Forest and Park Service of Finland	http://www.metsa.fi/
GEIXS	http://geixs.brgm.fr/
Geological Survey of Finland	http://www.gsf.fi/
GeoServe	http://www.ec-gis.org/geoserve.htm
German Metadata Service	http://www.atkis.de
GISEDI Europe	http://www.ec-gis.org/gisedi.htm
INSPIRE	http://www.jrc.cec.eu.int (latest accessed on October 20, 2004)
KMS-DK	http://www.geodata-info.dk
MADAME	http://www.ec-gis.org/madame.htm
MEGRIN	http://www.megrin.org
National Land Survey of Finland	http://www.nls.fi/
NCGI, Netherlands	http://www.ncgi.nl
Ravi, Netherlands	http://www.ravi.nl
Statistics Finland	http://www.stat.fi/
Swedish Metadata Service	http://www.megi.lm.se/

URL's in this chapter are accessed in May, 2005, unless otherwise stated in the text

World Spatial Metadata Standards
H. Moellering, H.J.G.L. Aalders & A. Crane (Editors)

North American Metadata Standards Developments in Canada and the United States of America

Kian Fadaie [a], Julie Binder Maitra [b], Valerie E. Hume [c], Harold Moellering [d], Henri J.G.L. Aalders [e]

[a] *Senior Adivisor, Science and Development, Agriculture and Agrifood Canada, Canada*
e-mail: fadaie@agr.gc.ca
[b] *FGDC Standards Co-ordinator, Federal Geographic Data Committee, U.S.A.*
e-mail: jmaitra@usgs.gov
[c] *Indian and Northern Affairs Canada, Northern Program, Canada*
e-mail: humev@inac.gc.ca
[d] *Professor of Geography & Geodetic Science, Ohio State University*
e-mail: geohal+@osu.edu
[e] *Delft University of Technology, Research Institute for Housing Urban and Mobility Studies, The Netherlands*
e-mail: h.j.g.l.aalders@otb.tudelft.nl

1. Regional SDI activities

GeoConnections Canada (on-line see http://www.geoconnections.org/) and the Federal Geographic Data Committee (U.S.A.) are governmental organisations that develop the Spatial Data Infrastructure (SDI) in their respective countries. These organisations have also collaborated with other U.S.A. and Canadian partners on several cross-border demonstration projects in recent years.

Canada and the U.S.A. are members of the Permanent Committee on Spatial Data Infrastructure for the Americas (on-line see http://www.cpidea.org.co/). PC IDEA was established in 2000 to co-ordinate development of SDI's in the Americas (for the foundation and goals of PC IDEA, see also the introduction of the Central and South American and the Caribbean chapter of this book). Mexico is also a member of PC IDEA, as are many countries in Central and South America. In 2002, PC IDEA became a liaison to ISO/TC 211, Geographic Information/Geomatics. PC IDEA also contributes to the development of the Global Spatial Data Infrastructure (GSDI, on-line see http://www.gsdi.org/), which itself has evolved into a formal organisation that promotes SDI Capacity building around the world.

The Pan American Institute of Geography and History (PAIGH, on-line see http://www.ipgh.org.mx/default_en.htm) is an older organisation, having been established in 1928 by the Sixth International Conference of American States. PAIGH provides technical assistance, conducts training at research centres, distributes publications, and organises technical meetings in the areas of cartography, geography, history and geophysics. The U.S.A. (but not Canada) is a member, as are Mexico and many other countries in Central and South America.

2. Overview of national metadata and SDI activities

2.1. *United States of America*

Federal

The Federal Geographic Data Committee (FGDC, on-line see http://www.fgdc.gov/) is an interagency committee organised to promote the co-ordinated use, sharing, and dissemination of geo-spatial data on a national basis. Executive Order 12906, signed by President Clinton in 1994 (EO, on-line see http://www.fgdc.gov/publications/documents/geninfo/execord.html), charged the FGDC with co-ordinating the Federal Government's development of the National Spatial Data Infrastructure (NSDI). EO 12906 also charged the FGDC to develop standards for implementing the NSDI. Member agencies were directed to document all new geo-spatial data it collected or produced using the FGDC Content Standard for Digital Geospatial Metadata (on-line see http://www.fgdc.gov/metadata/metadata.html) and to make metadata electronically accessible to the National Geospatial Data Clearinghouse (on-line see http://www.fgdc.gov/clearinghouse/clearinghouse.html).

The FGDC's collaboration with GeoConnections was noted above. FGDC and the National Imagery and Mapping Agency (NIMA, on-line see http://www.nima.mil/) also provide support to the GSDI Secretariat. The GSDI Secretariat promotes building SDI capacity around the world through technology transfer and training in topics that include metadata and clearinghouse.

The FGDC Content Standard for Digital Geospatial Metadata was one of the base documents for development of ISO 19115 (formerly ISO 15046-15), Geographic information – Metadata. It is expected that the ISO 19115 and planned ISO 19115-2 Imaginary Metadata and a Joint US / Canada Profile of 19115 will replace the FGDC Content Standard for Digital Geospatial Metadata.

State

States are under no compulsion to adopt FGDC-endorsed standards; however, several states have adopted the FGDC-endorsed Content Standard for Digital Geospatial Metadata and have established NSDI Clearinghouse nodes to enable users to search FGDC-compliant metadata to locate state-maintained geographic datasets. In 1995, the FGDC awarded a grant to the National States Geographic Information Council to implement an educational and research program on the FGDC metadata standard.

Federal/State

EO 12906 directs the FGDC to involve state, local, and tribal governments in the development and implementation of NSDI initiatives. The GeoData Alliance (on-line see http://www.geoall.net/) and the I-Team Initiative bring federal, state, local, and tribal organisations together to advance the NSDI.

Geospatial One-Stop (on-line see http://www.fgdc.gov/geo-one-stop/index.html) is a part of The Office of Management and Budget (OMB, on-line see http://www.whitehouse.gov/omb/) E-Government initiative to improve the effectiveness, efficiency, and customer service throughout the Federal Government. The Geospatial Information One-Stop Project will support 'one-stop' access by citizens and government to Federal Government and other geospatial data assets. The project will accelerate the development and implementation of NSDI. Organisations representing state, local, and tribal entities will participate as equal partners with the Federal Government on the Geospatial One-Stop Board of Directors.

2.2. *Canada*

Since the early work of Roger Tomlinson (on-line see http://www.esri.com/news/arcnews/fall01articles/drroger.html) on Geographic Information Systems, over 25 years ago, for the Canadian Land Inventory Database, Canada has been actively involved in the development and establishment of standards for geographic information. The earliest systems, twenty years ago, were intended to use computer systems to aid in digital cartography. In the mid 1980's the Canadian Department of National Defence, the Canadian Hydrographic Service, and the Canadian Department of Natural Resources (then the Department of Energy Mines and Resources), were active in the development of geographic information standards to address their respective domains of responsibility. In recent years the work of these agencies and others, in both the government and private sector, have been focused on the development of public domain standards through the formal national and international standards development bodies. The federal government in Canada has been active in the development of international standards for geographic information and for geographic information metadata through ISO/TC 211 since the inception of this international committee. The intent has been to develop generic standards internationally, and then to use them nationally as the basis for developing more detailed national standards that are localised to national conditions. With respect to metadata this localisation means that Canada needs to add support for both the French and English languages so that the metadata standard can operate in a multi-lingual environment, and of course needs to populate code tables of national information, such as provincial and territory administrative divisions.

Federal

The federal government long has been a supporter of the formal national standards system. The Standards Council of Canada (SCC, on-line see http://www.scc.ca/) has the mandate from parliament for the establishment of standards within Canada. The SCC acts as the contact for international standards development for Canada, and it delegates its actual standards writing tasks to different national standards writing bodies. The Canadian Standards Association (CSA, on-line see http://www.csa.ca/) is responsible for, among other things, information technology standards, and the Canadian General Standards Board (CGSB, on-line see http://www.pwgsc.gc.ca/cgsb/home/index-e.html) is responsible for geographic information. The CGSB itself is a division of the Canadian federal department of Public Works and Government Services.

In the early 1990's the Canadian General Standards Board – Committee on Geomatics (CGSB-CoG), the committee responsible for developing national standards in this area, developed four national standards. Two of these were data interchange standards, and the other two were related to the description of geographic information (metadata). These standards are:

- CAN/CGSB-171.1-95 – CGIS-SAIF – Canadian Geomatics Interchange Standard – Spatial Archive and Interchange Format: Formal Definition;
- CAN/CGSB-171.2-94 – Geomatic Data Sets Cataloguing Rules;
- CAN/CGSB-171.3-95 – Directory Information Describing Digital Geo-Referenced Data Sets;
- CAN/CGSB-171.4-94 – CGIS-DIGEST Canadian Geomatics Interchange Standard – Digital Geographic Information Exchange Standard.

The CGIS-SAIF is a modelling standard for geographic information, developed by a provincial agency, and the CGIS-DIGEST is the NATO STANAG 7074 DIGEST, which is needed for support of Canada's military internationally.

The standard for 'Geomatic Data Sets Cataloguing Rules' applies to the identification of maps in map libraries, and the 'Directory Information Describing Digital Geo-Referenced Data Sets' is a national metadata standard. This metadata standard was not widely implemented in Canada, due to Canada's need for compatibility internationally and with the U.S.A. Canada has actively worked in the ISO/TC 211 international standards committee to allow for an international solution.

It is expected that the older national metadata standard CAN/CGSB-171.3 will be replaced by a profile of the ISO/TC 211 metadata standard (ISO 19115), and that this profile will be developed in co-operation with the U.S.A. through the formal standards development process in both countries. In the meantime, a number of provinces and organisations have either adopted or adapted the existing standard to fit their own purposes.

The principal work related to metadata is currently determining how best to implement the new ISO/TC 211 standard in the context of the Canadian Geospatial Data Infrastructure (CGDI, on-line see http://cgdi.gc.ca), metadata being only one of the standards that will be reviewed for this purpose. This is an important government initiative to foster the growth of the geomatics industry in Canada and to make government data available on-line. It aims to develop co-ordinated national access to geo-spatial information thereby contributing to the knowledge economy, new job growth, value added commercialisation, better community decision-making and advanced applications.

It is Canada's intention to adopt international standards or profiles of the suite of ISO standards pertaining to geomatics, and to localise them as required. At the same time Canada and the U.S.A. are working together, bilaterally between government agencies, and through the formal national standards development processes in each country so that any profiles of standards in the two countries will be compatible, so that there will be common data interchange across the region. A joint standardisation agreement has been established between the SCC and the ANSI (on-line see http://www.ansi.org/). The hope has been expressed that Mexico will want to participate in this arrangement in the near future to address all of North America. This co-operation will extend beyond metadata to other areas of the standardisation of geographic information and to compatible data holdings. For example, the GeoConnections Secretariat for the CGDI and the US FGDC (on-line see http://www.fgdc.gov/) have announced a joint project to develop framework data for the CGDI and US NSDI.

Provincial

Within Canada, work related to metadata (the development of metadata standards, or the use of metadata) is being carried out at all levels of government: federal government, provincial and territorial governments, municipal governments, First Nations governments and institutions of public government (e.g., resource management boards, land use planning commissions). Some provinces have based their metadata standard on the older national standard. Others have adopted something that is more akin to the US FGDC standard or have developed a system for describing unique-like data. It is expected that the development of a joint Canada/U.S.A. bi-national metadata standard based on international standards will converge these initiatives.

Federal/Provincial

The Canadian Council on Geomatics (CCOG) is a federal–provincial–territorial consultative body for geomatics between the mapping agencies of the federal and provincial governments. Its aims are to provide a consultative forum to discuss common issues, exchange information on programs, the discussion of legislation of general interest, develop, promote and promulgate national and international geomatics standards and work together to support a Canadian geo-spatial data infrastructure. In the past, the CCOG has been involved in the standardisation process, but now it supports the development of standards through the formal public standards process through the SCC and jointly with the US ANSI.

3. National organisations for NSDI's in North America

Although mostly the infrastructure on standardisation in Canada and the U.S.A. is national oriented, also bi-lateral relations exists between the two countries (see Figure 1). Because of the ISO/TC 211 developments in the standardisation of geo-spatial information and also through the OGC specifications developments, proposals for a formal Memorandum of Understanding (MoU) between the two countries have been established in September 2003, aiming at a North American profile for ISO standards.

3.1. *United States of America*

3.1.1. *National standards organisations*

National standards in the U.S.A. come under the purview of the American National Standards Institute (ANSI). ANSI is the official national standards body of the U.S.A., and is the US member of the International Organisation on Standardisation (ISO). It accredits the US Technical Advisory Group (TAG) to ISO Technical Committee 211 (ISO/TC 211), Geographic information/Geomatics, which developed ISO 19115.

ANSI itself does not develop American National Standards (ANS's); rather, it accredits other organisations that follow its rules and principles as standards development organisations.

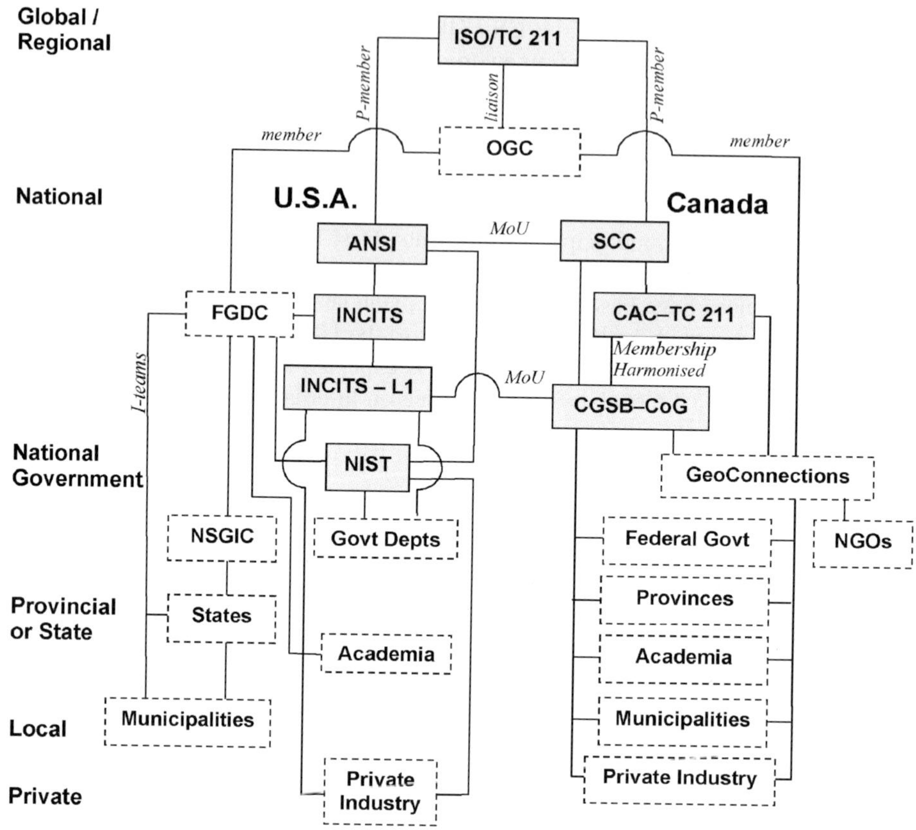

Abbreviations:

ANSI	American National Standards institute
CAC	Canadian Advisory Committee
CGOC	Canadian Council on Geomatics
Depts	Departments
ISO/TC 211	International Organisation for Standardisation / Technical Committee 211on Geographic Information - Geomatics
FGDC	Federal Geographic Data Committee
GCSB-CoG	Canadian General Standards Board-Committee on Geomatics
Govt	Government
I-team	Geospatial Information Initiative, I-Team Initiative
MoU	Memorandum of Understanding
NGO	National Governmental Organisation
NIST	National Institute of Standards and Technology
NSGIC	National States Geographic Information Council
OGC	OpenGIS® Consortium
P-member	Participating member
SCC	Standards Council of Canada

Legend:

formal

informal

Fig. 1. General geo-spatial standards infrastructure in Canada and the U.S.A.; with formal (in full line presentation) and informal (in dotted line representation) organisations and relationships.

ANSI accredited the InterNational Committee for Information Technology Standards (INCITS, on-line see http://www.incits.org), sponsored by the Information Technology Industry Council (ITI) trade association for the creation and maintenance of formal information technology standards. INCITS Technical Committee L1, Geographic Information Systems (GIS), focuses on national standards for digital geographic data. INCITS L1 oversees maintenance projects for existing ANSI standards and develops strategy for adoption and implementation of ISO standards as ANS's (see Figure 2).

Federal standards organisations

The FGDC develops geo-spatial data standards needed by the federal government through its Sub-committees and Working Groups. The FGDC is not accredited by ANSI, and uses its independent status to expedite the development standards that are urgently needed for use by the federal government. The FGDC Content Standard for Digital Geospatial Metadata (US FGDC STD-001-1998, version 2.0) is a leading example of a standard that was urgently needed by the federal government. In 1994, it became the first standard that was endorsed by the FGDC, soon after EO 12906 was issued.

OMB Circular A-119 (on-line see http://www.whitehouse.gov/OMB/circulars/a119/a119.html) directs federal agencies to use 'voluntary consensus standards' such as ANS's in procurement and regulatory activities, except where 'inconsistent with law or otherwise impractical'. To that end, the FGDC and several member agencies participate as members of INCITS L1 to identify appropriate voluntary consensus standards for the NSDI.

3.1.2. *Stakeholders – Government*
On-line see http://www.fgdc.gov/stakeholders/stakeholders.html

Federal

The NSDI Framework is a collaborative effort to create a widely available source of basic geographic data. Seven data themes have been identified as needed by a wide variety of geo-spatial applications. These data themes are elevation, ortho-imagery, hydrography, geodetic control, cadastral registration, government units, and transportation. One of the tasks of Geospatial One-Stop is to develop national geo-spatial data content standards for all seven NSDI Framework data themes.

The revised draft OMB Circular A-16 on Coordination of Geographic Information and Related Spatial Data Activities (on-line see http://www.fgdc.gov/publications/a16intro.html) designates lead FGDC-member agencies for each of the seven NSDI Framework data themes:

- US Geological Survey (USGS), Department of the Interior, is the lead federal agency for elevation (terrestrial), ortho-imagery, and hydrography data;
- National Oceanic and Atmospheric Administration (NOAA), Department of Commerce, and the US Army Corps of Engineers are co-leads for bathymetric elevation data;
- Bureau of Land Management (BLM), Department of the Interior, is the lead federal agency for cadastral data;
- Minerals Management Service (MMS), Department of the Interior, is the lead federal agency for Offshore Cadastre, the land management system used on the Outer Continental Shelf of the U.S.A.;

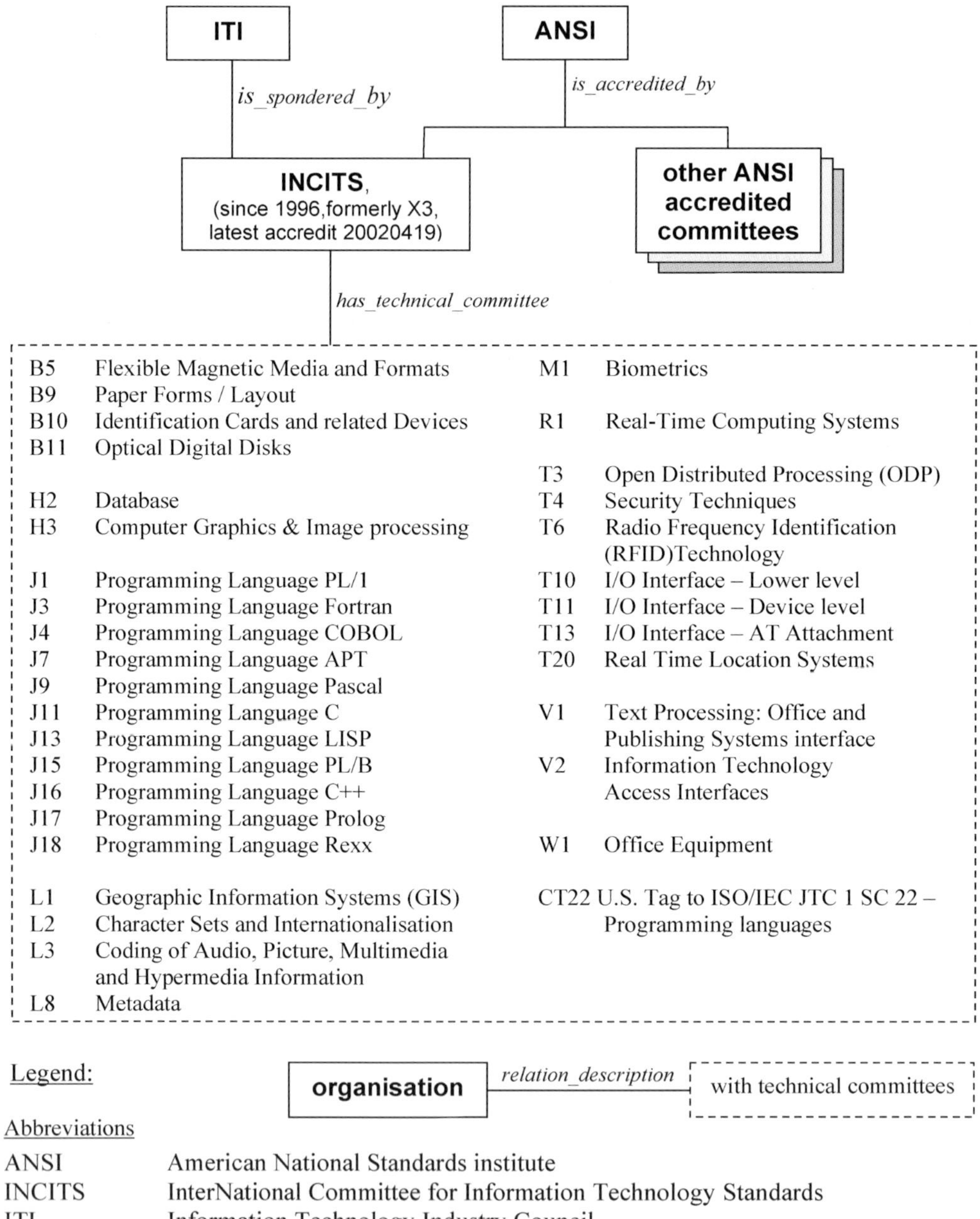

Fig. 2. ANSI – INCITS role in geo-spatial standards infrastructure in the U.S.A.

- National Geodetic Survey, which is part of NOAA, is the lead federal agency for geodetic control data. The National Spatial Reference System is the fundamental geodetic control for the U.S.A.;
- US Census Bureau, Department of Commerce, is the lead federal agency for government units that describe the official boundary of federal, state, local, and tribal governments;
- Bureau of Transportation Statistics, Department of Transportation, is the lead agency for transportation geo-spatial data.

State

State organisations that participate in developing the NSDI include the National States Geographic Information Council (NSGIC, on-line see http://www.nsgic.org/) and co-ordinating state GIS Councils (see Figure 1). NSGIC describes itself as 'an organisation of states committed to efficient and effective government through the prudent adoption of information technology', particularly geographic data and systems. NSGIC will be an equal partner on the Geospatial One-Stop Board of Directors.

In 1995, the FGDC awarded NSGIC a grant to implement a proposal for an educational and research program in support of Content Standards for Digital Geospatial Metadata (on-line see http://www.nsgic.org/papers/METAHMPG.HTM). The first phase of the project was the execution of case studies on the implementation of the standard in nine NSGIC member states. The second phase involved the preparation of a metadata primer to provide a practical explanation of the metadata standard for state and local governments and a compilation of the case studies produced in the first phase.

Regional, local, and tribal

Various regional, local, and tribal entities participate in developing the NSDI. Organisations representing regional, local, and tribal entities include the National Association of Counties (NACo), National League of Cities (NLC), the International City/County Management Association (ICMA), and the Intertribal GIS Council. These organisations will be equal partners with federal organisations on the Geospatial One-Stop Board of Directors.

The FGDC Cooperative Agreements Program (CAP, on-line see http://www.fgdc.gov/funding/funding.html) provides assistance to regional, local, and tribal organisations in creating metadata for data discovery through the clearinghouse. No comprehensive survey of use of metadata by regional, local, and tribal organisations has been done.

3.1.3. *Stakeholders – Cross-sector*
Open GIS® Consortium

The OpenGIS® Consortium (OGC, on-line see http://www.opengis.org/) is an international, not-for-profit member consortium that develops software interface specifications for geo-spatial services. Membership includes vendors, governmental bodies, academia, and non-governmental organisations. The OGC specifications are made public after approval and are free for download and implementation. OGC interfaces are implemented in a variety of non-commercial and commercial software programs worldwide.

OGC Software interfaces relevant to SDI's include:

- Catalog Services Specification for discovery of digital geo-spatial data through metadata;
- Web Map Service specification for requesting and displaying mapped data from remote data stores;
- Geography Markup Language, a dialect of XML for encoding spatial features and their attributes and geometries;
- Web Feature Service for supporting request and delivery of formatted 'vector' geographic information over a Web Service.

The OGC has promoted SDI concepts in multi-organisational 'test bed' and 'pilot' initiatives that apply approved or candidate specifications in operational environments relevant to FGDC membership and affiliations. The OGC serves as a source for tested specifications that FGDC members can adopt and implement to improve interoperability.

I-teams

The I-Team Geospatial Information Initiative (I-Team Initiative, on-line see http://www.fgdc.gov/I-Team/) is a joint project of the FGDC, OMB, NSGIC, NACo, and other strategic partners. The I-Team Initiative aims to offer a coherent set of institutional and financial incentives to enable all levels of government and the private sector to collaborate in building NSDI framework data.

The I-Team Initiative relies on locally formed, interdependent partnerships (**I-Teams**) of federal, state, local, and tribal authorities, academia, and the private sector to implement state and regional portions of the NSDI in accordance with interoperability specifications and data standards, including metadata standards, as part of their ordinary business processes.

GeoData Alliance

The FGDC convened an inter-sector team to conceive an open, inclusive and democratic process to foster the development of the NSDI, as the geo-spatial data community recognised the need for a neutral forum in which all members could gather to debate and resolve important crosscutting issues. In 2000, the GeoData Alliance was established as a non-profit organisation 'to foster trusted and inclusive processes to enable the creation, effective and equitable flow, and beneficial use of geographic information'. A multi-sector council of trustees governs the GeoData Alliance, which is open to all individuals and institutions that subscribe to the purpose and principles.

3.2. *Canada*

3.2.1. *National standards*

National standards in geomatics in Canada come under the purview of the Canadian General Standards Board through the Committee on Geomatics (CGSB-CoG) and the Canadian Advisory Committee (CAC-TC 211) to ISO/TC 211. Much of the activity has been concentrated on participation in the ISO/TC 211 (on-line see http://www.isotc211.org) and in the anticipation of adopting some of its standards in support of the new Canadian Geospatial Data Infrastructure (CGDI) and industrially within Canada.

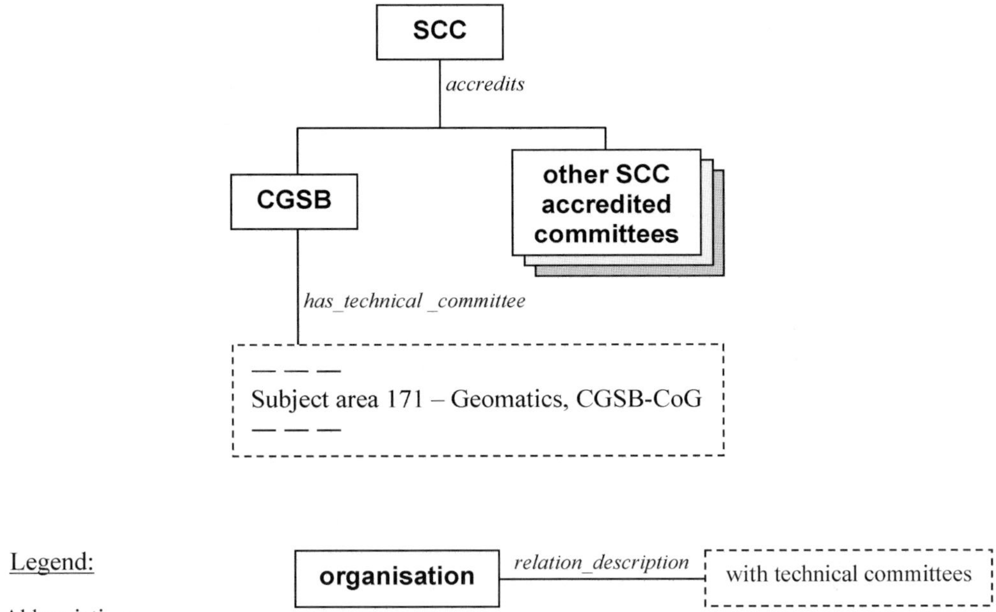

Abbreviations

CGSB-CoG Canadian General Standards Board – Committee on Geomatics
SCC Standards Council of Canada

Fig. 3. SCC – CGSB role in geo-spatial standards infrastructure in Canada.

Canadian Geo-spatial Data Infrastructure

One of the significant activities in Canada is the work on the CGDI by Technology Advisory Panel (CGDI-TAP). Standards are crucial to the development of a national infrastructure and are the responsibility of the CGDI-TAP that has developed a Technology Vision and Implementation Plan for the CGDI. A Report on the Development of a Plan and Process Model for the Standards Component of the CGDI has been written in readiness for this work, which must be implemented within the national Information Technology infrastructure. This discussion document points to the need for common practices to be identified and captured as standards as well as the need to address and resolve horizontal issues both nationally and internationally through standards that are not necessarily unique to geomatics.

Canadian Advisory Committee

The CAC-TC 211 Canadian Advisory Committee for ISO/TC 211 reviews draft standards from the international organisation. Several of its members are active participants in the technical committee, itself.

Canadian Committee on Geomatics

The CGSB-CoG Canadian Committee on Geomatics brings together representatives of the provincial governments that are involved in geomatics. This committee provides for the sharing of information about federal and provincial activities, including work on standards.

3.2.2. *Stakeholders – Government*
Federal

Within the federal government, several departments are using standards that are either national or international in scope. Geomatics Canada (part of the Natural Resources Canada) produces topographic maps and aeronautical charts, legal surveys of Canada lands, geodesy for precise positioning and applications or remotely sensed earth observation data. The Canada Centre for Remote Sensing (CCRS, on-line see http://www.ccrs.nrcan.gc.ca/ccrs/data/standards/welcom_e.html) is the national agency responsible for the development of remote sensing applications and related methodologies. The Federal government (Previously Natural Resources Canada and now Agriculture and Agrifood Canada) has led and participated in the development of standards for Imagery and Gridded Data in ISO/TC 211 (on-line see http://www.isotc211.org/). One of the areas of this work of the ISO/TC 211 Working Group 6 on Imagery has been the definition of new metadata elements for imagery and gridded data.

The Canadian Hydrographic Service (CHS, part of Fisheries and Oceans Canada, on-line see http://www.chs-shc.dfo-mpo.gc.ca/chs/) is responsible for surveying Canadian navigable waterways and for producing a number of nautical publications, including nautical charts, sailing directions, bathymetric maps and tide and current tables. It is implementing the revised edition of the International Hydrographic Organisation standard (IHO, on-line see http://www.iho.shom.fr/) standard (S57) for electronic nautical charts. To date, more than 400 electronic nautical charts have been produced, and an additional over 200 charts will complete its coverage within the next two years. The IHO has indicated that it is also revising its standards to align with the ISO/TC 211 metadata standards, and the CHS has been participating in fostering this alignment.

Environment Canada has responsibilities related to the natural environment. It is currently developing a Canadian Information System for the Environment, which will include descriptions of maps pertaining to its mandate. The Canadian Ice service has sponsored the extension of the IHO standard (S57) to cover ice objects. This is also being brought to the appropriate international forums through IHO and the World Meteorological Organisation (WMO, on-line see http://www.wmo.ch/index-en.html).

The Department of National Defence (DND, on-line see http://www.dnd.ca/) is responsible domestically for national emergency measures and homeland defence, and internationally for Canada's role in coalition forces. It has been an active participant in the NATO related DGIWG (on-line see http://www.digest.org/) and in the alignment of the NATO geographic information STANAGS with the ISO/TC 211 suite of standards. It has also been the promoter of the Canadian Geomatics Interchange Standard (CGIS-DIGEST-

CGSB 171.4-94). DND has been active in the efforts of the DGIWG to restructure the next generation of the DIGEST standard so that it aligns with the ISO/TC 211 suite of standards. In fact, DGIWG has adopted the ISO/TC 211 metadata standard and is currently adapting it to military requirements, adding security and intelligence metadata and additional details, such as code lists, to insure interoperability and data access services.

Provinces

Provinces as stakeholders have shown considerable independence in their adoption, creation and use of standards. Some of these have been based upon existing Canadian or U.S.A. standards. The value of true standardisation of metadata or any other tool has not been fully comprehended by some, and because there is no urgency given to standardisation, the number of unique systems is proliferating.

Municipalities

At recent workshops conducted for GeoConnections in different parts of the country, it was evident that municipalities also have an extraordinary mix of metadata and other tools for their geographic information systems. Although the value of harmonisation was made clear, resources are rarely readily provided for such activities by municipal governments. On their own, a number of the major municipalities are nevertheless moving ahead.

3.2.3. *Stakeholders – Non-government*

Open GIS® Consortium

A number of the leading Canadian software developers are participating in the activities of OGC and are strong supporters of this work and the resulting implementation specifications. OGC has indicated that its policy is to develop its implementation specifications based on international standards such as the ISO/TC 211 standards, although sometimes practical implementation considerations have limited this, or have resulted in feedback to the international standards development process. Expectations are high that standardised, open and practical solutions for reducing the technical barriers to finding, obtaining and using geo-spatial data will result.

The Geomatics Industry Association of Canada

The Geomatics Industry Association of Canada (GIAC, on-line see http://www.giac.ca/) is the only national business association dedicated to servicing the geomatics industry in Canada and it encompasses the entire range of geomatics disciplines. Its prime role is informing all aspects of the Geomatics industry of the developments in the standards arena.

4. Metadata activities in North America

4.1. *National geomatics standards development*

4.1.1. *United States of America*

The FGDC endorsed Version 1.0 of the FGDC Content Standard for Digital Geospatial Metadata in 1994. The current Version 2.0, which the FGDC endorsed in 1998, provides for the definition of profiles and extensibility through User Defined Metadata Extensions. Version 2.0 also modifies some production rules to ease implementation.

There is a close relationship between the FGDC Content Standard for Digital Geospatial Metadata and metadata elements in the Spatial Data Transfer Standard (SDTS, on-line see http://mcmcweb.er.usgs.gov/sdts/), which has been approved as an ANS (INCITS 320). Metadata content in the SDTS is used to determine the fitness of the dataset for the user's purpose. The Metadata Standard contains metadata used to search for digital spatial datasets through a clearinghouse (metadata for locating, describing access, use, and distribution), which may not be contained in the SDTS transfer set.

The FGDC Clearinghouse activity selected ANSI Z39.50-1995 (ISO 10163-1995) as the search and retrieval protocol for geo-spatial metadata on different host computers (or servers). The Z39.50 protocol was initially developed by the library community to discover bibliographic records using a standard set of attributes. On the server, Z39.50 software, typically communicates with an appropriate search engine to process the query and formulate the results. A server may be compliant with Z39.50 but may not be capable of handling structured metadata.

Z39.50 maintenance agencies can register specific sets of attributes, operators, and rules of implementation as Application Profiles. The Global Information Locator Service profile (GILS, on-line see http://www.gils.net/) selects a minimum number of Z39.50 features but requires those features on every compliant server. GILS adopts MAchine-Readable Cataloging (MARC, on-line see http://www.loc.gov/marc/) semantics for the representation and communication of bibliographic and related information in machine-readable form for elements used in locator records, and the GILS profile maintains a one-to-one correspondence between GILS elements and MARC tagged elements.

The Content Standard for Spatial Metadata inherits the metadata content of the GILS Profile and provides further elaboration peculiar to spatial data. The FGDC has developed a profile for geo-spatial metadata, called 'GEO', which provides guidance on how to implement FGDC metadata elements within a Z39.50 service. The GEO Profile is intended to be a superset (contain all the elements) of the GILS Profile. Therefore, servers compliant with the GEO Profile should also be GILS-compliant.

Profiles

Version 2.0 of the FGDC Content Standard for Digital Geospatial Metadata enables geospatial data communities to create profiles of the base standard and extensions (new metadata elements) for their requirements. The FGDC endorsed the Content Standard for Digital Geospatial Metadata, Part 1: Biological Data Profile in 1999 and the Metadata Profile for

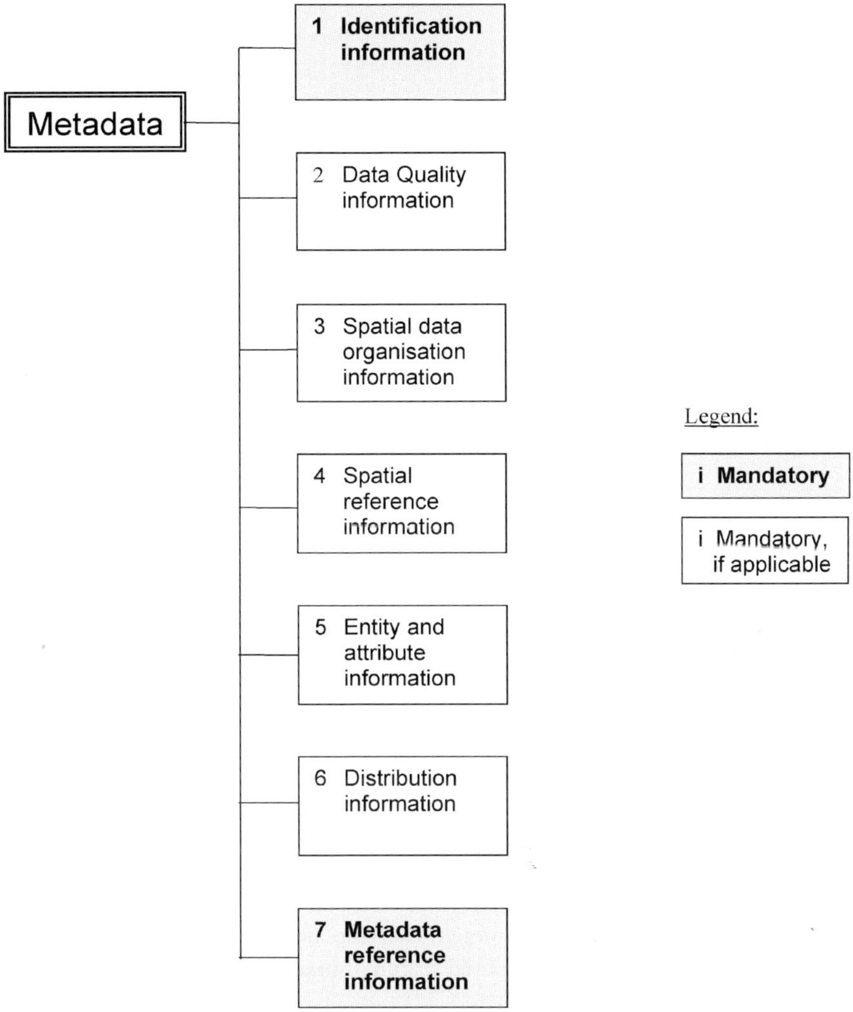

Fig. 4. Graphical representation of the Federal Geographic Data Committee's Contents Standard for Digital Geospatial Metadata, FGDC STD-002, June 1998. (After: Susan Stitt, USGS Center for Biological Information.)

Shoreline Data in 2001. The Content Standard for Digital Geospatial Metadata: Extensions for Remote Sensing Metadata is pending FGDC endorsement as of July 2002. A FGDC standards project to develop a Thematic Supplement for Geospatially Referenced Cultural and Demographic Data Metadata was withdrawn by the FGDC Cultural and Demographic Data Sub-committee, which had responsibility for that project.

Once the U.S.A. adopts a profile of ISO 19115 as a national standard, it is expected that FGDC-endorsed profiles and extensions will be added to the national standard.

4.1.2. *Canada*

Canadian Metadata Standard

A Canadian Metadata Standard was formally approved by the Canadian General Standards Board in 1992 and implemented by agencies. However, some of the lead agencies for earth observation data with strong international involvement overrode national interests and committed to the US FGDC standards. The differences between the two standards are being resolved as part of the joint Canada/U.S.A. standardisation work and the various stakeholders have committed to adoption of the TC211 specifications on its release.

National Canadian profile of the TC/211 Metadata

A national profile of the TC211 Metadata specification will be influenced and implemented by GeoConnections for data discovery and data access under the CGDI.

5. North American geo-spatial metadata – Services/Query

5.1. *United States of America*

NSDI Clearinghouse

The National Geospatial Data Clearinghouse enables search across a network of geo-spatial data collections based on metadata compliant with the FGDC standard held by over 100 institutions in the U.S.A. (and more internationally). The Clearinghouse includes a search interface or portal hosted by the FGDC and mirrored at five other FGDC member organisations around the U.S.A. Through search forms, one can compose a simple or complex query to locate documented geo-spatial data of all types and formats. The data and metadata are maintained and served by the hosting institutions to promote ownership and currentness. To search the Clearinghouse, one actually performs a distributed search of many metadata collections at the same time.

The implementation of OGC Web Map Services by FGDC member organisations means that not only can data resources be located, but can be displayed as maps in conventional browsers. The Clearinghouse search has recently included the ability, where Web Map Services exist, to launch a map viewer window that lets the user find and display and overlay multiple maps of data from multiple servers.

5.2. *Canada*

Canadian Geo-spatial Data Infrastructure

CEONET forms an important component of the Canadian Geo-spatial Data Infrastructure (CGDI). The CGDI is a distributed system with an emphasis on connecting and providing access to geo-spatial data and services. Data custodians manage their own resources. Led by the

Access Advisory Node of GeoConnections their objectives are to provide discovery, evaluation, and access to Canadian public geo-spatial data and services by national and international clients through a common window based on interoperable, reusable tools and services and provide mechanisms for suppliers to advertise and distribute Canadian geo-spatial data and services, from base data to value-added services.

Northern Information Network

The Northern Information Network (NIN, on-line see http://esd.inac.gc.ca/nin/) contains a directory that describes geo-spatial datasets covering Canada's northern territories and adjacent areas. The metadata uses the national standard and currently has about 500 maps described. Every effort is being made to use the metadata standard for describing both traditional knowledge and western science.

Atlantic Coastal Zone Information Steering Committee

The Atlantic Coastal Zone Information Steering Committee (ACZISC, on-line see http://www.dal.ca/aczisc/) provides a focus and forum for the development and co-ordination of a regional coastal zone information infrastructure and the promotion of integrated coastal zone management initiatives. It encourages the implementation of appropriate mechanisms to manage information and data effectively and to facilitate its exchange and dissemination. It also promotes the awareness of initiatives relating to coastal information and management in order to minimise duplication and to maximise collaborative opportunities. It is establishing linkages with other committees and organisations in order to collaborate on activities of mutual interest.

6. Summary and conclusion

6.1. *Regional plans*

Possible U.S.A./Canada Spatial Metadata profile

The standards committee structures in Canada and the U.S.A. are very similar, and past co-operation between the ANSI (American National Standards Institute) and the SCC (Standards Council of Canada) provides a good model for future co-operation for developing common or compatible profiles. In the past, joint work went through a working level joint committee of the two national groups. This provides a forum to develop consensual positions, which then can be ratified by the main groups of users and providers.

This process is being discussed between the Canadian CGSB Committee on Geomatics and the US ANSI INCITS L1 Committee to develop in parallel compatible Canadian and U.S.A. national metadata standard and have it approved in parallel in both countries as a joint North American profile of ISO 19115 (see Figure 5). In addition, there have been informal discussions between the U.S.A. and Canada on a common North American strategy to adopt and implement other ISO/TC 211 standards.

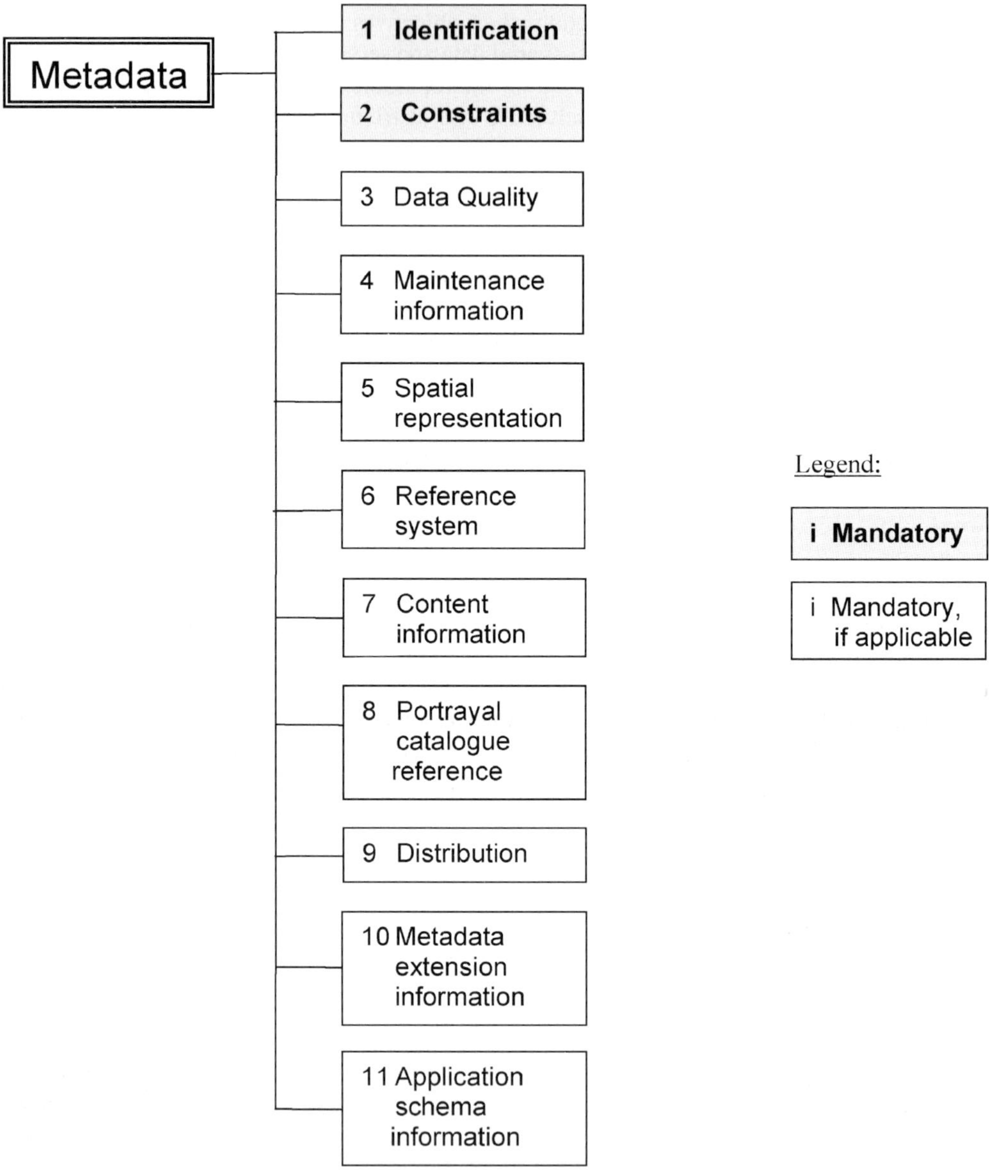

Fig. 5. Graphical representation of the possible outcome of the North American Profile for the ISO 19115 Metadata Standard.

Now, the U.S.A. is discussing a working document with Canada to make the general existing Memorandum of Understanding (MoU) between the two countries effective. This document endeavours to outline procedures for ANSI/INCITS-L1 and SCC/CGSB-CoG to develop common profiles of ISO/TC 211 standards. In addition, numerous Federal mapping agencies in the U.S.A. and Canada have joint mapping projects with each other, and with Mexico. Plans may include developing regional and international clearinghouse nodes.

6.2. *Future metadata standards plans*

6.2.1. *United States of America*

The U.S.A. will adopt ISO 19115 as an ANS after ISO 19115 is finalised as an IS. A proposal has been submitted to INCITS L1 for changes to obligations of entities/elements required to bring ANSI's adoption of ISO 19115 in line with obligations required in the FGDC v.2.0 Content Standard for Digital Metadata. The amendments in the proposed standard are intended to allow systems that are currently compliant with the FGDC Content Standard for Digital Metadata to be compliant with ISO 19115. The impact of this amended standard would be to align U.S.A. metadata standards with the ISO metadata standard, which, in turn, would permit/support queries on metadata globally and to open markets to metadata tool developers. Future amendments to the ANSI standard might include incorporation of FGDC community profiles and Intelligence Community requirements.

6.2.2. *Canada*

GeoConnections has developed an international clearinghouse for geo-spatial data. Originally called CEONet, it is similar to ten other clearinghouses in the U.S.A. and other parts of the world. The metadata specifications for data discovery and access are currently based upon the FGDC metadata standards, thereby facilitating the sharing of metadata information with Committee on Earth Observation Satellites International Direction (CEOS IDN) and Global Change Master Directory (NAS GCMD).

Future support for ISO/TC 211 is planned. The ISO/TC 211 metadata specifications are also being implemented by other agencies in Canada since they are compatible with those used by the U.S.A. and can be used as the basis of the U.S.A./Canada national profiles.

Acknowledgements

Richard Hogan of the US Geological Survey and Richard Pearsall of the US Federal Geographic Data Committee are recognized for their efforts to help design the outline for this chapter.

C. Douglas O'Brien of Idon Technologies contributed to the development of some of the Canadian portion of this chapter.

Cameron Wilson of Natural Resources Canada provided valuable assistance to help finalize Figure 1.

References

Organisation	For on-line contact see:
ACZISC, Canada	http://www.dal.ca/aczisc/
American National Standards Institute	http://www.ansi.org/

Organisation	For on-line contact see:
Canadian Geospatial Data Infrastructure	http://cgdi.gc.ca
Canadian General Standards Board	http://www.pwgsc.gc.ca/cgsb/home/index-e.html
Canadian Standards Association	http://www.csa.ca/
CCRS, Canada	http://www.ccrs.nrcan.gc.ca/ccrs/data/standards/welcom_e.html
GIAC, Canada	http://www.giac.ca/
DGIWG	http://www.digest.org/
DND, Canada	http://www.dnd.ca/
Executive Order (U.S.A.)	http://www.fgdc.gov/publications/documents/geninfo/execord.html
FGDC	http://www.fgdc.gov/
FGDC Cooperative Agreements Program	http://www.fgdc.gov/funding/funding.html
FGDC Content Standard Metadata	http://www.fgdc.gov/metadata/metadata.html
GeoConnections, Canada	http://www.geoconnections.org/
GeoData Alliance, U.S.A.	http://www.geoall.net/
GILS, U.S.A.	http://www.gils.net/
GSDI	http://www.gsdi.org/
IHO	http://www.iho.shom.fr/
Int. Com. for Information Technology Standards	http://www.incits.org
ISO/TC 211	http://www.isotc211.org/
I-Team Initiative	http://www.fgdc.gov/I-Team/
MARC	http://www.loc.gov/marc/
National Geospatial Data Clearinghouse, U.S.A.	http://www.fgdc.gov/clearinghouse/clearinghouse.html
National States Geographic Information Council	http://www.nsgic.org/
NSGI Metadata	http://www.nsgic.org/papers/METAHMPG.HTM
NSDI Geospatial One-Stop	http://www.fgdc.gov/geo-one-stop/index.html
NGA, U.S.A.	http://www.nga.mil/
NIN, Canada	http://esd.inac.gc.ca/nin/
Office of Management and Budget, U.S.A.	http://www.whitehouse.gov/omb/
OpenGIS® Consortium	http://www.opengis.org/
OMB Circular A-16	http://www.fgdc.gov/publications/a16intro.html
OMB Circular A-119	http://www.whitehouse.gov/omb/circulars/a119/a119.html
PAIGH	http://www.ipgh.org.mx/default_en.htm
PC-IDEA	http://www.cpidea.org/
Roger Tomlinson	http://www.esri.com/news/arcnews/fall01articles/drroger.html
SDTS, U.S.A.	http://mcmcweb.er.usgs.gov/sdts/
Standards Council of Canada	http://www.scc.ca/
U.S.A. Stakeholders – Government	http://www.fgdc.gov/stakeholders/stakeholders.html
WMO	http://www.wmo.ch/index-en.html

URL's in this chapter are accessed in May, 2005, unless otherwise stated in the text.

H. Moellering, H.J.G.L. Aalders & A. Crane (Editors)

Metadata Standards Development Activities in the Asia – Pacific Region

Craig Macauley, lead author, editor [a]

contributing authors:
Du Dao-sheng [b], Kazuhiko Akeno [c], Tae-jung Min [d]

[a] *Government of South Australia, Australia*
e-mail: macauley.craig@saugov.sa.gov.au
[b] *State Laboratory for Information Engineering in Surveying, Mapping and Remote Sensing, Wuhan University, China*
e-mail: dsdu@dns.liesmars.wtusm.edu.cn
[c] *Information Access Division, Geo-information Department, Geographical Survey Institute of Japan*
e-mail: akeno@gsi.go.jp
[d] *National Geography Institute, Korea*
e-mail: tjmin@moct.go.kr

1. General regional structure for spatial data organisations

This section provides an overview of two regional spatial data organisations. The first, the Permanent Committee on GIS Infrastructure for Asia and the Pacific (PC GIAP), covers the whole region, while the second, ANZLIC the Spatial Information Council is a joint initiative between Australia and New Zealand.

1.1. *PC GIAP*

The Permanent Committee on GIS Infrastructure for Asia and the Pacific provides the framework for spatial data infrastructure (SDI) initiatives in the region.

The Committee was established pursuant to Resolution 16 of the 13th United Nations Regional Cartographic Conference for Asia and the Pacific, Beijing 1994. It reports to the United Nations Regional Cartographic Conference for Asia and the Pacific (UNRCC-AP).

The aims of the Committee are to maximise the economic, social and environmental benefits of geographic information in accordance with Agenda 21 by providing a forum for nations from Asia and the Pacific to:

1. co-operate in the development of a regional geographic information infrastructure;
2. contribute to the development of the global geographic information infrastructure;

3. share experiences and consult on matters of common interest;
4. participate in any other form of activity such as education, training, and technology transfer.

PC-GIAP has 55 member countries, extending from Armenia, Iran, Azerbaijan and Kazakhstan in the northwest to New Zealand, French Polynesia, American Samoa and Cook Islands in the southeast.

PC GIAP has the following structure:

- Working Group 1 – Regional Geodesy;
- Working Group 2 – Fundamental Data;
- Working Group 3 – Cadastre;
- Working Group 4 – Institutional Strengthening.

Working Group 2's 2000–2003 work plan has nine key strategies, one of which is: 'Develop a specification and implementation plan for a Data Node network for the region, with priority given to metadata, data dictionary and clearing house issues'. (See Section 4.8 for more on the PC GIAP Data Node Project; on-line http://www.pcgiap.org/.)

1.2. *ANZLIC – the Spatial Information Council*

ANZLIC (on-line see http://www.anzlic.org.au/infrastructure_metadata.html) is the peak council for public sector spatial data management in Australia and New Zealand.

ANZLIC provides a framework for other national bodies to contribute to ANZLIC objectives. Those bodies include the Public Sector Mapping Agencies (PSMA) and the Intergovernmental Committee on Surveying and Mapping (ICSM).

ANZLIC's vision is that economic, social and environmental decision-making is supported by spatial information. ANZLIC will achieve its vision through:

- national leadership and advocacy;
- partnerships between the community, industry and governments;
- development and adoption of standards, policies and guidelines.

To achieve its vision, ANZLIC has established the following goals for 2000 to 2005:

- development of a comprehensive framework of policies and standards;
- enhance the availability of accessible and SDI-compliant datasets;
- recognition, at all levels of government, industry and the community, of the necessity for quality spatially referenced information;
- create a competitive, innovative and robust spatial information industry.

The ten members of ANZLIC represent six states, two territories and the federal government in Australia, and New Zealand.

Two Standing Committees report to ANZLIC:

1. Spatial Data Infrastructure;
2. Industry Development.

The ANZLIC SDI Standing Committee co-ordinates the ANZLIC Australian Spatial Data Infrastructure (ASDI) technical implementation projects as well as ASDI-related activities of ICSM and ANZLIC sub-committees and working groups. The Standing Committee also co-ordinates the development of standards and protocols required to implement the ASDI.

Please note that this information was accurate in May 2002. Readers are encouraged to go to the ANZLIC Web site at http://www.anzlic.org.au for the most recent information and details of current activities in this area (contact details: Executive Officer, ANZLIC – the Spatial Information Council, E-mail: info@anzlic.org.au, http://www.anzlic.org.au/).

2. Identification of major national spatial data organisations

There are many countries actively pursuing innovative metadata initiatives in the Asia – Pacific region. This section contains a selection of countries in the region and identifies the major spatial data organisations in those countries.

2.1. *Australia*

Geoscience Australia – National Mapping Division

Geoscience Australia – National Mapping Division (formerly AUSLIG, on-line see http://www.auslig.gov.au/) is Australia's national mapping agency and provides essential geographic information as well as the geodetic framework that underpins all accurate mapping and global positioning services in Australia.

The National Mapping Division is responsible for:

- providing small and medium scale topographic maps of Australia;
- acquiring, cataloguing, processing and distributing satellite imagery;
- providing accurate geodetic control networks and geoid models;
- administering the Office of Spatial Data Management and the Australian Spatial Data Directory.

Public Sector Mapping Agencies

The Public Sector Mapping Agencies (PSMA Australia Ltd, on-line see http://www.psma.com.au/) is a consortium of Australian government mapping agencies from the six states, two territories and the federal government. It was formed in 1993 to lead a project which integrated the best available cadastral and topographic data to produce a national digital map base of Australia.

Since then, PSMA has been involved in developing other national mapping products formed by combining datasets from PSMA member organisations.

2.2. *China*

National Technical Commission of Geographic Information Standardisation

The Chinese National Technical Commission of Geographic Information Standardisation (SBTS/TC 230) was founded in 1997. The secretariat is located at the National Geomatics

Centre of China (NGCC). The main members are the National Bureau of Surveying and Mapping, the Chinese Academy of Sciences, the Ministry of Construction and the China National Institute of Standardisation. In addition, thirty other member organisations are involved in the production and application of GI.

The scope of the Technical Commission (TC) is standardisation in the field of GI at the national level. It is responsible for researching and making suggestions for guiding principles and policies, drawing up and revising national standards, implementing issued national standards, studying issues related to standardisation and exchanging experiences of standardisation with foreign countries in the field of GI-standardisation. The TC will ensure that GI standards development will be replaced by a developing structured set of standards in China, guided by applications.

Geo-Spatial Information Co-ordination Commission of China

The State Department formally founded the Commission in 1998. One of its major functions is to organise, study and formulate the planning, standards, regulations, policies, laws and development strategies for the construction of the national spatial information infrastructure and GIS development.

China ISO/TC 211 domestic activities network

China was one of the observing members of ISO/TC 211 when the Committee was founded in June 1994. China became a participating member in 1995 and has since than sent several experts to attend the Working Group meetings and plenary meetings. More than ten Chinese experts work in the five Working Groups of ISO/TC 211. They participate in activities of the Committee including discussions for the forty-one projects which are being developed, reviewing Working Group drafts, providing technical material, proposing suggestions, voting for new project proposals, etc. Within China, there are more than forty organisations taking part in the China domestic activities network of ISO/TC 211. Expert groups relevant to the nine Working Groups of the ISO/TC 211 have been formed. These experts will contribute in the near future. The 7th Plenary and Working Group meetings of ISO/TC 211 were held in Beijing during September 1998. All these efforts will enable the acceleration of Chinese GI standards to comply with international standards.

National Geomatics Centre of China

The National Geomatics Centre of China (NGCC) was founded in 1995. It is also the National Archives of Surveying and Mapping of China. NGCC is one of the organisations that comprise the State Bureau of Surveying and Mapping (SBSM) of China. The main responsibilities of NGCC are as follows:

- National Fundamental Geographic Information System (NFGIS) maintenance and information service including data distribution;
- National Geodetic Control Network and Global Positioning System Network;
- administrative boundary survey;

- technical archives and results of surveying and mapping;
- thematic GIS developing;
- digital mapping and digital image mapping;
- technical consultancy in application projects.

The secretariat of the National Technical Commission of Geographic Information Standardisation (CSBTS/TC 230) is at NGCC. The ISO/TC 211 national body of China is also at NGCC.

Standardisation Institution of SBSM in China

The Standardisation Institution of Surveying and Mapping was formally founded by the State Bureau of Surveying and Mapping (SBSM) in 1984 and is responsible for developing and implementing technical standards of the traditional products of surveying and mapping in order to ensure the quality of the products. In the last ten years, the Institution has developed national and trade standards for GIS and NSDI.

2.3. *Fiji*

Fiji Land Information System

The Fiji Land Information System (FLIS) Program is a co-operative venture of all governmental and non-governmental land-related agencies for the establishment of an integrated land information system for Fiji. The Fiji Land Information Council (FLIC) is the executive arm of the Program. The Council's main role is the formulation of policies. It also administers and manages FLIS projects. Membership of the Council consists of Permanent Secretaries (Chief Executives) of all governmental land related organisations, including the General Manager of the Native Land Trust Board (NLTB) and the Managing Director of Telecom Fiji Ltd.

Native Land Trust Board

The Native Land Trust Board (NLTB) is a statutory organisation responsible for the administration of all land owned by the natives, which comprises more than 80% of the land in Fiji.

Telecom Fiji Ltd

Telecom Fiji Ltd is the leading GIS agency in the utility sector responsible for the development of telecommunication infrastructure, an important component of FLIS.

2.4. *Japan*

Geographical Survey Institute

The Geographical Survey Institute (GSI, on-line see http://www.gsi.go.jp/) is the governmental organisation engaged in survey administration and basic survey activities based on the

Survey Act and the Placement Law of the Ministry of Land, Infrastructure and Transport. GSI is also engaged in technical research and development, and international co-operation in the field of survey and mapping.

GSI provides fundamental geographic information for SDI programs and technical standards on GI.

Association of Precise Survey and Applied Technology (APA)

APA is a non-profit organisation for research on geomatics by members of private companies. APA is also the Secretariat of the ISO/TC 211 Domestic Committee.

2.5. *Korea*

National Geography Institute

The National Geography Institute (NGI, on-line see http://www.ngi.go.kr/) is the central surveying and mapping organisation under the Ministry of Construction and Transportation of Korean government. NGI provides fundamental geographic information and technical standards on surveying and mapping.

NGI is responsible for:

- policy, standards and co-ordination of surveying and mapping including GIS;
- management of the national geodesy, surveying and mapping programs;
- implementation of the National Geographic Information System (NGIS) of Korea.

National Computerisation Agency

The missions of the National Computerisation Agency (NCA, on-line see http://www.nca.or.kr/) are to:

- provide expertise in preparing and implementing initiatives and plans for information promotion;
- construct and promote the Korea Information Infrastructure (KII);
- develop and establish IT standards;
- evaluate and audit IT projects and systems.

Korean Agency for Technology and Standards

The Korean Agency for Technology and Standards (KATS, on-line see http://www.ats.go.kr/) is the national body of Korea for the ISO. KATS is responsible for the industrial standards in Korea.

2.6. *Malaysia*

National Infrastructure for Land Information System

The National Infrastructure for Land Information System (NaLIS, on-line see http://www.nalis.gov.my/) was established in 1997. It is a spatial data infrastructure initiative to facilitate sharing of geographic information among land related agencies. At present, NaLIS activities focus on three main areas:

- GI standardisation;
- development of core datasets;
- implementation of prototypes to enable data to be shared on-line.

Department of Survey and Mapping Malaysia

The Department of Survey and Mapping Malaysia (DSMM, on-line see http://www.jupem.gov.my/) is a government agency responsible for cadastral surveys, topographic mapping, geodetic surveys and boundary delimitation.

2.7. *New Zealand*

Land Information New Zealand

Land Information New Zealand (LINZ, on-line see http://www.linz.govt.nz/) is a government department with roles and responsibilities in the following key areas:

- National Spatial Reference System and cadastral survey infrastructure;
- topographic and hydrographic information;
- land titles;
- standard setting for rating valuation;
- crown property;
- assisting the government address land related aspects of Treaty of Waitangi issues.

The main role of the department is a regulatory one, to set guidelines and standards and manage contracts for carrying out the day-to-day business associated with each of the key areas.

Terralink

Terralink (on-line see http://www.terralink.co.nz) is a provider of information about land, property and seabed. The organisation specialises in mapping, land information systems and spatial data capture.

2.8. *The Philippines*

National Mapping and Resource Information Authority

The National Mapping and Resource Information Authority (NAMRIA) is the central mapping and resource information agency of the national government and is attached to the Department of Environment and Natural Resources. It is responsible for topographic mapping, geodetic surveys, hydrographic survey, land classification survey, remote sensing data analysis, and geographic information management.

NAMRIA heads an Inter-Agency Task Force on Geographic Information (IATFGI), which is carrying out activities to build standards for data sharing and dissemination of geographic information. Among these activities is a World Bank assisted project to prepare a National Geographic Information Infrastructure Framework Plan (2001–2005).

2.9. *Thailand*

Geo-Informatics and Space Technology Development Agency

The Geo-Informatics and Space Technology Development Agency (GISTDA, on-line see http://www.gistda.or.th/) is a new public organisation established to co-ordinate all activities for space technology development and geo-informatics in Thailand. Its major roles include:

- application development;
- satellite database and natural resources information centre development;
- provision of data services;
- provision of technical services and human resources development;
- research and development;
- standardisation in Remote Sensing and geo-informatics technology.

This new agency is also expected to have major roles in co-ordinating the development of GI databases and applications among all Thai organisations.

Thai Industrial Standards Institute/Technical Committee 904: Geographic Information

The Thai Industrial Standards Institute/Technical Committee 904: Geographic Information (TISI/TC904) is the national body of Thailand for ISO. TISI/TC904 is responsible for the standards related to GI in Thailand.

Royal Thai Survey Department

The Royal Thai Survey Department (RTSD, on-line see http://www.rtsd.mi.th/) is responsible for topographic maps of Thailand. It has started the production of digital version of its 1 : 50 000 and 1 : 250 000 topographic map series. The new version of the 1 : 50 000 digital topographic map of the whole country became available in 2002.

Department of Lands

The Department of Lands (DOL, on-line see http://www.dol.go.th/) is a governmental agency responsible for:

- land registration;
- statutory actions;
- surveying and mapping of land parcels;
- valuation of land.

DOL has computerised their land registration system in 40 land offices and has implemented LIS/GIS in four major land offices. Projects funded by the World Bank will be implemented in the next few years to upgrade the Department's processes and systems to meet its future requirements and to follow international and Thailand standards.

3. Identification of specific national/regional metadata standards development activities taking place

This section provides a description of individual national metadata standards development activities in selected countries.

3.1. *Australia and New Zealand*

3.1.1. *ANZLIC's Metadata Guidelines*

In 1996, ANZLIC published a set of core metadata elements to be used for data discovery and determining data fitness for use. Since then, these elements (contained in ANZLIC's Metadata Guidelines) have gained widespread acceptance as a simple set of 30 metadata elements that can be easily collected, stored and distributed (on-line see http://www.anzlic.org.au/asdi/metaelem.htm).

The principal vehicle for the distribution of ANZLIC metadata is through the Australian Spatial Data Directory (ASDD), a distributed Internet system. The ASDD was launched in 1998 and has since steadily grown in content to become the key source of spatial information in Australia. See Section 4.1 for more on ASDD.

Review of ANZLIC's Metadata Guidelines

Since being published in 1996, some inconsistencies and deficiencies have been identified in ANZLIC's Metadata Guidelines and some additional metadata elements have been defined. Consequently, in November 1999, the ANZLIC Metadata Working Group (AMWG) initiated a minor review of the ANZLIC Guidelines. The review is an interim measure pending the completion of the international metadata standard ISO 19115, Geographic information – Metadata.

The review had seven aims. These were:

- correct inconsistencies and deficiencies in the current core elements;
- describe elements that have been subsequently defined but are not included in the guidelines;
- harmonise the ANZLIC Document Type Definition (DTD) with the Metadata Guideline Elements;
- describe the current working environment of ANZLIC, the Working Group and implementation of Australian Spatial Data Infrastructure (ASDI) including the Australian Spatial Data Directory and the ANZLIC Metadata Entry Tool (MET);
- describe the current environment for the transfer of ANZLIC metadata core elements;
- publish all the above, ensuring the guidelines are consistent;
- provide the foundation for the addition of new metadata elements to the ANZLIC profile.

The review was completed in February 2001 and Version 2 of the Guidelines was published shortly afterwards.

3.1.2. *ANZLIC Profile of ISO 19115*

ANZLIC's vision for metadata is:

Users of spatial data in Australia and New Zealand will have on-line access to information directories that are accurate and current and are in an internationally compatible format to better enable them to identify and access the information they require.

ANZLIC and Standards Australia have sponsored the ANZLIC Metadata Working Group's (AMWG) involvement in ISO/TC 211's development of ISO 19115 to ensure harmonisation between ANZLIC's core metadata elements and the new international metadata standard. Given the success of this harmonisation work, it is expected there will be minimal change between the current ANZLIC elements and the new ANZLIC profile of ISO 19115.

The purpose of the ANZLIC profile is to support the development of the ASDI Distribution Network (ASDI DN – see Section 4.1 for more on this initiative). This will be done by including metadata elements in the ANZLIC profile, which extend the capability of ASDD to enable access to data, products and services that comprise the ASDI. Development of the profile commenced during 2001 and were finished in 2002.

3.2. *China*

3.2.1. *Discussion of national spatial metadata standards*

With the development of information systems and the separation between data production and applications, data sharing has become one of the important objectives for the construction of spatial data infrastructures, among which the metadata for spatial data is the key to realising spatial data sharing. The demand for applications includes organisation and maintenance of spatial data, providing information for data catalogues and data transfer centres, and managing and explaining the requirements of spatial data.

In 1999, the 21st Century Agenda Management Centre of China edited and published a book named 'Study on Chinese Geographic Information Metadata Standards'. This book has had a deep analysis and study on the metadata standards in countries as U.K. and U.S.A. and the metadata standards being formulated by ISO/TC 211. The definition, effects, significance, common formats and matters on standardisation of metadata have been discussed. The content, level, structure, grading, property, characteristics of metadata in addition with the principles and methods for metadata extension of information sharing of China has been proposed. The basic models, operation tools of metadata standard implementation, metadata collection and establishment, maintenance, updating and quality guarantee of metadata databases have been studied.

This book also provides 'Metadata Standard of China's Sustainable Development of Information Sharing' (discussion draft). The standard regulates the metadata content of China's sustainable development information sharing, including ID, content, quality, status and other relevant characteristics of data. It can be used for comprehensive description on the sustainable development information or other information dataset, dataset catalogue and information exchange network service. The operated object may be dataset, dataset series, essential entity and attribute.

The organisational structure of this metadata standard is as follows:

(a) thematic content and applicable range;
(b) referenced standard;
(c) technical terms;
(d) hierarchical structure and property of metadata;
(e) grading and characteristics of metadata;

(f) metadata content:
  (f.1) first level metadata content: catalogue information of metadata;
  (f.2) second level metadata content:
    – basic information: dataset ID and explanations on spatial and temporal range, status, security, limitation, etc;
    – evaluation information on data quality;
    – data mark information: application, data resources, productive technical, etc;
    – spatial data presentation and spatial information presentation, etc;
    – reference system information with explanations on spatial and temporal reference systems;
    – essential classification information with definitions and explanations of the essential types;
    – data distribution information, i.e.: data provider, dealing purview, etc;
    – metadata reference information: actualisation and the responsible unit information;
    – literature cited;
    – responsible unit information: unit or individual ID related to the dataset;
    – address information: the communication information of dataset provider;

(g) Principle and method for metadata extension are:
  – extension type;
  – extension;
  – extension method.

This standard draft is a research achievement but not the formal national standard. The spatial data metadata standard of China has been formally established as a project and is in the process of being regulated.

3.2.2. *Published standards*

The following list provides some examples of national GI standards that have been published:

- geographical grid;
- data classification and codes for National Core GI;
- data classification and codes for Forest Resources;
- coding system of River Names within China;
- classification and codes for Features of Topographic Maps at the Scales of 1 : 500, 1 : 1000 and 1 : 2000;
- coding structure and rules for Urban Geographic Features – City Roads, Road Intersections, Blocks and Municipal Piping System;
- codes for the Administrative Sub-divisions for the Peoples Republic of China;
- rules for the code representation of Administrative Divisions under Counties;
- technical interface between Topographic Database and Geographic Names Database;
- code for Highway Classification;
- name and number of National Trunk Highway Routes;
- naming and the coding Rules for Highway Bridges.

3.2.3. *Standards awaiting approval*
The following list provides examples of draft standards that are waiting for approval or examination:

- codes for Mountain Range and Peak Names of China;
- data transfer format for Information of Surveying and Mapping;
- quality standard for Digital Elevation Model Products.

3.2.4. *Standards being developed*
The following list provides some examples of national standards that are being developed:

- Geographic Information – Metadata Completed, May 2005;
- Geographic Information – Terminology;
- Data Quality Control and Evaluation;
- classification and coding system for Spatial Information at the highest level;
- Geographic Information – Conformance and testing;
- specification for Electronic Navigational Map.

The national standard, 'Guide to Standardisation for Urban Geographic Information System (UGIS)', has been developed and was published in 1998. The contents of this standard are:

- introduction;
- basic terms;
- management of UGIS standardisation;
- list of standards related to UGIS;
- general design for UGIS and detailed design for its subsystems;
- spatial positioning in UGIS;
- contents of urban geographic information, data classification and coding system;
- spatial data structure in UGIS;
- environment of software and hardware for UGIS;
- data quality control for UGIS; implementation and maintenance of UGIS;
- network of UGIS;
- security of UGIS.

It is a unified and basic guideline document for all organisations, institutes, companies who research on, design, develop or use UGIS in China.

3.2.5. *Relationship of standards work to other regional efforts*
The development of China's spatial metadata standard will reference standards developed by U.S.A., New Zealand and Australia as well as ISO 19115.

The spatial data standards that have been developed by other countries in the Asia – Pacific region were referenced when China developed its own spatial data standards. Generally, these standards were used in one of three ways: adopted, adapted or created from a profile.

3.2.6. *Planned or future spatial standards activities*
The developments of following standards are planned soon:

- Geographic Information – Reference Model;

- standards for location-based services (LBS);
- qualifications and certification of personnel.

3.3. *Fiji*

The Fiji Land Information Council (FLIC) has introduced a policy, which ensures that a directory of metadata is maintained. Its purpose is to maximise community access to land and geographic information and avoid duplication.

The development of a metadata standard is being considered. As part of this work, FLIS is reviewing current international metadata initiatives.

3.4. *Japan*

Metadata was standardised as a part of Japanese National Standards for Geographic Information (JNSGI), which was developed by GSI and private companies under a joint research program and approved as governmental standard. Generally, the standardisation was based on the first Committee Draft (CD) of ISO 19115.

The Japanese metadata standard is almost compliant with the first CD of ISO 19115. The differences are listed below:

- changing order of address notation to meet Japanese custom;
- adding elements for designating extensive character code set in distribution information (in Japan, extensive code sets are occasionally needed for specific fields).

Based on this standard, a profile for clearinghouse was developed as Japanese Metadata Profile (JMP). ISO 19115 is now a International Standard (IS). The Japanese standard and JMP are being revised to conform to it.

3.5. *Korea*

3.5.1. *National NGIS metadata interim standard*

The GIS Standard Sub-committee under the Steering Committee of NGIS develops the NGIS Metadata Interim Standard. As mentioned in Section 2.5 above, the metadata standard was based on the ISO/TC 211 15046-15 Metadata Draft Version 4.4 (now known as ISO 19115).

This NGIS metadata standard defines the schema required for describing geographic information and services. It provides information about the identification, the extent, the quality, the spatial and temporal schema, spatial reference, and distribution of digital geographic data. This standard is applicable to the cataloguing of datasets, clearinghouse activities, and the full description datasets.

3.5.2. *Review of NGIS metadata interim standard*

Since being published in 1999, some inconsistencies and deficiencies have been identified in the NGIS Metadata Standard and the ISO/TC 211 Metadata standard has changed. Therefore, in December 1999, the NGI initiated a review of the NGIS Metadata Standard and developed a new metadata standard based on the international metadata standard ISO 19115.3.

The new metadata standard of NGI is almost compliant with the third Committee Draft of ISO 19115. The NGI is responsible for the establishment of the national geographic database and the management of digital mapping. NGI will issue the new NGI metadata standard as NGIS Metadata Standard through the Geographic Information Sub-committee of NGIS. The NGIS Metadata Interim Standard will be superseded by the new NGI metadata standard in 2001.

Presently a profile for specific fields will be developed based on the NGI metadata standard. Now, that ISO 19115 is issued as an International Standard, the NGI metadata standard will be revised to conform to it.

3.6. *Malaysia*

In order to ensure spatial data that is available at various agencies can be shared, it has been decided under the NaLIS activities that all spatial data needs to be documented and thus there is a need to develop a Malaysian metadata standard.

Malaysia is actively participating in the ISO/TC 211 activities and using the metadata documentation developed by ISO 19115 to develop the Malaysian Metadata Standard.

3.7. *The Philippines*

A draft geographic data standard has been prepared by the IATFGI for final completion.

Concerning metadata standards, the World Bank Study includes a review of metadata standards in several countries and a recommendation to actively participate in the development of the ISO/TC 211 standards and adopt these to meet local requirements.

Draft policy documents have been completed for presidential action to pursue the development of the NGII, including the establishment of a national geographic information clearinghouse network.

3.8. *Thailand*

The Sub-committee on GIS Data Standard under the National Committee on GIS Coordination and Development has identified the Metadata standard to be the highest priority. A study project on Thai GIS Metadata standard has been conducted based on the second Committee Draft of ISO 19115. The following and final version of ISO 19115 had also been monitored and scrutinised.

Although there has been no formal Thai GIS Metadata standard declared yet, it is quite apparent that the ISO 19115 standard will be followed. Several organisations constructing geo-spatial databases have been advised to produce their metadata based on ISO 19115.

4. Description of metadata query/distribution services that are in operation or planned

This section describes metadata services that are planned or in operation in selected countries.

4.1. *Australia*

The Australian Spatial Data Directory (ASDD) provides a metadata discovery service for spatial information. The ASDD is located at http://www.auslig.gov.au/asdd/ and is maintained and developed by Geoscience Australia on behalf of ANZLIC.

In conjunction with the ASDD, ANZLIC is promoting the development of the Australian Spatial Data Infrastructure Distribution Network (ASDI DN). The aim of the ASDI DN is to provide access to spatial data. With its focus on data discovery, the Australian Spatial Data Directory (ASDD) will provide the foundation for ASDI DN. New directory features that provide information about products, services and solutions will complement the ASDD. New metadata elements will be developed to describe these features.

The development of distribution networks is also underway in the states, territories and the federal government. Some of these initiatives (such as in Northern Territory and South Australia) are already using ISO 19115 as the basis for sophisticated metadatabases. Given this trend, it is likely that a number of detailed ISO profiles will emerge during the next two years to support particular SDI initiatives in Australia and New Zealand.

4.2. *China*

As mentioned above, experts from NGCC and other organisations have developed Geographic Information – Metadata of China, the national standard. It was published at the end of the year 2001.

There are two levels for the content of the standard: the first level (or core metadata) and the second level (or full metadata).

The first level metadata for the database of the National Fundamental Geographic Information System of China (NFGIS) can be found at the website: http://nfgis.nsdi.gov.cn/

4.3. *Fiji*

In 1999, a high-level metadata application named 'FLIS Metadata' was developed which enables the storage of the data required to create 'Geographic Information Directory', a report that is generated periodically by FLIS.

A lower level entity and attribute level metadata database is being designed.

4.4. *Japan*

In order to promote accessibility of spatial data, it is important to create metadata of spatial data and make it available to the public through a metadata search system such as a clearinghouse.

GSI was implemented a clearinghouse gateway system in the Japanese language environment. This can deal with multi-byte character code sets including Japanese characters, as well as being compliant with the International Standard for information retrieval – ISO 23950. This is the first implementation in the world of a clearinghouse dealing with multi-byte character code sets using the GEO profile based on ISO 23950. By using the gateway, users can search not only GSI's metadata, but also metadata of other agencies all at once. The GSI's gateway is also harmonised with the FGDC clearinghouse in the U.S.A., so those users can search both clearinghouses at the same time.

Some clearinghouse nodes were implemented in Japanese with assistance from GSI in accordance with the Japanese SDI program. As of September 1, 2001, metadata in eight governmental clearinghouse nodes and two non-profit organisation's clearinghouse nodes can be retrieved through GSI clearinghouse gateway system.

4.5. *Korea*

The Korean government has initiated a full-scale implementation of the NGIS (National Geographic Information System) Master Plan in 1995. The success of NGIS Master Plan marks a turning point in the physical management system of our land. The NGIS Master Plan includes generation of spatial databases for NGIS, establishment of data standardisation, assistance of GIS-related technical development and development of framework for utilisation and application of NGIS.

The goal of the NGIS master plan is to develop environmentally healthy and pleasant land as well as to promote national competitiveness and productivity. We expect the success of the NGIS Master Plan to mark a turning point in the management system of our land.

The major issues of the NGIS Master Plan can be categorised as follows:

- establishment of basic database for spatial information:
  - database design;
  - digital mapping of the topographical data;
- development of GIS-related technology and training of GIS specialists:
  - development of GIS fundamental technology;
  - development of GIS application;
  - training of GIS specialists;
- standardisation of spatial data:
  - standard for basic spatial data;
  - spatial data transfer format;
- financial support for development of GIS application systems:
  - spatial Decision Support System;
  - administration support system;
- management and distribution of spatial information:
  - implementation and operation of information database clearing house;
- development of spatial data-related legislative law and framework;
- joint funding between public and private sector;
- update and modification of the NGIS Master Plan, rolling annual plans.

The first phase of the NGIS Master Plan was completed in 2000. The main purpose of the first phase is to establish a basic GIS infrastructure to produce various kinds of digital maps. In the second phase of the NGIS (2001 through 2005), the aim is to spread GIS application and utilisation, and then maintain the digital maps.

In order to promote access and sharing of spatial data, the steering committee of NGIS on behalf of Korean government commenced studying clearinghouse implementation in 1999. The aim of this pilot project is to improve access and use to spatial data. The pilot project planed to implement a clearinghouse into the NGI in 2001.

4.6. *Malaysia*

The NaLIS Clearinghouse is now developing a metadata discovery service for spatial data in Malaysia.

4.7. *New Zealand*

The New Zealand Officials Committee for Geo-spatial Information (OCGI) has proposed a collaborative approach to establishing a geo-spatial metadata directory. To this end, the Ministry for the Environment is undertaking a project to collect geo-spatial metadata in conjunction with their Environmental Indicators Programme. New Zealand intends adopting ISO 19115 and the ANZLIC Metadata Guidelines for geo-spatial metadata.

The geo-spatial metadata project is complementary to the E-government Portal Implementation Project, which is underpinned by the New Zealand Government Locator Service (NZGLS) discovery-level metadata standard. (NZGLS is based upon the Dublin Core metadata element set.)

4.8. *PC GIAP APSDI Data Node Project*

In order to promote GIS development in Asia and the Pacific, PC GIAP has proposed that the Asia – Pacific Spatial Data Infrastructure (APSDI) be developed and implemented. APSDI is planned to be a network of databases, located throughout the Asia – Pacific region, that together provide the fundamental data needed to achieve the region's economic, social, human resources development and environment objectives.

APSDI Data Node is a network of nodes established to distribute, manage, and maintain fundamental data in each member country of the APSDI (on-line see http://www.gsi.go.jp/PCGIAP/98wg/98wg2/nodes/nodes.htm). These data nodes will provide the means to:

- advertise data collection, requirements, inventory, and quality;
- support documentation of basic spatial datasets.

According to the project plan, Australia, China, Japan and Iran will act as demonstration data nodes in the first stage. In each country, a sub-node of the APSDI is planned and they will be linked with each other.

4.9. *The Philippines*

No metadata services are currently in operation. NAMRIA has initiated the establishment of a national geographic information clearinghouse network and has completed a draft proposal.

The plan is to set up a portal to be operated by NAMRIA or a proposed National Geographic Information Council (NGIC). To be known as NGICNet, it will have nodes on lands and surveys, agriculture and environment, socio-economic, and infrastructure/utilities. Metadata standards will be developed and installed, based or adopted from international standards, for the search, browsing, and access/downloading of geographic information.

4.10. *Thailand*

A pilot system for on-line spatial data searching is under development by the Department of Survey Engineering, Chulalongkorn University, as part of the Government Data Infrastructure (GDI) study project. Although this pilot system will be developed based on the FGDC metadata standard, a plan to upgrade to support the ISO 19115 Metadata standard has already been prepared. The result of this study project concerning policy and implementation plan for Thai National Spatial Data Infrastructure will be proposed to the Thai government.

At present, various government organisations that are constructing or plan to construct geo-spatial databases are being directed by the Bureau of the Budget to produce metadata for their geo-spatial databases. The availability of digitally standard metadata and the data transfer supporting policies will result in an on-line metadata query/distribution service in Thailand.

5. Summary and conclusions with prognosis for the future

5.1. *Summary*

Three main organisations encourage regional collaboration on metadata initiatives in the Asia – Pacific region:

- ANZLIC, the Spatial Information Council;
- The Permanent Committee on GIS Infrastructure for Asia and the Pacific (PC GIAP);
- ISO/TC 211, Geographic information/Geomatics.

ANZLIC joins the two nations of New Zealand and Australia in SDI work in general and metadata work in particular. Even though the Council comprises just two countries, ANZLIC's Metadata Guidelines have been applied in a number of other countries in the region.

PC GIAP has a focus on SDI development in the region. The metadata component of that work is a part of the Data Node Project that began in 2000.

ISO/TC 211 has brought together countries world-wide in the development of the ISO 191XX series of standards. It has also provided a forum for nations in the Asia – Pacific region to meet and exchange information about metadata projects.

However, apart from the work that takes place under the auspices of the three organisations above, most countries in the region tend to work on metadata standards and directory systems that are country-specific. It is envisaged that the work of ISO 19115 will redress this situation by providing a common basis for development work.

5.2. *Conclusions with prognosis for the future*

The Asia – Pacific region is geographically large and culturally diverse. For logistical reasons, this summary of regional metadata developments has been confined to a sample of countries. However, it must be said that there are many countries actively pursuing innovative metadata initiatives that have not been reported in this chapter.

Accepting this limitation, it is nevertheless evident that a common thread is woven through the developments in the region. That thread is ISO 19115. The introduction of an international metadata standard is timely given the number of metadata initiatives that are underway or planned at the state, national and regional levels. The provision of a common dictionary of metadata elements will promote the development of interoperable directory systems during the coming years.

Acknowledgement

The Asia Pacific chapter is based on contributions from the following experts:

Australia: Mr Craig Macauley, Government of South Australia, mailto: macauley.craig@saugov.sa.gov.au

China: Prof. Du Dao-sheng, The State Laboratorium for Information Engineering in Surveying, Mapping and Remote Sensing, Wuhan University, mailto: dsdu@dns.liesmars.wtusm.edu.cn

Prof. Jiang Jing-tong, National Geomatics Center of China, mailto: rmliu@public3.bta.gnet.cn

Fiji: Mr Ilaitia Navunisaravi, Fiji Land Information Systems Support Centre, Department of Lands & Surveys, mailto: inavunisaravi@lands.gov.fj

Japan: Mr Kazuhiko Akeno, Head of Information Access Division, Geoinformation Department, Geographical Survey Institute of Japan, mailto: akeno@gsi.go.jp

Republic of Korea: Mr Tae-jung Min, Director, National Geography Institute, mailto: tjmin@moct.go.kr

Mr Heungmuk Cho, National Geography Institute, mailto: hmcho@moct.go.kr

Malaysia: Mr Hasan Jamil, Director of Survey (Geodetic Division), Department of Survey and Mapping, Malaysia, mailto: hasanj@jupem.gov.my

New Zealand: Mr Bryan Teahan, Consultant, Terralink International Ltd, mailto: bryan.teahan@terralink.co.nz

Mr Mack Thompson, Cadastral Survey Adviser, Office of the Surveyor-General, Land Information New Zealand, mailto: mthompson@linz.govt.nz

PC GIAP Data Node Project: Mr Glenn Johnstone, Geodesy Program, National Mapping Division, Geoscience Australia, mailto: glennjohnstone@auslig.gov.au

The Philippines: Ms. Francisca N. Dayrit, GIS Consultant, mailto: dayrits@skyinet.net

Ms. Linda San Diego Papa, Director, Information Management Department, National Mapping and Resource Information Authority, mailto: lsdp@psdn.org.ph

Thailand: Mr Chanin Tinnachote, Assistance Professor, Department of Survey Engineering Chulalongkorn University, mailto: chanin.ti@chula.ac.th

References

Organisation	For on-line contact see:
ANZLIC, Australia / New Zealand	http://www.anzlic.org.au/infrastructure_metadata.html
ANZLIC's Metadata elements	http://www.anzlic.org.au/asdi/metaelem.htm
APSDI, PC GIAP	http://www.gsi.go.jp/PCGIAP/98wg/98wg2/nodes/nodes.htm
AUSLIG, Australia	http://www.auslig.gov.au/
DOL, Thailand	http://www.dol.go.th/
DSMM, Malaysia	http://www.jupem.gov.my/
GISTDA, Thailand	http://www.gistda.or.th/
GSI, Japan	http://www.gsi.go.jp/
KATS, Korea	http://www.ats.go.kr/
LINZ, New Zealand	http://www.linz.govt.nz/
NaLIS, Malaysia	http://www.nalis.gov.my/
NCA, Korea	http://www.nca.or.kr/
NGI, Korea	http://www.ngi.go.kr/
PC GIAP Data Node Project	http://www.pcgiap.org/
PSMA Australia Ltd	http://www.psma.com.au/
RTSD, Thailand	http://www.rtsd.mi.th
Terralink, New Zealand	http://www.terralink.co.nz

URL's in this chapter are accessed in May, 2005, unless otherwise stated in the text.

H. Moellering, H.J.G.L. Aalders & A. Crane (Editors)

Metadata Standards Development Activities in Central and South America and the Caribbean

Tatiana Delgado-Fernández[a], Dora Inés Rey-Martinez[b], Martha Ivette Chaparro-Dominguez[c]

[a]*Director, Computation Division, GEOCUBA, Republic of Cuba*
e-mail: geocuba@teleda.get.tur.cu
[b]*Executive Secretary, PC IDEA, Colombia*
e-mail: direy@igac.gov.co
[c]*Technical Secretary, Colombian Standardisation Committee on Geographic Information, Colombia*
e-mail: mchaparr@igac.gov.co

1. General regional structure for spatial data organisations

1.1. *Permanent Committee on SDI for the Americas – PC IDEA*

The Permanent Committee on Spatial Data Infrastructure for the Americas (PC IDEA, on-line see www.cpidea.org.co) was established pursuant to Resolution 3 of the Sixth United Nations Regional Cartographic Conference for the Americas held in New York in 1997. After a process of consolidation, 21 American countries signed an agreement on February 29, 2000 in Bogotá, Colombia, during the Pan American Seminar on National Spatial Data Infrastructure.

The goals of the Permanent Committee are comprised within the principles of the 21st Agenda of the United Nations Conference on the Environment and Development in order to maximise the economic, social and environmental benefits derived from the use of geo-spatial information.

The Permanent Committee shall endeavour to achieve the following objectives:

- establish and co-ordinate policies and technical rules for the development of the regional geo-spatial data infrastructure for the Americas;
- promote the establishment and development of the national geo-spatial data infrastructures of each member of the Permanent Committee;
- encourage the exchange of geo-spatial data among all the members of the community of the Americas, respecting their autonomy, pursuant to their national laws and policies;
- promote the co-operation, research, complementation, and exchange of experiences in fields related to the geo-spatial discipline;
- define guidelines and strategies to include cadastral information in the geo-spatial data infrastructure for the Americas.

The Committee shall promote the participation of all countries from the continent in the development of a regional infrastructure, as well as the establishment of sub-regional committees to stimulate the development of permanent databases in the sub-ordinate regions of North America, Central America, the Caribbean, and South America.

As an emerging initiative, PC IDEA is taking advantage from other regional and national SDI's established initiatives around the world. The experience from PC GIAP, EUROGI and several NSDI's worldwide is offering real lessons learned to developing countries in the Americas.

PC IDEA's work is based on respect for cultural and political differences and the specific conditions facing each country. The possibility to establish sub-regional committees under the PC IDEA umbrella also facilitates consensus. So far, two sub-regional committees are conformed: for the Caribbean Islands and for Central American nations.

After the first Meeting in Bogotá, PC IDEA has organised two committee meetings more. The second meeting took place along with the 7th United Nations Regional Cartographic Conference for the Americas on January 25, 2001 and the third meeting was held along with the 5th Global Spatial Data Infrastructure Conference in Cartagena, Colombia, May 21, 2001.

PC IDEA has implemented two regional surveys executed by the International Centre on Tropical Agriculture (CIAT, Centro Internacional de Agricultura Tropical) through its Project PROCIG and by the Agustin Codazzi Geographic Institute of Colombia (IGAC, on-line see http://www.igac.gov.co), who acts as Executive Secretary. These activities have been an excellent opportunity to identify key players in the community and to motivate them to join PC IDEA and its working groups. Nowadays PC IDEA counts 24 member countries.

In order to carry out the commitments, various Working Groups have been set up; they deal with: Legal and Economic matters, Communications to promote development of the SDI, and Technical Matters. The priority technical matters PC IDEA is facing are: inter-institutional policies and agreements, fundamental data, geographic information standards, clearinghouses, cadastral aspects, geographical names and capacity building (training and education).

1.2. *Central American GIS project for development – PROCIG*

The Central American GIS Project for Development (PROCIG) is integrated by several Central American organisations that seek to advance research and development of GIS in the region. This project was lead by USGS/EROS Data Center. PROCIG completed its activities in June 2001 and during the two years of the project its principal objective was to promote the integration of statistical and census data with other geographic information in order to generate products for public dissemination. An additional objective of PROCIG was to support initiatives for national spatial data infrastructure in each country. In this sense, the project not only aimed at capacity building, but also to implement metadata and clearinghouses nodes in each participating country. Many national organisations of Central America have received training, consultancy and other kind of support as part of PROCIG, in order to improve their knowledge of Spatial Data Infrastructure and to implement it.

PROCIG developed important working relationship with the Agustin Codazzi Geographic Institute in Bogotá, Colombia, as the leader of PC IDEA during its activities. Their efforts to integrate organisations working in geographic information management in Latin America benefited from strong representation from Central America [Hyman 2001].

1.3. *Pan American Institute of Geography and History (PAIGH)*

The Pan American Institute of Geography and History (PAIGH, on-line see http://www.ipgh.org.mx) was officially created at the Sixth Conference of American States held in Havana, Cuba, February 7, 1928. In 1930, the Government of the United Mexican States placed a building at the disposition of PAIGH, located in Mexico City.

The objectives of PAIGH are to:

- encourage, co-ordinate and disseminate Cartographical, Geographical, Historical and Geophysical studies, as well as other related scientific studies of interest to the Americas;
- promote and carry out studies, work and training in the mentioned fields of activity;
- promote co-operation among organisations interested in these fields of activity in the Americas, and with related international organisations.

1.3.1. *Inter-American Geo-spatial Data Network*

Inter-American Geo-spatial Data Network (IGDN, on-line see http://edcsnw3.cr.usgs.gov/igdn/mission.html) was one of the first regional efforts developing a regional spatial data infrastructure for the Americas. Introduced in 1994, IGDN is in prototyping stage under the responsibility of the USGS/EROS Data Center in partnership with the USAID [Borrero 2000].

IGDN is a group of Latin American and Caribbean organisations applying geo-spatial data, which are participating in the use of Federal Geographic Data Committee (FGDC) clearinghouse technology. Each participating member has a clearinghouse node that uses the FGDC Content Standard for Digital Geospatial Metadata (US FGDC STD-001-1998, version 2.0) and the Z39.50 protocol (ANSI Z39.50-1995, ISO 10163-1995). These clearinghouses are connected to the clearinghouse system for worldwide searching, discovery, and retrieval.

IGDN has a group of partners that have established NSDI Clearinghouse nodes. The partners are listed below:

- USGS/EROS Data Center (on-line see http://edc.usgs.gov/), containing Digital Elevation Model data of the western hemisphere and PAIGH Mapping of the Americas Data;
- PAIGH (on-line see http://spin.com.mx/~ipgh/);
- INEGI (on-line see http://www.inegi.gob.mx/);
- Kansas University (on-line see http://www.gemlab.ukans.edu) responsible for Biodiversity and Natural Resource Maps of Costa Rica;
- CIAT, Colombia (on-line see http://www.ciat.cgiar.org) covering data for Latin America and Africa;
- Caribbean Environment Program (on-line see http://www.cep.unep.org/). The 33 states and territories of the Wider Caribbean Region have joined in pursuit of a common goal, protection of the marine and coastal environment through the promotion of balanced and sustainable economic development;
- Uruguay National Clearinghouse (on-line see http://www.clearinghouse.com.uy).

1.3.2. *Atlas of the Americas pilot project*

The Atlas of the Americas (on-line see http://edcsnw3.cr.usgs.gov/igdn/mission.html) is a PAIGH two-year project to co-finance the execution of a pilot project aimed at providing technical transfer to participating Latin American countries to develop tools for sustainable development decision making. The main components of the project include:

1. creation of a clearinghouse network;
2. creation of an electronic atlas of Latin America;
3. creation of individual national electronic atlases;
4. capacity building;
5. electronic maps for the blind/visually impaired;
6. dissemination of materials via the Internet.

The project-executing agency is PAIGH in Mexico City, Mexico. Other participating agencies include: Carlton University (Canada), the Canadian Government, the USGS/EROS Data Center, and the Geographic Institutes of Argentina, Brazil, Chile, Costa Rica, Mexico, Panama, Peru, and El Salvador.

As part of this project, the following countries have developed clearinghouses and metadata within their countries: Argentina, Costa Rica, El Salvador, Mexico and Peru.

1.4. *DIGSA*

Created in 1978, the Directory of Geographic Institutes from South America, Spain and Portugal (DIGSA, on-line see http://www.digsa.com.pe) is the summit of national mapping institutions in South America, Spain and Portugal. It is a meeting of policy-makers. DIGSA eases exchange of information between members and stimulates formation of common policy in specific areas for mutual benefit. Since the XIth DIGSA meeting, held in Bogotá in 1995, the issue of spatial data infrastructures has been included in its agenda. Although DIGSA concedes priority to aspects like licensing, copyright and training policies, also digital data production, maintenance and quality are considered issues [Borrero 2000].

2. Identification of major national spatial data organisations

2.1. *Argentina*

The Argentinean Military Geographic Institute (IGM-Argentina, on-line see http://www.igm.gov.ar) is creating a database for the national basic and boundaries cartography and maintains the National Geodetic Network. They have the following digital data available:

- digital cartography at 1 : 250 000 scale;
- GIS at 1 : 250 000 scale (SIG 250);
- satellite image cartography.

The Cadastral and Territorial Information Direction manages the Territorial Information Systems based on cadastre and co-ordinates the inter-institutional actions related with cadastre and territorial information.

2.2. *Brazil*

Brazil's Bureau of Census (IBGE, on-line see http://www.ibge.gov.br) is responsible for a large number of spatially-bound data collections, including the population census, and a multitude of socio-economical indicators.

Brazil's Geological Survey (CPRM, on-line see http://www.cprm.gov.br/ingles/index.html) is very active in collection and dissemination of spatial datasets and is committed to the use of FGDC metadata standards to the extent of having developed special software for data entry.

The Brazil National Institute for Space Research (INPE, on-line see http://www.inpe.br) is responsible for different activities including:

- operation of a remote sensing ground station (receives LANDSAT data since 1974 and SPOT data since 1986), forming one of the largest archives of RS data in the world;
- development of software technology for GIS and RS, called the SPRING system, possibly the most complete freeware available worldwide;
- graduate program in RS and GIS, including associated research;
- numerical weather prediction and climate centre, using a supercomputer, which issues forecast and climate analysis and has many other products available on-line;
- development of remote sensing satellites, including the CBERS series of satellites, built in co-operation with China.

2.3. *Chile*

The Military Geographic Institute of Chile (IGM-Chile, on-line see http://www.igm.cl) is the official body responsible for geographic information in Chile. It has focused one of its efforts on the developing of 'Geo-Spatial National Data Infrastructure' [Gran 2001].

2.4. *Colombia*

The Colombian Geographic Institute Agustin Codazzi (IGAC, on-line see http://www.igac.gov.co) develops policy and produces, analyses and publishes geo-referenced information on cartography, agrology (soil studies), cadastre and geography.

Since 1996, IGAC acts as Technical Secretariat promoting standards, metadata and clearinghouse for the Colombian Spatial Data Infrastructure (ICDE, on-line see http://www.icde.org.co) and stimulates ICDE and PC IDEA creation.

2.5. *Costa Rica*

The Costa Rican National Geographic Institute (IGN-Costa Rica, on-line see http://www.ign.cr.com) is the official institution dedicated to provide geographic information and basic cartography to the country.

2.6. *Cuba*

The Hydrography and Geodesy Service of Cuba (including the National Office of Hydrography and Geodesy and GEOCUBA, on-line see http://www.cmw.inf.cu/ciget/consultora/tecnogest/patrocinio/geocuba/geocuba.htm) is responsible for national cartography, geodesy, hydrography, cadastre, photogrammetry, remote sensing and other geomatics activities in Cuba. Since 2000, it is leading the National Geo-spatial Data Infrastructure in Cuba.

2.7. *El Salvador*

National Geographic Institute (IGN-El Salvador, on-line see http://www.cnr.gob.sv) belongs to the National Registry Centre and brings information on geography, cartography, hydrography and geodesy for projects related to human and natural environment. IGN makes general and thematic maps on these topics.

2.8. *Guatemala*

The National Geographic Institute 'Ingeniero Alfredo Obiols Gómez' (IGN, on-line see http://www.ign.gob.gt) is the public entity in Guatemala dedicated to generate, publish and distribute geographic information and basic cartography in Guatemala. It is the national representative institution in cartography, geography and geodesy international organisations.

2.9. *Guyana*

Iwokrama International Centre for Rain Forest Conservation and Development (on-line see http://www.iwokrama.org) is an autonomous non-profit research and development institute established by Guyana and the Commonwealth to demonstrate how tropical forest biodiversity may be conserved and sustainably utilised for ecological, social and economic benefits.

2.10. *Honduras*

National Geographic Institute (IGN-Honduras, on-line see http://www.ns.sdnhon.org.hn/~ignhon) is responsible for production and dissemination of the national cartography. IGN-Honduras participates jointly with the Geographic Information Centre, the National Natural Resources and Environment Secretary in the implementation of national clearinghouse (on-line see http://www.usgs.unitec.edu).

2.11. *Mexico*

The National Institute of Statistics, Geography and Informatics (INEGI, Instituto Nacional de Estadística, Geografía e Informática, on-line see http://www.inegi.gob.mx/) is the agency responsible for integrating Mexico's systems of statistical and geographic information; in addition it promotes and co-ordinates the development of informatics in the country and stimulates the development of a National Spatial Data Infrastructure in Mexico.

2.12. *Nicaragua*

Nicaraguan Institute for Territorial Studies – INETER (on-line see http://www.ineter.gob.ni) established a Geo-spatial Metadata Committee in Nicaragua as part of PROCIG project (http://www.clearinghouse.gob.ni) together with:

- MARENA (Ministerio de Ambiente y Recursos Naturales, Ministry of Environment and Natural Resources);
- INEC (Instituto Nacional de Estadística y Censos, Statistics and Census National Institute);

- MAGFOR (Ministerio de Agricultura-Ganadería y Forestaría, Ministry of Agriculture-Livestock Farming and Forestry);
- ALISTAR (on-line see http://www.alistar.org.ni/), a non-profit organisation.

INETER provides the geographic data related to geodesy and cartography, geophysics, geology, meteorology, hydraulic resources, cadastre, photogrammetry and other geographic data. The committee will be dedicated to establish and co-ordinate the creation and maintenance of geo-spatial metadata in Nicaragua.

2.13. *Panama*

Tommy Guardia National Geographic Institute (IGN – Tommy Guardia, on-line see http://www.mop.gob.pa/igntg) is the official institute dedicated to provide information on geography, cartography, geodesy, photogrammetry, hydrographic surveys and GIS in Panama. IGN – Tommy Guardia has also to establish technical norms in these fields. On behalf of the CTC (Comité Técnico Consultivo para Información Geográfica, Technical Committee for Consultancy on Geographic Information), IGN – Tommy Guardia is leading the activities of the National Spatial Data Infrastructure in Panama.

2.14. *Uruguay*

Military Geographic Institute (IGM – Uruguay, on-line see http://www.sgm.ejercito.gub.uy) is the official institute to provide geographic information and basic cartography to the country.

The National Clearinghouse for Geographic Data (CNDG, Clearinghouse Nacional de Datos Geográficos within the Dirección Nacional de Topografía, Ministerio de Obras Públicas, on-line see http://www.clearinghouse.gub.uy) consists of a network of providers, administrators and users of geographic information; it is has the following functions:

- elaborate metadata standards;
- organise metadata to disseminate them into the public;
- allow access to the available geographic information by means of Uruguay/Net/Internet;
- foster inter-institutional co-operation projects (public and private) to the production and distribution of geographic information.

2.15. *Venezuela*

Geographic Institute of Venezuela Simón Bolívar (IGVSB, on-line see http://www.igvsb.gov.ve) co-ordinates and develops policies and execute national plans of geography, geodesy, geophysics, cartography, remote sensing, cadastre, and soils. IGVSB is also leading the development of the National Spatial Data Infrastructure in Venezuela and acted as PC IDEA Presidency for the period 2001–2002.

3. Identification of specific national/regional metadata standards development activities taking place

This section contains a summary of metadata standards development activities taking place in a selection of countries in the region. The selection was made in accordance with the responses of the general SDI questionnaires for the PC IDEA countries before the 5th GSDI Conference in Cartagena, Colombia. The authors also used the updated-information provided by some countries to this report (see paragraph 5 of this chapter: List of contributors). The current work of PC IDEA Standards Working Group is also described.

3.1. *Argentina*

Since 1998, the Military Geographic Institute (IGM – Argentina) has implemented, with the support of PAIGH, a clearinghouse node based on the FGDC Content Standard for Digital Geospatial Metadata (US FGDC STD-001-1998, version 2.0). The metadata were loaded by using CORPS MET 95 software, also provided by FGDC.

Two workshops have taken place in Argentina, sponsored by the PAIGH, in order to disseminate the information about 'Metadata and Clearinghouse' and to facilitate the implementation of metadata.

3.2. *Brazil*

The Brazilian Government, through its Ministry for Science and Technology, launched in 2000 a national initiative called 'Sociedade da Informacao' (Information Society) for improving the nation's use of information technology. This program considers the use and dissemination of spatial information as well as the establishment of a NSDI as an important part of its goals.

On the other hand, there are also other government agencies and committees, which are active in different aspects of establishing an NSDI for Brazil such as IBGE, DATASUS, CPRM and INPE.

3.3. *Chile*

In 2001, a Working Group led by the Ministry for Public Assets and Property and with the representation of various sectors among geo-spatial data providers and users was set up. Within this working group four specialised sub-committee were created to study the following aspects [Gran 2001]:

- digital cartography;
- provision of standards for territorial information;
- policies for the financing and transfer of territorial information;
- design and structuring of a clearinghouse.

The Working Group carries out a preliminary study of the existing standards in the world (e.g.: OpenGIS® Consortium, ISO/TC 211, FGDC) for the adaptation to the situation in Chile.

The IGM – Chile concentrates the efforts on the creation of a clearinghouse node, where metadata are already available for the regular cartographic datasets at 1 : 250 000, and

1 : 500 000 scales. Currently the metadata for the rest of the 1 : 50 000 cartographic dataset is under construction, to be followed by those for ortho-photos and large-scale cartographic datasets. The National Property Ministry have also implemented spatial metadata information.

3.4. *Colombia*

The Colombian Technical Norm (NTC4611), initially approved in May 1999 and reviewed in April 2002, was made under the join effort of public and private entities following the international standards. Starting in 1997, the reference material was international available documents of FGDC, ISO/TC 211, CEN/TC 287 and ANZLIC. These proposals showed common elements and remarkable differences. Regarding the general core of the documents, their main features and the Colombian conditions, it was possible to adapt a frame feasible for the national context.

The Colombian version of geographic metadata NTC4611, written in Spanish, is a subset that complies with the specifications of the FGDC standard and ISO/TC 211 and offers two conformity levels that can be applied according to the purpose of the obtained information.

The geographic metadata, included in the NTC4611, has a hierarchy of relationships and information organisation. It comprises seven main sections: identification, data quality, spatial representation of data, spatial reference, entities and attributes, metadata distribution and reference; and two support sections: citation and contact.

So, NTC4611 is applicable to datasets of individual data items, or series of datasets, geographic objects or attributes. It lets:

- identification of datasets (Conformity Level 1), to be used only for datasets and series of datasets and supports cataloguing purposes and the creation of metadata clearinghouses to facilitate the data discovery;
- complete documentation of a dataset (Conformity Level 2) defines the rank of the elements in the metadata required to identify, evaluate, access, use and manage geographic information. It is recommended to the data producers, to generate metadata at this level.

Criteria and guidelines for metadata development

There are as many ways to create metadata, as there are users in an organisation, while each organisation chooses the most convenient one. In Colombia it has been stated that any way will be valid if, at least, it guarantees the information discovery in a minimum level (comparable to the Conformity Level 1 in NTC4611). It is recommended to define a common criterion for the organisation. Once the scheme has been chosen, a list of the standard fields has to be generated, and they are mandatory for the organisation.

In Colombia, for the Metadata Development Plan, the following key topics included are:

- general inventory for geo-spatial data lists of the organisation;
- selection of the documentation scheme for the organisation: Conformity Level 1 or 2;
- selection of the organisational profile for the adopted scheme;
- priority list of the data, documented with metadata;
- selection of criteria and guidelines for metadata development, both for data in progress and existing data, with their availability, validity, accessibility form and access to the information;

- personnel selection to define whether the process will be done by the organisation's staff or through outsourcing;
- personnel preparation, implying to use appropriate people that transmit the technical aspects of metadata and the definitions of the organisation to use it;
- query forms by catalogues and lists on-line through a clearinghouse;
- schedule of activities.

The process is complex and its creation and management may be expensive. The justification of this effort relates to new opportunities; if a plan for adoption of metadata is consciously prepared and executed, it enables:

- increase of accessibility. The effectiveness and efficiency of the searches can be improved by a standardised and consistent documentation;
- institutional memory. To keep and document the relationships between the data and the producers, indicates authenticity, integrity and complexity of the information;
- extension of use. Well-structured documentation makes eases a huge number of possibilities for searching, localisation and manipulation of data;
- assurance of use of the most recent or appropriate available information;
- clarification of legal aspects as copyrights and conditions for reproduction or publication of the data;
- preserving accessibility and understanding of the data. There are many formats and storage media of digital data. Metadata keeps technical records and descriptions of the data from their creation throughout all updates;
- to avoid data duplication. Metadata reduce efforts in data acquisition and collection, and are a powerful tool to optimise the investments of the organisation.

According to the guidelines and structure of NTC4611, the main organisations of the Colombian Spatial Data Infrastructure – ICDE have been preparing their geographic product documentation, since 2000. Furthermore, they are developing the Colombian Spatial Data Directory – Clearinghouse whose main objective is to install a system for searching and accessing geographic information on Internet (on-line see http://codazzi10.igac.gov.co/gateway.html). Because Colombia has its own metadata standard, it was necessary to adapt and configure the program Isite, developed by CNIDR (supplied by FGDC), to create satisfactorily performance with the NTC4611 standard structure. This was accomplished successfully in the 2000. Now there are four available clearinghouse nodes in the main ICDE producer organisations.

3.5. *Costa Rica*

The IGN – Costa Rica, in collaboration of CIAT, developed a characterisation survey for the development of a National Geo-Spatial Data Infrastructure and interviewed 21 Costa Rican institutes that develop and manage geographic information [Elizondo 2001]. This survey is seen as the first phase in assessing GIS in Costa Rica and IGN – Costa Rica plans to continue with a more extensive survey that will aid in the development of NSDI.

Between June and August of 2001, a project pilot was undertaken to foster the developing of metadata in Costa Rica with the consultancy of CIAT and national institutes such as the IGN – Costa Rica, Ministry of Environment and Energy, Ministry of Agriculture and the National

Statistics and Census Institute. The project includes an inventory of spatial data from each participant, the creation of metadata and a web page per institute for data publication. Finally, the metadata are registered and available in the IGN-CATIE Clearinghouse Node.

The metadata standard chosen by Costa Rica is the FGDC Content Standard for Digital Geospatial Metadata (US FGDC STD-001-1998, version 2.0).

3.6. *Cuba*

During the II International Congress GEOMATICA 2002 held in Havana in May 2000, the Hydrographic and Geodetic Service of Cuba launched the National Spatial Data Infrastructure (IDERC, Infraestructura de Datos Espaciales de la República de Cuba) initiative sponsored by the Information Society Program of the Ministry of Informatics and Communications. Some local experiences in using metadata facilitate this spatial data sharing.

Local Experience: Biodiversity and Conservation Project in the Sabana-Camaguey Coastal Ecosystem

In the Sabana-Camaguey Coastal Ecosystem, comprising 465 square kilometres, a GEF/UNDP project oriented to the biodiversity conservation and the establishment of sustainable development is being carried out. Several multi-disciplinary studies were performed since the project's beginning in 1994 and have generated a significant amount of geo-spatial data, being integrated in databases to be used for all participants to improve the environmental decision-making process in the area. In order to achieve its goal, the 63 participating institutes organised their metadata to describe the diverse datasets provided by different sources and to share the information among them. So far, training courses about procedures and internationally accepted (IOC, UNESCO) standards have been developed. There is a program to implement the metadata based on ISO 19115 standard. They expect to create a clearinghouse node for metadata and spatial data discovery the following years.

National Experience: The Spatial Data Infrastructure of the Cuban Republic (IDERC)

IDERC is following the experience of several countries, as well as the international experiences in the development of Spatial Data Infrastructure, as described in the SDI Cookbook [Nebert 2001]. In regard with metadata, the FGDC approach and ISO 19115 are being considered, as well as the OGC Specifications to manage them.

Since the inclusion of Cuba as member of PC IDEA in May 2001 during the 3rd PC IDEA Meeting held in Cartagena de Indias, Colombia, several actions have been carried out to take advantage of the regional experience. The general plan, regardless institutional framework activities, to undertake the Cuban Spatial Data Infrastructure include the following steps:

1. feasibility study (completed in July, 2001);
2. capacity building (dissemination of SDI information, courses, workshops, pilots projects);
3. Metadata Development Project (questionnaire for SDI actors, inventory of geo-spatial data, creation of metadata);
4. establishment of, at least, a clearinghouse node.

Currently, a pilot project addressed by Hydrographic and Geodetic Service to implement the first ISO 19115-based metadata in Cuba is being developed.

3.7. *El Salvador*

Information acquired by IGN and the Ministry of the Environment during the PC IDEA's founding workshop in February 2000, directors of CNR, MARN (Ministerio del Medio Ambiente y Recursos Naturales, Ministry of Environment and Natural Resources) called a meeting for the main users and providers of geographic information, to establish a committee in charge of creating the National Spatial Data Infrastructure in El Salvador.

So far, the committee has been working on training, metadata development and the creation of the first clearinghouse node (on-line see http://clearinghouse.cnr.gob.sv/metadatos/index.htm). As part of this strategy, in March 2001 the 1st Symposium of National Spatial Data Infrastructure of El Salvador was held, among others organised by the National Registry Centre (CNR), Vice Ministry of Housing, Ministry of the Environment and Natural Resources. It also relied on the participation of:

- US Geological Survey (USGS);
- National Imaging and Mapping Agency (NIMA);
- US Agency for International Development (USAID);
- International Institute on Tropical Research (CIAT).

El Salvador applies the FGDC Content Standard for Digital Geospatial Metadata (US FGDC STD-001-1998, version 2.0). The following institutions are participating in the creation of geographic information metadata in El Salvador:

- Ministry of the Environment;
- National Registry/IGN-El Salvador;
- Vice-Ministry of Housing and Urban Development;
- San Salvador Metropolitan Area Planning Office.

3.8. *Guatemala*

Since October 1999, as part of the countries harmed by hurricane Mitch, Guatemala received training, technical assistance and equipment for implementing a clearinghouse node from a Project supported by USGS (US Geological Service) to help to reduce and mitigate natural disasters. This project is co-ordinated by the National Geographic Institute (IGN, Ingeniero Alfredo Obiols Gómez), who centralises and validates all the information, with the participation of:

- Secretariat of Planning and Programming for the Presidency (responsible of the administration of the clearinghouse node);
- National Co-ordinator for Disasters Reduction;
- National Institute of Vulcanology, Meteorology and Hydrology;
- Del Valley University;
- other providers and users.

Since the beginning of the project, each participating institute developed their spatial data inventory, metadata development plan and the metadata development activity, modelled according to the IGN experience. In addition, they received training in metadata development, clearinghouse technologies and web page design.

The metadata standard used by Guatemala is the FGDC Content Standard for Digital Geospatial Metadata (US FGDC STD-001-1998, version 2.0). So far, there is a clearinghouse node with approximately 345 available metadatasets (on-line see http://www.clearinghouse.segeplan.gob.gt/).

3.9. *Guyana*

A metadata standard for Guyana is being discussed at present, and the Iwokrama International Centre for Rain Forest Conservation and Development is contributing. The main activities that have take place in regard with the metadata in Guyana are:

1. participation in a national forum (the Guyana Integrated Natural Resources Information System) for GIS;
2. formulation of a National Policy on GIS (recommendations for the Government of Guyana), that includes an initial recommendation for a metadata standard (FGDC/ISO-based).

3.10. *Honduras*

After the Mitch hurricane disaster, a rebuilding project in Honduras began to be developed with the support of USGS in January 2000. As the result, the Geographical Information Centre (CIGEO, Centro de Información Geográfica) was created, aimed to strengthen the capacity building in GIS and metadata themes, as well as the creation and maintenance of a clearinghouse node.

Since March 2001, with the support of PROCIG-CIAT, a new complementary project of metadata developing is being executed [Buck 2001]. The main objectives of this project are:

1. create an inventory of participants in spatial data;
2. produce the metadata and to publish them in each institutional web page;
3. deliver updated metadata to SINIA (Sistema Nacional de Información Ambiental – National Environmental Information System) monthly;
4. make metadata and spatial data on-line available in a clearinghouse node.
5. The metadata standard implemented in Honduras is the FGDC Content Standard for Digital Geospatial Metadata (US FGDC STD-001-1998, version 2.0).

3.11. *Mexico*

National Spatial Data Infrastructure in Mexico is co-ordinated by the National Institute of Statistics, Geography and Informatics (Instituto Nacional de Estadística, Geografía e Informática, INEGI) aiming at geo-spatial data sharing among the users and producers in the nation.

With regard to metadata, they use the FGDC Content Standard for Digital Geospatial Metadata (US FGDC STD-001-1998, version 2.0). Metadata are prepared for datasets produced in several spatial themes and scales as DEM's, orthophotos data, topographic and toponymic data, natural resources and other datasets. These metadata could be accessed through several gateways. The query system is based on the Isite software and the ANSI Z39.50-1995 (ISO 10163-1995) protocol, on a Linux platform. So far, several basic web-mapping services have been implemented; they will be extended and improved in the future.

3.12. *Nicaragua*

In Nicaragua, there is not yet a NSDI initiative structured under a government policy; however there are two important efforts with regard to metadata:

1. Riabin–Pnuma Project (Red Interamericana de Información sobre la Biodiversidad, Inter Americas Biodiversity Information Network). This project includes training and installation of a web server with more than 100 metadata records. The ministries of Natural Resources and Agriculture, Universities and NGO's participated in the project, as well as agencies from Guatemala, Honduras, El Salvador and Nicaragua;
2. Mitch Clearinghouse Project (USGS-USAID). This project was started responding the heavy impact of the hurricane Mitch in October 1998. The disaster showed decision-makers and users in general the lack of integrated information related to natural disasters, in particular geographic information in manageable formats.

3.13. *Panama*

The National Geographic Institute Tommy Guardia (IGN – Tommy Guardia) and SENACYT (Secretaría Nacional de Ciencia y Tecnología – National Secretariat for Science, Technology and Innovation of the Panama Republic) have organised courses of Metadata and Clearinghouse development, with the support of CIAT, PROCIG and the USGS/EROS Data Center as part of the national strategy to create the National Spatial Data Infrastructure in Panama.

Following these training activities, IGN – Tommy Guardia is developing an institutional metadata work plan that will result in a functional, registered metadata clearinghouse.

3.14. *Uruguay*

Uruguay was one of the pioneer using metadata and clearinghouses in the region. They have also carried out an important role to the dissemination of information related to this theme by translating into Spanish publications on metadata tools and standards produced by the FGDC (on-line see http://www.clearinghouse.gub.uy). The metadata standard used by Uruguay is the FGDC Content Standard for Digital Geospatial Metadata (US FGDC STD-001-1998, version 2.0). So far, they have over 3000 metadata records (both English and Spanish) in their server, covering digital and paper cartography, imagery, land use, etc. They have also carried out courses and workshops to disseminate the knowledge among the main producers and users.

3.15. *Venezuela*

Venezuela is using the FGDC Content Standard for Digital Geospatial Metadata (US FGDC STD-001-1998, version 2.0). At present, IGVSB consolidates existing metadata catalogues and prepares to launch the institute's clearinghouse during 2002.

3.16. *PC IDEA*

The Standards Working Group of PC IDEA has a specific task to gather, develop and promote, and apply the efficient management of standards for regional geographic information through its members. This task will include responsibility for subjects with regard to geographic information interoperability.

Active participation links with initiatives like ISO/TC 211 Geographic information and Geomatics, as well as other ISO Technical Committees close to the geo-spatial information subjects. Amongst them are ISO/TC 10 – Products technical information, ISO/TC 204 – Control and information transport system. Also other organisations are considered such as DGIWG – Digital Geographic Information Working Group (on-line see http://www.digest.org/) with DIGEST, IHO with the Transfer Standard for Digital Hydrographic Data (S57), OGC – OpenGIS® Consortium (on-line see http://www.opengis.org) and SEDRIS (on-line see http://www.sedris.org) and others. ISO/TC 211 approved the request for the Class A Liaison Status for PC IDEA at the Plenary of ISO/TC 211 held in Bangkok, May 2002.

4. Description of metadata query/distribution services that are planned or in operation

The general in Central and South America and the Caribbean the ANSI Z39.50 standard for Metadata search/distribution or clearinghouse services is used, following the FGDC approach.

ANSI Z39.50 (or ISO 23950) is a search and retrieval protocol developed initially in the library community for access to virtual catalogues. Key features of the ISO 23950 protocol include [Nebert 2001]:

- support of registered public 'field' attributes for query across multiple servers where they may be mapped to private attributes;
- platform-independent implementation over TCP/IP using ASNI encoded protocol data units;
- ability to request both content (known as Element Sets) and presentation format (Preferred Syntax);
- GEO (Geo-spatial Metadata) Profile with registered implementation guidance for current FGDC and ANZLIC metadata and soon to include ISO 19115 metadata elements.

The use of a generalised query protocol as ISO 23950 permits a migration from national forms of metadata to future forms being developed through international consensus under ISO/TC 211 guidance and their draft metadata standard 19115. The countries of Latin America that have implemented the Z39.50 protocol for their clearinghouses will take advantage of its flexibility with GEO profile to migrate to future standards.

As illustrative national experience, briefly the Colombian approach of metadata query/distribution service, that is in operation and planned, will be described. Finally, the PC IDEA activities are also shown.

Colombian approach:

According to Lilia Patricia Arias, the IGAC's Clearinghouse Co-ordinator, the implementation of the Clearinghouse for ICDE can be summarised as follows:

Configuration and installation the Colombian clearinghouse nodes

In order to promote the clearinghouse service in the region, FGDC supplies training materials and the program Isite developed by CNIDR, which must be installed on the nodes, operating in conformity with the FGDC Content Standard for Digital Geospatial Metadata (US FGDC STD-001-1998, version 2.0).

After the user defines the search criteria, the clearinghouse nodes are in charge of processing information queries, and returning hits.

Because Colombia has its own metadata standard, it was necessary to adapt and configure the program to perform satisfactorily with the NTC4611 standard structure. This was accomplished successfully in 2000. Now, four clearinghouse nodes are available in the main ICDE producer organisations.

Install the National Spatial Data Gateway

Clearinghouse Information searching is performed through a web form, in which the user can specify different criteria like: geographic co-ordinates, date, text string, and query nodes. This form is only a part of a system called Gateway that manages simultaneous searches, using a conventional web browser. Gateway receives hits produced by the nodes, and groups them together to be listed in the user browser, who accesses to the metadata by clicking hyperlinks.

ICDE's goal for the 2001 is to assess the most appropriate technological solution for the national gateway installation, and clearinghouse nodes connection. This solution facilitates an independently information management from organisations, without endangers of shared query systems. As result of the technological assessment to improve the National Clearinghouse, IGAC will implement a Metadata Management System based on a robust database engine. Additionally, the application and web servers and the software resources required to install the National Spatial Data Gateway and the institutional clearinghouse node has been acquired. This implementation is focused on providing access mechanism to IGAC's fundamental data and guarantees compatibility between the NTC4611 standard and the global search and retrieval requirements for geo-spatial information.

Colombian experience will be documented and a subsequent process of knowledge transfer will be carried out at national and regional levels, as a contribution to the technical implementation of SDI in developing countries.

Define and develop the on-line Geographic Information Services

This phase becomes the convergence point of ICDE components, for creating a broad array of geographic information products and services, which will be accessed by users through Internet. Carrying out this phase implies the building of National Data Framework and the definition of policies and standards for information management that must be developed jointly by organisations.

In the middle term, it will accomplish activities related to the ICDE's type of services, focused on the user requirements; the assessment of existing information and updating needs; strategies for the development of data production and transfer, and the advanced technology implementation for achieving the database interoperability and the on-line delivery.

From technological point of view, ICDE institutions must develop an architectural model based on a distributed or network system of data servers that will be accessed by users through web interfaces. The initial goal is to provide catalogue services to query metadata records. Later, geo-spatial datasets will be linked to metadatabases in order to enable visualisation, downloading and printing operations of geo-spatial products in image formats. Finally, geo-spatial information will be available to the user in standard transfer formats, via FTP and dynamic mapping applications, which will be implemented according to the OpenGIS® specifications, national implementation of ISO/TC 211 standards, the On-line Government guidelines, ICDE policies and agreements, and the institutional definitions about E-commerce and E-business.

Setting up on-line services will allow optimising resources investment, increasing the market of geographic products and services, and the most relevant issue, enable access to strategic geographic information for supporting decision-making and sustainable development.

4.1. *PC IDEA activities*

The last survey carried out by PC IDEA shows countries that have already developed Z39.50 clearinghouses nodes. Countries responding affirmatively to Kate Lance (USGS/EROS Data Center) were: Argentina, Chile, Colombia, Costa Rica, El Salvador, Guatemala, Guyana, Honduras, Mexico, Nicaragua and Uruguay.

Douglas Nebert, co-ordinator for the US Federal Geospatial Data Committee (FGDC) and chairperson of the Technical Working Group for Global Spatial Data Infrastructure (GSDI), heads the Clearinghouses Working Group of PC IDEA. The general objective of the Clearinghouse Working Group is to install a clearinghouse node in each member country of PC IDEA, to establish the query system of geographic metadata with the support of FGDC and Colombia to facilitate the query and retrieval of geographic information by Internet among the countries of the region.

5. Summary and conclusions with prognosis for the future

After analysing the multiple efforts developing of metadata applications in the Central, South American and the Caribbean Region, it is possible to conclude the following issues:

1. at present, the region could be characterised as being in a stage of capacity building (wide dissemination of SDI information, short intensive courses, workshops, pilot projects, etc.) and in the beginning of implementations of metadata and clearinghouses;
2. work on the spatial data standards activities tends to follow the US FGDC-NDSI and ISO/TC 211 approach. Most of the countries in the region use the FGDC Content Standard for Digital Geospatial Metadata (US FGDC STD-001-1998, version 2.0);
3. in general, Central and South America and the Caribbean countries use of ANSI Z39.50 as standard of metadata search/distribution or clearinghouse services following the FGDC approach;
4. metadata and clearinghouses activities are addressed frequently by the National Geographic Institutes or the official institutions for Cartography and Geodesy, who are the

major geographic information providers. However, the number of inter-institutions relationships at national level created during this process, is significant;
5. some of the initiatives in the countries in the region have an important linkage with Information Society vision (Brazil, Colombia, Cuba);
6. main application scenarios are oriented towards the Environmental Decision-Making and Disaster Management in the region;
7. following PROCIG and other similar experiences, the importance Pilot Projects that allow to foster the introduction of new technologies and their generalisation becomes apparent. The multiple tasks carried out by Central America countries as part of PROCIG in the developing of metadata applications and clearinghouses are good examples;
8. member institutions and countries of the PC IDEA have committed to work together in the future to develop the Regional Spatial Data Infrastructure for the Americas to maximise the economic, social and environmental benefits from geographic information.

List of contributors

Argentina	Ricardo César Millet, IGM Director mailto: dir@mapas.igm.gov.ar, public@mapas.igm.gov.ar Ricardo Antonio Torchetti, IGM Technical Area Head mailto: jcart@mapas.igm.gov.ar, sigra@mapas.igm.gov.ar
Chile	Maria Loreto Advis Neira, Research and Development Department, Military Geographic Institute, mailto: mladvis@igm.cl Alvaro Medina Aravena, Ministry of National Property, mailto: amedina@mbienes.cl
Columbia	Lilia Patricia Arias, Clearinghouse Co-ordinator of Agustín Codazzi, Geographic Institute of Colombia, mailto: lparias@igac.gov.co
Cuba	Pedro Martínez Fernández, GEOCUBA, mailto: geohab@geohab.get.cma.net
El Salvador	Roberto López Meyer, National Registry Centre, National Geographic Institute, mailto: lopezmeyer@ign.cnr.gob.sv
Guatemala	Tanya Aguayo, National Co-ordinator of Metadata, mailto: taguayo@segeplan.gob.gt
Guyana	Vijay Datadin, Iwokrama International Centre for Rain Forest Conservation and Development, mailto: vdatadin@solutions2000.net
Mexico	José Luis Ornelas de Anda, INEGI, mailto: jornelas@antares.inegi.gob.mx Rafael Arrioja Urzúa, INEGI, mailto: rarrioja@dgg.inegi.gob.mx
U.S.A.	Kate Lance, USGS/EROS Data Center, mailto: klance@usgs.gov
Uruguay	Carlos López-Vázquez, National Clearinghouse of Geographic Data, mailto: carlos@clearinghouse.gub.uy, mdmaster@clearinghouse.gub.uy
Venezuela	Marlys García S, Co-ordinator of Institutions Relationships and International Co-operation, mailto: marlysgar@hotmail.com

References

[Borrero 2000] *Case Study of Trans-national initiatives: Latin-America* S. Borrero, 2000. GSDI-4 Conference, 13–15 March 2000, Cape Town, South Africa.

[Buck 2001] *Proyecto Centroamericano de Información Geográfica – resumen del Trabajo en Honduras* M. Buck, 2001. On-line see: http://www.procig.org

[Elizondo 2000] *Caracterización Preliminar de la Infraestructura Nacional de Datos Espaciales en Costa Rica* C. Elizondo, M. Aguilar, M. Buck, 2000. On-line see: http://www.procig.org

[Fuentes 2001] *Entrenamiento e Implementación de Metadatos y Clearinghouses en Tommy la Guardia* D. Fuentes, 2001. On-line see: http://www://procig.org

[Gran 2001] *Chilean Geo-Spatial National Data Infrastructure (INDE)*
P. Gran, et al., August, 2001. Proceedings of the 20th International Cartographic Conference, ICC2001 Beijing, China, Volume 4, pages 2188–2196.

[Hyman 2001] *Final Report to InfoDev Program of PROCIG – El Proyecto Centroamericano de Información Geográfica*
G. Hyman, et al., July, 2001. On-line see: http://www.procig.org

[Nebert 2001] *Developing Spatial Data Infrastructure: The SDI Cookbook. Version 1.1*
D. Nebert, May, 2001. On-line see: http://www.gsdi.org

Organisation	**For on-line contact see:**
ALISTAR, Nicaragua	http://www.alistar.org.ni/
Caribbean Environment Program	http://www.cep.unep.org/
CIAT, Colombia	http://www.ciat.cgiar.org
CNDG, Uruguay	http://www.clearinghouse.gub.uy
CPRM, Brazil	http://www.cprm.gov.br/
DGIWG	http://www.digest.org/
DIGSA, South America	http://www.digsa.com.pe
U.S.A. Geographic Information Centre,	http://www.ncgia.ucsb.edu
National Natural Resources and Environment	
Guatemala clearinghouse node	http://www.ign.gob.gt/clearing.htm
IBGE, Brazil	http://www.ibge.gov.br
ICDE, Colombia	http://www.icde.org.co
IGM – Argentina	http://www.igm.gov.ar
IGM – Chile	http://www.igm.cl
IGM – Uruguay	http://www.sgm.gub.uy
IGN – El Salvador	http://www.cnr.gob.sv
IGN, Guatemala	http://www.ign.gob.gt
IGN – Tommy Guardia, Panama	http://www.mop.gob.pa/igntg
IGAC, Colombia	http://www.igac.gov.co
IGVSB, Venezuela	http://www.igvsb.gov.ve
INEGI, Mexico	http://www.inegi.gob.mx/
INETER, Nicaragua	http://www.ineter.gob.ni
INPE, Brazil	http://www.inpe.br
Iwokrama, Guyana	http://www.iwokrama.org
OGC, OpenGIS® Consortium	http://www.opengis.org
PAIGH	http://www.ipgh.org.mx
PC IDEA	www.cpidea.org/
SEDRIS	http://www.sedris.org
Uruguay National Clearinghouse (node)	http://www.clearinghouse.com.uy
USGS/EROS Data Center	http://edc.usgs.gov/

URL's in this chapter are accessed in May, 2005, unless otherwise stated in the text.

World Spatial Metadata Standards
H. Moellering, H.J.G.L. Aalders & A. Crane (Editors)

Spatial Metadata in Africa and the Middle East

Antony Cooper [a], Elizabeth J.O. Gavin [b]

[a] *Centre for Logistics and Decision Support, CSIR, Pretoria, South Africa*
e-mail: acooper@csir.co.za
[b] *Statistics Sout Africa, Pretoria, South Africa*
e-mail: lizg@statssa.gov.za

1. Introduction

With well over 60 different countries in Africa and the Middle East, a review such as this can sketch only the background for understanding the use of spatial data across the region, especially the use of standards for spatial data, including metadata. The region stretches from the Caucus Mountains in the north down to Antarctica, incorporating a diverse range of terrain and environment: alpine to ocean, tropical to temperate, desert to jungle, rural to urban, land-locked countries to island states. There is also a wide diversity of cultures and languages and varying levels of information and communications infrastructure, resulting in a heterogeneous environment.

This chapter attempts to paint a broad picture of the spatial metadata situation in Africa and the Middle East, through describing briefly the activities in countries and regional bodies across the region, and providing more detail in one or two countries as examples. We describe first some activities that span the whole of the African continent and then, for convenience, we have grouped the summaries by region: western, eastern, northern and southern Africa, and the Middle East. Recording and making available metadata is crucial for increasing access to spatial datasets relevant for policy and decision makers at all levels, for addressing issues such as land reform and administration, environmental management, ensuring food security, alleviating poverty, combating diseases and implementing global conventions.

This chapter draws heavily on the survey of spatial data infrastructures (SDIs) in Africa [Gavin 2000], conducted by the National Spatial Information Framework (NSIF) Directorate in South Africa on behalf of the SDI in Africa Initiative (on-line see http://geoinfo.uneca.org/sdiafrica/). In addition, an attempt has been made to solicit information on their spatial metadata activities from all the member countries of the International Cartographic Association (ICA) in Africa and the Middle East (Algeria, Ghana, Guinea, Iran, Israel, Jordan, Madagascar, Morocco, Nigeria, Qatar, South Africa, Sudan, Tanzania, Tunisia, Uganda, Yemen). We have also used the best practices studies of the Environmental Information System Programme for Sub-Saharan Africa (EIS) [EIS 1999].

In addition to considering spatial data organisations, we will also consider specific once-off projects or initiatives, as many of these have a better resource than national organisations and produce legacy base datasets (hopefully with metadata), and consequently have a significant impact. It should be noted that there are also many global initiatives in which countries in Africa are involved, such as Global Map (on-line see http://www.iscgm.org), including 19 participating countries in Africa. The resources involved in these initiatives may be significant in comparison with resources invested by countries in their own spatial data collection and management.

2. General regional structure for spatial data organisation

There are a number of organisations addressing the need for spatial metadata across Africa, including the United Nations Economic Commission for Africa, the Spatial Data Infrastructure in Africa initiative, EIS-AFRICA and the United Nations Environmental Programme.

2.1. *United Nations Economic Commission for Africa*

In June 1999 in Addis Ababa, the first meeting was held of the Committee on Development Information (CODI) of the United Nations Economic Commission for Africa (UN ECA). CODI, as one of the seven subsidiary bodies of the Commission for Africa, was established in 1997 by the twenty-third meeting of the Conference of African Ministers responsible for Economic and Social Development and Planning. (For further details, including the report from the first meeting, refer to the UN ECA web site, http://geoinfo.uneca.org/). CODI provides policy and technical guidance to the ECA's sub-programme 'harnessing information for development'. At the June 1999 meeting, CODI's three sub-committees met in parallel, bringing together several previously separate initiatives to improve co-ordination in the area of development information:

- Sub-committee on Statistics: this replaced the Co-ordinating Committee on African Statistical Development (CASD), the primary task of which was developing the strategy of the Implementation of the Addis Ababa Plan of Action for Statistical Development in Africa in the 1990s;
- Sub-committee on Information and Communication Technologies: this sub-committee focuses on the development of information and communications technology (ICT) infrastructure, embodied by the African Information Society Initiative (AISI), which was adopted at the ECA Conference of Ministers in 1996;
- Sub-committee on Geo-information: known as CODI-GEO, this replaces the previous Regional Cartographic Conference for Africa. The terms of reference agreed upon for CODI-GEO at its first meeting include those of:
  - facilitating capacity building with regard to spatial information capture, management, dissemination and utilisation;
  - promoting the development of standards in these areas and of resources.

Currently, CODI-GEO is compiling a list of successful GIS projects in Africa, with the intention of identifying a few to use as case studies for developing best practices. At its September

2001 meeting, CODI-GEO focused on SDI development, including metadata capturing and publishing.

2.2. *The Spatial Data Infrastructure in Africa initiative*

On 15 July 1999, immediately following the Earth Data Information Systems (EDIS) 1999 Conference held in Pretoria, South Africa, a workshop was convened to consider the development of a Spatial Data Infrastructure (SDI) for Africa. It was organised by the Program on Environment Information System in Sub-Saharan Africa (EIS) and the South African National Spatial Information Framework (NSIF) Directorate.

An interim task team was established, which is conducting an on-going survey of SDI in Africa [Gavin 2000], co-ordinated by the National Spatial Information Framework Directorate in South Africa. The interim task team has been liaising with CODI-GEO about forming an African SDI, and organised a meeting on SDI in Africa in Cape Town, South Africa, on 15 March 2000, in association with the 4th Global Spatial Data Infrastructure (GSDI) Conference. At that meeting, two Working Groups were established to address the most critical issues inhibiting the development of SDI in Africa:

- Metadata Working Group: led by the Regional Remote Sensing Unit in Harare, Zimbabwe. This group will address issues such as standardisation and the publishing of metadata records;
- Geodetic Working Group: led by the South African Chief Directorate of Surveys and Mapping in Mowbray, Cape Town. This working group will focus on the development of a sound geodetic network for Africa, with the rekindling of the ADOS project by the IAG.

The interim task team aims at stimulating the sharing of ideas and participating in other initiatives world-wide (especially GSDI), through establishing and maintaining a network of interested individuals. Significant incentives for an African SDI are to promote investment opportunities (through providing data to assist potential investors), more-informed decision making, trans-border decision support systems (especially for natural resource issues), aligning regional projects, developing a common understanding amongst countries and disaster mitigation and prevention. The SDI in Africa mailing list is being used to debate these issues.

Major obstacles include the perceived control of SDI by national mapping agencies, different political systems and legacies between countries, lack of awareness of the value of SDI's and spatial information, varying stages of information technology development and the availability of resources, especially money.

2.3. *EIS-Africa*

Since its inception in 1990, the goal of the Program on Environment Information Systems (EIS) in Sub-Saharan Africa has been the promotion of information usage in decision-making, particularly within the context of environmental management and sustainable development. The Program is now entering a new phase, in which it will in future operate as a non-profit organisation, under the new name 'EIS-AFRICA: Network for the Co-operative Management of Environmental Information' (on-line see http://www.eis-africa.org/). EIS-AFRICA has been registered as an 'association', that is, a non-governmental organisation (NGO), in South Africa.

Over the past decade, the program has assisted Sub-Saharan African countries to develop operational EISs in order to meet the most important needs of decision-makers, planners and users of resources, through networking and capacity building efforts. The need to harmonise geo-spatial information collection efforts has been emphasised. EIS-AFRICA aims to play a leading role in continuing to facilitate the strategic development and use of geo-information in support of the effective management of, including monitoring and the reporting on, environmental resources for furthering development.

An aspect identified as critical to success in achieving the aim of increased information use in decision-making, is the increased availability and accessibility of policy-relevant information to policy and decision-makers at all levels. The information required for effective environmental management is pertinent also to a variety of other programmes, such as land reform and administration, ensuring food security and poverty alleviation, as well as the implementation of global conventions at the local, national sub-regional levels, all require the same basic datasets. Central to accomplishing the goal of readily accessible and relevant information, is the promotion of a unified approach to the development of inter-linked national, sub-regional and continental geo-spatial data infrastructure for Africa, supporting the sharing and reuse of data by different users for a variety of applications. Thus EIS-AFRICA will in future be emphasising the need to align standards for digital spatial data, especially with respect to the capturing of metadata and it's publishing.

In association with the African Organisation for Cartography and Remote Sensing (AOCRS), EIS is involved in organising the Africa GIS meeting, held biennially. This presents an important platform for exchange of knowledge and experience within the geographic information community in Africa.

2.4. *UNEP-GRID and UNEP.Net*

The United Nations Environmental Programme (UNEP) is, through UNEP-GRID (its Global Resource Information Database), in the process of producing sub-regional State of Environment (SoE) reports for Southern, Western and Central Africa. All of these reports will have associated SoE metadatabases. The most advanced SoE initiative is that for Southern Africa. UNEP's collaborating institution on this initiative is the Regional Remote Sensing Unit of the SADC Food Security Programme in Harare. The unit has already drafted a strategy to develop the SoE metadata, for which UNEP will soon provide the necessary funding. The SoE metadata for West Africa will be developed later this year. CEDARE (Centre for Environment and Sustainable Development in the Arab Region and Europe) in Cairo represents another Collaborating Centre of UNEP on environmental assessment and reporting.

UNEP is aiming to establish UNEP.net (http://www.unep.net/) as the authoritative portal for environmental information. A portal for Africa will be developed, and, as with all regional portals, will include metadata and associated spatial datasets.

3. African regions

West Africa

3.1. *West African regional initiative*

The AGRHYMET Centre in Niamey, Niger focuses on hydrological and meteorological data.

AGRYHYMET has been involved in capturing a considerable amount of metadata for the Sahelian countries of West Africa.

3.2. *Benin*

In Benin, the governmental authorities responsible for the management and distribution of digital spatial data (from remote sensing sources), on behalf of the public sector, are:

- ASECNA: for data with low resolution (e.g.: NOAA and METEOSAT); and
- CENATEL: for data with high resolution (e.g.: SPOT and LANDSAT).

Both, ASECNA and CENATEL, act as agents for the producers of the data, funnelling requests through to them. CENATEL is involved in regional initiatives for establishing SDIs. The private sector, universities and non-governmental organisations use pre-processed digital spatial data for their applications. Currently, the Internet is not used to provide access to spatial data in Benin.

There is no central budgeting or co-ordination for acquiring digital spatial data of Benin, or for the development of a spatial data infrastructure. Development projects funded by international organisations continue to be a major source of data. However, progress is being made in sensitising the Executive and Legislative branches of the government about the need for a well-established spatial data infrastructure for Benin.

3.3. *Ghana*

The National Framework for Geo-spatial Information Management (NAFGIM) co-ordinates the production and exchange of compatible spatial data. The NAFGIM co-ordinator is based in the Environmental Protection Agency. Five datasets (topography, meteorology, land suitability, land use/land cover and land ownership) are being produced under a government-funded project on Environmental Resources Management Project (GERMP).

Though some institutions in Ghana have been recording metadata for their data holdings, they have not been doing this in a co-ordinated manner. The result is that much of the data of Ghana that could be shared is either undocumented, or has not been documented to any widely used standard. Data is then distributed mainly in response to requests to suppliers, often in hard copy format. Using the freeware metadata tool CorpsMet95, which complies with the current FGDC metadata standard (on-line see http://www.fgdc.gov/metadata/), the Remote Sensing Applications Unit (RSAU) of the University of Ghana has captured the metadata for the land-cover and land-use coverages resulting from the Ghana Environmental Resource Management Project.

There is a keen awareness of the need to implement metadata standards in Ghana, and with this in mind, the Environmental Protection Agency arranged for the FGDC to run a workshop on metadata in Ghana in August 2000. It is anticipated that metadata will be published through the web site of NAFGIM, presently under construction, before the end of 2001.

3.4. *Ivory Coast*

Within the Ivory Coast, government institutions, researchers and academics play an important role in gathering spatial data. There is a degree of co-ordination by governmental institutions in this arena. Spatial data are transferred between individuals in response to requests, and are published in reports. A project to develop a water resource database in order to predict well productivity in the crystalline basement in Ivory Coast is capturing metadata, along with hydrological and rainfall data, as well as crop statistics. However, there is no focus on making this metadata readily available to other potential users of the datasets under development. Spatial data standards are being addressed within the context of the project.

East Africa

3.5. *East African regional initiative*

Plans for a clearinghouse in East Africa were discussed at the Regional Integrated Information Systems (RIIS) Strategy workshop, organised by the Intergovernmental Authority on Development (IGAD), which was held in Jinja, Uganda, 26–29 July 1999. IGAD's initiative to develop a regional network of data producers in Eastern Africa is still at a formative stage.

3.6. *Ethiopia*

In Ethiopia, the main government organisations that are involved in the capture of digital spatial data are the Ethiopian Mapping Authority (EMA, on-line see http://www.telecom.net.et/), the Ministry of Water Resources and the Ministry of Agriculture. The spatial information captured includes data from satellite imagery, as well as that used in topographic and thematic mapping. The Geophysical Observatory and Geology Department of the Addis Ababa University also capture spatial data. There is no known private sector involvement, and NGO involvement in this area is limited to capacity building.

By proclamation, the EMA co-ordinates public programmes involving spatial information. The EMA distributes digital data topographic and thematic data using floppy diskettes, CD-ROM and magnetic tapes. Spatial data are also distributed as part of research reports.

In June 1999, the Environmental Support Project (ESP), which aims at enhancing data management, was initiated. This project involves the collection of metadata, making the data available to users, developing spatial data standards, developing a legal framework for accessing and distributing data and developing a pricing policy. Currently, the ANZLIC spatial metadata standard is being used. The metadata are not confined to only the spatial data, but also encompass projects, experts, reports, etc. The ESP will not attempt to create basic datasets of Ethiopia, but rather to enable access to existing spatial data, by providing metadata on CD-ROM's, in catalogues or over the Internet. The project currently involves the Ministry of Water Resources, the Ministry of Mines and Energy (in particular the Institute for Geological Surveys), the Ministry of Agriculture, the Environmental Protection Authority, the EMA and the Ethiopian Science and Technology Commission. The regional offices will be included at a later stage. A Protocol for Inter-Agency Co-operation has been drafted and is in the process of being signed by the participants. By the end of 2000, the Environmental Support Project will have established their Web site. The project is scheduled to finish in February 2003.

Ethiopia, through the EMA, is participating in the Global Map programme (on-line see http://www.iscgm.org).

3.7. *Kenya*

There are many government departments in Kenya that deal with spatial data, though none is mandated to be the co-ordinator of such activities. They include the Department of Resource Survey and Remote Sensing (DRSRS), the Kenya Wildlife Service (KWS), the Forestry Department, Kenya Marine Fisheries Research Institute (KMFRI), National Museums of Kenya (NMK), Central Veterinary Laboratory (CVL), Department of Natural Resources, and Surveys of Kenya. Three major providers of spatial data within Kenya are Surveys of Kenya, DRSRS and the United Nations Environmental Programme (UNEP), which is based in Nairobi and which provides Internet access to digital data. Few private-sector organisations are involved in digital data capture or application development; they mainly specialise in aerial surveys and map production. There is little effort to incorporate other non-digital information on to the maps.

Academic institutions involved in managing spatial data include the University of Nairobi, MOI University and Egerton University. The Kenya Polytechnic presents several training courses that include components on GIS and Remote Sensing, and as such, is involved in capturing spatial data, as well as associated metadata, which is made available to potential data users. NGO's that need to use spatial data generally contract universities or private firms to do the work.

The Kenya wetlands conservation and training programme aims at conserving Kenya's wetlands and building capacity for the management and sustainable utilisation of wetland resources, which includes the mapping of marine and terrestrial wetlands. The Nairobi City Council captures and documents data needed for managing the city. The Nairobi Informal Settlements Co-ordinating Committee (NISCC) co-ordinates a wide spectrum of stakeholders.

In Kenya, there is no agency specifically tasked with the distribution of spatial data, so in general users have to search for what they need. There is also no forum, wherein users can voice their needs for data. Many organisations do not yet capture metadata, but awareness of the need to do so is growing. Kenya's National Environment Secretariat (NES) and the Forest Department were funded to develop a clearinghouse mechanism for Kenya, but this has not yet been implemented. Organisations in Kenya have in the main not yet embraced the need for spatial data standards, and at present, no organisation is mandated to develop standards.

3.8. *The Seychelles*

The Seychelles GIS Centre was established in 1996 by the Ministry of Land Use and Habitat (MLUH), for the purpose of capturing, manipulating and distributing spatial data, as well as linking with other spatial data distributors and users. Other major users and distributors include other divisions in the Ministry, other Ministries, Departments and private firms, including utilities. A publicly funded orthophoto-mapping project began in 1998, and reached completion in February 2000. A dataset called the National Inventory for Seychelles, including roads, rivers, land use, topography and cadastre, is under development.

Spatial data are distributed in hard copy as well as via the MLUH's Intranet to MLUH users. Metadata is collected and made available to spatial data users. Web pages for the GIS Centre are being developed. The MLUH is not currently involved in developing standards or policy for spatial data other than a pricing policy. MLUH has links with MAPS in the United Arab Emirates.

3.9. *Uganda*

In Uganda, the main government Departments involved in the capture of digital spatial data are the Departments of Surveys and Mapping, Planning, Petroleum Exploration and Water Development. There is no known private sector involvement in digital spatial data capture in Uganda. The United Nations High Commissioner for Refugees (UNHCR) has been collecting data on feeder roads in areas of their operations but maintenance plans for these data are not clear. From the academic sector, Makerere University Institute of Environment and Natural Resources (MUIENR) has a database on Uganda bio-diversity.

Concerning co-ordination of the capture and maintenance of spatial data in Uganda, the Information Section of the National Environment Management Authority (NEMA) has taken the initiative to co-ordinate the capture, maintenance and management of spatial information through providing financial and material support. This arrangement, called the Horizontal Environment Information Network, covers only a few governmental organisations at present and should ideally be broadened should funds become available for this.

The Surveys and Mapping Department sells topographic maps from their offices in Entebbe. The National Biomass Study has a website (http://www.imul.com/forestry) for viewing and downloading less-detailed digital data. Digital and paper copies with more-detailed information can also be bought from their office or ordered by other means. The aim of the National Biomass Study, which operates nationally in Uganda, is the provision of reliable and regularly updated data, information and knowledge about land cover and biomass for use by key stakeholders. The National Biomass Study has six datasets covering the whole country: land cover (use), rivers, and roads, gazette areas, administrative units and contours.

The metadata concept is new to many organisations in Uganda and no formal or *de facto* standards exist. However, data capturing organisations have records concerning lineage and data quality.

The National Biomass Study has a liaison with a regional initiative IGAD's Regional Integrated Information Systems (RIIS), although the RIIS is still in a formative stage.

3.10. *Sudan*

The Sudan Survey Department will have been transformed by the end of 2000 into the Sudan National Survey Corporation, under the supervision of the Ministry of Survey and Physical Development. It will be the major supplier of spatial data in the Sudan, together with the National Remote Sensing Centre. Sudan is a member of the Regional Centre for Surveying, Mapping and Remote Sensing, which is based in Nairobi, Kenya, and of the East African Section of the Africover Project.

Several Ministries and Departments in the Sudan are in the process of setting up the hardware and software needed for storing their data digitally. By the end of 2000, it is expected that the Sudan National Survey Corporation and several other organisations in Sudan will be able to organise their metadata in accordance with the international standards specifications.

North Africa

3.11. *Morocco*

Several departments are users of spatial data, mostly as hard copy, but increasingly, in a digital format. The two most significant organisations with respect to the creation, management and distribution of digital spatial data at a national level are the Administration for Land Conservation, the Cadastre and of Cartography (L'administration de la Conservation Foncière, du Cadastre et de la Cartographie) and the Royal Centre for Remote Sensing (Centre Royal de Télédétection Spatiale, CRTS). Other departments involved in the creation of base digital spatial data are the Geology Directorate (Direction de la Géologie) and the Statistics Directorate (Direction de la Statistiques). The limited involvement of the private sector focuses largely on the development of specific applications.

The Department of the Prime Minister is creating a National Council for Geographic Information (Comité National de l'Information Géographique, CNIG), that will develop the digital geographic information sector and put in place an institutional framework for the co-ordination of exchange procedures and the dissemination of digital spatial data. Currently the National Council for Cartography (Comité National de Cartographie) deals with aspects relating to mapping, while the National Council for Remote Sensing (Comité National de Télédétection Spatiale), established by the CRTS in 1993, has recorded information concerning existent programs and base data developed by different departments. Both these committees' research users' needs and initiate programmes in response to these needs.

Each institution is responsible for disseminating the data it produces. The CRTS provides information on the availability of digital spatial data and how data may be accessed via its website (http://www.crts.gov.ma). A further project underway is an archiving and access system for digital data, which become operational in September 2000.

A study is underway regarding metadata and procedures for access and utilisation of digital spatial data. There is an awareness of the role that the availability of metadata can play in minimising duplication in data capture and ensuring appropriate use of existing data. While standards in general are addressed by a component within the Ministry of Commerce and Industry, which by and large adopts international standards, it is anticipated that the CNIG will form a working group to deal specifically with digital spatial data.

3.12. *Algeria*

In Algeria, the National Council for Geographic Information (Comité National de l'Information Géographique, CNIG), acts as a co-ordinator with respect to the capture and processing of spatial data, digital and analogue. National institutions, which are major spatial data users or producers, are represented on this council. The National Institute for Cartography and Remote Sensing is a major data producer, while the National Centre for Spatial Techniques is also involved with processing digital spatial data. There are many other users of spatial data, including universities, research centres and other government departments, such as Agriculture and Geology.

The National Institute for Cartography and Remote Sensing distributes spatial data. The Internet is not used for distributing spatial data at this stage in Algeria. The NCGI is responsible for standard development, through its Commission for GIS and standardisation. Currently, there is no liaison with international standard development initiatives.

Southern Africa

3.13. *The Southern African Development Community (SADC)*

There are currently 14 member states of SADC, several of which do not, from a geographical perspective, really form part of 'southern' Africa. The member states are: Angola, Botswana, Democratic Republic of Congo (DRC), Lesotho, Malawi, Mozambique, Mauritius, Namibia, Seychelles, South Africa, Swaziland, Tanzania, Zambia and Zimbabwe. There are many SADC programmes, one of which is the Food Security Programme. There are also other regional initiatives covering subsets of countries within SADC, such as the Regional Tsetse and Trypanosomosis Control Programme (RTTCP), which involves the countries of Malawi, Mozambique, Zambia and Zimbabwe, with the Zimbabwean Ministry of Agriculture as the lead agency.

The secretariat for the SADC Food Security Programme is based at the Regional Remote Sensing Unit (RRSU) in Harare, Zimbabwe. The RRSU is a centre of technical expertise, which provides training and technical support in the fields of remote sensing and GIS, to advance both early warning systems for food security and natural resources management. The RRSU operates largely on funding provided by the SADC Member States, which will continue for an indefinite period. In addition to SADC funding, the RRSU will receive some funding from the Netherlands Government, until June 2001. The RRSU has links with South Africa's National Spatial Information Framework (NSIF), and has also been active in GSDI's Technical Working Group, contributing to the SDI cookbook it is publishing.

There are many role-players with respect to digital data within the SADC region. In general, one would find the following major role players within each country within the region:

- governmental departments: typically the Office of the Surveyor General, Central Statistical Offices, National Meteorological Department (Weather Bureau), Department of Agriculture or equivalents;
- universities and other research institutes;
- organisations, especially a range of UN organisations such as FAO, WFP and UNEP; but also WWF, FEWS (USAID) and the Red Cross;
- national and international NGO's.

Spatial data would most often be distributed within the region in hard-copy format, often as a component of a report. In general, there is no centralised system for distributing data in a digital format, although a number of initiatives do exist. Data are distributed mainly on CD-ROM, with the provision of some information via the Internet.

The RRSU maintains comprehensive metadata on datasets it compiles, which cover the entire SADC region. In 1999 these metadata records were captured using MetaLite, an FGDC-compliant metadata-capturing tool (on-line see http://www.fgdc.gov/metadata/), and published through the clearinghouse administered by NSIF in South Africa. The RRSU is also involved in developing spatial data standards and framing policy relating to the accessing and distribution of spatial data. The RRSU regards the development of standard data (transfer) formats and operation data policies as critical factors for the successful development of SDI in the region.

The Environment Resource Centre for Southern Africa also based in Harare, Zimbabwe, runs the Environment Reporting Programme for Southern Africa, covering the SADC region. There is a focus on raising awareness of environmental issues and influencing policy and decision-making. The Centre is affiliated to UNEP-GRID, and interacts with the RRSU, SIRDC, CSIR (based in South Africa), State of Environment reporting practitioners and the offices of the Surveyors General within the region. Data are collected at the SADC, national and river basin level. Metadata is also captured and is made available to spatial data users.

3.14. *SAFARI 2000*

SAFARI 2000 is a regional science initiative in Southern Africa to study land-atmosphere processes and emissions (biogenic, pyrogenic or anthropogenic), and the consequences of their deposition to bio-geophysical and bio-geo-chemical systems. It builds on a number of existing activities undertaken by the National Aeronautical and Space Administration (NASA), the international community and African countries. Data for SAFARI 2000 are being gathered from ground-based and airborne platforms, as well as earth observing satellites such as the NASA EOS Terra platform. The project covers: terrestrial ecology, land use and land cover change, aerosols and trace gas chemistry and transport, surface radiation, cloud characterisation and radiation effects and modelling (on-line see http://www.wits.ac.za/fac/engineering/civil/Safari/index.html).

The key outputs from the project are a database of all known emissions in the region for several key pollutants and with models that can predict where the pollutants go and how they are chemically altered. The project team has recognised how crucial it will be to record metadata for all the datasets, to prepare data for synthesis and analysis and to archive and distribute data after the completion of the SAFARI 2000 project. A metadata editor on the Internet has been developed to allow participants to capture their metadata, critical to the design of this tool has been the realisation that not all researchers will have access to high-speed Internet connections. The tool also allows researchers to keep their metadata private until their datasets are ready for publication. Once captured, the metadata records are loaded onto NASA's metadata search facility hosted by the Oak Ridge National Laboratories [SAFARI 2001].

3.15. *South Africa*

In 1997, resources were set aside by the Department of Land Affairs to develop the National Spatial Information Framework (NSIF), South Africa's SDI building programme. The aim of the programme is the establishment of a technical and policy framework, enabling unimpeded access to and utilisation of spatial information for effective and efficient governance, planning and decision making, through all spheres of government. This includes the provision of systems supporting access to key spatial information. Like similar initiatives elsewhere, along with standard development and the framing of policy and institutional arrangements, there is a strong focus on the development of a clearinghouse, and as part of this, the capturing and publishing of standardised metadata.

Currently an FGDC-type clearinghouse, termed the Spatial Data Discovery Facility (SDDF, on-line see http://www.nsif.org.za/), makes available close to 3000 metadata records, including records pertaining to datasets covering the whole SADC region. These records correspond to spatial data holdings within both the public and private sectors in South Africa. A metadata standard was *de facto* imposed through the provision of free metadata capturing tools, both MetaLite and a web-based system for capturing metadata. South Africa has been a pioneer in the development of standards for spatial data [Cooper 1991], and work is proceeding on developing a national standard for metadata, through the profiling of the international standard, ISO 19115 [ISO 19115], amongst other standards [Gavin 2001].

3.16. *Lesotho*

The major spatial data producers and users in Lesotho include several government departments and agencies, such as the Departments of Environment, Statistics, Lands and Survey, the Lesotho Highlands Development Authority and the National Environment Secretariat. Other role players are the SADC Water Sector and SADC Environment and Land Management Sector, both of which are based in Lesotho.

Currently spatial data are distributed using CD-ROM and floppy diskettes, as well as in the form of hard-copy maps. While there is not yet any formal policy concerning spatial data access and distribution, informal agreements between users and suppliers are usually entered into ensuring that information concerning any updating or modification to the datasets will be provided to the parties concerned.

The Committee on Environmental Data Management (CEDAMA) was established in February 1999 to advise the National Environmental Secretariat on issues pertaining to environmental data management. These include fostering institutional arrangements that will support co-ordination of efforts in gathering and managing spatial data, as well as researching policies and standards to promote the sharing of spatial information and raising an awareness of the importance of spatial data for Lesotho. CEDAMA is involved in collecting metadata and making metadata available to potential data users. South Africa's NSIF participated in a workshop in Lesotho in 1999 to share its experiences, and a follow up workshop is planned to establish a clearinghouse for spatial data, following the capturing of metadata on spatial data.

3.17. *Botswana*

In Botswana the major producers, distributors and users of spatial data include several government departments such as:

- Department of Surveys and Mapping (DSM);
- Department of Town and Regional Planning;
- Department of Lands, Housing and Local Government;
- Department of Geological Survey and Mining;
- Department of Agriculture and Water Affairs;
- Department of Wildlife, Conservation and Tourism,

as well as parastatals, i.e., quasi-governmental organisations, such as:

- the Botswana Power Corporation;
- Water Utilities Corporation;
- Botswana Telecommunications.

DSM is responsible for creating the Botswana National Atlas and the National Integrated GIS (to be completed in 2004). As the major producer and user of spatial information, DSM plays an informal co-ordinating role with respect to publicly funded projects relating to spatial information.

DSM does collect metadata, and makes this available to potential spatial data users. DSM is also involved in developing standards for digital spatial data in general. Botswana, through DSM, is also participating in the Global Mapping Project.

3.18. *Namibia*

The Surveyor General and the National Remote Sensing Centre, falling within the Ministry of the Environment and Tourism, are the major data producers in Namibia. Metadata is captured as a matter of course. The private sector is also active in producing spatial data. Often spatial data are distributed in the form of paper maps. A body to co-ordinate the gathering and management of spatial data within the public sector was established recently.

The State of Environment Reporting Programme (SOER) operates from the Department of Environmental Affairs and promotes sustainable development. Spatial data are distributed largely in response to individual requests. There is liaison with other SOER programmes in the SADC region. Metadata is collected and made available to data users, as is the case for similar programmes running throughout the region.

3.19. *Zambia*

The role-players in spatial data production, distribution and maintenance, as well as the major users within Zambia, include national planning units, local government institutions, pilot communities for the Environmental Support Programme (ESP), donors, researchers and academic institutions. The Environmental Information Network and Monitoring System, which operates within the Ministry of Environment and Natural Resources, aims to provide action-oriented environmental information system and decision support tools to various stakeholders, providing them with better access to information on environmental issues and plans (on-line see: http://www.menr.gov.zm). Metadata on datasets such as those pertaining to air and water quality, wildlife distribution and soil degradation, is collected and made available to users.

An Environmental Information System Forum is involved in co-ordinating publicly funded programmes involving spatial information. Spatial data are distributed using websites and on CD-ROM, but also through publications such as newsletters and conference proceedings, meetings and electronic billboards.

Middle East

3.20. *Saudi Arabia*

Today, the scope of geographic information activities in the Kingdom of Saudi Arabia encompasses: strategic applications, the planning and management of infrastructure, agriculture, environment, industries, health, education, communication, telecommunication, electricity, transportation, roads and highway development, census, coastal studies, engineering, natural and human resources, national economy, general welfare, geology, land use analysis, urban and rural planning, mineral exploration, sand control (desertification), water resources assessment, hydrographic studies, tourism, oil spill monitoring and many other applications and future planning. Therefore, the Kingdom of Saudi Arabia became a member of the ISO/TC 211 in 1998 and supports several research and development projects. One of these projects is the development of Geographic Information Infrastructure. This project revolves around the development of long-term strategy for national geographic information infrastructure (NSGII) design. It includes the implementation of geographic information standards, data documentation, metadata standards, national geographic information clearinghouses, procedures, policy institutional, technical issues, and other issues effecting the collection, processing, display, storage, distribution, availability, reliability and accessibility of correct, up-to-date geographic information in the Kingdom of Saudi Arabia.

In the near future, the Kingdom of Saudi Arabia hopefully will create the infrastructure necessary for providing rapid access to accurate and current geographic information at the national level. The Data documentation and metadata standards, which will be used in the Kingdom of Saudi Arabia, will follow very closely the ISO/TC 211 standards (with certain modifications to suite the country).

3.21. *Israel*

Geo-related standards in Israel are one of the responsibilities of the Survey of Israel, under governmental resolutions in 1989 and 1993. Work on the Israeli metadata standard started in 1991, as part of the then newly developed data transfer standard (IEF'91). The basic idea was to supply each of the features in the databases and transferred datasets with 'source' information. Thus, each of the features (objects, entities) within the National GIS databases was provided with an additional attribute: 'Source-ID'. This was an integer value serving as a pointer to a relate table. In this table, for each entry is stored metadata including the date of acquisition, method of acquisition, positional accuracy and supplier ID. Thus, when most of the data were acquired by photogrammetric mapping, all features in a large area received the same source-code entry. It is important to notice that all metadata was focused on spatial information.

In 1995, when National GIS was established, some non-spatial data were added to the basic spatial data. In addition, some government ministries began to develop their own value-added layers, based on the 11 topographic layers generated by the Survey. In order to co-ordinate all these databases into the envisaged national GIS, the Survey began to produce a Spatial Data Index. At this time, it was realised that the next step should be the adoption of an independent metadata standard. The idea was to wait for the ISO standard and to generate an appropriate Israeli profile. By 1998, it was clear that ISO would not finish the work in time and work began in October 1998 to develop an Israeli metadata standard.

The Israeli Metadata Standard (Format) IMF2000 is aimed at establishing the framework and mechanism for transferring information related to spatial data. Originally, it was launched to supplement spatial data offered, by the Survey of Israel, from the National GIS databases. Finally, the proposed standard was developed to accommodate any spatial data regardless of their origin, format or domain (vector, raster or textual). In addition, the standard was devised as an instrument of spatial data marketing and distribution. Thus, the standard was divided into separate chapters that each accommodate various aspects of the description and quality definition of the spatial data in terms of spatial accuracy, completeness and currentness, and information related to availability of information and its accessibility. The major motivation in the development and the emphasis of the separate structure of individual chapters was to accommodate many optional models and a variety of data without using separate profiles, but rather various parts (chapters) of the common standard.

The standard is in its final stage of approval. The final draft has been tested and is adopted officially in December 2000, with a mandatory review in October 2001. The additional options embedded in the metadata standard required the Survey to upgrade its transfer standard (after 10 years of usage), again, to accommodate metadata information within the framework of the transfer standard. Thus, the later will enable the transfer of free text attributes and encoded metadata information, as well as the spatial and other non-spatial information.

4. Summary and conclusions

Africa and the Middle East are entering a period in which one may expect rapid growth in the capturing and publishing of spatial metadata. Many countries are poised to formalise a co-ordinated approach towards the development of their spatial data infrastructures. As a focus on metadata, almost invariably forms a part of the strategy towards developing spatial data infrastructures. It is inevitable that this will reinforce the need to capture and publish spatial metadata in a standardised way. The growth in international co-ordination through regional and even global initiatives will reinforce the progress towards the standardisation of metadata capture and publishing.

It should be noted that there are a number of challenges impeding the implementation of spatial data infrastructures in many countries in the region. These include political priorities not supporting spatial data infrastructure development, a lack of national mapping or SDI policies. But also, national mapping agencies steeped in tradition whose products are still paper-based, difficulties with obtaining inter-agency co-operation, few standards and poor compliance to them, and the lack of resources and skilled personnel [Clarke 2001]. However, these impediments are hardly unique to the region and are also to be found, to some degree at least, in many countries in more developed regions.

These trends towards the co-ordination of efforts to develop spatial data infrastructure, coincide with the completion of the ISO metadata standard. Nevertheless, this fact augurs well for the development of systems supporting the meaningful comparison of spatial metadata held within different metadata repositories, as these are likely to be developed in alignment with ISO guidelines. Further, many current metadata repositories in Africa have already been developed within the ambit of FGDC standards. This can, in no small measure, be attributed to the availability of free, easy-to-use tools to capture metadata to FGDC standards.

The provision of a tool (or platform-dependent tools) to capture metadata to ISO standards would, in all likelihood, greatly encourage the wide usage of this standard, especially by nations or regional initiatives lacking resources to develop their own standards and capturing tools. Work has already begun on the development of an ISO equivalent of MetaLite

[FGDC/NSIF 2001]. There is also likely to be a tool available in the future to migrate legacy metadata, captured according to FGDC standards, to the format of ISO 19115 [Pearsall 2001].

Acknowledgement

We would like to acknowledge the many scientists in Africa and the Middle East who provided information on metadata activities in their countries and regions, which we have used in compiling this overview. We would also like to acknowledge the support of our organisations, which made this work possible.

We would are appreciative of the efforts of Orlando Nino-Fluck of UN ECA, Jacob Gyamfi-Aidoo, then of EIS Africa, Amashani Reddy and Dirk Boon both then of the NSIF Directorate, who collected information, arranged the translations of the questionnaire into French, distributed the questionnaires, collated and analysed the responses, and translated the French responses into English. Without their contribution, this chapter could not have been written and would not have been the same.

References

[Clarke 2001]
Global spatial information issues and challenges and opportunities for developing nations
D.G. Clarke, August 2001, 20th International Cartographic Conference, Beijing.

[Cooper 1991]
The South African standard for the exchange of digital geo-referenced information
A.K. Cooper and Clarke, D.G. 1991.
in: H. Moellering (ed.), Spatial database transfer standards: current international status, International Cartographic Association and Elsevier Applied Science, pp. 154–168, ISBN 1-85166-677-X.

[EIS 1999]
Best practices of Environmental Information Systems (EIS): the case of Zimbabwe
N. Mbudzi, Y. Jairosi, C. Vogel and D. Bohnet, 1999. http://www.grida.no/eis-ssa/products/zimbabwe/index.htm (Accessed 29 March 2001).

[Sylla 1999]
Les expériences en matière de Systèmes d'Information sur l'Environnement en Afrique Sub-Saharienne Cas du Sénéga C.I. Sylla, I.A.Wade, P. Hengue and P. Gerbe, 1999.
http://www.grida.no/eis-ssa/products/senegal/index.htm (Accessed 29 March 2001).

[FGDC/NSIF 2001]
An overview of Spatial Data Infrastructure in Africa,
D.D. Nebert and C. Crawford, March 2001, private communication.

[Gavin 2000]
E.J.O. Gavin, March 2000, GSDI 4, Cape Town; On-line see: web site: http://www.gsdi.org/conferences (Accessed: March 29, 2001).

[Gavin 2001]
Developing South African Standards Based on ISO 19100
E.J.O. Gavin, March 2001. ISO/TC 211 Workshop on Standards in Action, Lisbon.

[ISO 19115]
Geographic Information/Geomatics: Metadata
ISO/TC 211.

[Mensah 2000]
The Need For Developing A Standardised Meta-Data For Spatial Datasets In Ghana
F.K. Mensah, March 2000, GSDI 4, Cape Town.

[Pearsall 2001]
US Federal Geographic Data Committee ISO 19115 Harmonization Activities
R. Pearsall, March 2001, ISO/TC 211 Workshop on Standards in Action, Lisbon.

[SAFARI 2001]
SAFARI 2000: A Southern African Regional Science Initiative
On-line see: web page: http://www.wits.ac.za/fac/engineering/civil/Safari/index.html (Accessed 14 March, 2001).

Organisation	For on-line contact see:
CRTS, Morocco	http://www.crts.gov.ma
EIS-AFRICA	http://www.eis-africa.org/
Environmental Information Network and Monitoring System, Zambia	http://www.menr.gov.zm (last accessed in July, 2002)
EMA Ethiopia	http://www.telecom.net.et/
FGDC metadata standard, U.S.A.	http://www.fgdc.gov/metadata/
Global Map (programme)	http://www.iscgm.org
National Biomass Study	http://www.imul.com/forestry (last accessed in July, 2002)
SAFARI 2000	http://www.wits.ac.za/fac/engineering/civil/Safari/index.html
SDDF, South Africa	http://www.nsif.org.za/
SDI in Africa Initiative	http://geoinfo.uneca.org/sdiafrica/
UN ECA	http://geoinfo.uneca.org/
UNEP.net	http://www.unep.net/

URL's in this chapter are accessed in May, 2005, unless otherwise stated in the text.

World Spatial Metadata Standards
H. Moellering, H.J.G.L. Aalders & A. Crane (Editors)

Global Spatial Metadata Activities in the ISO/TC211 Geographic Information Domain

Olaf Østensen [a], David M. Danko [b]

[a] *Chairman of ISO/TC 211, Geographic Information/Geomatics, Chairman Joint Steering Group on Spatial Standardisation and Related Interoperability, Norwegian Mapping Authority, Norway*
e-mail: olaf.ostensen@statkart.no

[b] *Editor ISO 19115, Project Team Leader ISO 19115-2, Senior Consultant – GIS Standards Environmental Systems Research Institute, Inc., U.S.A.*
e-mail: ddanko@esri.com

1. Introduction to ISO

Standards are documented agreements containing technical specifications or other precise criteria to be used consistently as rules, guidelines, or definitions of characteristics, to ensure that material, products, processes and services are fit for their purpose. The International Organisation for Standardisation (ISO, on-line see http://www.iso.ch) is a worldwide federation of national standards bodies from some 130 countries, one from each country.

The existence of non-harmonised standards for similar technologies in different countries or regions can contribute to so-called 'technical barriers to trade'. Export-minded industries have long sensed the need to agree on world standards to help rationalise the international trading process. This was the origin of the establishment of ISO.

Work in the fields of environmental management and geographic information is among the newer activities, taken up mid 1990's.

International standardisation is well-established for many technologies in such diverse fields as information processing and communications, textiles, packaging, distribution of goods, energy production and utilisation, shipbuilding, banking and financial services. It will continue to grow in importance for all sectors of industrial activity for the near future.

Development agencies are increasingly recognising that a standardisation infrastructure is a basic condition for the success of economic policies aimed at achieving sustainable development. Creating such an infrastructure in developing countries is essential for improving productivity, market competitiveness, and export capability.

1.1. *ISO and the market*

The World Trade Organisation (WTO) is the international organisation dealing with the global rules of trade between nations. Its main function is to ensure that trade flows as smoothly, predictably and freely as possible.

ISO – together with IEC (International Electrotechnical Commission) and ITU (International Telecommunication Union) has built a strategic partnership with WTO. The political agreements reached within the framework of WTO require underpinning by technical agreements. ISO, IEC and ITU, as the three principal organisations in international standardisation, have the complementary scopes, the framework, the expertise and the experience to provide this technical support for the growth of the global market.

The agreement with WTO recognises the important contribution that international standards and conformity assessment systems can make to improving efficiency of production and facilitating international trade. Where international standards exist or their completion is imminent, the code of Good Practice says that standardising bodies should use them, or the relevant parts of them, as a basis for standards they develop. It also aims at the harmonisation of standards on as wide a basis as possible, encouraging all standardising bodies to play as full a part as resources allow in the preparation of international standards by the relevant international body, including the ISO and IEC.

1.2. *Geographic information market and standards*

All businesses that produce, distribute, or utilise spatial information alone or in conjunction with non-spatial information will benefit from spatial standards. Environments supported include geographic information, decision support, data mining, data warehousing, modelling and simulation. Application areas include but are not limited to automated mapping, geo-engineering, computer-aided drafting and design, and entertainment. These span the planning, design, construction, operation, and maintenance of facilities and their supporting infrastructure such as communications, transportation, and utilities.

A common way to describe the market is by dividing it into segments: the traditional geographic information systems (GIS) market, business support systems (BSS), and personal productivity (PP). A description can be given as:

- GIS:
  - spatial information contributes the most value;
  - traditional market for spatial technology;
- BSS:
  - spatial information does not contribute the most value;
  - spatial technology embedded in business applications;
- PP:
  - users want to communicate with maps/ geographic information;
  - follows Office Suite market; a new emerging market is location-based mobile services (LBMS).

Many market places will benefit significantly from interoperable access to spatial information and services. Industry sectors include such areas as the travel and tourism industries, the

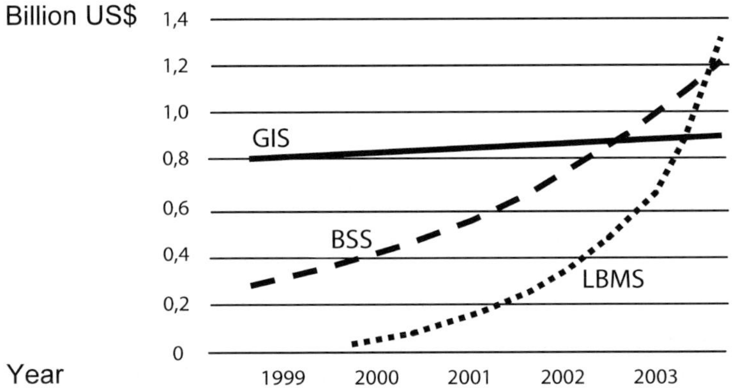

Fig. 1. Revenues in the traditional market for Geographical Information Systems (GIS), for business support systems (BSS) and the emerging technology, location based mobile services (LBMS) according to several sources.

mapping and routing industries, communications, utilities, transportation, national defence, agriculture, disaster management and public safety, location/mobile services, inventory management, real and synthetic environmental modelling and gaming, and the emerging needs of electronic commerce for spatial information.

Achieving more interoperability requires proactive co-ordination of spatial standards at both the abstract and implementation levels of detail. Proactive co-operation between spatial standards activities should also help to utilise, more efficiently, available resources by minimising technical overlap, where appropriate. Such co-ordination and co-operation should lead to more market relevant spatial standards and could serve as a useful road map for all interested parties.

1.3. *Scope of the ISO/TC 211*

The formal scope of ISO/TC 211 Geographic information/Geomatics (on-line see http://www.isotc211.org/tc2) is as follows:

Standardisation in the field of digital geographic information aims to establish a structured set of standards for information concerning objects or phenomena that are directly or indirectly associated with a location relative to the Earth.

These standards may specify, for geographic information, methods, tools and services for data management (including definition and description), acquiring, processing, analysing, accessing, presenting and transferring such data in digital/electronic form between different users, systems and locations. The work shall link to appropriate standards for information technology and data where possible, and provide a framework for the development of sector-specific applications using geographic data.

Beyond the needs within traditional applications of digital geographic information, there is a growing recognition among users of information technology that indexing by location is a fundamental way to organise and to use digital data. Increasingly, digital data from a wide variety of sources is being referenced to locations for use in a diversity of applications. Conse-

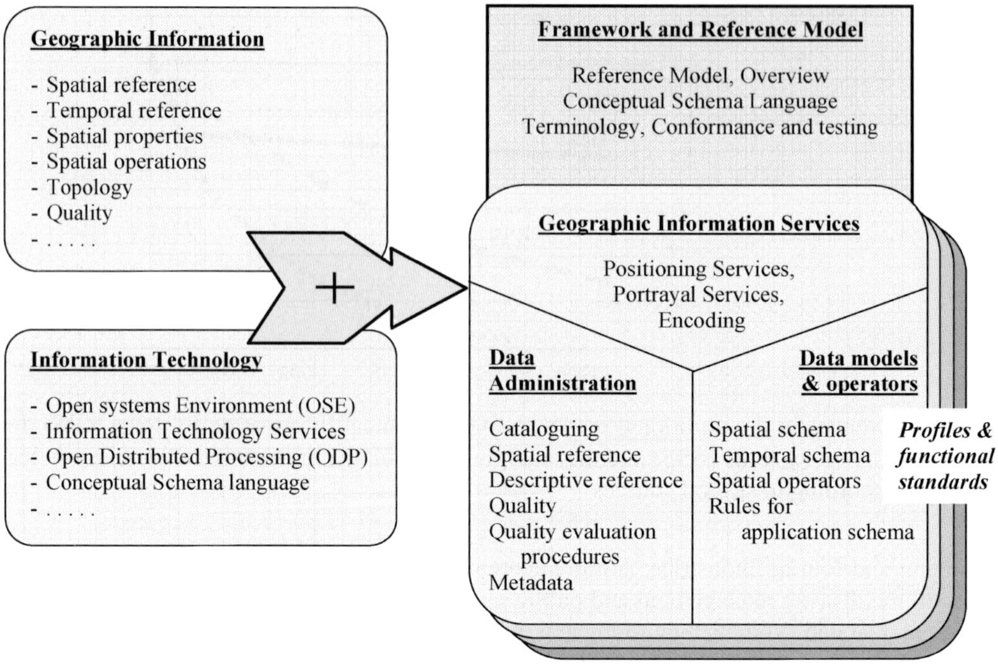

Fig. 2. 'Map' of the ISO/TC 211 approach to standardisation.

quently, there is an increasing need for standardisation of geographic information and services for processing this information. To meet this need, the ISO 191XX series (i.e., the standards developed by ISO/TC 211) standardises relevant aspects of the description and management of geographic information and geographic information services. This standardisation will:

- increase the understanding and usage of geographic information;
- increase the availability, access, integration, and sharing of geographic information;
- promote the efficient, effective, and economic use of digital geographic information and associated hardware and software systems;
- contribute to a unified approach to addressing global ecological and humanitarian problems.

The ultimate benefits of standardisation are based on the use of widely recognised and accepted international voluntary standards developed to the highest technical level by an open consensus process that includes all those affected. Beyond standardisation of traditional geographic functionality: innovative, new, and unknown technology and application domains present challenges transcending the established process of geographic standardisation. Previously, standardisation was a process for recognising and codifying the status quo of technology. Standardisation is now beginning to define the requirements and implementation of new technology.

The implied mandate for ISO/TC 211 is to develop an integrated set of standards for geographic information. Equally important, if not more so, is the non-stated strategic direction for the international deployment of such standards. Accordingly, the strategic directions for

ISO/TC 211 can be viewed in terms of development, deployment, and the underlying co-ordination/consensus process that integrates both these phases for successful standardisation.

For development, the major issues include: standards technical development, organisations developing geographic or related standards, priorities of standards, standards and interoperability testing, and speed of developing technical specifications. For deployment, the key issues are implementation of standards, standards education/training, and user communities supporting ISO/TC 211 standards.

Inherent and pervasive through standards development, deployment, and their co-ordination/consensus process are considerations for the implementers and users of geographic standards. Such as data transfer standards that are implemented by vendors or data cataloguing standards implemented by data producers, or metadata standards implemented by vendors, data producers, and general users of geographic information. Implementers and user requirements need to be considered in conjunction with the standards development and deployment and the process of integrating such requirements.

Traditionally, geographic information was produced and used by the geographic community. Increasingly, geographic information is being created and used by everyone else, especially, in the business community. Hence, the once all-important technical issues for experts are now being sub-ordinated to the business issues confronting government and commercial organisations. Previously, the cost of standardisation was minimal because of the number of users and requirements. Because geographic information has changed, in many countries, from being the essence of national mapping organisations to being the common commodity of consumers in the electronic/Internet/wireless communities – the diverse requirements, costs, and complexity for geographic standardisation has increased dramatically.

1.4. *The work programme of ISO/TC 211*

- ISO/TC 211 is currently accomplishing a very challenging task in developing – mostly in parallel – the set of standards as illustrated in the table below. The first standard was published in 2000, and almost a dozen standards and technical reports have been published by mid-2005.

A brief description of the various items:

ISO 19101 Reference model describes the overall requirements for standardisation and the fundamental principals that apply in developing and using standards for geographic information. Information system analysts, program planners, and developers of geographic information standards that are related to Geographic information standards intend ISO 19101 Geographic information – Reference model for use. In addition, others may use the ISO 19101 Reference model in order to understand the basic principals of the series of standards and the overall requirements for standardisation of geographic information.

ISO 19102 Overview provides a reference to the ISO 191XX suite of geographic information standards. It allows potential users to understand the purpose of each standard in the suite and how they relate to each other. It enables all users of geographic information to identify quickly which of the ISO 191XX standards are required for their purposes.

ISO 19103 Conceptual schema language. This technical specification (ISO/TS) is concerned with the adoption and use of a conceptual schema language (CSL) for developing computer-interpretable models, or schemas, of geographic information. Standardisation of

Table 1. The current ISO/TC 211 work programme

ISO 19101	Geographic information – Reference model
ISO 19102	Geographic information – Overview
ISO 19103	Geographic information – Conceptual schema language
ISO 19104	Geographic information – Terminology
ISO 19105	Geographic information – Conformance and testing *(published)*
ISO 19106	Geographic information – Profiles
ISO 19107	Geographic information – Spatial schema
ISO 19108	Geographic information – Temporal schema
ISO 19109	Geographic information – Rules for application schema
ISO 19110	Geographic information – Feature cataloguing methodology
ISO 19111	Geographic information – Spatial referencing by co-ordinates
ISO 19112	Geographic information – Spatial referencing by geographic identifiers
ISO 19113	Geographic information – Quality principles
ISO 19114	Geographic information – Quality evaluation procedures
ISO 19115	Geographic information – Metadata
ISO 19116	Geographic information – Positioning services
ISO 19117	Geographic information – Portrayal
ISO 19118	Geographic information – Encoding
ISO 19119	Geographic information – Services
ISO/TR 19120	Geographic information – Functional standards *(new revision started)*
ISO/TR 19121	Geographic information – Imagery and gridded data
ISO/TR 19122	Geographic information – Qualifications and certification of personnel
ISO 19123	Geographic information – Schema for coverage geometry and functions
ISO 19124	Geographic information – Imagery and gridded data components
ISO 19125-1	Geographic information – Simple feature access – Common architecture
ISO 19125-2	Geographic information – SFA Geographic information – SQL option
ISO 19126	Geographic information – Profile Geographic information – FACC Data Dictionary
ISO 19127	Geographic information – Geodetic codes and parameters
ISO 19128	Geographic information – Web map server interface
ISO 19129	Geographic information – Imagery, gridded and coverage data framework
ISO 19130	Geographic information – Sensor and data models for imagery and gridded data
ISO 19131	Geographic information – Data product specifications
ISO 19132	Geographic information – Location based services possible standards
ISO 19133	Geographic information – Location based services tracking and navigation
ISO 19134	Geographic information – Multi-modal location based services for routing and navigation
ISO 19135	Geographic information – Procedures for registration of geographical information items
ISO 19136	Geographic information – Geography Markup Language
ISO 19137	Geographic information – Generally used profiles of the spatial schema and of similar important other schemas
ISO 19138	Geographic information – Data Quality Measures
ISO 19139	Geographic information – Metadata – Implementation Specification
ISO 19140	Geographic information – Technical amandment to the ISO 191** Geographic information series of standards for harmonisation and enhancements

geographic information requires the use of a formal CSL to specify unambiguous schemas that can serve as a basis for data interchange and the definition of interoperable services. An important goal of ISO 191XX suite of standards is to create a framework in which data interchange and service interoperability can be realised across multiple implementation environments. The adoption and consistent use of a CSL to specify standardised geographic information is of fundamental importance in achieving this goal.

ISO 19104 Terminology. This document gives definitions for terms used within the 191XX family of standards. Terms have been harmonised within the 191XX family of standards. The format of the terminology conforms to ISO 10241 International terminology standards – Preparation and layout. A short guide for definition writing is included. The terminology will probably be available on the web.

ISO 19105 Conformance and testing provides the framework, concepts and methodology for testing and criteria to be achieved to claim conformance to this family of standards. It is based on material in ISO 9646-1, ISO 10303-31 and ISO 10641. In order to facilitate testing, all testable parts of the ISO 191XX suite contain a conformance clause, which explicitly specifies all the requirements that have to be satisfied to claim conformance to that part of the standard.

ISO 19106 Profiles provides the guidelines for preparation of a profile, product specification or implementation specification. Many of the standards in the 191XX series are abstract in nature. Each domain application implementation will require all the generic options to be resolved for the standards for which it complies. A profile documents the selection of options from a generic or abstract standard or the realisation of the rules or models in a rule-based standard. A profile provides the limited scope and functionality for effective specialist implementations of data and systems. The registration mechanism for profiles, product specifications and implementation specifications is defined to allow international, national and private profiles. Taxonomy of profiles is maintained so that a users or implementers can know which profiles exist and are applicable.

ISO 19107 Spatial schema provides a conceptual schema for describing aspects of the spatial characteristics of geographic features. Components of this schema may be specialised within an application schema to describe specific feature types. This part of the standard also describes a set of operators that, if correctly implemented on this schema, will ensure total consistency in the result.

A feature may have associated with it one or more spatial attributes. The value of a spatial attribute is a spatial object, either a geometric object or a topological object, that describes one or more characteristics such as location, size, shape, and spatial relationships to other spatial objects in the same 'world'. A spatial object can consist of a single geometric or topological primitive of 0, 1, 2 or 3 dimensions or of a set of these as prescribed in an application schema.

ISO 19108 Temporal schema defines standard concepts needed to describe the temporal characteristics of geographic information. These include both metadata elements that describe temporal characteristics of datasets and feature attributes that describe temporal characteristics of features. This standard identifies elements to be included in the definition of such characteristics whether in metadata, feature and attribute catalogues, or application schemas. It depends upon existing information technology standards for the interchange of temporal information. It therefore specifies the Gregorian calendar and Universal Co-ordinated Time as a preferred basis for interchanging temporal information. However, it also includes methods for describing calendars other than the Gregorian calendar when the use of such calendars is appropriate.

This standard is limited to describing temporal characteristics of features independently of space. There is increasing interest in describing the spatial behaviour of features as a function of time for applications such as simulations and predictive modelling. Although ISO 19108 only standardises the use of time in feature attributes, it has been written so as not to preclude the combination of the spatial and temporal dimensions in future ISO/TC 211 standards.

ISO 19109 Rules for application schema show how to develop schemas, which identify how the various parts of the 191XX suite of standards shall be applied for particular application domains. This family of standards is generic and must be tailored for specific application domains. The use of common rules for this process will allow for data and systems to be interchanged both within and across different application domains. The core of this process is the General Feature Model, which acts as a platform for the parts of this family of standards, particularly Metadata and Feature cataloguing. Specific feature catalogues will need to be developed for each application domain and the content of metadata will be specific to each application domain.

The use of a formal application schema for representing geographic information will enable the development of physical implementations and data interchange facilities to be at least procedural if not, in the short term, automated and specific this use of an application schema that will allow specific physical applications to inter-operate and share data.

ISO 19110 Feature cataloguing methodology. This part provides a standard framework for organising and reporting the classification of real world phenomena in a set of geographic data. Geographic features are the representations of real world phenomena associated with a location relative to the Earth, about which data are collected, maintained and disseminated. Feature catalogues defining the types of features represented in geographic data enable the dissemination, sharing and use of geographic data through a better understanding of the contents and meaning of the data. The feature catalogue for a given dataset is included or referenced in its metadata. Reference may be made to a general feature catalogue for an application area and then the metadata should identify the subset of feature types occurring in the particular dataset. The establishment of general feature catalogues for specific application domains will enable an understanding of data across domains and thus promote the sharing of data. Applications may use a number of general feature catalogues in order to allow sharing of data from different application domains.

In addition to identifying feature types, feature catalogues define feature types by identifying their associated feature operations, feature attributes, and feature relationships. Explanations of feature cataloguing concepts and an example feature catalogue are provided.

ISO 19111 Spatial referencing by co-ordinates. Any co-ordinate-based usage of geographic information needs a full definition of the underlying reference system. This International Standard defines the conceptual schema for the description of spatial referencing by co-ordinates. It establishes a common requirement for describing 1-, 2- and 3-dimensional co-ordinate reference systems (CRS's) including the datum giving the relation of the co-ordinate system to the Earth. It allows additional descriptive information to be provided. The standard stipulates that CRS's do not alter with time. Changes with time are catered for by specification of new CRS's that identify the epoch of their realisation. Compound co-ordinate reference systems (CCRS's) are included for situations when the components of position come from two different CRS's, one of the CRS is for vertical co-ordinates.

The requirements for transformation of co-ordinates with reference to different datums are given, as well as the requirements for conversion of co-ordinates between different co-ordinate systems including map projection.

It is applicable to producers and users of geographic information. Although it is applicable to digital geographic data, its principles can be extended to many other forms of geographic data such as maps, charts, and text documents.

ISO 19112 Spatial referencing by geographic identifiers. Spatial references may be provided using geographic identifiers, instead of co-ordinates, where an association is made to a known geographic feature. The association may be containment, such as within a country, based on a local co-ordinate, such as a given distance along a street, or loosely related, such as adjacent or near to a building.

This standard defines a consistent manner for spatial referencing by geographic identifiers. Spatial referencing systems using geographic identifiers comprise:

- one or more location types to be used for reference, such as street;
- reference scheme for uniquely identifying the location instances, such as street name; and
- one or more gazetteers recording all location instances that occur within a reference space, such as a street gazetteer.

ISO 19113 Quality principles. One of the main objectives of this family of standards is to enable geographic data to be shared and widely available in and across application domains. As this objective becomes realised it will increasingly be important to report the quality of data so that data users may use quality criteria in selecting a dataset best suited to their application requirements. ISO 19113 establishes the principles for describing the quality of geographic data and specifies components for reporting quality information. It also provides an approach to organising knowledge about data quality.

ISO 19113 identifies two components for describing data quality. Data quality overview elements provide informative non-quantitative information that may be subjectively evaluated and data quality elements and their data quality sub-elements provide quantitative quality information on how well a dataset meets the criteria set forth in its product specification. A non-exhaustive listing of data quality overview elements in ISO 19113 includes purpose, lineage and usage; user-defined data quality overview elements may be created to extend the listing. A non-exhaustive listing of data quality elements in ISO 19113 includes completeness, logical consistency, positional accuracy, temporal accuracy and thematic accuracy; user-defined data quality elements may be created to extend the listing. Each data quality element is comprised of several aspects called data quality sub-elements. Data quality information for each applicable data quality sub-element is reported using descriptors of a data quality sub-element, consisting of a data quality scope, data quality measure, data quality evaluation procedure, data quality result, data quality value domain, and data quality date.

The metadata schema of 19115 Metadata is the mandatory method for reporting both non-quantitative and quantitative data quality information. The 19114 Quality Evaluation Procedures' Quality Evaluation Report may be used to more extensively report quantitative quality information.

ISO 19114 Quality evaluation procedures. This part establishes a framework of quality evaluation procedures for a dataset of geo-spatial data so that data producers can define how well their products meet their product specification and users can define their requirements and how well they are met. The product specification or user requirements should allow the acceptable quality levels to be determined for each quality metric. An estimation of a datasets' quality is made by sampling, computer processing and/or indirectly by deduction for comparison with the acceptable quality level. This allows reporting on quality evaluation results.

ISO 19115 Metadata. The objective of this standard is to provide an international dictionary of terms and a structure for describing digital geographic data. This standard defines metadata elements and a schema, which, when properly implemented by a data producer, will enable users to locate, access, evaluate and employ geographic data. Users will be able to discover, select, and purchase geographic data; determine whether data in a holding will be of use to them, and be able to apply geographic data in the most efficient way. By establishing a common set of metadata terminology, definitions, and extension procedures, this standard will promote the proper use and effective retrieval of geographic data. Supplementary benefits of this standard for metadata are to facilitate the organisation and management of geographic data. This standard for the implementation and documentation of data furnishes data producers with appropriate information so they can properly characterise their geographic data. It facilitates dataset cataloguing enabling data discovery, retrieval and reuse.

This standard is described much more deeply in paragraph 2–4 in this chapter.

ISO 19116 Positioning services. This part defines a standard interface data structure for use between positioning devices and geographic information application systems. Modern electronic positioning technology is making available a wide range of positioning instruments and devices that can measure the co-ordinates of a location on or near the Earth dynamically with great speed and accuracy. A variety of geographic information system applications can make use of these facilities including surveying, navigation and intelligent transport systems. Clearly, the application of these devices will be able to be taken up more readily with a standard interface.

ISO 19117 Portrayal. This standard concerns portraying geographic information as an image understandable by humans, including the methodology for describing symbols. The portrayal standard will provide applications with a common interface to supported standard symbol sets. Thus, this standard does not include standardisation of cartographic symbols but provides a standard interface for such standard symbol sets.

The portrayal schema allows for portrayal rules and specifications to be included in the application Feature Catalogue, the metadata of a dataset and/or in individual geographic features included in the data. Otherwise, default generic portrayal specifications are included at type level (line, area) and not the instance level (road, lake). The use of alternative symbol libraries allows the datasets to be portrayed according to purpose.

ISO 19118 Encoding specifies the encoding rules that shall be used for data interchange purposes. The encoding rule allows geographic information defined in an application schema to be coded into a system independent data structure suitable for transport or storage. The encoding rule specifies the type of data to be encoded, the syntax and structure and coding schemes used in the resulting data structure. The encoding rule defined shall be used to implement encoding services.

The application schemas shall be defined using the standardised conceptual schema language specified in ISO 19103 – Conceptual schema language and shall be in conformance with the rules mandated in ISO 19109 – Rules for application schemas.

The encoding rule specified shall be compatible with the conceptual schema language chosen, i.e., the Universal Modelling Language (UML). Since there are no encoding rules associated with UML, this standard defines an encoding rule based on the Extensible Markup Language (XML). Even though several encoding rules exist as ISO standards, the emerging Extensible Markup Language seems best fit for defining one internationally adopted encoding rule for geographic information. XML is system and computing platform independent, it has a large market push and it is designed to be interoperable with the World Wide Web.

ISO 19119 Services provides identification and definition of the service interfaces used for geographic information and definition of the relationships to the Open System Environment model. The definition of service interfaces allows a variety of applications with different levels of functionality to access and use of geographic information. While specialised services will appropriately remain an area for proprietary products, the interfaces to those services will be standardised. Geographic information system and software developers will use these standards to provide general and specialised services that can be used for all geographic information. It is important that work in this area is integrated with the approaches being developed within the more general world of information technology.

ISO/TR 19120 Functional standard is an informative technical report that identifies and reviews a number of international geographic standards currently in use. These functional standards provide a valuable input to the standardisation process as they are in active use within the global user community. As such, they can provide real-world examples of how the ISO 191XX series of standards could be applied in different user communities.

The technical report describes a number of specific issues that the functional standards currently address. These examples are highlighted as test cases to exercise the ISO 191XX standards.

The report also considers how the separate functional standards could be harmonised with each other by using the ISO 191XX family of standards. The report explores the concept of granularity within the ISO 191XX standards, and considers how the development of standard modules could promote harmonisation between existing geographic standards.

ISO/TR 19121 Imagery and gridded data is an informative technical report that provides information to assist ISO/TC 211 for including the topic of image and gridded data within the suite of ISO 191XX standards. This technical report reviews the manner in which raster and gridded data is currently being handled in the Geomatics community in order to propose how the 191XX suite of standards should support this type of data.

ISO/TR 19122 Qualifications and certification of personnel. Geographic information science/geomatics is an evolving discipline involving the acquisition, transformation, management and distribution of geo-spatially referenced data. It is the result of a merger between several independent and older disciplines, including geodesy, photogrammetry, land management and cartography, and more recently, geographic information systems and remote sensing. Geographic Information Science/Geomatics embraces a broad community including surveying and geomatics engineers, geographers, planners, landscape architects, computer scientists and geo-scientists (adapted from GEOIDE 1999).

Given this complex definition, the establishment of standards for the qualification and certification of personnel is a difficult challenge. Several of the external liaisons to ISO/TC 211 (e.g., ICA, ISPRS, FIG) have a mandate to address these issues for their membership. Within ISO/TC 211 member countries, specific initiatives have been completed or are underway, e.g., the work of the AGI in Great Britain on Professional Development; the certification debate at URISA and the model curriculum effort by UCGIS are pertinent in the U.S.A.

ISO 19123 Coverage geometry and functions defines a conceptual schema for the spatial characteristics of coverages. Coverages support mapping from a spatial domain to attribute values where attribute types are common to all geographic positions within the spatial domain. A spatial domain consists of a collection of direct positions in a co-ordinate space; the collection may be defined in terms of geometric objects, such as curves or surfaces, which contain the direct positions. Examples of coverages include raster, triangulated irregular networks, point coverages, and polygon coverages. Coverages are the prevailing data structures in a number of application areas, such as remote sensing, meteorology, and bathymetric, elevation, soil, and vegetation mapping.

ISO 19124 – Imagery and gridded data components is a preparatory work that aims to standardise concepts for the description and representation of imagery and gridded data in the context of the ISO 191XX suite of standards. This includes new work on the following aspects of such data: Rules for application schemas, Quality principles and Quality evaluation procedures, Spatial reference systems, Visualisation, and Exploitation services. The work will also identify aspects of existing parts of the family of standards that need to be expanded to address imagery and gridded data. The work has resulted in proposals for new work described later in this chapter.

ISO 19125-1 – Simple feature access – Part 1: Common architecture is a two part standard; part 1 provides a common architecture.

ISO 19125-2 – Simple feature access – Part 2: SQL Option provides an implementation specification for the SQL environment that:

- specifies an SQL schema that supports storage, retrieval, query and update of simple geospatial feature collections;
- establish an architecture for the implementation of feature tables;
- define terms to use within the architecture;
- apply to both SQL Components and SQL with Geometry Types Components;
- describe a set of SQL Geometry Types together with the SQL functions on those types;
- not attempt to standardise any part of the mechanism by which the Geometry Types are added to and maintained in the SQL environment.

ISO 19126 – Profile – FACC Data Dictionary is a profile. It is based on rules and methods defined in ISO 19110 Geographic information – Feature cataloguing methodology, in the context of DGIWG. It defines a Data Dictionary and includes the definition of Features and Attributes only, which may be of use to the wider international community.

ISO/TC 211 has worked with DGIWG since its establishment, with respect to the development of the ISO/TC 211 suite of standards. Since the beginning, the intent of DGIWG has been to develop profiles of the ISO Geographic Information standards that correspond to the components of the DIGEST standard. The structure of DIGEST and the FACC Data Dictionary are described in the ISO/TR 19120 Geographic information – Functional standards.

ISO 19127 – Geodetic codes and parameters is a Technical Specification on geodetic codes and parameters that defines rules for the population of tables of geodetic codes and parameters and identifies the data elements required within these tables, in compliance with ISO 19111, Geographic information – Spatial referencing by co-ordinates. The Specification also makes recommendations for use of the tables. These recommendations should address the legal aspects, the applicability to historic data, the completeness of the tables, and a mechanism for maintenance.

ISO/TC 211 has recognised the need to develop unified lists of geodetic codes and parameters. There currently exist many lists of geodetic codes and parameters in national standards, the standards of liaison organisations, and industrial specifications and software products. None of these complies with the conceptual schema in ISO 19111. The development of a Technical Specification would provide the opportunity for the creation of a comprehensive list in compliance with ISO 19111. This list should be available electronically and allow for dynamic entry of additional geodetic codes and parameters.

ISO 19128 – Web map server interface. This standard describes a Web Map Server (or just Map Server). A Map Server can do three things. It can:

- produce a map (as a picture, as a series of graphical elements, or as a packaged set of geographic feature data);
- answer basic queries about the content of the map;
- tell other programs what maps it can produce and which of those can be queried further.

To first order, a standard web browser can ask a Map Server to do these things just by submitting requests in the form of Uniform Resource Locators (URLs). The content of such URL's depends on which of the three tasks is requested. All URL's include a Web Mapping Technology specification version number and a request type parameter and in addition:

- to produce a map, the URL parameters indicate which portion of the Earth is to be mapped, the co-ordinate system to be used, the type(s) of information to be shown, the desired output format, and perhaps the output size, rendering style, or other parameters;
- to query the content of the map, the URL parameters indicate what map is being queried and which location on the map is of interest.

1.5. *Co-operation and co-ordination in geomatics standards*

The increasing recognition for the value of spatial data and geographic information has spawned the entry of new players into the spatial standardisation arena, both from within ISO and externally. This has resulted in the formation of a co-operative agreement with the OpenGIS® Consortium (OGC) and a Joint Steering Group on Spatial Standardisation and Related Interoperability, chaired by the ISO/TC 211 Chairman. Consequently, a new agenda is emerging for international spatial standardisation that includes traditional and new innovative applications across a spectrum of disciplines. For ISO/TC 211, these developments are resulting in new strategic directions.

The co-operative agreement with the OpenGIS® Consortium was established in 1999. The OpenGIS® Consortium is a seven-year-old not-for-profit membership organisation dedicated to the development of interface specifications that support open access to geographic information and geo-spatial processes. Membership is open to all entities (a list of the current over 200 members may be found at http://www.opengis.org).

The OpenGIS® Consortium develops and provides, through a membership submission and consensus process, implementation-level technical specifications for interfaces to geo-spatial processes and geo-spatial information. The specifications are informed by emerging standards in the ISO/TC 211 work groups and work items. These interfaces deliver a corresponding level of interoperability in geo-spatial solutions.

The co-operation between OGC and ISO/TC 211 so far has led to four standards under development that are direct adoption of corresponding OGC specifications. These are ISO 19125 Part 1 – Part 3 and ISO 19128. The other way around, OGC has adopted ISO 19107 Spatial Schema as their abstract specification for geometry and topology. In addition, a large number of co-ordination and harmonisation efforts have resulted from the close collaboration. This can for instance be found within the service architecture, conceptual schema language, imagery and gridded data, and other work.

Also important are the discussions and influences at strategic level. Especially is there a fruitful co-operation in activities concerning emerging technologies like location-based services. These considerations led to the establishment of a Joint Steering Group on Spatial Standardisation and related Interoperability (on-line see http://www.jsgspatial.org/) in May 2000.

The purpose of the steering group is to share information in order to manage co-ordination activities better, and not to assign or manage specific spatial standards projects. The intent is to foster the rapid insertion of spatial technology into mainstream information technology while also working to optimise standards development resources.

As an advisory group, the forum will respect the different culture, practices and rules of the member organisations. Such a forum will serve to draw attention to potential loss of interoperability at an early stage and identify opportunities for synergistic co-operation.

1.6. *International standards and technical reports*

The following standards or technical reports have been published by ISO:

- ISO 19101: 2002 Geographic information – Reference model;
- ISO 19105: 2000 Geographic information – Conformance and testing;
- ISO 19106: 2003 Geographic information – Profiles;
- ISO 19107: 2003 Geographic information – Spatial Schema;
- ISO 19108: 2002 Geographic information – Temporal schema;
- ISO 19111: 2003 Geographic information – Spatial referencing by coordinates;
- ISO 19112: 2003 Geographic information – Spatial referencing by geographic identifiers;
- ISO 19113: 2002 Geographic information – Quality principles;
- ISO 19114: 2003 Geographic information – Quality evaluation procedures;
- ISO 19115: 2003 Geographic information – Metadata;
- ISO/TR 19120: 2001 Geographic information – Functional Standards;
- ISO/TR 19121: 2000 Geographic information – Imagery and gridded data;
- ISO 19125-1: 2004 Simple feature access – Part 1: Common architecture;
- ISO 19125-2: 2004 Simple feature access – Part 2: SQL Option.

1.7. *Draft international standards*

A draft international standard (DIS) is a draft standard published by ISO and referable externally as are Final Draft International Standards (FDIS), the last step before reaching International Standard (IS) status. Often standards at these stages of development start to be implemented. The following ISO/TC 211 documents are currently (as of September 2004) at these stages:

- ISO/DIS 19104 Geographic information – Terminology;
- ISO/DIS 19109 Geographic information – Rules for application schema;
- ISO/FDIS 19110 Geographic information – Feature cataloguing methodology;
- ISO/DIS 19116 Geographic information – Positioning services;
- ISO/DIS 19117 Geographic information – Portrayal;
- ISO/DIS 19118 Geographic information – Encoding;
- ISO/DIS 19119 Geographic information – Services;
- ISO/DIS 19123 Geographic information – Schema for coverage geometry and functions;
- ISO/DIS 19128 Geographic information – Web Map server interface.

Most other documents are at a late committee draft (CD) stage, soon to progress into DIS.

1.8. *Future work of ISO/TC 211*

New work is almost continuously proposed to ISO/TC 211. Examples of new work items are within imagery like Geographic information – Imagery, gridded and coverage data framework and Geographic information – Sensor and data model for imagery and gridded data. Other new work items are Data product specifications, Procedures for registration of geographical information items, and Geography Markup Language.

Perhaps the largest area of new growth in the nearby future will be the emerging technology called location based services. Location based services are services (through combination of hardware devices, communication networks, often wireless, and software applications) that access, provide or otherwise act upon location information. We distinguish between mobile position determination systems that determine the location of a mobile terminal and application-oriented location service, which exploit device location in some application service sought by a client. Examples include:

- Traffic Information, e.g., 'You are about to join a ten kilometres traffic queue, turn right on the A3 ahead';
- Emergency Services, e.g., 'Help, I'm having a heart attack!';
- Roadside Emergency, e.g., 'Help, my car has broken down!'.

A special advisory group on location-based services has been set up to define requirements for new work in this field.

2. ISO 19115, Metadata standard

As we move into the age of global spatial data infrastructures, knowledge about widely distributed and dissimilar geographic data is essential to universally allow users to locate, evaluate, extract, and employ the data. Varied and wide spread communities with a common understanding of metadata will be able to manage, share, and reuse each other's geographic data, making global interoperability a reality. The ISO Standard for Geographic Information – Metadata (ISO 19115) is designed to provide this common understanding.

Many geographic metadata standards have been in existence prior to the development of this ISO standard. In many cases these separate information community, regional, and national standards evolved in separate niches and are incompatible. Several general metadata standards that do provide minimal global interoperability are insufficient to support geographic information. This incompatibility and insufficiency was the motivation for the development of ISO 19115. The ISO metadata standard was designed to:

- support geographic information;
- work with wider information technology standards and practices;
- serve the global community, in a multi-national, multi-language environment;
  based on a foundation of national, regional, and special information community standards and experiences; a thorough requirements analysis, and implementation testing.

The ISO 19115 Metadata standard is being developed within ISO/TC 211 Working Group 3. It defines and standardises a comprehensive set of metadata elements and their characteristics, along with the schema necessary to fully, and extensively, document geographic data. The standard applies to all geographic data: it is applicable to datasets in series, datasets, individual geographic features, and their attributes. The standard defines the minimum set of metadata required to serve the wide range of metadata applications, as well as optional metadata elements to support a more extensive description of geographic data. Because of the diversity of geographic data, no single set of metadata elements will satisfy all requirements; for this reason the ISO metadata standard provides a standardised way for users to extend their metadata and still ensure interoperability allowing other users to comprehend and exploit this extended metadata.

2.1. *Development history*

2.1.1. *Requirements analysis*

Development of the ISO Metadata standard began in late 1995. Working draft 1.0 was released for review within the ISO/TC 211 Work Item 15 Project Team in March 1996. Prior to developing the working draft a requirements-analysis was performed to ensure the standard would support the full range of requirements for which metadata is needed and to ensure the selected metadata elements served a purpose.

The Project Team's requirements-analysis established metadata elements supporting four major applications for metadata:

1. *locate*. Metadata elements were selected which enable users to locate geographic information and allow producers to 'advertise' their data. These metadata elements help organisations locate external data and find partners to share in data collection and maintenance. Metadata elements were selected to help organisations identify data, facilitate better management, storage, retrieval, and reuse;
2. *evaluate*. Once having located a dataset, users require metadata to determine its 'fitness for an intended use'. Understanding the quality and accuracy, the spatial and temporal schema,

the content and geographic feature definition, and the spatial reference system used, users are able to determine if a dataset fills their needs;

3. *extract*. After locating and determining if a dataset will fill their needs, users require metadata elements describing how to access the dataset and transfer it to their site. Metadata elements were established to enable a person or organisation to know where a dataset is located, its size, format, media, price, and restrictions on use. Once it has been transferred, users need to understand how to interpret the data and incorporate it into their holdings;
4. *employ*. Metadata elements were included which facilitate appropriate use of data allowing users to properly merge and combine data with their own, apply it correctly, and have a full understanding of its properties and limitations.

In order to understand the requirements for metadata elements, the Project team established four primary circumstances in which the above applications would be used. These included:

1. within a catalogue, to enable discovery and locating geographic data. Metadata for cataloguing purposes should be in a form not unlike a library card catalogue or on-line catalogue. Catalogue Metadata supports searches by subject matter/area coverage/theme, author/producer, detail/resolution/scale, currency/date, data structure/form, and physical form/media;
2. within a database management system with historical records: metadata should support the documentation of data holdings to support storage, updates, production management, and maintenance of geo-spatial data. Historical records will provide legal documentation to protect an organisation if conflict arises over the use or misuse of geo-spatial data;
3. within a geo-spatial dataset to be used by application software operating on geo-spatial data. Metadata should accompany a dataset and be in a form to support the proper application of geo-spatial data. GIS and other application software using data will need to evaluate data as it applies it to a situation. In this form the metadata may be incorporated into the structure of the data itself;
4. as a text report in a human readable form to provide users an understand of their data. Metadata in a form in which a computer can locate and sort information, or manage the warehousing, production, and use by application software will greatly enhance the use of geo-spatial data but eventually a human must understand the data. One persons' or organisations' geo-spatial data is a subjective abstract view of the real world, it must be understood by others to ensure the data is used correctly. Metadata needs to be in a form that can be readily and thoroughly understood by users. Perhaps metadata in one or two forms will support all of these areas.

Defining metadata elements to fulfil the above requirements did not start from scratch. Geographic information has been in use for many years and experience has been growing in this field. Many national, regional, and special use groups have developed standards and methods for transferring and handling this type of data. The experience gained in the development and use of these standards was invaluable in the development of the ISO metadata standard. The initial ISO metadata standard was based on:

- *ANZLIC Working Group on Metadata: Core Metadata Elements*, Australia and New Zealand Land Information Council, November 1995, Sydney;

- Canadian Directory Information Describing Digital Geo-referenced Data Sets, Canadian General Standards Board, July 1994, Ottawa;
- *Standard for Geographic Information – Metadata,* European Committee for Standardisation (CEN), September 1996, Brussels;
- *US Federal Geographic Data Committee (FGDC)* Content Standard for Geo-spatial Metadata, Federal Geographic Data Committee, June 8, 1994, Washington DC.

Many transfer/exchange standards also carry metadata. These transfer standards were also examined and provided input to the ISO standard. Some of the transfer standards that provided metadata elements were:

- *Digital Geographic Information Exchange Standard (DIGEST)*, Digital Geographic Information Working Group, January 1994;
- *International Hydrographic Organisation Special Publication 57*, International Hydrographic Bureau, October 1995, Monaco;
- *Spatial Data Transfer Standard (SDTS)*, National Institute for Standards and Technology (NIST), US Department of Commerce, August 1992, Gaithersburg, MD.

Experienced users of these standards also added input as to what worked, what did not work and what was important. Project Experts representing countries from around the world reviewed early drafts of the metadata standard. These individuals represented the producer and user communities; so, the metadata requirements and experiences of both of these communities were incorporated into the developing standard.

As each metadata element was developed, and/or derived from the sources above, and incorporated into the standard, it was examined as to its worth in fulfilling the requirements outlined in a metadata usage reference matrix (see Table 2). If the ISO metadata standard is to be successful, it must fulfil the requirements of these transfer and metadata standards and provide a common vehicle providing interoperability across these diversified standards and uses.

2.1.2. *Formal development process*

The ISO Metadata standard was developed under strict ISO rules using an iterative process. ISO/TC 211 national experts and Class A Liaison organisations performed the development. During the Working Draft stage, the ISO/TC 211 Project Team 15 established a consensus on the design of the document, the project Team Leader/Editor then updated the document and created a new version, which was again reviewed by the Project Team. As the document reached Committee Draft status, the reviews became more formal. Each Committee Draft was officially reviewed by the entire ISO/TC 211 membership. Reviewers were allowed at least 3

Table 2. Metadata usage reference matrix

	Catalog	**Within dataset**	**Historical record**	**Textual report**
Locate	X		X	X
Evaluate	X	X	X	X
Extract	X	X		
Employ		X		X

months to examine the document in detail. Comments, complete with a recommended change, were formally submitted to the ISO/TC 211 Secretariat and officially published. After each review/comment period, an Editing Committee was formed to officially respond to the comments. The Editing Committees consisted of 16–18 national (limited to one representative from each nation) and 3–4 Class A Liaison (limited to one per organisation) members. The committees met for a one to two weeks' period in order to adequately address the comments. Each comment was fully addressed – either the requested change was incorporated, or a detailed explanation was provided as to how the comment was resolved. A detailed schedule of the ISO metadata development cycles is provided in Table 3.

A large number of comments were received on each Committee Draft; this shows the avid and broad interest in metadata and the desire for an international standard. The comments ranged from editorial comments, to comments on new technical approaches to metadata or to comments on the inconsistency of the metadata standard with the other ISO/TC 211 standards, which were in various different stages of development. Comments also drove the change from a metadata schema based on an entity-relationship model to a schema based on a UML object model. The design of the standard evolved through the versions of the Committee Drafts. The standard evolved from providing 'levels of conformance'; to using an abstract schema based comprehensive and essential profiles and to the final design that includes an abstract schema and a single scaleable comprehensive profile. Of the 33 ISO/TC 211 member nations the following were most involved in the metadata standard development: Australia, Canada, China, Finland, France, Germany, Japan, Korea, Malaysia, Norway, South Africa, Spain, Sweden, United Kingdom, and the U.S.A. The Digital Geographic Information Working Group,

Table 3. ISO 19115 metadata standard development cycles

Version name	Date
Working drafts	
Version 1.0	March 1996
Version 1.1	October 1996
Version 1.2	December 1996
Version 2.0	January 1997
Version 2.1	June 1997
Version 3.0	September 1997
Version 3.1	December 1997
Version 4.0	February 1998
Committee drafts	
Version 1.0	July 1998 (962 comments)
Version 2.0	November 1999 (1422 comments)
Version 3.0	June 2000 (655 comments)
Draft International Standard	
Version 1.0	September 2001
Final Draft International Standard	
Version 1.0	September 2002
International Standard	
Version 1.0	May 2003

the International Cartographic Association, the International Hydrographic Bureau, and the OpenGIS® Consortium actively participated as Class A liaisons.

3. Description of ISO 19115

The ISO Metadata Standard uses an abstract object model and a data dictionary to provide a complete schema and definition for metadata. The object model is based on the Object Management Group's (OMG) *Unified Modelling Language (UML)*. The data dictionary is based on ISO 11179 (all parts) *Information Technology – Specification and standardisation of data elements*. As with the development process described above, the metadata standard document follows the conventions prescribed by the International Organisation for Standardisation, Geneva, Switzerland.

3.1. *Scope*

The official Scope of the document (Clause 1) was established by ISO/TC 211 members through a ballot procedure at the very beginning of the development process. This Scope has served as the objective and goal for the development project team.

3.2. *Conformance*

Clause 2 contains the Conformance clause, which (along with an Abstract test suite in the ISO Metadata standards' Annex D) provides the requirements that any product, claiming conformance to the standard, must pass.

3.3. *Normative references*

Normative references (Clause 3) lists documents that contain provisions, which are conditions of the ISO metadata standard, through reference in the text of the metadata standard.

3.4. *Terms and definitions*

The standard defines terms along with their definitions (Clause 4), which allow the user to understand the metadata and are applicable when employing the standard.

3.5. *Symbols and abbreviated terms*

Clause 5, Symbols and abbreviated terms, provides definitions for the abbreviations used in the document, describes the UML notation and modelling techniques, and provides a table of the relationships between the UML and ISO 11179 terminology.

3.6. *Requirements*

The fundamental core of the standard is the Requirements clause (Clause 6). Three primary parts of the Requirements clause are:

- Metadata Application information, that defines the classes of information to which metadata applies;
- Metadata packages, that defines the geographic metadata;
- Core metadata for geographic datasets, a list of the minimum essential elements required describing a dataset.

3.6.1. *Metadata Application*
The ISO metadata standard specifies that metadata must be provided for every geographic dataset, and may be provided for aggregations of datasets (datasets in series or a collection of datasets belonging to a common activity or collection), individual geographic features, feature types, feature attributes, and feature attribute types.

3.6.2. *Metadata packages*
The standard uses the concept of UML packages to combine collections of like metadata classes and show relationships between these collections. Each package contains one or more classes of metadata, which may be specialised (sub-classed) or generalised (super-classed). Each class of metadata contains attributes, which identify the discrete units of metadata. The standard defines the following packages:

- metadata entity set information (which is a mandatory package that is instantiated by the remaining packages);
- identification information;
- constraint information;
- data quality information;
- maintenance information;
- spatial representation information;
- reference system information;
- content information;
- portrayal catalogue information;
- distribution information;
- metadata extension information;
- application schema information;
- extent information;
- citation information;
- responsible party information.

3.6.3. *Core metadata for geographic datasets*
The metadata standard defines an extensive list of metadata elements; a user may select a subset to suite their purposes. However, if certain key metadata elements are not included in the subset, the metadata may not be of any value. For this reason the ISO 19115 Metadata standard provides a list of 'core' metadata, which is recommended as a minimum for geographic datasets. These are listed below (M = mandatory, C = mandatory under certain conditions, and O = optional). The mandatory and conditional metadata must be included in a metadataset, the additional optional are highly recommended to provide users with a very basic understanding of a dataset. This core set of metadata elements must be included in any community profiles that conform to the ISO standard:

- dataset title M;
- dataset reference date M;
- dataset responsible party O;
- geographic location of the dataset M;
- dataset language M;
- dataset character set C;
- dataset topic category M;
- scale of the dataset O;
- abstract describing the dataset M;
- dataset format name O;
- dataset format version O;
- additional extent information O;
- spatial representation type O;
- reference system O;
- lineage statement O;
- on-line resource O;
- metadata file identifier O;
- metadata standard name O;
- metadata standard version O;
- metadata language C;
- metadata character set C;
- metadata point of contact M;
- metadata date stamp M.

3.7. *UML diagrams*

The ISO 19115 Metadata standard provides the complete abstract UML model (in Annex A) for metadata. This model is an integral part of the entire abstract model for geographic information developed by ISO/TC 211. The metadata model diagrams are presented in 16 views of metadata corresponding to the metadata packages defined above. The model diagrams provide a good illustration of the classes and their attributes, and the type and cardinality of the relationships between classes for geographic metadata.

3.8. *Data dictionary for geographic metadata*

The UML abstract model provides a well-defined graphic description of metadata, however it does not provide formal textual definitions for each class, attribute, and relationship: the data dictionary (Annex B of ISO 19115) defines this. The UML model and the data dictionary must be used together to define completely geographic metadata. The data dictionary is closely linked to the UML model. It defines a name, short name, definition, obligation, condition, maximum occurrence, data type, and domain for each item in the model. The precise definition for each metadata item translated into the many languages of users throughout the world will provide global interoperability, the goal of an ISO standard.

3.9. *Metadata extensions and profiles*

The ISO metadata standard defines almost 300 metadata elements; most of these are listed as 'optional'. They are explicitly defined in order to allow users with diverse backgrounds understand exactly what is being described. Individual communities, nations, or organisations will develop 'community profiles' of the ISO standard. They will establish a subset using the 'core' and selected optional metadata elements from the ISO standard for their standardised community profile. As an example a user community that believed the price of a dataset was an important part of their business model would make certain the metadata element 'standardOrderProcess.fees' was included in their profile.

A community of users may want to establish additional metadata elements that are not in the ISO standard. For example, a community may want to develop metadata elements for the status of datasets within their own system to help manage production. However, these added elements would not be known outside the community unless they are published. A community profile must also establish field sizes and domains for all metadata elements. If one system within a community uses 32 characters for the title of a dataset and another system handles 8 characters interoperability will not be achieved. Standardising selected domains within a community is important allowing searches that are more efficient and better system control.

3.10. *Comprehensive dataset metadata application profile*

The abstract model provided in the standard must be implemented through profiles. The standard provides a comprehensive basic profile that will be adequate for a wide variety of information communities. The use of this profile by a wide variety of users will provide interoperability across information communities. It enables immediate use of the standard and provides a guide to other communities in the development of their own profiles.

3.11. *Informative annexes*

The standard also provides numerous informative annexes that provide examples and guides to help implement the standard. Informative annexes do not provide standard provisions, which stipulate conformance. Informative annexes include:

- *metadata extension methodology* which provides advice on stages to follow when creating additional metadata elements;
- *metadata implementation* which provides guidelines for the management of metadata;
- *implementation examples* are provided to help increase an understanding of the standard;
- *multilingual support for free text metadata element* provides a method for identifying the language of a free text metadata element when the language for that element is different then the language specified for the dataset, or when providing a metadata element in multiple languages.

4. Future development

4.1. *Metadata for services*

The scope of the metadata standard ISO 19115 specifies that it defines metadata for services as well as data. A service is a capability, which a service provider entity makes available to a service user entity through a set of interfaces that define behaviour, such as a use case. A service is software that enables users to access and process geographic data from a variety of sources across a generic computing interface within an open information technology foundation. Accessing and processing geographic data means to geographic data users that they can query remote databases and control remote processing resources. They can take advantage of other distributed computing technologies such as software delivered to the user's local environment from a remote environment for temporary use. The metadata standard provides the very basic metadata that allows a service to be identified. ISO/TC 211 Project Team 19, dealing with *ISO 19119 – Services* is developing more detailed service metadata. This Services standard presents a list of identified and defined geo-spatial services. It identifies and defines the service interfaces used for geographic information and defines their relationship to the Open Systems Environment model. The Services standard defines an approach to service discovery by defining a metadata model for service instances. Metadata records can then be managed and searched using a catalogue service as is done for dataset metadata. The metadata elements for a service provide sufficient information to allow a client to invoke the service based on the metadata records. The ISO 19119 Services standard is at the Draft International Standard stage of development and is expected to become an International Standard in approximately the same period as the Metadata Standard.

4.2. *Metadata for imagery and gridded data*

ISO 19115 currently does not provide detailed metadata about imagery and gridded data types such as metadata about the collection systems used for imagery. Examples are platform identification and orientation and sensor orientation and geometric and radiometric calibration information. ISO/TC 211 established an Imagery and Gridded Data Project Team to address this type of data. It has produced a report (ISO/TC 211 N1017, Dec 2000), detailing issues concerning imagery and gridded data and recommending approaches to address these issues. One recommendation from the report is to develop ISO 19115 – Metadata Part 2, which will contain the more detailed imagery and gridded metadata. Because of this report, several new work items have been initiated within ISO/TC 211. These include Project number 19129 – *Geographic information – Imagery, gridded and coverage data framework* and Project number 19130 – *Geographic information – Sensor and data model for imagery and gridded data.* Findings from these new work items will provide input to ISO 19115 – Metadata Part 2.

5. Concluding remarks

The goal of international standardisation in the field of geographic information is to develop a family of standards that will:

- support the understanding and usage of geographic information;
- increase the availability, access, integration, and sharing of geographic information, enable inter-operability of geo-spatially enabled computer systems;
- ease the establishment of geo-spatial infrastructures on local, regional and global level;
- and thus contribute to a unified approach to addressing global ecological and humanitarian problems and geo-spatial data infrastructures.

ISO/TC 211 is about to finalise its first massive, parallel effort in establishing the fundamental standards in the field. The development has attracted the resources of more than 500 individuals on a global basis. A large number of organisations have followed, influenced, and participated in the work. ISO/TC 211 has acted as a huge meeting point for expertise from users, academia, producers and vendors in the field of geographic information and geomatics.

The work will continue in the traditional and fundamental field, but will also be extended to new application areas and new technologies. Geomatics is still a young application field!

References

[ISO/TC 211-N1017, 2000]
Draft review summary from stage 0 of project 19124, Geographic information – Imagery and gridded data components
ISO/TC 211 N1017, 2000-12-01.

[SDI Cookbook 1999]
The SDI Cookbook, version 1.0
GSDI. Contact: www.gsdi.org.

[ISO 2002]
International Organisation for Standardisation. Contact: www.iso.ch;
ISO/TC 211 Geographic information/Geomatics. Contact: http://www.isotc 211.org/.

[JSGSSRI 2000]
Joint Steering Group on Spatial Standardisation and Related Interoperability
Contact: http://www.jsgspatial.org/.

Organisation	For on-line contact see:
GSDI	www.gsdi.org
ISO	http://www.iso.ch
Joint Steering Group on Spatial Standardisation and related Interoperability	http://www.jsgspatial.org
OGC, OpenGIS® Consortium	http://www.opengis.org

URL's in this chapter are accessed in May, 2005, unless otherwise stated in the text.

PART 3

SCIENTIFIC AND TECHNICAL CHARACTERISTICS FOR ASSESSING METADATA STANDARDS FOR GEOGRAPHIC DATASETS

Edited by

Harold Moellering and Aaron Crane

H. Moellering, H.J.G.L. Aalders & A. Crane (Editors)

Introduction to the Scientific and Technical Characteristics for Assessing Metadata Standards for Geographic Datasets

Harold Moellering [a], Aaron Crane [b]

[a]*Department of Geography, Ohio State University, Columbus, Ohio, United States of America*
e-mail: geohal+@osu.edu
[b]*Navteq, Chicago, Illinois, United States of America*
e-mail: aaron.crane@navteq.com

1. Introduction

Background

This chapter contains a set of scientific and technical characteristics for the independent scientific assessment of metadata standards for spatial and geographic datasets. The fundamental definition of spatial metadata used by the ICA Standards Commission for this work is "data that allow for the description of content, spatial extent, quality (accuracy, complexity, consistency, and currentness) and other characteristics of spatial data" (ICA, 1996, 1997, 1998, 1999, 2000, 2001, 2002). As such this is the kind of information which is provided with geographical information, and which allows for a better understanding of geographical data (geographical information is defined as any information which has an earth location of any form). Spatial metadata information also encompasses transfer format templates, counts the number of data items available in the transfer, conceptual data model and catalogs, positioning references of the geographical data, and a quality report among other things. It also can include the logical description of a spatial data transfer metafile (after ICA, 1994).

The spatial metadata standard Characteristics Questionnaire template presented in this section is based on discussions held by members of the ICA Commission on Spatial Data Standards initially held in Costa Brava, Spain, 1995, and by the Commission meeting the next summer in Den Haag, Netherlands, 1996, work had progressed to Version 4.0 on the scientific and technical Characteristics Questionnaire. Plans were made for prototype field testing of the Questionnaire on two of the mature metadata standards. By the next year at the Commission meetings at Uppsala, Sweden, 1997, the scientific and technical Characteristics Questionnaire was well developed and fairly mature, and reports were presented on empirical field tests of the Characteristics Questionnaire on one of the National standards: the North American USA

FGDC metadata standard reviewed by David Danko, and on one Regional standard: the European CEN TC287 metadata standard reviewed by Paula Ahonen of Finland. The results of those two field tests revealed that the Characteristics Questionnaire document was sound, but that some rewording, reordering of the questions, and polishing were in order to make the document smoother and more widely understandable by nonnative English speakers.

Plans were then made to polish the Characteristics Questionnaire document, and to further refine the glossary of terms being used in it. Upon completion of each draft of the Characteristics Questionnaire, the document was circulated to the Commission members for review and comments. The comments received were then discussed and incorporated into the document at the next meeting of the Commission.

During the following year this work was carried out along with the early assessment of several more National and International standards. This work was reported on in the 1998 Commission meeting at Aix-en-Provence, France. Additional discussion focused on refining the format of the Characteristics Questionnaire, and the concept of Obligation of certain metadata items was added to the questionnaire. The levels of obligation are consistent with most of the metadata standards in the world and are usually: Mandatory, Conditional, and Optional. At this Commission meeting, all preliminary testing of the Characteristics Questionnaire had been completed, and Version 6.6 became the final version of the Questionnaire for the formal assessment of the twenty National and International spatial metadata standards and profiles to be examined.

At the conclusion of the 1998 Commission meeting, copies of the Characteristics Questionnaire were sent to the Commission member representatives of each of the known metadata standards organizations in the world. The Questionnaires were accompanied by a glossary and set of instructions for completing the Characteristics Questionnaire. At the August 1999 Commission meeting in Ottawa, Canada the responses were reviewed by the Commission for consistency and completeness. Feedback was then provided to each of the Part Four chapter authors. The Commission was very aware that their respective standards groups were then just completing their national metadata standards, and hence those remaining few would be assessed as soon as they were finalized. This follow up process continued until the list for candidate National and International metadata standards, and a few example subject matter profiles, was formally closed a couple of years later.

Objectives of the Spatial Metadata Standards Assessment Project

These scientific and technical characteristics are presented in the form of a Characteristics Questionnaire in template form, designed to obtain information in a systematic and standardized format regarding critical characteristics of a particular National or International standard, or example subject matter profile of one of them. The intention here is to publish both the final assessment characteristics, and responses to these characteristics questions for all current National and International standards in terms of the scientific and technical characteristics presented here independent of the native language(s) of that standard or profile. All of this material has been organized, collated, and edited into the ICA publication presented here.

The objectives for this process were initially adopted at the 1995 Costa Brava Commission meeting. These Commission objectives are the same as for the work of the Commission on spatial database transfer standards (Moellering, H. and R. Hogan (eds), 1997, p. 63), and are intended to:

- support education regarding spatial metadata standards;
- standardize the terminology and structure used for describing spatial metadata standards;
- facilitate the description and interpretation of spatial metadata standards for users;
- provide characteristics independent of the native language(s) of that standard or profile;
- facilitate the comparison of alternative available standards;
- provide for 'value-free' comparisons of selected standards;
- enable users to apply their own value judgements to the comparisons;
- enable short listing of potential standards by users for particular applications; and
- assist standards implementers to identify international commonalities between the various National and International standards.

2. Terminology for Scientific and Technical Spatial Metadata Standards Characteristics

In order that this Characteristics Questionnaire document can be understood in a very broad international setting, it is important to clearly explain the terms that are used within the Characteristics Questionnaire itself. This glossary was developed in conjunction with the Characteristic Questionnaire and resulting template document. This glossary was developed and extensively reviewed by the Commission members over a series of regular Summer Commission meetings, and by further email communication. The resulting Glossary was distributed with the Characteristics Questionnaire to each of the Part 4 chapter authors for their assessment of their own National or International metadata standard, for which they serve as a member on that specific metadata standard development Committee. The existence of such a glossary does not guarantee consistency in the way each Part 4 chapter author answers each query in the Characteristics Questionnaire, because some terminology is used differently by different organizations. In addition, interpretation of terms within the context of other languages and different National and International metadata standards adds complexity to the task of ensuring a consistent understanding of the resulting responses in each Part 4 chapter. The glossary was very helpful in the assessment of each spatial metadata standard evaluated, but more importantly it was the carefully organized peer review by members of the Commission of all Part 4 chapters submitted in the regular Summer Commission meetings that ensures the consistency of the assessments presented by the author of each chapter.

The Glossary of Terms and Definitions, V. 7.0 – Final Version

In order that the resulting spatial metadata standards Characteristics Questionnaire can be understood more uniformly and completely, a glossary of Terms and their definitions is provided here. This is designed to facilitate the effective reading and understanding of the queries in the Characteristics Questionnaire itself, as well as the resulting Part 4 chapters on the National and International standards, and the example subject matter profiles. The terms and definitions listed here are directly germane to the Characteristics Questionnaire itself, and are not intended to serve as a general reference for spatial metadata concepts, and associated terms and definitions. During the development of the Characteristics Questionnaire, members of the ICA Standards Commission carefully selected and considered the terms and definitions presented here. It should be pointed out that even such a systematic effort to provide and utilize consistent terms and definitions, every country has its own national language(s) which are used in their spatial metadata standard. This situation is further compounded where in a few cases some terms are officially defined differently in various National or International standards. Perhaps the clearest example of such a variation of defined usage is for the terms Feature, Entity, and Object, where one, or perhaps more of these terms can relate to things that exist in the physical human reality in which all of us live, and other of these terms can relate to the digital representation(s) of such things in a spatial database. Such conflicts between officially defined terminology in specific official standards cannot be resolved here, but must be recognized when reading these following chapters. The ISO/TC211 Committee on Geomatics/Geographic Information has a Working Group on Terminology striving to come up with a set of world spatial terms and definitions that are universally applicable (TC211, 2005). Meanwhile, the work presented here recognizes these terminology differences, and has diligently strived to achieve systematic and uniform terms and definitions in this spatial metadata Characteristics Questionnaire. The sources for terms and definitions used in this ICA work and its resulting book are as follows:

(TC211) These definitions come from ISO/TC211 Committee on Geomatics/Geographic Information Terms Repository found at: http://www.isotc211.org/. The reference document is ISO/TC211 document N1485, 2003, and ISO/TC211 document N1758, 2005. The reader should note that this terminology glossary is being updated every three months.

(ICA/TR) Taken from the ICA Standards Commission Technical Report, Moellering, H., and C. Clement (eds), Technical Characteristics for Assessing Standards for the Transfer of Spatial Data and Brief International Descriptions, Columbus: ICA Standards Commission, 1994, 104 p.

(ICA) Terms defined at the Commission meetings in: Den Haag, Netherlands, 1996, Uppsala, Sweden, 1997, Aix-en-Provence, France, 1998, Ottawa, Canada, 1999, Monaco, 2000, Beijing, China, 2001, and Brno, Czech Republic, 2002. Many of these definitions are based on the Commission book on spatial data transfer standards: Moellering, H. and R. Hogan (eds), Spatial Database Transfer Standards 2: Characteristics for Assessing Standards and Full Descriptions of the National and International Standards in the World, Elsevier Science, 1997, 373 pp. + Crosstable, ISBN 0080424333.

(ESDQ Pxx) Taken from the ICA book: Guptill, S., and J. Morrison (eds), Elements of Spatial Data Quality, Elsevier Science, 1995, 250 pp., ISBN 0080424325. Pxx represents the page number for the reference definition.

(IHO) Terms provided by the International Hydrographic Organization.

In cases where appropriate, it has been strongly felt by the members of the ICA Standards Commission that use of the terms and definitions developed by the ISO/TC211 Project 19104 Terminology Group (TC211, 2003, 2005) should take preference in use in this Characteristics Questionnaire. This has been an evolutionary process because Commission work on this spatial metadata Characteristics Questionnaire began in 1995, shortly after the ISO/TC211 Committee was founded. In all cases the sharpening of the terms and definitions by the TC211 Terminology Group has served to make the Commission work more effective and in harmony with broader spatial standards efforts. In cases where a TC211 definition does not exist for an important term used herein, the definition for that term has come from four additional sources: the Commission Technical Report on Spatial Transfer Standards (Moellering and Clement (eds), 1994), the Commission itself, the book edited by Guptill and Morrison (1995) on spatial data quality, and a term defined by the International Hydrographic Organization.

The Terms and Definitions Used in this Characteristics Questionnaire

Accuracy – closeness of a measured value set or values to the accepted reference value(s). (**TC211**)
Attribute – a characteristic of a feature, object, entity, or phenomenon. (**ICA**)
Completeness – the presence and absence of feature, object, entity and phenomenon instances; relationship instances; and attribute instances. (**ICA**, after CEN 12656)
Conformance – fulfillment of specified requirements. (**TC211**)
Currentness – "up-to-datedness" of a dataset. (**ESDQ** p. 149)
Data – reinterpretable representation of information in a formalized manner suitable for communication, interpretation, or processing. (**TC211**)
Data class – group of feature(s)/object(s), entity(ies), phenomenon(a) with common characteristic(s) or relationship(s). (**ICA**)
NOTE: Some standards refer to "data type" instead of "data class".
Data dictionary – an information resource that lists and defines all relevant data elements. (**ICA**)
Data element – discrete unit of data; it may contain class, and/or characteristics, and/or relationships. (**ICA**)
Data type – kind of value that can be assigned to a data element. (**ICA**)
NOTE: Examples include: "integer", "real", "string", "date", etc.
Database schema – rules and procedures to represent the structure of a database. (**ICA/TR**)
Dataset – identifiable collection of related data. (**ICA**)
DEM – Digital Elevation Model.
Document – a written item, as a book, article or letter of factual or informative nature. (**TC211**)
Domain – a defined set of values for an attribute. (**ICA/TR**)
DTM – Digital Terrain Model.
NOTE: Elevation model that contains topological surface structure lines, and sometimes the Warntz Network.
Encoding – process of representing information content and structure in a format suitable for a computer. (**ICA**)
Entity – 1. Representation of a collection of data elements in a conceptual schema. 2. Class of features/objects with common properties. (**ICA**)
FACC – Feature Attribute Coding Catalog.
NOTE: After DGIWG feature and attribute catalog. (**ICA**)
Feature/object – an entity or phenomenon as it exists or appears in the real world represented in digital form, including semantics, metrics, and symbology. (Please note that due to various international uses of the terms, Feature, Entity and Object, this definition cannot be made unambiguously clear.) (**ICA/TR**)

NOTE: The following is also used: **Feature** – abstraction of a real world phenomenon. (**TC211**); **Object** – entity with well-defined boundary and identity that encapsulate state and behavior.

Geodetic datum – a set of parameters specifying the reference surface or the reference coordinate system used for geodetic control. (after **IHO**)

Geoid – level surface which best fits mean sea level either locally or globally. (**TC211**)

Geometric primitive – geometric object representing a single, connected, homogeneous element of space. (**TC211**)

Grid – network composed of two or more sets of curves in which the member of each set intersect the members of the other sets in a systematic way that divides a space. (**TC211**)

Horizontal datum – geodetic datum for horizontal positioning (longitude and latitude on the ellipsoid). (**ICA**)

Image – a permanent record of the likeness of any natural or manmade features, objects and activities. (**ICA**)

ISBD – International Standard Bibliographic Description.

ISBN – International Standard Book Number.

ISSN – International Standard Serials Number.

Lineage – description of the process and basic information used to obtain the data contained in the transfer.

NOTE: Lineage includes information about source material, source scale and accuracy when appropriate, methods of quality control, dates, methods of derivation, etc. (**ICA/TR**)

Logical consistency – description of the level of compliance to the relations defined in the data model.

NOTE: Logical consistency may be expressed in numerals or by providing a description of the quality control process, tests performed, etc. (**ICA/TR**)

Metadata – data about data. (**ICA, TC211**)

NOTE: The following are also used:

Metadata – data that allow the description of content, spatial extent, quality (accuracy, complexity, consistency and currentness) and other characteristics of spatial data. (**ICA**)

Metadata – data about a geographic dataset or geographic datasets. (after **CEN 287**)

Metadata element – discrete unit of metadata (**ICA, TC211**)

Normative – required in order to be compliant with a particular clause of a standard. (**ICA**)

Object – see feature/object.

OSKA – Objekt Schlüssel Katalog.

Positional accuracy – the nearness of a feature/object's position to the estimated true position. (**ICA**, after **ESDQ** p. 32)

Product specification – a specification for a dataset, comprised of a set of profiles; specifically including the schema, metadata, quality information, reference system, structure primitives and encoding. (**ICA/TR**)

Profile – a clearly defined and limited subset/superset of a standard. (**ICA**)

NOTE: It is designed for use within an application area or information community.

Quality – totality of characteristics of a product that bear on its ability to satisfy stated and implied needs. (**TC211**)

Quality assurance methods – procedures to verify and evaluate quality. (**ICA**)

Query – a structured statement to extract information from a dataset. (**ICA/TR**)

Raster – one or more regular 2-dimensional arrays of pixels or grid cells covering an area. (**ICA**)

S-57 OC – Object Catalogue of the International Hydrographic Organization (IHO) Transfer Standard for Digital hydrographic Data. (After **IHO**)

Schema – formal description of a model. (**TC211**)

Spatial data model – rules, specifications, procedures and a geometric schema to represent real world geographic features in digital form. (**ICA**)
NOTE: The term spatial schema also used.

Spatial data type – the type of primitive used in spatial data models. (**ICA**)
NOTE: For example: Vector, Raster, and Object, etc. primitives. See also (**TC211**).

Spatial referencing – the convention by which spatial location can be specified in a standard. (**ICA/TR**)
NOTE: For example: using coordinates (latitude/longitude, UTM, ...) or by other means (e.g., postal codes).

Spatial schema – rules and procedures to represent the geometric and/or topological structure of the real world. (**ICA**)
NOTE: The term spatial data model is also used.

Temporal accuracy – the degree to which the information adequately describes geographic phenomena, in terms of temporal precision, frequency, and process history. (**ICA**, After **ESDQ** p. 164)

Theme – subject matter set of spatial data. (**ICA**)
Note: Can be layer, level, coverage, etc.

Transfer – moving data from one location to another. (**ICA**)

Transfer mechanism – a set of procedures using a minimal data model in a highly flexible environment to transfer data from one system to another. (**ICA/TR**)

Transfer standard – statement of the procedures used to structure data for transfer between two environments. (**ICA/TR**)

Triangular Irregular Network (TIN) – a tessellation using triangles (often based on the Delunay algorithm).
NOTE: The vertices of thc triangles form irregularly spaced nodes. Unlike the grid, the TIN allows dense information in complex areas, and sparse information in simpler or more homogeneous areas. (**ICA/TR**)

URL – Uniform Resource Locator.

Vector – composition of directed lines.
NOTE: A directed line (or vector) is a quantity that has both magnitude and direction. (**TC211**)

Vertical datum – datum describing the relation of gravity-related heights to the Earth. (**TC211**)

3. Spatial Metadata Characteristics of the National and International Spatial Metadata Standards

Organizational Structure of the Characteristics Questionnaire

The scientific and technical characteristics have been organized into 13 categories. In order to more effectively facilitate understanding of a particular National or International standard, or a subject matter profile of one. The categories begin with the more general characteristics and then move toward increasingly specific and detailed characteristics of the standard being analyzed. The 13 categories of characteristics are as follows:
Section I. Introduction to the Standard

0. Assessment Information;
1. Administration of Metadata Standard;
2. Use and Implementation of the Metadata Standard;
3. Linkage and Coordination of the Metadata Standard.

Section II. Categories of Kinds of Spatial Metadata

4. Dataset Identification;
5. Status of Dataset;
6. Extent of the Dataset;
7. Data Content of the Dataset;
8. Data Quality of the Dataset;
9. Spatial Data Structure of the Dataset;
10. Spatial Reference of the Dataset;
11. Availability and Distribution of the Dataset;
12. Authorization and Verification of the Metadata.

Each of these categories of scientific and technical characteristics is an essential part of any metadata standard or profile for spatial or geographic data. Integrated into these categories of characteristics are 58 secondary levels of characteristics, and more than 200 tertiary characteristics about each standard or profile. This provides a wealth of scientific and technical information about each standard or profile in a very systematic fashion.

In order for the user to understand a particular National or International standard, or profile of one of them, and its characteristics with respect to other standards or profiles, it must be assessed systematically and uniformly in light of all of the above categories. The Characteristics Questionnaire itself now follows.

References

[Guptill, S., and J. Morrison (eds), 1995]
Elements of Spatial Data Quality, Elsevier Science, 250 pp., ISBN 0080424325.

[ICA, 1996]
Meeting of the ICA Standards Commission, Den Haag, Netherlands.

[ICA, 1997]
Meeting of the ICA Standards Commission, Uppsala, Sweden.

[ICA, 1998]
Meeting of the ICA Standards Commission, Aix-en-Provence, France.

[ICA, 1999]
Meeting of the ICA Standards Commission, Ottawa, Canada.

[ICA, 2000]
Meeting of the ICA Standards Commission, Monaco, 2000.

[ICA, 2001]
Meeting of the ICA Standards Commission, Beijing, China.

[ICA, 2002]
Meeting of the ICA Standards Commission, Brno, Czech Republic.

[ISO/TC2211 – N1485, 2003]
Text for 19104 Geographic Information – Terminology, as sent to the ISO Central Secretariat for publication.

[ISO/TC2211 – N1785, 2005]
ISO/TC211 Terminology Spreadsheet – Edition 5.

[Moellering, H., and C. Clement (eds), 1994]
Technical Characteristics for Assessing Standards for the Transfer of Spatial Data and Brief International Descriptions, Columbus: ICA Standards Commission, 104 pp.

[Moellering, H., and R. Hogan (eds), 1997]
Spatial Database Transfer Standards 2: Characteristics for Assessing Standards and Full Descriptions of the National and International Standards in the World, Elsevier Science, 373 pp. + Crosstable, ISBN 0080424333.

Scientific and Technical Characteristics of Metadata Standards for Geographical Datasets

Edited by
Harold Moellering, Aaron Crane

Developed by
ICA Spatial Data Standards Commission 1995–1999
Version 7.0 (September, 2002)

I. Information about the Metadata Standard Itself

SECTION 0 – Assessment Information

This section provides the user with a general description of the scope and intended use of the standard. Information on the particular who prepared the assessment is also provided.

0.1 Respondent

Assessment prepared by:
Date assessment completed:
Relationship of respondent to standard:

0.2 Brief Summary Statement

A. Summary of the scope and intent of the standard:

SECTION 1 – Administration of Metadata Standard

This section provides a general description of the administrative framework, within which the standard was developed, tested and currently resides. Significant details are also provided in terms of the developmental and managerial history as well as information on the existence of documentation, software tools and training materials. This section also provides a point of contact for further information on the standard.

1.1 Name of Standard

A. Name of standard:

1. In original language(s):
2. English translation, if appropriate:

B. Version/Edition:
C. Language(s) of documentation:
D. Acronym(s):
E. Official ID:
F. Name(s) of recognizing standards authority(ies):

1.2 Responsible Institutions

Contact information about those institutions that are (or were) responsible for the processes involved in the evolution of the standard:

A. Production and/or development:
B. Testing:
C. Conformance:
D. Maintenance:
E. Distribution:
F. Help Desk/User Support:

1.3 Development History

A. Beginning of work on metadata standard:
B. Milestones:
C. Anticipated completion:
D. Update cycle:

1.4 Status of the Standard

A. Is the standard officially recognized:
B. Date(s) of recognition:
C. Anticipated date of recognition:
D. Current stage in recognition process:
E. Steps still required to achieve recognition:

1.5 Access to Documentation of the Standard

A. Is the standard copyrighted?:
B. Name and address of copyright owner:
C. Do restrictions on the use of the standard apply?:
D. Is the standard documentation currently available?:
E. List the form/media of standard documentation:
F. Is the standard available on the Internet?:
G. Price:
H. Contact details for obtaining the standard:
I. Does the standard have ISBN/ISSN numbers?:

1.6 Available Software Tools

A. Is software available to implement the standard as a database?:
B. Is software available to test compliance of metadata with the standard?:
C. Is software available to produce metadata compliant with the standard?:
D. Other relevant software:
E. Is there a test dataset available to test software tools for implementing the standard?:
F. For each product available above, please list contact information:

1.7 Available Training Materials

A. Are organized training sessions available?:
B. Contact details for training sessions:
C. Training documentation available:

SECTION 2 – Use and Implementation of the Metadata Standard

This section provides information on the way in which the standard is used and the purpose of its use. It also provides descriptions about the level of abstraction of the standard.

2.1 Intended Use of the Standard
Can the standard be used:

A. Within a transfer data structure?:
B. To produce hardcopy catalog of datasets?:
C. To catalog data in a digital on-line catalog?:
D. Other:

2.2 Intended Purpose of the Standard
Is the intended purpose of the standard:

A. To support discovery, identification, location of data?:
B. To support how to access data?:
C. To support the determination of dataset fitness for use?:
D. To support transfer of dataset?:
E. To support an organization's data management?:
F. Other:

2.3 Table of Contents

A. Provide the standard's Table of Contents:

2.4 Level of Abstraction of the Metadata Standard Definition

A. Does the standard define metadata elements?:
B. Does the standard define a model for the relationships between the metadata elements?:
C. Provide an overview diagram of the model of the metadata, if available.
D. Does the standard define domains and/or data types of the metadata elements?:

E. Does the standard define how the metadata should be encoded?:
F. Does the standard define a database schema for the automatic generation of a database?:

2.5 Extensibility of the Standard

A. Is the metadata description extensible (e.g., can user add metadata domains, elements, thematic profiles, etc.)?:

2.6 Availability of Metadata
Does the standard specify:

A. How metadata are to be accessed?:
B. A mechanism for querying the metadata?:
C. Particular standardized query protocols?:
D. That the metadata should be encoded for computer access?:
E. That the metadata should be encrypted for security?:

2.7 Examples

A. Does the standard provide examples of implementations?:
B. How many examples are available?:

SECTION 3 – Linkage and Coordination of the Metadata Standard

This section provides information on how the metadata standard is linked to other standards, particularly transfer standards, and how such linkage is implemented. If a significant linkage does exist, detailed technical information is provided on the precise nature of the linkage to the standard, and how the linkage affects implementation of the metadata standard.

3.1 Associated Transfer Standards

A. Is the metadata standard associated with a transfer standard(s)?:
B. If A is yes, provide references and contact details for each linked standard:
C. If A is yes, are both (all) standards coordinated?:
D. If A is yes, describe how the coordination works, both from a technical and conformance (normative) viewpoint:

3.2 Incorporated into transfer mechanism

A. Is the metadata standard incorporated into a transfer mechanism?:
B. If A is yes, provide references and contact details for each incorporated mechanism:
C. If A is yes, describe how:

3.3 Other linked, coordinated or associated standards

A. Is the metadata standard linked, coordinated, or associated with any other standard(s) not listed above?:
B. If A is yes, provide references and contact details for each standard:
C. If A is yes, describe how for each relevant standard:

II. Categories the Metadata Standard Uses to Describe Datasets

The following sections are categories of metadata that may be used by the metadata standard to describe data. Not all categories are relevant to different datasets and may not necessarily be present in the standard. The presence or absence of a category should not be used for judging one standard against another.

Note: For each of the questions in the following sections please provide the obligation status information. Valid obligation values are:

	Obligation Level	
M	Mandatory	This information must be entered to comply with the metadata standard.
C	Conditional	This information is entered based on other information in the metadata.
O	Optional	This information is optionally entered in the metadata.

If the answer is 'NO', the obligation status should be left blank.

SECTION 4 – Dataset Identification

This section examines how the standard provides for the specification of the various sorts of titles and identifiers for a dataset.

4.1 The title(s) of a dataset
Does the standard allow/require the provision for:

Obligation

A. __ Previous title(s)?:
B. __ Current title(s)?:
C. __ Possible future title(s)?:
D. __ Official titles in different languages?:
E. __ Abbreviated title(s)?:
F. __ Alternative Title(s)?:
G. __ Does the standard support effective dates of titles?:
H. __ Other identifiers (e.g., ISBN, National Stock Number, Uniform Product Code)?:

4.2 Parties responsible for the dataset
Does the standard allow/require the provision for:

Obligation

A. __ Author?:
B. __ Owner?:
C. __ Other?:

4.3 ISBD Information

Obligation

A. __ Does the metadata standard specify inclusion of mandatory ISBD information (see glossary) within the metadata?:

SECTION 5 – Status of the Dataset

This section examines how the standard provides for the specification of the current status and updating of the data set.

5.1 Status and Progress of Datasets
Does the metadata standard allow/require the provision for:

Obligation

A. __ Information on the status of the entire dataset?:
B. __ Information on the status of the particular subset (e.g. observation time, inauguration time, extraction time)?:

Does the metadata standard specify:

C. __ A structure for updating or maintenance information?:
D. __ A structure to describe historical information (see also Section 8.3 G)?:

5.2 Information on Incremental Updating
Does the metadata standard specify:

Obligation

A. __ A structure for information on incremental updates of the entire dataset?:
B. __ A structure for information on incremental updates of particular portions of the dataset?:
C. __ A structure to describe the update process for the dataset?:

SECTION 6 – Extent of the Dataset

This section examines how the standard provides for the specification of the spatial and temporal extent of the dataset.

6.1 The spatial extent of the dataset

Obligation

A. __ Does the metadata standard allow/require for the description of horizontal extent (e.g., bounding rectangle/polygon, named location (i.e., Stockholm, etc.))?:
B. __ Does the metadata standard allow/require for the description of vertical extent (e.g., minima/maxima heights, etc.)?:
C. __ Does the standard require specific units for describing extent?:
D. __ Does the standard allow specification of the units for describing extent?:
E. __ Does the standard require specific spatial reference system for describing extent?:
F. __ Does the standard allow specification of the spatial reference system for describing extent?:

6.2 The temporal extent of the dataset

Obligation

A. __ Does the standard allow/require for the description of time period for which the dataset is relevant?:
B. __ Does the standard allow/require specification of the temporal reference system?:

SECTION 7 – Data Content of the Dataset

This section examines how the standard provides for the specification of the data content of the dataset in terms of overview elements, themes, data definitions and product specifications.

7.1 Dataset overview
Does the standard allow/require for the provision of the following information:

Obligation

A. __ Dataset overview?:
B. __ Abstract – brief summary of the content of the dataset?:
C. __ Purpose – summary of the intentions with which the dataset was developed?:
D. __ Usage – list of activities for which the dataset has been used? (see also Section 8.5):
E. __ Spatial schema – (e.g., vector, raster)? (see also Section 9):
F. __ Spatial reference system? (see also Section 12):
G. __ Language?:
H. __ Character set (e.g., ISO8859-1, JISX0208)?:
I. __ List of documents providing further information about the dataset?:
J. __ Sample/preview dataset?:
K. __ Related datasets?:
L. __ Original size of the dataset?:
M. __ Compression? (see also Section 11.5):
N. __ Other?:

7.2 Theme(s)

Obligation

A. __ Does the standard allow/require the provision for the description of theme(s) (e.g., topography, land cover, transport, etc.) in the dataset?:
B. __ If A is yes, does the standard allow the referencing of an outside dictionary or thesaurus?:
C. __ If A is yes, does the standard itself include a dictionary or thesaurus within the metadata?:
D. __ If A is yes, does the standard allow the use of unauthorized user-defined keywords?:

7.3 Data Definition

Obligation

A. __ Does the standard allow/require the provision for the description of the data definition?:

B. If A is yes, does the standard allow/require the provision for the description of:
__ 1. Features/Objects?:
__ 2. Attributes?:
__ 3. Relations?:

C. __ Can attributes be described separately from feature/object classes?:

D. __ Does the standard allow/require the specification of the domain of the attributes?:

E. __ Does the standard allow/require the specification of data type of the attributes?:

F. __ Does the standard allow/require the specification of particular classification coding schemes (e.g., FACC, OSKA, S-57 OC)?:

7.4 Product Specification

Obligation

A. __ Does the standard allow/require for the provision of information about whether the dataset meets a particular product specification?:

B. __ If A is yes, can it be internal to the Metadata?:

C. __ If A is yes, can it be external to the Metadata?:

SECTION 8 – Data Quality of the Dataset

This section examines how the standard deals with the question of data quality. This includes homogeneity and metrics for data quality, quality assurance, and usage.

8.1 Overall Data Quality Statement

Obligation

A. __ Does the standard allow/require for the provision of quality to be described? If yes, please answer the questions below. (*If no, all following answers in Section* 8 *are N/A, please skip to Section* 9.):

B. __ Does the standard allow/require the provision for the quality information referencing an external quality standard, quality assurance standard, or specification?:

8.2 Homogeneity of Data Quality

Obligation

A. __ Does the standard allow/require for the variation of the quality of the dataset to be described?:

B. What type of aggregation of quality information is allowed?
__ 1. Entire database?:
__ 2. Feature/Object instance?:

__ 3. Feature/Object class?:
__ 4. Attribute instance?:
__ 5. Attribute class?:
__ 6. Layer?:
__ 7. Theme?:
__ 8. Geographic extent?:
__ 9. Other?:

8.3 Metrics for Data Quality
Does the standard allow/require the provision of values for:

Obligation

A. __ Positional accuracy?:
1. Standard error?:
2. Maximum error?:
3. Error correlation?:
4. Confidence level (e.g., 90%, 95%)?:
5. Other?:

B. __ Attribute accuracy?:
1. Standard error?:
2. Maximum error?:
3. Error frequency?:
4. Error correlation?:
5. Confidence level (e.g., 90%, 95%)?:
6. Other?:

C. __ Temporal accuracy?:
1. Standard error?:
2. Maximum error?:
3. Error correlation?:
4. Confidence level? (e.g., 90%, 95%):
5. Other?:

D. __ Completeness?:
1. Error of omission?:
2. Error of commission?:
3. Other?:

E. __ Currentness?:
1. Date?:
2. Maximum age?:
3. Temporal extent?:
4. Other?:

F. __ Logical consistency?:

G. __ Lineage?:
1. List of processing steps?:
2. List of values of processing parameters?:
3. List of printouts of statistics, etc.?:

4. Source material?:
5. Other?:

H. __ Additional Quality Information?:

8.4 Quality Assurance Methods

Does the standard allow/require the provision of information about the methods used to determine:

Obligation

A. __ Positional accuracy?:
1. Standard error in adjustments?:
2. Repeated measurements?:
3. Independent measurements?:
4. Subjective evaluations?:
5. Other?:

B. __ Attribute accuracy?:
1. Standard error?:
2. Error classification matrix?:
3. Independent measurements?:
4. Subjective evaluation?:
5. Other?:

C. __ Temporal accuracy?:
1. Independent measurements?:
2. Subjective evaluation?:
3. Other?:

D. __ Completeness?:
1. Field checks?:
2. Repeated measurements?:
3. Independent measurements?:
4. Subjective evaluation?:
5. Error of commission?:
6. Other?:

E. __ Currentness?:
1. Date of source material?:
2. Repeated measurements?:
3. Independent measurements?:
4. Subjective evaluation?:
5. Other?:

F. __ Logical consistency?:
1. Verification results of consistency checks?:
2. Repeated processing?:
3. Comparison with independent sources?:
4. Subjective evaluation?:
5. Other?:

G. __ Lineage?:

1. Checking results of processing steps:
2. Subjective evaluation:
3. Other:

H. __ Additional Quality Information?:

8.5 Usage

Obligation

A. __ Does the metadata standard allow/require the provision of information about usage?:
1. Who used it:
2. When it was used:
3. For what application it was used:
4. The effectiveness of use:
5. Other:

8.6 Alternative Quality Description:

SECTION 9 – Spatial Data Structure of the Dataset

This section examines how the standard provides for the specification of the spatial data model and spatial data types.

9.1 Spatial Data Model
Does the standard allow/require the provision for:

Obligation

A. __ The description of the geometrical and/or topological primitives used to model the real world?:

B. __ Information about an external standard for geometrical and/or topological primitive specification?:

9.2 Spatial Data Types
Does the standard allow/require the provision for:

Obligation

A. __ Information about the spatial data type(s) used in the dataset?:

B. __ If A is yes, which of the following generic spatial data types are supported? For each, please list specific data primitive levels supported by each data type.
1. Vector (e.g., spaghetti, chain-node/node-edge-face, planar graph, full topology)?:
2. Areal Raster (e.g., thematic raster-scanned and classification maps, image raster – satellite images, scanned aerial photographs)?:
3. Grid/Matrix (e.g., Digital Elevation Model)?:
4. Triangular Irregular Network (TIN)?:
5. Other?:

9.3 Other Data Types

Does the standard allow the provision of information about:

Obligation

A. __ Video?:
B. __ Sound?:
C. __ Other?:

SECTION 10 – Spatial Reference of the Dataset

This section examines how the dataset provides for the specification of the spatial reference characteristics of the data set.

Does the standard allow/require the provision for the following information:

10.1 Geoidal Model

Obligation

A. __ Geoidal model name?:
B. __ Geoidal model definitions?:

10.2 Geodetic datum

Obligation

A. __ Geodetic datum name?:
B. __ Geodetic datum definitions?:

10.3 Reference Ellipsoid

Obligation

A. __ Reference ellipsoid name?:
B. __ Reference ellipsoid definitions?:

10.4 Projection

Obligation

A. __ Projection name?:
B. __ Projection definitions?:

10.5 Coordinate Systems

Obligation

A. __ Coordinate system type?:
B. __ Coordinate system definitions?:

10.6 Height Reference System

Obligation

A. __ Vertical datum name?:
B. __ Types of heights?:

10.7 Ancillary Geodetic Information

Obligation

A. __ Does the standard allow/require the provision for ancillary geodetic, gravimetric, or magnetometric information?:

10.8 Transformation Parameters
Does the standard specify a method to provide information for coordinate transformation(s):

Obligation

A. __ Model name?:
B. __ Control points?:
C. __ Transformation parameters?:

10.9 Non-Coordinate Spatial Reference
Does the standard allow/require the provision for information on non-coordinate spatial reference systems (e.g., postal systems, street address)?:

Obligation

A. __ Reference system name?:
B. __ Reference system definitions?:
C. __ Other?:

SECTION 11 – Availability and Distribution of the Dataset

This section examines how the standard provides information on the availability of the dataset?

Does the standard allow/require the provision for:

11.1 The Distributor(s) of the Dataset

Obligation

A. __ Information about the distributor(s) of the dataset?:
B. __ Contact information for the distributor(s)?:

11.2 Restrictions to Access and Usage

Obligation

A. __ Information about copyright on the dataset (e.g., date, period, copyright owner, etc.)?:

B. __ Information about restrictions on the access of the dataset?:
C. __ Information about restrictions on the usage of the dataset?:

11.3 Pricing Details

Obligation

A. __ Information about pricing?:

11.4 Distribution of the Dataset

Obligation

A. __ Information about distribution media?:
B. __ Information about on-line access?:
C. __ Information about units of distribution (e.g., tiles, layers, polygons, elements)?:

11.5 Distribution Format

Obligation

A. __ Information about distribution formats (e.g., commercial CAD format, commercial GIS formats, National/International transfer standard)?:
B. __ Information about the encoding of the dataset (e.g., binary, ASCII)?:
C. __ Information about compression method?:
D. __ Information about compressed dataset size?:
E. __ Information about the distribution mechanism?:
F. __ Information about the encapsulation of data (e.g., ISO 8211, ISO 8824)?:

Ordering Procedures

Obligation

A. __ Information about ordering procedures for the dataset?:

11.7 Support Services

Obligation

A. __ Information about support services (e.g., additional data transformations, format or coordinate conversions)?:

SECTION 12 – Authorization and Verification of the Metadata

This section examines how the standard provides for the description of the authorization and verification on the metadata of a dataset (e.g., how is the metadata itself authorized and verified?).

12.1 Metadata reference

Obligation

A. __ Does the standard allow/require the provision of information about reference dates of the metadata (e.g., last check date, last update date, review date)?:

12.2 Verification Authority

Obligation

A. __ Does the standard allow/require the provision for information about who verified/authorized the metadata?:

B. __ Does the standard specify the method for providing this information (e.g., ISO 9000):

12.3 Statistical Methods Used for Verification

Obligation

A. __ Does the standard specify the statistical methods to be used for verification of the metadata?:

B. __ Does the standard allow/require the provision for the description of the statistical methods used to verify the metadata?:

END OF CHARACTERISTICS

FIN

PART 4

SCIENTIFIC AND TECHNICAL ASSESSMENTS WITH FULL DESCRIPTIONS OF THE SPATIAL METADATA STANDARDS

Edited by

Aaron Crane, Charles Ruhl, and Harold Moellering

Introduction to the Scientific and Technical Assessments of Spatial Metadata Standards

Harold Moellering

Department of Geography, Ohio State University, Columbus, Ohio 43210, United States of America
e-mail: geohal+@osu.edu

Introduction

Part 4 of this book is an assessment of each of the National and International spatial metadata standards in the world with the scientific and technical Questionnaire developed by the Commission during the years 1995 to 1999. These ICA scientific and technical characteristics were field tested in 1998 and 1999 to ensure the efficacy and effectiveness of the queries asked of each standard.

Hence, each National and International Spatial Metadata standard can be assessed with a consistent Questionnaire of items developed, and agreed to, by the International Community as represented by members of the ICA Spatial Data Standards Commission, that has developed these spatial metadata standards. Therefore, the reader will see a consistent structure of queries assessed against each of the National and International spatial data standards in the world. The reader then has a wonderful scientific resource by which one can learn about each spatial metadata standard in the world. Beyond that, the reader has a consistent basis for comparing these various standards one against another for a specific purpose as defined by the user. If desired, the reader also has the basis for choosing a specific standard for a particular purpose that is influenced solely by the scientific and technical characteristics of that standard. This material then helps the reader, or user, to make a more rational decision when trying to decide which standard might be used for a particular purpose.

In the later part of 1999 the Commission began to write the first drafts of the Questionnaire chapters for the National and International spatial metadata standards that were mature enough to begin this process. In subsequent Commission meetings at Ottawa, 1999, Monaco, 2000, Beijing, 2001, Brno, 2002, and later evaluations, a major part of each of the four day meetings was devoted to peer review of the Part 4 metadata questionnaires for the standards that were available at that time. Some Questionnaire chapters were peer reviewed more than once as their corresponding metadata standards were updated.

Each of the Questionnaires contains 13 Primary categories of characteristics, 58 Secondary subcategories of characteristics, and about 278 Tertiary characteristics. This provides a wealth of scientific and technical information about each spatial metadata standard. Each standard is assessed with this Questionnaire in the original language of the standard by a Commission member from the original standards body from that Country or organisation. The resulting answers in English were put into the Questionnaire template framework to format the replies systematically, make them easier to understand, and facilitate comparisons between various standards. This very systematic and highly structured form of information about these various standards and profiles promotes a quicker understanding of an individual standard, and improves the comparison process. Please note that Part 5 provides a summary Crosstable of all 22 standards and profiles by the 70 crucial characteristics that may be an initial beginning point for neophyte readers of such standards.

Please note that these chapters on the various spatial metadata standards are intended to serve as a companion set of characteristics that work in tandem with the analogous kinds of characteristics pertaining to the various spatial data transfer standards assessed in our recent 1997 ICA book [Moellering and Hogan (eds), 1997].

References

[Moellering and Hogan (eds), 1997]
Spatial Database Transfer Standards 2: Characteristics of Assessing Standards and Full Descriptions of the National and International Standards in the World, London: Pergamon/Elsevier Science, 373 pp. + Crosstable, ISBN 0080424333.

NATIONAL SPATIAL METADATA STANDARDS

World Spatial Metadata Standards
H. Moellering, H.J.G.L. Aalders & A. Crane (Editors)

Australia and New Zealand ANZLIC Metadata Guidelines

Craig Macauley

I. Information about the Metadata Standard Itself

SECTION 0 – Assessment Information

This section provides the user with a general description of the scope and intended use of the standard. Information on the respondent who prepared the assessment is also provided.

0.1 Respondent

Assessment prepared by:

Craig Macauley
Principal Consultant, ASDI Program
ANZLIC
PO Box 550
MARLESTON SA 5033
Australia
Fax: + 61 8 8293 4898
Phone: + 61 8 8226 4601
E-mail: macauley.craig@saugov.sa.gov.au

Date assessment completed:

July 1999, updated November 2000 and May 2002, Reviewed September, 2004

Relationship of respondent to standard:

Convenor, ANZLIC Metadata Working Group

0.2 Brief Summary Statement

A. Summary of the scope and intent of the standard:
ANZLIC is the peak body for spatial information coordination in Australia and New Zealand. One of ANZLIC's key strategies is to maximise community access to land and geographic information. If community access to land information is to be maximised, adequate descriptions of the characteristics of all geographically

referenced data sets must be available and accessible to the community at large.
A Working Group was formed by ANZLIC in April 1995 to work on the following tasks to improve community access to data:

- **produce a guidelines document which identifies and defines the mandatory metadata elements of a national land and geographic data directory system, discussing creation, maintenance and directory custodianship issues;**
- **using the guidelines document, promote the concept of a national data directory system to help determine the priorities and issues for implementing a national directory;**
- **develop an implementation plan for a national data directory system, including procedures for transfer of metadata between jurisdictions and the national directory system.**

In July 1996, the ANZLIC Metadata Guidelines were released. The Guidelines define a core set of metadata elements for the purpose of data discovery. Since 1996, the Guidelines have been widely adopted throughout Australia and New Zealand. The Australian Spatial Data Directory (ASDD) was recently implemented as a Web-based national metadata system. It is based upon ANZLIC's core metadata elements.
(Refer to the ASDD web site at http://www.environment.gov.au/net/asdd/)

SECTION 1 – Administration of Metadata Standard

This section provides a general description of the administrative framework, within which the standard was developed, tested and currently resides. Significant details are also provided in terms of the developmental and managerial history as well as information on the existence of documentation, software tools and training materials. This section also provides a point of contact for further information on the standard.

1.1 Name of Standard

A. Name of standard:
   1. In original language(s):
      ANZLIC Metadata Guidelines
   2. English translation, if appropriate:
      N/A

B. Version/Edition:
   Version 1.1, January, 1997
   Version 2, February, 2001

C. Language(s) of documentation:
   English

D. Acronym(s):
   N/A

E. Official ID:
   N/A

F. Name(s) of recognizing standards authority(ies):
   N/A

1.2 Responsible Institutions

Contact information about those institutions that are (or were) responsible for the processes involved in the evolution of the standard:

A. Production and/or development:
ANZLIC Metadata Working Group
PO Box 2
BELCONNEN ACT 2616
Australia
Fax: + 61 2 6201 4366
Phone: + 61 2 6201 4299
E-mail: anzlic@auslig.gov.au
Web: http://www.anzlic.org.au/

B. Testing:
ANZLIC Metadata Working Group

C. Conformance:
ANZLIC Metadata Working Group

D. Maintenance:
Executive Officer, ANZLIC the Spatial Information Council
E-mail: info@anzlic.org.au
Web: http://www.anzlic.org.au/

E. Distribution:
Executive Officer, ANZLIC

F. Help Desk/User Support:
Executive Officer, ANZLIC

1.3 Development History

A. Beginning of work on metadata standard: **April 1995**

B. Milestones:
Version 1 published July 1996
Version 1.1 with revised attachment 2 published January 1997
Version 2, February, 2001

C. Anticipated completion: **Completed**

D. Update cycle:
Revised as required (minor review in August 2000)

1.4 Status of the Standard

A. Is the standard officially recognized: **NO**
B. Date(s) of recognition: **N/A**
C. Anticipated date of recognition: **June 2001**
D. Current stage in recognition process:
Guidelines are being harmonised with ISO 19115.
E. Steps still required to achieve recognition:

ANZLIC Metadata Guidelines are not endorsed by Standards Australia but have been widely adopted. The objective is to progress the Guidelines to formal standard status once compliance with ISO 19115 is achieved.

1.5 Access to Documentation of the Standard

A. Is the standard copyrighted?: **YES**
B. Name and address of copyright owner:
ANZLIC
PO Box 2
BELCONNEN ACT 2616
Australia
C. Do restrictions on the use of the standard apply?: **NO**
D. Is the standard documentation currently available?: **YES**
E. List the form/media of standard documentation:
Paper
F. Is the standard available on the Internet?: **YES**
http://www.anzlic.org.au/asdi/metaelem.htm
G. Price:
Free of charge
H. Contact details for obtaining the standard:
Executive Officer, ANZLIC the Spatial Information Council
E-mail: info@anzlic.org.au
Web: http://www.anzlic.org.au/

I. Does the standard have ISBN/ISSN numbers?: **NO**

1.6 Available Software Tools

A. Is software available to implement the standard as a database?: **NO**
B. Is software available to test compliance of metadata with the standard?: **NO**
C. Is software available to produce metadata compliant with the standard?: **YES**
ANZLIC Metadata Entry Tool (Microsoft Access)
D. Other relevant software:
None available
E. Is there a test dataset available to test software tools for implementing the standard?: **NO**
F. For each product available above, please list contact information:
Executive Officer, ANZLIC the Spatial Information Council
E-mail: info@anzlic.org.au
Web: http://www.anzlic.org.au/

1.7 Available Training Materials

A. Are organized training sessions available?: **YES**
B. Contact details for training sessions:

Executive Officer, ANZLIC the Spatial Information Council
E-mail: info@anzlic.org.au
Web: http://www.anzlic.org.au/

C. Training documentation available: **N/A**

SECTION 2 – Use and Implementation of the Metadata Standard

This section provides information on the way in which the standard is used and the purpose of its use. It also provides descriptions about the level of abstraction of the standard.

2.1 Intended Use of the Standard
Can the standard be used:

A. Within a transfer data structure?: **YES**
B. To produce hardcopy catalog of datasets?: **YES**
C. To catalog data in a digital on-line catalog?: **YES**
D. Other: **N/A**

2.2 Intended Purpose of the Standard
Is the intended purpose of the standard:

A. To support discovery, identification, location of data?: **YES**
B. To support how to access data?: **NO**
C. To support the determination of dataset fitness for use?: **YES**
D. To support transfer of dataset?: **NO**
E. To support an organization's data management?: **YES**
F. Other: **N/A**

2.3 Table of Contents

A. Provide the standard's Table of Contents

1. Background to ANZLIC
2. Why ANZLIC is Concerned with Metadata
3. Implementation Issues
4. Guidelines for Completing the Core Elements
5. INDEX – Worked Examples
Attachment 1: Reconciliation of Core Elements and Data Entry Tool Screens
Attachment 2: Recommended Guidelines for the Transfer of ANZLIC Metadata Core Element
Attachment 3: Members of the ANZLIC Working Group on Metadata
Attachment 4: Members of the ANZLIC Advisory Committee

2.4 Level of Abstraction of the Metadata Standard Definition

A. Does the standard define metadata elements?: **YES**
B. Does the standard define a model for the relationships between the metadata elements?: **NO**

C. Provide an overview diagram of the model of the metadata, if available. **N/A**

D. Does the standard define domains and/or data types of the metadata elements?: **YES**

Domains are defined for some elements (e.g., search words, maintenance & update frequency); the date is defined where relevant (e.g., beginning date); the remaining elements are free text.

E. Does the standard define how the metadata should be encoded?: **YES**

Two methods have been defined. Firstly, a document type definition (DTD) was published for transfer in XML format. Secondly, a character-delimited ASCII exchange format is in use that produces eight files when exporting form the Metadata Entry Tool (one for each table in the Metadata Entry Tool's Access database).

F. Does the standard define a database schema for the automatic generation of a database?: **NO**

Not explicitly, but a table structure is defined implicitly through the use of the Metadata Entry Tool.

2.5 Extensibility of the Standard

A. Is the metadata description extensible (e.g., Can user add metadata domains, elements, thematic profiles, etc.)?: **YES**

Extra elements can be added. The core elements exist at the "Page 0" level. At the jurisdiction or theme level, additional elements (i.e., Page 1) can be defined. At the agency level, more elements (i.e., Page 2) can be defined.

2.6 Availability of Metadata

Does the standard specify:

A. How metadata are to be accessed?: **NO**

B. A mechanism for querying the metadata?: **NO**

C. Particular standardized query protocols?: **NO**

Note that Z39.50 has been adopted by ANZLIC as the protocol to be used in the development of the Australian Spatial Data Directory (ASDD) which is based on ANZLIC's core metadata elements.

D. That the metadata should be encoded for computer access?: **NO**

E. That the metadata should be encrypted for security?: **NO**

2.7 Examples

A. Does the standard provide examples of implementations?: **YES**

B. How many examples are available?: **16**

SECTION 3 – Linkage and Coordination of the Metadata Standard

This section provides information on how the metadata standard is linked to other standards, particularly transfer standards, and how such linkage is implemented. If a significant linkage does exist, detailed technical information is provided on the precise nature of the linkage to the standard, and how the linkage affects implementation of the metadata standard.

3.1 Associated Transfer Standards

A. Is the metadata standard associated with a transfer standard(s)?: **NO**
Note that guidelines exist for the transfer of ANZLIC core elements and are included as an attachment in the Metadata Guidelines document.
B. If A is yes, provide references and contact details for each linked standard: **N/A**
C. If A is yes, are both (all) standards coordinated?: **N/A**
D. If A is yes, describe how the coordination works, both from a technical and conformance (normative) viewpoint: **N/A**

3.2 Incorporated into transfer mechanism

A. Is the metadata standard incorporated into a transfer mechanism?: **YES**
B. If A is yes, provide references and contact details for each incorporated mechanism:
There are two mechanisms (XML and ASCII). The contact person for both is:
Executive Officer, ANZLIC the Spatial Information Council
E-mail: info@anzlic@org.au
Web: http://www.anzlic.org.au/
C. If A is yes, describe how:
Two methods have been defined. Firstly, a document type definition (DTD) was published for transfer in XML format. Secondly, a character-delimited ASCII exchange format is in use that produces eight files when exporting form the Metadata Entry Tool (one for each table in Tool's Access database).

3.3 Other linked, coordinated or associated standards

A. Is the metadata standard linked, coordinated, or associated with any other standard(s) not listed above?: **NO**
However, harmonisation is occurring between the ANZLIC guidelines and ISO 19115 metadata standard.
B. If A is yes, provide references and contact details for each standard: **N/A**
C. If A is yes, describe how for each relevant standard: **N/A**

II. Categories the Metadata Standard Uses to Describe Datasets

The following sections are categories of metadata that may be used by the metadata standard to describe data. Not all categories are relevant to different datasets and may not necessarily be present in the standard. The presence or absence of a category should not be used for judging one standard against another.

Note: For each of the questions in the following sections please provide the obligation status information. Valid obligation values are:

	Obligation Level	
M	Mandatory	This information must be entered to comply with the metadata standard.
C	Conditional	This information is entered based on other information in the metadata.
O	Optional	This information is optionally entered in the metadata.

If the answer is 'NO', the obligation status should be left blank.

SECTION 4 – Dataset Identification

This section examines how the standard provides for the specification of the various sorts of titles and identifiers for a dataset.

4.1 The title(s) of a dataset
Does the standard allow/require the provision for:

Obligation

A. __ Previous title(s)?: **NO**
B. **M** Current title(s)?: **YES**
C. __ Possible future title(s)?: **NO**
D. __ Official titles in different languages?: **NO**
E. __ Abbreviated title(s)?: **NO**
F. __ Alternative Title(s)?: **NO**
G. __ Does the standard support effective dates of titles?: **NO**
H. __ Other identifiers (e.g., ISBN, National Stock Number, Uniform Product Code)?: **NO**

4.2 Parties responsible for the dataset
Does the standard allow/require the provision for:

Obligation

A. __ Author?: **NO**
B. **M** Owner?: **YES**
C. __ Other?: **NO**

4.3 ISBD Information

Obligation

A. __ Does the metadata standard specify inclusion of mandatory ISBD information (see glossary) within the metadata?: **NO**

SECTION 5 – Status of the Dataset

This section examines how the standard provides for the specification of the current status and updating of the data set.

5.1 Status and Progress of Datasets
Does the metadata standard allow/require the provision for:

Obligation

A. **M** Information on the status of the entire dataset?: **YES**
B. __ Information on the status of the particular subset (e.g., observation time, inauguration time, extraction time)?: **NO**

Does the metadata standard specify:

C. **M** A structure for updating or maintenance information?: **YES**
D. **M** A structure to describe historical information (see also Section 8.3 G)?: **YES**

5.2 Information on Incremental Updating
Does the metadata standard specify:

Obligation

A. __ A structure for information on incremental updates of the entire dataset?: **NO**
B. __ A structure for information on incremental updates of particular portions of the dataset?: **NO**
C. __ A structure to describe the update process for the dataset?: **NO**

SECTION 6 – Extent of the Dataset

This section examines how the standard provides for the specification of the spatial and temporal extent of the dataset.

6.1 The spatial extent of the dataset

Obligation

A. **M** Does the metadata standard allow/require for the description of horizontal extent (e.g., bounding rectangle/polygon, named location (i.e., Stockholm, etc.))?: **YES**
B. __ Does the metadata standard allow/require for the description of vertical extent (e.g., minima/maxima heights, etc.)?: **NO**
C. **M** Does the standard require specific units for describing extent?: **YES**
D. __ Does the standard allow specification of the units for describing extent?: **NO**
E. __ Does the standard require specific spatial reference system for describing extent?: **NO**
F. __ Does the standard allow specification of the spatial reference system for describing extent?: **NO**

6.2 The temporal extent of the dataset

Obligation

A. **M** Does the standard allow/require for the description of time period for which the dataset is relevant?: **YES**

B. __ Does the standard allow/require specification of the temporal reference system?: **NO**

SECTION 7 – Data Content of the Dataset

This section examines how the standard provides for the specification of the data content of the dataset in terms of overview elements, themes, data definitions and product specifications.

7.1 Dataset overview

Does the standard allow/require for the provision of the following information:

Obligation

A. __ Dataset overview?: **NO**
B. **M** Abstract – brief summary of the content of the dataset?: **YES**
C. __ Purpose – summary of the intentions with which the dataset was developed?: **NO**
Planned to be in the next version.
D. __ Usage – list of activities for which the dataset has been used? (see also Section 8.5): **NO**
Planned to be in the next version.

E. **O** Spatial schema – (e.g., vector, raster)? (see also Section 9): **YES**
F. **O** Spatial reference system? (see also Section 12): **YES**
G. __ Language?: **NO**
H. __ Character set (e.g., ISO8859-1, JISX0208)?: **NO**
I. **C** List of documents providing further information about the dataset?: **YES**
J. __ Sample/preview dataset?: **NO**
K. __ Related datasets?: **NO**
L. __ Original size of the dataset?: **NO**
M. __ Compression? (see also Section 11.5): **NO**
N. __ Other?: **N/A**

7.2 Theme(s)

Obligation

A. **O** Does the standard allow/require the provision for the description of theme(s) (e.g., topography, land cover, transport, etc.) in the dataset?: **YES**
B. __ If A is yes, does the standard allow the referencing of an outside dictionary or thesaurus?: **NO**
C. **M** If A is yes, does the standard itself include a dictionary or thesaurus within the metadata?: **YES**
D. __ If A is yes, does the standard allow the use of unauthorized user-defined keywords?: **NO**

7.3 Data Definition

Obligation

A. __ Does the standard allow/require the provision for the description of the data definition?: **NO**

B. If A is yes, does the standard allow/require the provision for the description of:
  __ 1. Features/Objects?: **N/A**
  __ 2. Attributes?: **N/A**
  __ 3. Relations?: **N/A**

C. __ Can attributes be described separately from feature/object classes?: **NO**

D. __ Does the standard allow/require the specification of the domain of the attributes?: **NO**

E. __ Does the standard allow/require the specification of data type of the attributes?: **NO**

F. __ Does the standard allow/require the specification of particular classification coding schemes (e.g., FACC, OSKA, S-57 OC)?: **NO**

7.4 Product Specification

Obligation

A. __ Does the standard allow/require for the provision of information about whether the dataset meets a particular product specification?: **NO**

B. __ If A is yes, can it be internal to the Metadata?: **N/A**

C. __ If A is yes, can it be external to the Metadata?: **N/A**

SECTION 8 – Data Quality of the Dataset

This section examines how the standard deals with the question of data quality. This includes homogeneity and metrics for data quality, quality assurance, and usage.

8.1 Overall Data Quality Statement

Obligation

A. **M** Does the standard allow/require for the provision of quality to be described? If yes, please answer the questions below. (*If no, all following answers in Section* 8 *are N/A, please skip to Section* 9.): **YES**

B. __ Does the standard allow/require the provision for the quality information referencing an external quality standard, quality assurance standard, or specification?: **NO**

8.2 Homogeneity of Data Quality

Obligation

A. **O** Does the standard allow/require for the variation of the quality of the dataset to be described?: **YES**

B. What type of aggregation of quality information is allowed?
  C 1. Entire database?: **YES**
  C 2. Feature/Object instance?: **YES**
  C 3. Feature/Object class?: **YES**
  C 4. Attribute instance?: **YES**
  C 5. Attribute class?: **YES**
  C 6. Layer?: **YES**

C 7. Theme?: **YES**
C 8. Geographic extent?: **YES**
__ 9. Other?: **NO**

8.3 Metrics for Data Quality
Does the standard allow/require the provision of values for:

Obligation

A. **M** Positional accuracy?: **YES**
1. Standard error?: **YES**
2. Maximum error?: **YES**
3. Error correlation?: **YES**
4. Confidence level (e.g., 90%, 95%)?: **YES**
5. Other?: **YES**
An intuitive estimate is allowed in the absence of a quantitative measurement.

B. **M** Attribute accuracy?: **YES**
Free text field may contain the five elements below.
1. Standard error?: **YES**
2. Maximum error?: **YES**
3. Error frequency?: **YES**
4. Error correlation?: **YES**
5. Confidence level (e.g., 90%, 95%)?: **YES**
6. Other?:
An intuitive estimate is allowed in the absence of a quantitative measurement.

C. __ Temporal accuracy?: **NO**
1. Standard error?: **NO**
2. Maximum error?: **NO**
3. Error correlation?: **NO**
4. Confidence level? (e.g., 90%, 95%): **NO**
5. Other?: **NO**

D. **M** Completeness?: **YES**
Free text field may contain 2 elements below.
1. Error of omission?: **YES**
2. Error of commission?: **YES**
3. Other?: **NO**

E. **M** Currentness?:
1. Date?: **YES**
2. Maximum age?: **YES**
3. Temporal extent?: **NO**
4. Other?: **NO**

F. **M** Logical consistency?: **YES**
Free text field with no procedures.

G. **M** Lineage?: **YES**
  Free text field with no check procedures.
  1. List of processing steps?: **YES**
  2. List of values of processing parameters?: **NO**
  3. List of printouts of statistics, etc.?: **NO**
  4. Source material?: **YES**
  5. Other?: **NO**

H. __ Additional Quality Information?: **NO**

8.4 Quality Assurance Methods

Does the standard allow/require the provision of information about the methods used to determine:

Obligation

A. **O** Positional accuracy?: **YES**
  Free text field may contain the four elements below.
  1. Standard error in adjustments?: **YES**
  2. Repeated measurements?: **YES**
  3. Independent measurements?: **YES**
  4. Subjective evaluations?: **YES**
  5. Other?: **NO**

B. **O** Attribute accuracy?: **YES**
  Free text field may contain the four elements below.
  1. Standard error?: **YES**
  2. Error classification matrix?: **YES**
  3. Independent measurements?: **YES**
  4. Subjective evaluation?: **YES**
  5. Other?: **NO**

C. __ Temporal accuracy?: **NO**
  1. Independent measurements?: **NO**
  2. Subjective evaluation?: **NO**
  3. Other?: **NO**

D. **O** Completeness?: **YES**
  Free text field may contain the five elements below.
  1. Field checks?: **YES**
  2. Repeated measurements?: **YES**
  3. Independent measurements?: **YES**
  4. Subjective evaluation?: **YES**
  5. Error of commission?: **YES**
  6. Other?: **NO**

E. __ Currentness?: **NO**
  1. Date of source material?: **NO**
  2. Repeated measurements?: **NO**
  3. Independent measurements?: **NO**
  4. Subjective evaluation?: **NO**

5. Other?: **NO**

F. **O** Logical consistency?: **YES**

Free text field may contain the four elements below.

1. Verification results of consistency checks?: **YES**
2. Repeated processing?: **YES**
3. Comparison with independent sources?: **YES**
4. Subjective evaluation?: **YES**
5. Other?: **NO**

G. __ Lineage?: **YES**

Free text field may contain the three elements below.

1. Checking results of processing steps: **YES**
2. Subjective evaluation: **YES**
3. Other: **NO**

H. __ Additional Quality Information?: **NO**

8.5 Usage

Obligation

A. __ Does the metadata standard allow/require the provision of information about usage?: **NO**

1. Who used it: **NO**
2. When it was used: **NO**
3. For what application it was used: **NO**
4. The effectiveness of use: **NO**
5. Other: **NO**

8.6 Alternative Quality Description: **NO**

SECTION 9 – Spatial Data Structure of the Dataset

This section examines how the standard provides for the specification of the spatial data model and spatial data types.

9.1 Spatial Data Model
Does the standard allow/require the provision for:

Obligation

A. __ The description of the geometrical and/or topological primitives used to model the real world?: **NO**

B. __ Information about an external standard for geometrical and/or topological primitive specification?: **NO**

9.2 Spatial Data Types
Does the standard allow/require the provision for:

Obligation

A. **O** Information about the spatial data type(s) used in the dataset?: **YES**
B. **O** If A is yes, which of the following generic spatial data types are supported? **YES**
For each, please list specific data primitive levels supported by each data type.
1. Vector (e.g., spaghetti, chain-node/node-edge-face, planar graph, full topology)?: **YES**
2. Areal Raster (e.g., thematic raster – scanned and classification maps, image raster – satellite images, scanned aerial photographs)?: **YES**
3. Grid/Matrix (e.g., Digital Elevation Model)?: **YES**
4. Triangular Irregular Network (TIN)?: **YES**
Other?: **NO**

9.3 Other Data Types

Does the standard allow the provision of information about:

Obligation

A. __ Video?: **NO**
B. __ Sound?: **NO**
C. **O** Other?: **YES**

SECTION 10 – Spatial Reference of the Dataset

This section examines how the dataset provides for the specification of the spatial reference characteristics of the data set.

Does the standard allow/require the provision for the following information:

10.1 Geoidal Model

Obligation

A. **O** Geoidal model name?: **YES**
B. **O** Geoidal model definitions?: **YES**

10.2 Geodetic datum

Obligation

A. **O** Geodetic datum name?: **YES**
B. **O** Geodetic datum definitions?: **YES**

10.3 Reference Ellipsoid

Obligation

A. **O** Reference ellipsoid name?: **YES**
B. **O** Reference ellipsoid definitions?: **YES**

10.4 Projection

Obligation

A. **O** Projection name?: **YES**
B. **O** Projection definitions?: **YES**

10.5 Coordinate Systems

Obligation

A. **O** Coordinate system type?: **YES**
B. **O** Coordinate system definitions?: **YES**

10.6 Height Reference System

Obligation

A. **O** Vertical datum name?: **YES**
B. **O** Types of heights?: **YES**

10.7 Ancillary Geodetic Information

Obligation

A. __ Does the standard allow/require the provision for ancillary geodetic, gravimetric, or magnetometric information?: **NO**

10.8 Transformation Parameters
Does the standard specify a method to provide information for coordinate transformation(s):

Obligation

A. __ Model name?: **NO**
B. __ Control points?: **NO**
C. __ Transformation parameters?: **NO**

10.9 Non-Coordinate Spatial Reference
Does the standard allow/require the provision for information on non-coordinate spatial reference systems (e.g., postal systems, street address)?:

Obligation

A. __ Reference system name?: **NO**
B. __ Reference system definitions?: **NO**
C. __ Other?: **NO**

SECTION 11 – Availability and Distribution of the Dataset

This section examines how the standard provides information on the availability of the dataset?

Does the standard allow/require the provision for:

11.1 The Distributor(s) of the Dataset

Obligation

A. **M** Information about the distributor(s) of the dataset?: **YES**
B. **M** Contact information for the distributor(s)?: **YES**

11.2 Restrictions to Access and Usage

Obligation

A. **O** Information about copyright on the dataset (e.g., date, period, copyright owner, etc.)?: **YES**
B. **O** Information about restrictions on the access of the dataset?: **YES**
C. **O** Information about restrictions on the usage of the dataset?: **YES**

11.3 Pricing Details

Obligation

A. **O** Information about pricing?: **YES**

11.4 Distribution of the Dataset

Obligation

A. **M** Information about distribution media?: **YES**
B. **O** Information about on-line access?: **YES**
C. **O** Information about units of distribution (e.g., tiles, layers, polygons, elements)?: **YES**

11.5 Distribution Format

Obligation

A. **M** Information about distribution formats (e.g., commercial CAD format, commercial GIS formats, National/International transfer standard)?: **YES**
B. **O** Information about the encoding of the dataset (e.g., binary, ASCII)?: **YES**
C. __ Information about compression method?: **NO**
D. __ Information about compressed dataset size?: **NO**
E. __ Information about the distribution mechanism?: **NO**
F. __ Information about the encapsulation of data (e.g., ISO 8211, ISO 8824)?: **NO**

11.6 Ordering Procedures

Obligation

A. __ Information about ordering procedures for the dataset?: **NO**

11.7 Support Services

Obligation

A. __ Information about support services (e.g., additional data transformations, format or coordinate conversions)?: **NO**

SECTION 12 – Authorization and Verification of the Metadata

This section examines how the standard provides for the description of the authorization and verification on the metadata of a dataset (e.g., how is the metadata itself authorized and verified?).

12.1 Metadata reference

Obligation

A. **M** Does the standard allow/require the provision of information about reference dates of the metadata (e.g., last check date, last update date, review date)?: **YES**

12.2 Verification Authority

Obligation

A. __ Does the standard allow/require the provision for information about who verified/authorized the metadata?: **NO**
B. __ Does the standard specify the method for providing this information (e.g., ISO 9000): **NO**

12.3 Statistical Methods Used for Verification

Obligation

A. __ Does the standard specify the statistical methods to be used for verification of the metadata?: **NO**
B. __ Does the standard allow/require the provision for the description of the statistical methods used to verify the metadata?: **NO**

END OF CHARACTERISTICS
FIN

World Spatial Metadata Standards
H. Moellering, H.J.G.L. Aalders & A. Crane (Editors)

Canada
Directory Information Describing Digital Geo-referenced Data Sets
Information de répertoire décrivant les ensembles de données numériques à référence spatiale
CAN/CGSB-171.3-95

Kian Fadaie, Valerie E. Hume

I. Information about the Metadata Standard Itself

SECTION 0 – Assessment Information

This section provides the user with a general description of the scope and intended use of the standard. Information on the respondent who prepared the assessment is also provided.

0.1 Respondent

Assessment prepared by:

Valerie E. Hume,
655 Richmond Rood, Unit #14,
Ottawa, Ontario K2A 3Y3
Canada
Phone: +1 613 728 8617
E-mail: vhume@rogers.com

Assessment revised by:

Dr. Kian Fadaie, Senior Advisor, Agriculture and Agri-Food Canada,
International Science & Development
930 Carling Avenue
Ottawa, Ontario K1A 0C5
Canada
Location: Sir John Carling Building,
8th Floor, Room 836C
Tel: 613-694-2315

Fax: 613-759-7504
E-mail: fadaiek@agr.gc.ca
Date assessment completed:
July 2000, updated August 2002, Reviewed April, 2005
Relationship of respondent to standard:
Chairman of drafting committee, Valerie Hume
Chair of the Canadian Advisory Committee to ISO TC211, Dr. Kian Fadaie

0.2 Brief Summary Statement

A. Summary of the scope and intent of the standard:
This standard defines metadata for describing geographic information and services. It provides information about the identification, the extent, the quality, the spatial and temporal schema, spatial reference, and distribution of digital geographic data. This standard is applicable to:

- **The cataloguing of datasets, clearinghouse activities, and the full description of datasets;**
- **Geographic datasets, dataset series, and individual geographic features and feature properties.**

This standard defines:

- **Mandatory and optional metadata sections, metadata entities, and metadata elements;**
- **The minimum set of metadata required to serve the full range of metadata applications such as discovery, access, transfer, and usage of digital data);**
- **Optional metadata elements – to allow for a more extensive standard description of geographic data, if required;**
- **A method for extending metadata to fit specialized needs.**

Though this standard is applicable to digital data, its principles can be extended to many other forms of geographic data such as maps, charts, and textual documents as well as non-geographic data.
NOTE: Certain mandatory metadata elements may not apply to these other forms of data.

SECTION 1 – Administration of Metadata Standard

This section provides a general description of the administrative framework, within which the standard was developed, tested and currently resides. Significant details are also provided in terms of the developmental and managerial history as well as information on the existence of documentation, software tools and training materials. This section also provides a point of contact for further information on the standard.

1.1 Name of Standard

A. Name of standard:
   1. In original language(s):
      Directory Information Describing Digital Geo-referenced Data Sets Information de répertoire décrivant les ensembles de données numériques à référence spatiale
   2. English translation, if appropriate:
      N/A

B. Version/Edition:
   1st edition, 2nd version April 1995

C. Language(s) of documentation:
   English, French

D. Acronym(s):
   N/A

E. Official ID:
   CAN/CGSB-171.3-95

F. Name(s) of recognizing standards authority(ies):
   Canadian General Standards Board

1.2 Responsible Institutions

Contact information about those institutions that are (or were) responsible for the processes involved in the evolution of the standard:

A. Production and/or development:
   Canadian General Standards Board
   Ottawa, Canada K1A 1G6
   Telephone: (819) 956-0894
   Fax: (819) 956-1634

B. Testing:
   Canadian General Standards Board
   Ottawa, Canada K1A 1G6
   Telephone: (819) 956-0894
   Fax: (819) 956-1634

C. Conformance:
   Canadian General Standards Board
   Ottawa, Canada K1A 1G6
   Telephone: (819) 956-0894
   Fax: (819) 956-1634

D. Maintenance:
   Committee on Geomatics
   Canadian General Standards Board
   Ottawa, Canada K1A 1G6
   Telephone: (819) 956-0894
   Fax: (819) 956-1634

E. Distribution:
Canadian General Standards Board
Ottawa, Canada K1A 1G6
Telephone: (819) 956-0894
Fax: (819) 956-1634

F. Help Desk/User Support:
N/A

1.3 Development History

A. Beginning of work on metadata standard: **1993**

B. Milestones:
1st version 1994
2nd version April 1995
Electronic version available in 1995
Review has been undertaken in the Fall of 2001 to develop the North American profile of ISO 19115

C. Anticipated completion:

D. Update cycle: **5 years**

1.4 Status of the Standard

A. Is the standard officially recognized: **YES**

B. Date(s) of recognition: **April 1995**

C. Anticipated date of recognition: **N/A**

D. Current stage in recognition process: **N/A**

E. Steps still required to achieve recognition: **N/A**

1.5 Access to Documentation of the Standard

A. Is the standard copyrighted?: **YES**

B. Name and address of copyright owner:
Canadian General Standards Board
Ottawa, Canada K1A 1G6
Telephone: (819) 956-0894
Fax: (819) 956-1634

C. Do restrictions on the use of the standard apply?: **YES**
No part of the publication may be reproduced in any form without the prior permission of the publisher

D. Is the standard documentation currently available?: **YES**

E. List the form/media of standard documentation:
paper, diskette

F. Is the standard available on the Internet?: **NO**

G. Price: **$41.50 CDN**

H. Contact details for obtaining the standard:
CGSB Sales Centre
Ottawa Canada K1A 1G6

Telephone: (819) 956-0425
1-800-665-CGSB (Canada only)
Fax: (819) 956-5644

I. Does the standard have ISBN/ISSN numbers?: **NO**

1.6 Available Software Tools

A. Is software available to implement the standard as a database?: **YES**
Ninman
Natural Resources and Environment Branch
Indian and Northern Affairs Canada
Ottawa, Ontario, Canada K1A 0H4

B. Is software available to test compliance of metadata with the standard?: **NO**

C. Is software available to produce metadata compliant with the standard?: **YES**
Canadian General Standards Board
Ottawa Canada K1A 1G6
Telephone: (819) 956-0894
Fax: (819) 956-1634

D. Other relevant software: **N/A**

E. Is there a test dataset available to test software tools for implementing the standard?:

F. For each product available above, please list contact information:
Any data set could be used.

1.7 Available Training Materials

A. Are organized training sessions available?: **NO**
B. Contact details for training sessions: **N/A**
C. Training documentation available: **N/A**

SECTION 2 – Use and Implementation of the Metadata Standard

This section provides information on the way in which the standard is used and the purpose of its use. It also provides descriptions about the level of abstraction of the standard.

2.1 Intended Use of the Standard
Can the standard be used:

A. Within a transfer data structure?: **YES**
B. To produce hardcopy catalog of datasets?: **YES**
1 page at a time
C. To catalog data in a digital on-line catalog?: **YES**
D. Other: **NO**

2.2 Intended Purpose of the Standard
Is the intended purpose of the standard:

A. To support discovery, identification, location of data?: **YES**

B. To support how to access data?: **YES**
C. To support the determination of dataset fitness for use?: **YES**
D. To support transfer of dataset?: **NO**
E. To support an organization's data management?: **YES**
F. Other: **NO**

2.3 Table of Contents

A. Provide the standard's Table of Contents **N/A**

2.4 Level of Abstraction of the Metadata Standard Definition

A. Does the standard define metadata elements?: **YES**
B. Does the standard define a model for the relationships between the metadata elements?: **NO**
C. Provide an overview diagram of the model of the metadata, if available. **N/A**
D. Does the standard define domains and/or data types of the metadata elements?: **NO**
E. Does the standard define how the metadata should be encoded?: **NO**
F. Does the standard define a database schema for the automatic generation of a database?: **NO**

2.5 Extensibility of the Standard

A. Is the metadata description extensible (e.g., Can user add metadata domains, elements, thematic profiles, etc.)?: **YES**

2.6 Availability of Metadata

Does the standard specify:

A. How metadata are to be accessed?: **NO**
B. A mechanism for querying the metadata?: **NO**
C. Particular standardized query protocols?: **NO**
D. That the metadata should be encoded for computer access?: **NO**
E. That the metadata should be encrypted for security?: **NO**

2.7 Examples

A. Does the standard provide examples of implementations?: **NO**
B. How many examples are available?: **N/A**

SECTION 3 – Linkage and Coordination of the Metadata Standard

This section provides information on how the metadata standard is linked to other standards, particularly transfer standards, and how such linkage is implemented. If a significant linkage does exist, detailed technical information is provided on the precise nature of the linkage to the standard, and how the linkage affects implementation of the metadata standard.

3.1 Associated Transfer Standards

A. Is the metadata standard associated with a transfer standard(s)?: **NO**
B. If A is yes, provide references and contact details for each linked standard: **N/A**
C. If A is yes, are both (all) standards coordinated?: **N/A**
D. If A is yes, describe how the coordination works, both from a technical and conformance (normative) viewpoint: **N/A**

3.2 Incorporated into transfer mechanism

A. Is the metadata standard incorporated into a transfer mechanism?: **NO**
B. If A is yes, provide references and contact details for each incorporated mechanism: **N/A**
C. If A is yes, describe how: **N/A**

3.3 Other linked, coordinated or associated standards

A. Is the metadata standard linked, coordinated, or associated with any other standard(s) not listed above?: **NO**
B. If A is yes, provide references and contact details for each standard: **N/A**
C. If A is yes, describe how for each relevant standard: **N/A**

II. Categories the Metadata Standard Uses to Describe Datasets

The following sections are categories of metadata that may be used by the metadata standard to describe data. Not all categories are relevant to different datasets and may not necessarily be present in the standard. The presence or absence of a category should not be used for judging one standard against another.

Note: For each of the questions in the following sections please provide the obligation status information. Valid obligation values are:

	Obligation Level	
M	Mandatory	This information must be entered to comply with the metadata standard.
C	Conditional	This information is entered based on other information in the metadata.
O	Optional	This information is optionally entered in the metadata.

If the answer is 'NO', the obligation status should be left blank.

SECTION 4 – Dataset Identification

This section examines how the standard provides for the specification of the various sorts of titles and identifiers for a dataset.

4.1 The title(s) of a dataset
Does the standard allow/require the provision for:

Obligation

A. **O** Previous title(s)?: **YES**
B. **O** Current title(s)?: **YES**
C. __ Possible future title(s)?: **NO**
D. __ Official titles in different languages?: **NO**
E. **O** Abbreviated title(s)?: **YES**
F. __ Alternative Title(s)?: **NO**
G. __ Does the standard support effective dates of titles?: **NO**
H. __ Other identifiers (e.g., ISBN, National Stock Number, Uniform Product Code)?: **NO**

4.2 Parties responsible for the dataset
Does the standard allow/require the provision for:

Obligation

A. __ Author?: **NO**
B. **O** Owner?: **YES**
C. x Other?: **YES**
Scientific contact, Data manager, contact for access

4.3 ISBD Information

Obligation

A. __ Does the metadata standard specify inclusion of mandatory ISBD information (see glossary) within the metadata?: **NO**

SECTION 5 – Status of the Dataset

This section examines how the standard provides for the specification of the current status and updating of the data set.

5.1 Status and Progress of Datasets
Does the metadata standard allow/require the provision for:

Obligation

A. **O** Information on the status of the entire dataset?: **YES**
Time coverage
B. __ Information on the status of the particular subset (e.g., observation time, inauguration time, extraction time)?: **NO**

Does the metadata standard specify:
C. __ A structure for updating or maintenance information?: **NO**
D. __ A structure to describe historical information (see also Section 8.3 G)?: **NO**

5.2 Information on Incremental Updating
Does the metadata standard specify:

Obligation

A. __ A structure for information on incremental updates of the entire dataset?: **NO**
B. __ A structure for information on incremental updates of particular portions of the dataset?: **NO**
C. __ A structure to describe the update process for the dataset?: **NO**

SECTION 6 – Extent of the Dataset

This section examines how the standard provides for the specification of the spatial and temporal extent of the dataset.

6.1 The spatial extent of the dataset

Obligation

A. **O** Does the metadata standard allow/require for the description of horizontal extent (e.g., bounding rectangle/polygon, named location (i.e., Stockholm, etc.))?: **YES**

bounding polygon, jurisdictional/administrative unit, and other geographical description

B. **O** Does the metadata standard allow/require for the description of vertical extent (e.g., minima/maxima heights, etc.)?: **YES**

identification of vertical coverage, areas where vertical coverage is complete abd uniform

C. __ Does the standard require specific units for describing extent?: **NO**
D. **O** Does the standard allow specification of the units for describing extent?: **YES**
E. __ Does the standard require specific spatial reference system for describing extent?: **NO**
F. **O** Does the standard allow specification of the spatial reference system for describing extent?: **YES**

latitude and longitude

6.2 The temporal extent of the dataset

Obligation

A. **O** Does the standard allow/require for the description of time period for which the dataset is relevant?: **YES**
B. **O** Does the standard allow/require specification of the temporal reference system?: **YES**

SECTION 7 – Data Content of the Dataset

This section examines how the standard provides for the specification of the data content of the dataset in terms of overview elements, themes, data definitions and product specifications.

7.1 Dataset overview
Does the standard allow/require for the provision of the following information:

Obligation

A. **O** Dataset overview?: **YES**
Description of the data set

B. **O** Abstract – brief summary of the content of the dataset?: **YES**
Description of the data set

C. **O** Purpose – summary of the intentions with which the dataset was developed?: **YES**
Current applications of the data set

D. __ Usage – list of activities for which the dataset has been used? (see also Section 8.5): **NO**
E. **O** Spatial schema – (e.g., vector, raster)? (see also Section 9): **YES**
F. **O** Spatial reference system? (see also Section 12): **YES**
G. x Language?: **YES**
H. __ Character set (e.g., ISO8859-1, JISX0208)?: **NO**
I. **O** List of documents providing further information about the dataset?: **YES**
J. __ Sample/preview dataset?: **NO**
K. **O** Related datasets?: **YES**
L. **O** Original size of the dataset?: **YES**
M. __ Compression? (see also Section 11.5): **NO**
N. **O** Other?: **YES**
Map projection, geodetic datums, scale at which information was collected, Positional accuracy, Seamess/discontinuous relation of adjoining maps, etc.

7.2 Theme(s)

Obligation

A. **O** Does the standard allow/require the provision for the description of theme(s) (e.g., topography, land cover, transport, etc.) in the dataset?: **YES**
B. __ If A is yes, does the standard allow the referencing of an outside dictionary or thesaurus?: **NO**
C. __ If A is yes, does the standard itself include a dictionary or thesaurus within the metadata?: **NO**
but it includes a list of categories

D. **O** If A is yes, does the standard allow the use of unauthorized user-defined keywords?: **YES**

7.3 Data Definition

Obligation

A. x Does the standard allow/require the provision for the description of the data definition?: **YES**
B. If A is yes, does the standard allow/require the provision for the description of:
__ 1. Features/Objects?: **NO**

O 2. Attributes?: **YES**
O 3. Relations?: **YES**
Variables

C. x Can attributes be described separately from feature/object classes?: **YES**
D. __ Does the standard allow/require the specification of the domain of the attributes?: **NO**
E. __ Does the standard allow/require the specification of data type of the attributes?: **NO**
F. __ Does the standard allow/require the specification of particular classification coding schemes (e.g., FACC, OSKA, S-57 OC)?: **NO**

7.4 Product Specification

Obligation

A. __ Does the standard allow/require for the provision of information about whether the dataset meets a particular product specification?: **NO**
B. __ If A is yes, can it be internal to the Metadata?: **N/A**
C. __ If A is yes, can it be external to the Metadata?: **N/A**

SECTION 8 – Data Quality of the Dataset

This section examines how the standard deals with the question of data quality. This includes homogeneity and metrics for data quality, quality assurance, and usage.

8.1 Overall Data Quality Statement

Obligation

A. **O** Does the standard allow/require for the provision of quality to be described? If yes, please answer the questions below. (*If no, all following answers in Section* 8 *are N/A, please skip to Section* 9.): **YES**
B. x Does the standard allow/require the provision for the quality information referencing an external quality standard, quality assurance standard, or specification?: **YES**
in the section inviting comments about quality

8.2 Homogeneity of Data Quality

Obligation

A. **O** Does the standard allow/require for the variation of the quality of the dataset to be described?: **YES**
B. What type of aggregation of quality information is allowed?
O Entire database?: **YES**
__ Feature/Object instance?: **NO**
__ Feature/Object class?: **NO**
__ Attribute instance?: **NO**
__ Attribute class?: **NO**

__ Layer?: **NO**
__ Theme?: **NO**
O Geographic extent?: **YES**
__ Other?: **NO**

8.3 Metrics for Data Quality
Does the standard allow/require the provision of values for:

Obligation

A. **O** Positional accuracy?: **YES**
1. Standard error?: **NO**
O 2. Maximum error?: **YES**
3. Error correlation?: **NO**
4. Confidence level (e.g., 90%, 95%)?: **NO**
5. Other?: **NO**
B. __ Attribute accuracy?: **NO**
1. Standard error?: **NO**
2. Maximum error?: **NO**
3. Error frequency?: **NO**
4. Error correlation?: **NO**
5. Confidence level (e.g., 90%, 95%)?: **NO**
6. Other?: **NO**
C. __ Temporal accuracy?: **NO**
1. Standard error?: **NO**
2. Maximum error?: **NO**
3. Error correlation?: **NO**
4. Confidence level? (e.g., 90%, 95%): **NO**
5. Other?: **NO**
D. **O** Completeness?: **YES**
1. Error of omission?: **NO**
2. Error of commission?: **NO**
3. Other?: **YES**
identified in evaluation of completeness and uniformity of coverage, Circumstances and other details associated with the data collection
E. x Currentness?: **YES**
O 1. Date?: **YES**
2. Maximum age?: **NO**
O 3. Temporal extent?: **YES**
4. Other?: **NO**
F. __ Logical consistency?: **NO**
G. **O** Lineage?: **YES**
1. List of processing steps?: **NO**
2. List of values of processing parameters?: **NO**
3. List of printouts of statistics, etc.?: **NO**
O 4. Source material?: **YES**

5. Other?: **NO**

H. x Additional Quality Information?: **YES**

Comments about quality. Circumstances and other details associated with the data collection

8.4 Quality Assurance Methods

Does the standard allow/require the provision of information about the methods used to determine:

Obligation

A. __ Positional accuracy?: **NO**
1. Standard error in adjustments?: **NO**
2. Repeated measurements?: **NO**
3. Independent measurements?: **NO**
4. Subjective evaluations?: **NO**
5. Other?: **NO**

B. __ Attribute accuracy?: **NO**
1. Standard error?: **NO**
2. Error classification matrix?: **NO**
3. Independent measurements?: **NO**
4. Subjective evaluation?: **NO**
5. Other?: **NO**

C. __ Temporal accuracy?: **NO**
1. Independent measurements?: **NO**
2. Subjective evaluation?: **NO**
3. Other?: **NO**

D. __ Completeness?: **NO**
1. Field checks?: **NO**
2. Repeated measurements?: **NO**
3. Independent measurements?: **NO**
4. Subjective evaluation?: **NO**
5. Error of commission?: **NO**
6. Other?: **NO**

E. __ Currentness?: **NO**
1. Date of source material?: **NO**
2. Repeated measurements?: **NO**
3. Independent measurements?: **NO**
4. Subjective evaluation?: **NO**
5. Other?: **NO**

F. __ Logical consistency?: **NO**
1. Verification results of consistency checks?: **NO**
2. Repeated processing?: **NO**
3. Comparison with independent sources?: **NO**
4. Subjective evaluation?: **NO**
5. Other?: **NO**

G. __ Lineage?: **NO**
  1. Checking results of processing steps: **NO**
  2. Subjective evaluation: **NO**
  3. Other: **NO**
H. __ Additional Quality Information?: **NO**

8.5 Usage

Obligation

A. **O** Does the metadata standard allow/require the provision of information about usage?: **YES**
  1. Who used it: **NO**
  2. When it was used: **NO**
O 3. For what application it was used: **YES**
  4. The effectiveness of use: **NO**
  5. Other: **NO**

8.6 Alternative Quality Description: **NO**

SECTION 9 – Spatial Data Structure of the Dataset

This section examines how the standard provides for the specification of the spatial data model and spatial data types.

9.1 Spatial Data Model
Does the standard allow/require the provision for:

Obligation

A. __ The description of the geometrical and/or topological primitives used to model the real world?: **NO**
B. __ Information about an external standard for geometrical and/or topological primitive specification?: **NO**

9.2 Spatial Data Types
Does the standard allow/require the provision for:

Obligation

A. __ Information about the spatial data type(s) used in the dataset?: **NO**
B. __ If A is yes, which of the following generic spatial data types are supported? **N/A**
  For each, please list specific data primitive levels supported by each data type.
  1. Vector (e.g., spaghetti, chain-node/node-edge-face, planar graph, full topology)?: **N/A**
  2. Areal Raster (e.g., thematic raster – scanned and classification maps, image raster – satellite images, scanned aerial photographs)?: **N/A**
  3. Grid / Matrix (e.g., Digital Elevation Model)?: **N/A**

4. Triangular Irregular Network (TIN)?: **N/A**
5. Other?: **N/A**

9.3 Other Data Types

Does the standard allow the provision of information about:

Obligation

A. **O** Video?: **YES**
B. __ Sound?: **NO**
C. **O** Other?: **YES**
about any form of documentation available

SECTION 10 – Spatial Reference of the Dataset

This section examines how the dataset provides for the specification of the spatial reference characteristics of the data set.

Does the standard allow/require the provision for the following information:

10.1 Geoidal Model

Obligation

A. __ Geoidal model name?: **NO**
B. __ Geoidal model definitions?: **NO**

10.2 Geodetic datum

Obligation

A. **O** Geodetic datum name?: **YES**
B. __ Geodetic datum definitions?: **NO**

10.3 Reference Ellipsoid

Obligation

A. __ Reference ellipsoid name?: **NO**
B. __ Reference ellipsoid definitions?: **NO**

10.4 Projection

Obligation

A. **O** Projection name?: **YES**
B. __ Projection definitions?: **NO**

10.5 Coordinate Systems

Obligation

A. **O** Coordinate system type?: **YES**
B. __ Coordinate system definitions?: **NO**

10.6 Height Reference System

Obligation

A. x Vertical datum name?: **YES**
B. __ Types of heights?: **NO**

10.7 Ancillary Geodetic Information

Obligation

A. __ Does the standard allow/require the provision for ancillary geodetic, gravimetric, or magnetometric information?: **NO**

10.8 Transformation Parameters

Does the standard specify a method to provide information for coordinate transformation(s):

Obligation

A. __ Model name?: **NO**
B. __ Control points?: **NO**
C. __ Transformation parameters?: **NO**

10.9 Non-Coordinate Spatial Reference

Does the standard allow/require the provision for information on non-coordinate spatial reference systems (e.g., postal systems, street address)?:

Obligation

A. __ Reference system name?: **NO**
B. __ Reference system definitions?: **NO**
C. x Other?: **YES**
Names of areas or features

SECTION 11 – Availability and Distribution of the Dataset

This section examines how the standard provides information on the availability of the dataset?

Does the standard allow/require the provision for:

11.1 The Distributor(s) of the Dataset

Obligation

A. __ Information about the distributor(s) of the dataset?: **NO**
B. x Contact information for the distributor(s)?: **YES**

11.2 Restrictions to Access and Usage

Obligation

A. x Information about copyright on the dataset (e.g., date, period, copyright owner, etc.)?: **YES**
B. x Information about restrictions on the access of the dataset?: **YES**
C. x Information about restrictions on the usage of the dataset?: **YES**

11.3 Pricing Details

Obligation

A. x Information about pricing?: **YES**

11.4 Distribution of the Dataset

Obligation

A. x Information about distribution media?: **YES**
B. x Information about on-line access?: **YES**
C. __ Information about units of distribution (e.g., tiles, layers, polygons, elements)?: **NO**

11.5 Distribution Format

Obligation

A. x Information about distribution formats (e.g., commercial CAD format, commercial GIS formats, National/International transfer standard)?: **YES**
B. __ Information about the encoding of the dataset (e.g., binary, ASCII)?: **NO**
C. __ Information about compression method?: **NO**
D. __ Information about compressed dataset size?: **NO**
E. x Information about the distribution mechanism?: **YES**
F. __ Information about the encapsulation of data (e.g., ISO 8211, ISO 8824)?: **NO**

11.6 Ordering Procedures

Obligation

A. __ Information about ordering procedures for the dataset?: **NO**

11.7 Support Services

Obligation

A. __ Information about support services (e.g., additional data transformations, format or coordinate conversions)?: **NO**

SECTION 12 – Authorization and Verification of the Metadata

This section examines how the standard provides for the description of the authorization and verification on the metadata of a dataset (e.g., how is the metadata itself authorized and verified?).

12.1 Metadata reference

Obligation

A. x Does the standard allow/require the provision of information about reference dates of the metadata (e.g., last check date, last update date, review date)?: **YES**

Date of last revision, Date of last scientific update, Frequency of updates

12.2 Verification Authority

Obligation

A. __ Does the standard allow/require the provision for information about who verified/authorized the metadata?: **NO**

Normally a user extension

B. __ Does the standard specify the method for providing this information (e.g., ISO 9000): **NO**

12.3 Statistical Methods Used for Verification

Obligation

A. __ Does the standard specify the statistical methods to be used for verification of the metadata?: **NO**

B. __ Does the standard allow/require the provision for the description of the statistical methods used to verify the metadata?: **NO**

END OF CHARACTERISTICS
FIN

World Spatial Metadata Standards
H. Moellering, H.J.G.L. Aalders & A. Crane (Editors)

China
地理信息 元数据
Geographic Information – Metadata
GB/T 19710–2005

Jing-tong Jiang, Ruo-mei Liu

I. Information about the Metadata Standard Itself

SECTION 0 – Assessment Information

This section provides the user with a general description of the scope and intended use of the standard. Information on the respondent who prepared the assessment is also provided.

0.1 Respondent

Assessment prepared by:

Name: Jiang Jingtong position: Professor
Name: Liu Ruomei position: Assisted professor
Organization: National Geomatics Center of China (NGCC)
Address: No.1 Baishengcun, Zi Zhu Yuan, Beijing 100044, China
Fax: +86-10-68424101
Phone: +86-10-68424105
E-mail: rmliu@public3.bta.net.cn

Date assessment completed:

August, 2001; Updated May 2002, Updated May, 2005

Relationship of respondent to standard:

Author: Jiang Jingtong, Liu Ruomei
Drafting committee member: Zhou X, Jia Yunpeng, Jiang Zuqin, Yao Yanming
User: GIS and spatial data producers, distributors and users, GIS software developers, etc.
Custodian: Standardization Administration of China (SAC)

0.2 Brief Summary Statement

A. Summary of the scope and intent of the standard:

It is a basic national standard for geographic information in China. It is applicable for all data producers and distributors to manage their data, including cataloguing of datasets, clearinghouse activities, and the full description of datasets; it is also useful for users to find data which they want and to evaluate data for their application purposes.

SECTION 1 – Administration of Metadata Standard

This section provides a general description of the administrative framework, within which the standard was developed, tested and currently resides. Significant details are also provided in terms of the developmental and managerial history as well as information on the existence of documentation, software tools and training materials. This section also provides a point of contact for further information on the standard.

1.1 Name of Standard

A. Name of standard:
   1. In original language(s): 地理信息 元数据
   2. English translation, if appropriate:
      Geographic Information – Metadata

B. Version/Edition:
   2005

C. Language(s) of documentation:
   Chinese

D. Acronym(s):
   N/A

E. Official ID:
   GB/T 19710–2005

F. Name(s) of recognizing standards authority(ies):
   Standardization Administration of China (SAC)

1.2 Responsible Institutions

Contact information about those institutions that are (or were) responsible for the processes involved in the evolution of the standard:

A. Production and/or development:
   Jiang Jingtong Professor
   Liu Ruomei, Assistant professor
   National Geomatics Center of China (NGCC)
   No.1 Baishengcun, Zi Zhu Yuan, Beijing 100044, China
   Fax: +86-10-68424101
   Phone: +86-10-68424105
   E-mail: rmliu@public3.bta.net.cn

B. Testing:

Jiang Jingtong Professor
Liu Ruomei, Assisted professor
National Geomatics Center of China (NGCC)
No. 1 Baishengcun, Zi Zhu Yuan, Beijing 100044, China
Fax: +86-10-68424101
Phone: +86-10-68424105
E-mail: rmliu@public3.bta.net.cn

C. Conformance:

Jiang Jingtong Professor
Liu Ruomei, Assisted professor
National Geomatics Center of China (NGCC)
No. 1 Baishengcun, Zi Zhu Yuan, Beijing 100044, China
Fax: +86-10-68424101
Phone: +86-10-68424105
E-mail: rmliu@public3.bta.net.cn

D. Maintenance:

National Technical Commision of Geographic Information of China (SAC/TC 230)
National Geomatics Center of China (NGCC)
No. 1 Baishengcun, Zi Zhu Yuan, Beijing 100044, China
Fax: +86-10-68424076
Phone: +86-10-68424076
E-mail: gist@nsdi.gov.cn

E. Distribution:

Standardization Administration of China (SAC)
Fax: +86-10-58811656
Phone: +86-10-58811666

F. Help Desk/User Support:

China Publish House for Standard
No. 16 Shanlihe Northern Street, Beijing 100045
Fax: +86-10-68522112

1.3 Development History

A. Beginning of work on metadata standard: **1996-01**

B. Milestones:

1995-12 – Started to attend developing ISO/TC 211 standard: Geographic Information – Metadata
1996-07 – Started to research on Metadata standard as the national key research project
1998-10 – Finished to develop the first level of Metadata standard and tested it through examples
1999-06 – Published the book ⟨Research on the Chinese Geographic Information Metadata

Standard⟩ for the first time in China
2000-01 – Started to develop the national standard
⟨Metadata for Geographic Information⟩
2002-06 – WD
2003-06 – CD
2004-09 – Voting

C. Anticipated completion: **2005-03: national standard**
D. Update cycle: **about every 5 years**

1.4 Status of the Standard

A. Is the standard officially recognized: **expected**
B. Date(s) of recognition: **N/A**
C. Anticipated date of recognition: **2002-06**
D. Current stage in recognition process: officially recognition
E. Steps still required to achieve recognition:
2004-10 ~ 2005-03 official recognition

1.5 Access to Documentation of the Standard

A. Is the standard copyrighted?: **YES**
B. Name and address of copyright owner:
Standardization Administration of China (SAC)
C. Do restrictions on the use of the standard apply?: **NO**
D. Is the standard documentation currently available?: **NO**
E. List the form/media of standard documentation:
Paper only
F. Is the standard available on the Internet?: **NO**
G. Price: **¥80 ~ 150 each copy**
H. Contact details for obtaining the standard:
China Publish House for Standard
No. 16 Shanlihe Northern Street, Beijing 100045
Fax: +86-10-68522112
I. Does the standard have ISBN/ISSN numbers?: **YES**
The standard will have ISBN/ISSN number when it is published.

1.6 Available Software Tools

A. Is software available to implement the standard as a database?: **YES**
It has been developed a set of software tools to implement profiles of the standard.

B. Is software available to test compliance of metadata with the standard?: **YES**
The software has been developed based on the content of the profile of the standard.

C. Is software available to produce metadata compliant with the standard?: **YES**

D. Other relevant software: **YES**

E. Is there a test dataset available to test software tools for implementing the standard?: **YES**

F. For each product available above, please list contact information:

Jiang Jingtong, Professor
Liu Ruomei, Assisted professor
National Geomatics Center of China (NGCC)
No.1 Baishengcun, Zi Zhu Yuan, Beijing 100044, China
Fax: +86-10-68424101
Phone: +86-10-68424105
E-mail: rmliu@public3.bta.net.cn

1.7 Available Training Materials

A. Are organized training sessions available?: **YES**

B. Contact details for training sessions:

Jiang Jingtong, Professor
Liu Ruomei, Assisted professor
National Geomatics Center of China (NGCC)
No.1 Baishengcun, Zi Zhu Yuan, Beijing 100044, China
Fax: +86-10-68424101
Phone: +86-10-68424105
E-mail: rmliu@public3.bta.net.cn

C. Training documentation available:
Training documentation will be created after the standard published. The book ⟨Research on the Chinese Geographic Information Metadata Standard⟩ is a reference as well.

SECTION 2 – Use and Implementation of the Metadata Standard

This section provides information on the way in which the standard is used and the purpose of its use. It also provides descriptions about the level of abstraction of the standard.

2.1 Intended Use of the Standard
Can the standard be used:

A. Within a transfer data structure?: **YES**
B. To produce hardcopy catalog of datasets?: **YES**
C. To catalog data in a digital on-line catalog?: **YES**
D. Other: **N/A**

2.2 Intended Purpose of the Standard
Is the intended purpose of the standard:

A. To support discovery, identification, location of data?: **YES**
B. To support how to access data?: **YES**
C. To support the determination of dataset fitness for use?: **YES**
D. To support transfer of dataset?: **YES**
E. To support an organization's data management?: **YES**
F. Other: **N/A**

2.3 Table of Contents

A. Provide the standard's Table of Contents

It is divided into two levels.
The contents of level one include core metadata entities and elements.
The contents of level two include full metadata entities and elements.
It includes:
Identification information
Constraint information
Data quality information
Maintenance information
Spatial representation information
Reference system information
Content information
Portrayal catalogue information
Distribution information
Metadata extension information
Application schema information
Extent information
Citation and responsible party information
etc.

2.4 Level of Abstraction of the Metadata Standard Definition

A. Does the standard define metadata elements?: **YES**
B. Does the standard define a model for the relationships between the metadata elements?: **YES**
C. Provide an overview diagram of the model of the metadata, if available. **N/A**
D. Does the standard define domains and/or data types of the metadata elements?: **YES**
E. Does the standard define how the metadata should be encoded?: **YES**
F. Does the standard define a database schema for the automatic generation of a database?: **NO**

2.5 Extensibility of the Standard

A. Is the metadata description extensible (e.g., Can user add metadata domains, elements, thematic profiles, etc.)?: **YES**

The user can add metadata domains, elements, thematic profiles, etc.
But the rules for metadata extension must be complied.

2.6 Availability of Metadata
Does the standard specify:

A. How metadata are to be accessed?: **YES**
 metadata could be accessed by visiting main website nodes and subnodes using key words of metadata.
B. A mechanism for querying the metadata?: **YES**
 Metadata could be queried by key words of metadata, or by domains, element names, etc.
C. Particular standardized query protocols?: **NO**
D. That the metadata should be encoded for computer access?: **YES**
E. That the metadata should be encrypted for security?: **NO**
 all of metadata are open to the whole world.

2.7 Examples

A. Does the standard provide examples of implementations?: **YES**
 examples are given in a informative annex of the standard.
B. How many examples are available?: **4**

SECTION 3 – Linkage and Coordination of the Metadata Standard

This section provides information on how the metadata standard is linked to other standards, particularly transfer standards, and how such linkage is implemented. If a significant linkage does exist, detailed technical information is provided on the precise nature of the linkage to the standard, and how the linkage affects implementation of the metadata standard.

3.1 Associated Transfer Standards

A. Is the metadata standard associated with a transfer standard(s)?: **NO**
B. If A is yes, provide references and contact details for each linked standard: **N/A**
C. If A is yes, are both (all) standards coordinated?: **N/A**
D. If A is yes, describe how the coordination works, both from a technical and conformance (normative) viewpoint: **N/A**

3.2 Incorporated into transfer mechanism

A. Is the metadata standard incorporated into a transfer mechanism?: **NO**
B. If A is yes, provide references and contact details for each incorporated mechanism: **N/A**
C. If A is yes, describe how: **N/A**

3.3 Other linked, coordinated or associated standards

A. Is the metadata standard linked, coordinated, or associated with any other standard(s) not listed above?: **YES**

B. If A is yes, provide references and contact details for each standard:
   1. **The Codes for the Administrative Divisions**
   2. **The Data Classification and Codes for the National Fundamental Geographic Information**
   3. **The Codes of River Names**
   4. **The Data Transfer Format for Geographic Information**
   5. **The Naming, Numbering and Coding Rules of Highway Route**
   6. **The Specification of Digital Products of the National Fundamental Geographic Information**
   7. **The Data Quality Control of Geographic Information**

C. If A is yes, describe how for each relevant standard:
   1. **The Codes for the Administrative Divisions – linked with the codelist of the national standard of Codes for the administrative divisions of the People's Republic of China.**
   2. **The Data Classification and Codes for the National Fundamental Information – linked with codelist and identification information of metadata.**
   3. **The Codes of River Names – linked with codelist and reference system information of metadata.**
   4. **The Data Transfer Format for Geographic Information – linked with codelist and Distribution information of metadata.**
   5. **The Naming, Numbering and Coding Rules of Highway Route – linked with reference system information and codelist of metadata.**
   6. **The Specification of Digital Products of the National Fundamental Geographic Information – linked with Distribution information of metadata.**
   7. **The Data Quality Control of Geographic Information – linked with Data quality information of metadata.**

II. Categories the Metadata Standard Uses to Describe Datasets

The following sections are categories of metadata that may be used by the metadata standard to describe data. Not all categories are relevant to different datasets and may not necessarily be present in the standard. The presence or absence of a category should not be used for judging one standard against another.

Note: For each of the questions in the following sections please provide the obligation status information. Valid obligation values are:

	Obligation Level	
M	Mandatory	This information must be entered to comply with the metadata standard.
C	Conditional	This information is entered based on other information in the metadata.
O	Optional	This information is optionally entered in the metadata.

If the answer is 'NO', the obligation status should be left blank.

SECTION 4 – Dataset Identification

This section examines how the standard provides for the specification of the various sorts of titles and identifiers for a dataset.

4.1 The title(s) of a dataset
Does the standard allow/require the provision for:

Obligation

A. __ Previous title(s)?: **NO**
B. **M** Current title(s)?: **YES**
C. __Possible future title(s)?: **NO**
D. __Official titles in different languages?: **NO**
E. __Abbreviated title(s)?: **NO**
F. **O** Alternative Title(s)?: **YES**
G. __ Does the standard support effective dates of titles?: **NO**
H. __ Other identifiers (e.g., ISBN, National Stock Number, Uniform Product Code)?: **NO**

4.2 Parties responsible for the dataset
Does the standard allow/require the provision for:

Obligation

A. **M** Author?: **YES**
B. **M** Owner?: **YES**
C. **O** Other?: **YES**
Data producer, manager and distributer

4.3 ISBD Information

Obligation

A. __ Does the metadata standard specify inclusion of mandatory ISBD information (see glossary) within the metadata?: **NO**

SECTION 5 – Status of the Dataset

This section examines how the standard provides for the specification of the current status and updating of the data set.

5.1 Status and Progress of Datasets
Does the metadata standard allow/require the provision for:

Obligation

A. **M** Information on the status of the entire dataset?: **YES**
Use an element to specify the status of the entire dataset.
B. **C** Information on the status of the particular subset (e.g., observation time, inauguration time, extraction time)?: **YES**
This kind of information is described in the entity of lineage.

Does the metadata standard specify:

C. **C** A structure for updating or maintenance information?: **YES**
D. **M** A structure to describe historical information (see also Section 8.3 G)?: **YES**

5.2 Information on Incremental Updating
Does the metadata standard specify:

Obligation

A. **C** A structure for information on incremental updates of the entire dataset?: **YES**
This kind of information is described in the entity of lineage.
B. **C** A structure for information on incremental updates of particular portions of the dataset?: **YES**
This kind of information is described in the entity of lineage.
C. **C** A structure to describe the update process for the dataset?: **YES**
This kind of information is described in the entity of lineage.

SECTION 6 – Extent of the Dataset

This section examines how the standard provides for the specification of the spatial and temporal extent of the dataset.

6.1 The spatial extent of the dataset

Obligation

A. **M** Does the metadata standard allow/require for the description of horizontal extent (e.g., bounding rectangle/polygon, named location (i.e., Stockholm, etc.))?: **YES**
Extent information includs the bounding polygon and named location of the dataset
B. **C** Does the metadata standard allow/require for the description of vertical extent (e.g., minima/maxima heights, etc.)?: **YES**
Minima/maxima heights
C. **C** Does the standard require specific units for describing extent?: **YES**
D. **C** Does the standard allow specification of the units for describing extent?: **YES**
E. **C** Does the standard require specific spatial reference system for describing extent?: **YES**
F. **C** Does the standard allow specification of the spatial reference system for describing extent?: **YES**

6.2 The temporal extent of the dataset

Obligation

A. **C** Does the standard allow/require for the description of time period for which the dataset is relevant?: **YES**
B. **C** Does the standard allow/require specification of the temporal reference system?: **YES**

The temporal reference system includes three common types of temporal reference systems: calendars (used with clocks for greater precision), temporal coordinate systems, and ordinal temporal reference systems.

SECTION 7 – Data Content of the Dataset

This section examines how the standard provides for the specification of the data content of the dataset in terms of overview elements, themes, data definitions and product specifications.

7.1 Dataset overview

Does the standard allow/require for the provision of the following information:

Obligation

A. **O** Dataset overview?: **YES**
B. **M** Abstract – brief summary of the content of the dataset?: **YES**
C. **M** Purpose – summary of the intentions with which the dataset was developed?: **YES**
D. **O** Usage – list of activities for which the dataset has been used? (see also Section 8.5): **YES**
E. **M** Spatial schema – (e.g., vector, raster)? (see also Section 9): **YES**
F. **M** Spatial reference system? (see also Section 12): **YES**
G. **M** Language?: **YES**
H. **M** Character set (e.g., ISO8859-1, JISX0208)?: **YES**
I. **O** List of documents providing further information about the dataset?: **YES**
J. __ Sample/preview dataset?: **NO**
K. __ Related datasets?: **NO**
L. **O** Original size of the dataset?: **YES**
M. **C** Compression? (see also Section 11.5): **YES**
Dependant on different datasets.
N. **M** Other?: **YES**
Projection used for the dataset, data quality, point of contact, constraints of using the data, and so on.

7.2 Theme(s)

Obligation

A. **O** Does the standard allow/require the provision for the description of theme(s) (e.g., topography, land cover, transport, etc.) in the dataset?: **YES**
B. **O** If A is yes, does the standard allow the referencing of an outside dictionary or thesaurus?: **YES**
C. **O** If A is yes, does the standard itself include a dictionary or thesaurus within the metadata?: **YES**
The standard itself includes a codelist.

D. **O** If A is yes, does the standard allow the use of unauthorized user-defined keywords?: **YES**

7.3 Data Definition

Obligation

A. **O** Does the standard allow/require the provision for the description of the data definition?: **YES**

B. If A is yes, does the standard allow/require the provision for the description of:
O 1. Features/Objects?: **YES**
O 2. Attributes?: **YES**
O 3. Relations?: **YES**

C. **O** Can attributes be described separately from feature/object classes?: **YES**

D. **O** Does the standard allow/require the specification of the domain of the attributes?: **YES**

E. **O** Does the standard allow/require the specification of data type of the attributes?: **YES**

F. **O** Does the standard allow/require the specification of particular classification coding schemes (e.g., FACC, OSKA, S-57 OC)?: **YES**

For example: the Data Classification and Codes for the National Fundamental Geographic Information.

7.4 Product Specification

Obligation

A. **O** Does the standard allow/require for the provision of information about whether the dataset meets a particular product specification?: **YES**

B. **O** If A is yes, can it be internal to the Metadata?: **YES**

C. __ If A is yes, can it be external to the Metadata?: **NO**

SECTION 8 – Data Quality of the Dataset

This section examines how the standard deals with the question of data quality. This includes homogeneity and metrics for data quality, quality assurance, and usage.

8.1 Overall Data Quality Statement

Obligation

A. **M** Does the standard allow/require for the provision of quality to be described? If yes, please answer the questions below. (*If no, all following answers in Section* 8 *are N/A, please skip to Section* 9.): **YES**

B. **O** Does the standard allow/require the provision for the quality information referencing an external quality standard, quality assurance standard, or specification?: **YES**

The national standard: the Data Quality Control of Geographic Information.

8.2 Homogeneity of Data Quality

Obligation

A. **C** Does the standard allow/require for the variation of the quality of the dataset to be described?: **YES**

B. What type of aggregation of quality information is allowed?

- **M** 1. Entire database?: **YES**
- **O** 2. Feature/Object instance?: **YES**
  when the metadata is for individual geographic feature instance.
- **O** 3. Feature/Object class?: **YES**
  when the metadata is for individual geographic feature type.
- **O** 4. Attribute instance?: **YES**
  when the metadata is for individual geographic attribute instance.
- **O** 5. Attribute class?: **YES**
  when the metadata is for individual geographic attribute type.
- **O** 6. Layer?: **YES**
- **O** 7. Theme?: **YES**
- **O** 8. Geographic extent?: **YES**
- **O** 9. Other?: **YES**
  Version, etc.

8.3 Metrics for Data Quality

Does the standard allow/require the provision of values for:

Obligation

A. **O** Positional accuracy?: **YES**

1. Standard error?: **YES**
2. Maximum error?: **YES**
3. Error correlation?: **YES**
4. Confidence level (e.g., 90%, 95%)?: **NO**
5. Other?: **YES**
   Absolute external positional accuracy, relative internal positional accuracy, gridded data positional accuracy, and so on.

B. **O** Attribute accuracy?: **YES**

1. Standard error?: **YES**
2. Maximum error?: **YES**
3. Error frequency?: **YES**
4. Error correlation?: **YES**
5. Confidence level (e.g., 90%, 95%)?: **YES**
6. Other?: **YES**
   Thematic classification correctness, non-quantitative attribute accuracy, quantitative attribute accuracy, etc.

C. **O** Temporal accuracy?: **YES**

1. Standard error?: **YES**

2. Maximum error?: **YES**
3. Error correlation?: **NO**
4. Confidence level? (e.g., 90%, 95%): **NO**
5. Other?: **YES**
Temporal consistency, temporal validity.

D. **O** Completeness?: **YES**
1. Error of omission?: **YES**
2. Error of commission?: **YES**
3. Other?: **YES**

E. **O** Currentness?: **YES**
1. Date?: **YES**
2. Maximum age?: **YES**
3. Temporal extent?: **YES**
4. Other?: **YES**
layer completeness, feature type completeness, etc.

F. **O** Logical consistency?: **YES**
Conceptual consistency, Domain consistency, Formal consistency, Topological consistency.

G. **M** Lineage?: **YES**
1. List of processing steps?: **YES**
2. List of values of processing parameters?: **YES**
3. List of printouts of statistics, etc.?: **NO**
4. Source material?: **YES**
5. Other?: **YES**
Time of data processing, updating date and method, etc.

H. **O** Additional Quality Information?: **YES**
Documents for dataset.

8.4 Quality Assurance Methods

Does the standard allow/require the provision of information about the methods used to determine:

Obligation

A. **O** Positional accuracy?: **YES**
1. Standard error in adjustments?: **YES**
2. Repeated measurements?: **NO**
3. Independent measurements?: **YES**
4. Subjective evaluations?: **YES**
5. Other?: **NO**

B. **O** Attribute accuracy?: **YES**
1. Standard error?: **YES**
2. Error classification matrix?: **YES**
3. Independent measurements?: **YES**
4. Subjective evaluation?: **YES**
5. Other?: **NO**

C. **O** Temporal accuracy?: **YES**
1. Independent measurements?: **NO**
2. Subjective evaluation?: **YES**
3. Other?: **NO**

D. **O** Completeness?: **YES**
1. Field checks?: **YES**
2. Repeated measurements?: **YES**
3. Independent measurements?: **YES**
4. Subjective evaluation?: **YES**
5. Error of commission?: **YES**
6. Other?: **YES**
Error of omission.

E. **O** Currentness?: **YES**
1. Date of source material?: **YES**
2. Repeated measurements?: **NO**
3. Independent measurements?: **NO**
4. Subjective evaluation?: **YES**
5. Other?: **YES**
Date of updating.

F. **O** Logical consistency?: **YES**
1. Verification results of consistency checks?: **YES**
2. Repeated processing?: **NO**
3. Comparison with independent sources?: **NO**
4. Subjective evaluation?: **YES**
5. Other?: **YES**
Using software tools.

G. **O** Lineage?: **YES**
1. Checking results of processing steps: **YES**
2. Subjective evaluation: **YES**
Other: **YES**
Checking relative documents.

H. **O** Additional Quality Information?: **NO**

8.5 Usage

Obligation

A. **O** Does the metadata standard allow/require the provision of information about usage?: **YES**
1. Who used it: **YES**
2. When it was used: **YES**
3. For what application it was used: **YES**
4. The effectiveness of use: **YES**
5. Other: **YES**
Demonstrations by data producers.

8.6 Alternative Quality Description: **YES**

Purpose of data produced, etc.

SECTION 9 – Spatial Data Structure of the Dataset

This section examines how the standard provides for the specification of the spatial data model and spatial data types.

9.1 Spatial Data Model

Does the standard allow/require the provision for:

Obligation

A. **O** The description of the geometrical and/or topological primitives used to model the real world?: **YES**

B. __ Information about an external standard for geometrical and/or topological primitive specification?: **NO**

9.2 Spatial Data Types

Does the standard allow/require the provision for:

Obligation

A. **M** Information about the spatial data type(s) used in the dataset?: **YES**

B. **C** If A is yes, which of the following generic spatial data types are supported? For each, please list specific data primitive levels supported by each data type.

1. Vector (e.g., spaghetti, chain-node/node-edge-face, planar graph, full topology)?: **YES**

   Full topology.

2. Areal Raster (e.g., thematic raster – scanned and classification maps, image raster – satellite images, scanned aerial photographs)?: **YES**

   Thematic raster – scanned and classification maps, image raster – satellite images, scanned aerial photographs.

3. Grid/Matrix (e.g., Digital Elevation Model)?: **YES**

   DEM

4. Triangular Irregular Network (TIN)?: **YES**

5. Other?: **YES**

   Audio, video.

9.3 Other Data Types

Does the standard allow the provision of information about:

Obligation

A. **O** Video?: **YES**

B. **O** Sound?: **YES**

C. **O** Other?: **YES**

Statistical data

SECTION 10 – Spatial Reference of the Dataset

This section examines how the dataset provides for the specification of the spatial reference characteristics of the data set.

Does the standard allow/require the provision for the following information:

10.1 Geoidal Model

Obligation

A. **O** Geoidal model name?: **YES**
B. **O** Geoidal model definitions?: **YES**

10.2 Geodetic datum

Obligation

A. **M** Geodetic datum name?: **YES**
B. **O** Geodetic datum definitions?: **YES**

10.3 Reference Ellipsoid

Obligation

A. **O** Reference ellipsoid name?: **YES**
B. **O** Reference ellipsoid definitions?: **YES**

10.4 Projection

Obligation

A. **C** Projection name?: **YES**
B. **O** Projection definitions?: **YES**

10.5 Coordinate Systems

Obligation

A. **C** Coordinate system type?: **YES**
B. **O** Coordinate system definitions?: **YES**

10.6 Height Reference System

Obligation

A. **C** Vertical datum name?: **YES**
when the referencing system is compound.
B. **O** Types of heights?: **YES**
when the referencing system is compound.

10.7 Ancillary Geodetic Information

Obligation

A. **C** Does the standard allow/require the provision for ancillary geodetic, gravimetric, or magnetometric information?: **YES**

Yes, when the data is a special theme, such as the gravity data, geodetic data.

10.8 Transformation Parameters

Does the standard specify a method to provide information for coordinate transformation(s):

Obligation

A. **C** Model name?: **YES**
B. **C** Control points?: **YES**
C. **C** Transformation parameters?: **YES**

10.9 Non-Coordinate Spatial Reference

Does the standard allow/require the provision for information on non-coordinate spatial reference systems (e.g., postal systems, street address)?:

Obligation

A. **C** Reference system name?: **YES**
B. **O** Reference system definitions?: **YES**
C. **O** Other?: **YES**

gazetteer, etc.

SECTION 11 – Availability and Distribution of the Dataset

This section examines how the standard provides information on the availability of the dataset?

Does the standard allow/require the provision for:

11.1 The Distributor(s) of the Dataset

Obligation

A. **M** Information about the distributor(s) of the dataset?: **YES**
B. **M** Contact information for the distributor(s)?: **YES**

11.2 Restrictions to Access and Usage

Obligation

A. **M** Information about copyright on the dataset (e.g., date, period, copyright owner, etc.)?: **YES**
B. **M** Information about restrictions on the access of the dataset?: **YES**
C. **M** Information about restrictions on the usage of the dataset?: **YES**

11.3 Pricing Details

Obligation

A. Information about pricing?: **NO**

11.4 Distribution of the Dataset

Obligation

A. **O** Information about distribution media?: **YES**
B. **O** Information about on-line access?: **YES**
C. **O** Information about units of distribution (e.g., tiles, layers, polygons, elements)?: **YES**

11.5 Distribution Format

Obligation

A. **M** Information about distribution formats (e.g., commercial CAD format, commercial GIS formats, National/International transfer standard)?: **YES**
B. **O** Information about the encoding of the dataset (e.g., binary, ASCII)?: **YES**
C. __ Information about compression method?: **NO**
D. __ Information about compressed dataset size?: **NO**
E. **O** Information about the distribution mechanism?: **YES**
F. __ Information about the encapsulation of data (e.g., ISO 8211, ISO 8824)?: **NO**

11.6 Ordering Procedures

Obligation

A. **O** Information about ordering procedures for the dataset?: **YES**

11.7 Support Services

Obligation

A. **O** Information about support services (e.g., additional data transformations, format or coordinate conversions)?: **YES**
additional data transformations, format or coordinate conversions, data selection from or adding to the original dataset.

SECTION 12 – Authorization and Verification of the Metadata

This section examines how the standard provides for the description of the authorization and verification on the metadata of a dataset (e.g., how is the metadata itself authorized and verified?).

12.1 Metadata reference

Obligation

A. **M** Does the standard allow/require the provision of information about reference dates of the metadata (e.g., last check date, last update date, review date)?: **YES**

Last update date.

12.2 Verification Authority

Obligation

A. **M** Does the standard allow/require the provision for information about who verified/authorized the metadata?: **YES**

Contact information includes name, organization, address, phone number, fax number, email, etc.

B. __ Does the standard specify the method for providing this information (e.g., ISO 9000): **NO**

12.3 Statistical Methods Used for Verification

Obligation

A. **O** Does the standard specify the statistical methods to be used for verification of the metadata?: **YES**

B. __ Does the standard allow/require the provision for the description of the statistical methods used to verify the metadata?: **NO**

END OF CHARACTERISTICS

FIN

World Spatial Metadata Standards
H. Moellering, H.J.G.L. Aalders & A. Crane (Editors)

Czech Republic
Standard ISVS pro strukturu a vymenny format metadat informacnich zdroju
Standard for Structure and Transfer Format of Metadata on Geo-data Sets
011/01.02

Petr Rapant, Bronislava Horakova, Jan Ruzicka

I. Information about the Metadata Standard Itself

SECTION 0 – Assessment Information

This section provides the user with a general description of the scope and intended use of the standard. Information on the respondent who prepared the assessment is also provided.

0.1 Respondent

Assessment prepared by:

Petr Rapant
Researcher
Joint Laboratory of CAGI and Institute of Geoinformatics
VSB – Technical University of Ostrava
17. listopadu 15
708 33 Ostrava – Poruba
Czech Republic
Fax: +420 59 691 8589
Tel.: +420 59 732 5470
E-mail: petr.rapant@vsb.cz

Bronislava Horakova
Head of Laboratory
Joint Laboratory of CAGI and Institute of Geoinformatics
VSB – Technical University of Ostrava
17. listopadu 15

708 33 Ostrava – Poruba
Czech Republic
Fax: +420 59 691 8589
Tel.: +420 59 732 5599
E-mail: bronislava.horakova@vsb.cz
http://www.cagi.cz/midas

Jan Ruzicka
Researcher
Joint Laboratory of CAGI and Institute of Geoinformatics
VSB – Technical University of Ostrava
17. listopadu 15
708 33 Ostrava – Poruba
Czech Republic
Fax: +420 59 691 8589
Tel.: +420 59 732 5472
E-mail: jan.ruzicka@vsb.cz

Date assessment completed:
July, 2002, Reviewed September, 2004

Relationship of respondent to standard:
1, user
2, author
3, author

0.2 Brief Summary Statement

A. Summary of the scope and intent of the standard:
Standard is based on provisional standard CEN ENV 12657 for metadata and it is extended by structures permitting description of data sets with no spatial reference and also structures for description of events, services, SW, and documents. These extended possibilities are based mainly on Dublin Core Metadata Initiative specification.

SECTION 1 – Administration of Metadata Standard

This section provides a general description of the administrative framework, within which the standard was developed, tested and currently resides. Significant details are also provided in terms of the developmental and managerial history as well as information on the existence of documentation, software tools and training materials. This section also provides a point of contact for further information on the standard.

1.1 Name of Standard

A. Name of standard:
1. In original language(s):
Standard ISVS pro strukturu a vymenny format metadat informacnich zdroju

2. English translation, if appropriate:
   Standard for Structure and Transfer Format of Metadata on Geo-data Sets

B. Version/Edition:
   01.02-22.2.2002

C. Language(s) of documentation:
   Czech

D. Acronym(s):
   N/A

E. Official ID:
   011/01.02

F. Name(s) of recognizing standards authority(ies):
   Ministery of Informatics
   Czech Association for Geoinformation (CAGI)

1.2 Responsible Institutions

Contact information about those institutions that are (or were) responsible for the processes involved in the evolution of the standard:

A. Production and/or development:
   Czech Association for Geoinformation
   office 229B
   Novotneho lavka 5
   116 68 Praha 1
   Czech Republic
   Tel./Fax: +420 221 082 374
   http://www.cagi.cz

B. Testing:
   Ministery of Informatics
   Havelkova 22
   130 00 Praha 3
   Czech Republic
   Tel.: +420 221 008 111
   Fax: +420 224 221 484
   E-mail: posta@micr.cz

C. Conformance:
   Ministery of Informatics
   Havelkova 22
   130 00 Praha 3
   Czech Republic
   Tel.: +420 221 008 111
   Fax: +420 224 221 484
   E-mail: posta@micr.cz

D. Maintenance:

Czech Association for Geoinformation
office 229B
Novotneho lavka 5
116 68 Praha 1
Czech Republic
Tel./Fax: +420 221 082 374
E-mail: cagi@cagi.cz
http://www.cagi.cz

E. Distribution:

Ministery of Informatics
Havelkova 22
130 00 Praha 3
Czech Republic
Tel.: +420 221 008 111
Fax: +420 224 221 484
E-mail: posta@micr.cz

F. Help Desk/User Support:

Joint Laboratory of CAGI and Institute of Geoinformatics
VSB – Technical University of Ostrava
17. listopadu 15
708 33 Ostrava – Poruba
Czech Republic
Fax: +420 69 691 8589
Tel.: +420 69 732 5599
E-mail: bronislava.horakova@vsb.cz
http://www.cagi.cz/midas

1.3 Development History

A. Beginning of work on metadata standard: **April, 2000**

B. Milestones:

September, 2000 – draft proposal
December, 2000 – final version

C. Anticipated completion: **January, 2001**

D. Update cycle: **1 year**

1.4 Status of the Standard

A. Is the standard officially recognized: **YES**

B. Date(s) of recognition: **February, 2002**

C. Anticipated date of recognition: **N/A**

D. Current stage in recognition process: **completed**

E. Steps still required to achieve recognition: **N/A**

1.5 Access to Documentation of the Standard

A. Is the standard copyrighted?: **NO**
B. Name and address of copyright owner: **N/A**
C. Do restrictions on the use of the standard apply?: **NO**
D. Is the standard documentation currently available?: **YES**
E. List the form/media of standard documentation:
paper, www

F. Is the standard available on the Internet?: **YES**
http://www.micr.cz/files/461/uvis-S011.01.02-20020222PDF.zip

G. Price: **FREE**
H. Contact details for obtaining the standard:
Ministery of Informatics
Havelkova 22
130 00 Praha 3
Czech Republic
Tel.: +420 221 008 111
Fax: +420 224 221 484
E-mail: posta@micr.cz
I. Does the standard have ISBN/ISSN numbers?: **YES**
ISSN 1213-225X

1.6 Available Software Tools

A. Is software available to implement the standard as a database?: **YES**
B. Is software available to test compliance of metadata with the standard?: **NO**
C. Is software available to produce metadata compliant with the standard?: **YES**
D. Other relevant software: **N/A**
E. Is there a test dataset available to test software tools for implementing the standard?: **NO**
F. For each product available above, please list contact information:
Ad A – MIDAS (Metainformation System of Geodata Sets):
Bronislava Horakova
head of laboratory
Joint Laboratory of CAGI and Institute of Geoinformatics
VSB – Technical University of Ostrava
17. listopadu 15
708 33 Ostrava – Poruba
Czech Republic
Fax: +420 69 691 8589
Tel.: +420 69 732 5599
E-mail: bronislava.horakova@vsb.cz
http://www.cagi.cz/midas
Ad C – Metainformation system for zoning:

Ministry for Regional Development
Staromestske nam. 6
110 15 Prague 1
Czech Republic
Tel.: +420 224 861 111
Fax: +420 224 861 333
http://www.mmr.cz

1.7 Available Training Materials

A. Are organized training sessions available?: **YES**
B. Contact details for training sessions:
Bronislava Horakova
Head of Laboratory
Joint Laboratory of CAGI and Institute of Geoinformatics
VSB – Technical University of Ostrava
17. listopadu 15
708 33 Ostrava – Poruba
Czech Republic
Fax: +420 69 691 8589
Tel.: +420 69 732 5599
E-mail: bronislava.horakova@vsb.cz
http://www.cagi.cz/midas
C. Training documentation available:
late 2002 at www.cagi.cz/midas

SECTION 2 – Use and Implementation of the Metadata Standard

This section provides information on the way in which the standard is used and the purpose of its use. It also provides descriptions about the level of abstraction of the standard.

2.1 Intended Use of the Standard
Can the standard be used:

A. Within a transfer data structure?: **YES**
B. To produce hardcopy catalog of datasets?: **NO**
C. To catalog data in a digital on-line catalog?: **YES**
D. Other: **N/A**

2.2 Intended Purpose of the Standard
Is the intended purpose of the standard:

A. To support discovery, identification, location of data?: **YES**
B. To support how to access data?: **YES**
C. To support the determination of dataset fitness for use?: **YES**
D. To support transfer of dataset?: **NO**

E. To support an organization's data management?: **YES**
F. Other: **N/A**

2.3 Table of Contents

A. Provide the standard's Table of Contents

Introduction
1 Scope
2 Normative references
3 Terms and definitions
4 Description of content and structure of metadata
4.1 Metadata classes and associations
4.2 Codelists
5 Description of metadata transfer format
5.1 Introductory notes
5.2 Definition of transfer format
6 Annexes
Annex A (informative) Codelists and classifications used in standard
Annex B (normative) Document type definition (DTD) for description of metadata transfer format
Annex C (informative) Example of transferred metadata file
Annex D (informative) Possible interpretation of transferred metadata file from Annex C

2.4 Level of Abstraction of the Metadata Standard Definition

A. Does the standard define metadata elements?: **YES**
B. Does the standard define a model for the relationships between the metadata elements?: **YES**
C. Provide an overview diagram of the model of the metadata, if available. **NO**
D. Does the standard define domains and/or data types of the metadata elements?: **YES**
E. Does the standard define how the metadata should be encoded?: **YES**
F. Does the standard define a database schema for the automatic generation of a database?: **NO**

2.5 Extensibility of the Standard

A. Is the metadata description extensible (e.g., Can user add metadata domains, elements, thematic profiles, etc.)?: **NO**

2.6 Availability of Metadata

Does the standard specify:

A. How metadata are to be accessed?: **NO**
B. A mechanism for querying the metadata?: **NO**
C. Particular standardized query protocols?: **NO**
D. That the metadata should be encoded for computer access?: **NO**
E. That the metadata should be encrypted for security?: **NO**

2.7 Examples

A. Does the standard provide examples of implementations?: **YES**
example dataset of transferred metadata including example of interpretation of it

B. How many examples are available?: **1**

SECTION 3 – Linkage and Coordination of the Metadata Standard

This section provides information on how the metadata standard is linked to other standards, particularly transfer standards, and how such linkage is implemented. If a significant linkage does exist, detailed technical information is provided on the precise nature of the linkage to the standard, and how the linkage affects implementation of the metadata standard.

3.1 Associated Transfer Standards

A. Is the metadata standard associated with a transfer standard(s)?: **NO**
B. If A is yes, provide references and contact details for each linked standard: **N/A**
C. If A is yes, are both (all) standards coordinated?: **N/A**
D. If A is yes, describe how the coordination works, both from a technical and conformance (normative) viewpoint: **N/A**

3.2 Incorporated into transfer mechanism

A. Is the metadata standard incorporated into a transfer mechanism?: **NO**
B. If A is yes, provide references and contact details for each incorporated mechanism: **N/A**
C. If A is yes, describe how: **N/A**

3.3 Other linked, coordinated or associated standards

A. Is the metadata standard linked, coordinated, or associated with any other standard(s) not listed above?: **YES**

B. If A is yes, provide references and contact details for each standard:
CEN ENV 12657 – Geographic Information – Data description – Metadata

C. If A is yes, describe how for each relevant standard:
Standard for structure and transfer format of metadata on geo-data sets is based on CEN ENV 12657 – Geographic Information – Data description – Metadata and will periodically accommodate to any changes in this one

II. Categories the Metadata Standard Uses to Describe Datasets

The following sections are categories of metadata that may be used by the metadata standard to describe data. Not all categories are relevant to different datasets and may

not necessarily be present in the standard. The presence or absence of a category should not be used for judging one standard against another.

Note: For each of the questions in the following sections please provide the obligation status information. Valid obligation values are:

Obligation Level		
M	Mandatory	This information must be entered to comply with the metadata standard.
C	Conditional	This information is entered based on other information in the metadata.
O	Optional	This information is optionally entered in the metadata.

If the answer is 'NO', the obligation status should be left blank.

SECTION 4 – Dataset Identification

This section examines how the standard provides for the specification of the various sorts of titles and identifiers for a dataset.

4.1 The title(s) of a dataset
Does the standard allow/require the provision for:

Obligation

A. __ Previous title(s)?: **NO**
B. **M** Current title(s)?: **YES**
C. __ Possible future title(s)?: **NO**
D. **O** Official titles in different languages?: **YES**
E. **O** Abbreviated title(s)?: **YES**
F. **O** Alternative Title(s)?: **YES**
G. __ Does the standard support effective dates of titles?: **NO**
H. **O** Other identifiers (e.g., ISBN, National Stock Number, Uniform Product Code)?: **YES**
version

4.2 Parties responsible for the dataset
Does the standard allow/require the provision for:

Obligation

A. **O** Author?: **YES**
B. **O** Owner?: **YES**
C. **O** Other?: **YES**
provider, administrator, copyright owner

4.3 ISBD Information

Obligation

A. __ Does the metadata standard specify inclusion of mandatory ISBD information (see glossary) within the metadata?: **NO**

SECTION 5 – Status of the Dataset

This section examines how the standard provides for the specification of the current status and updating of the data set.

5.1 Status and Progress of Datasets
Does the metadata standard allow/require the provision for:

Obligation

A. __ Information on the status of the entire dataset?: **NO**
B. __ Information on the status of the particular subset (e.g., observation time, inauguration time, extraction time)?: **NO**
Does the metadata standard specify:
C. __ A structure for updating or maintenance information?: **NO**
D. __ A structure to describe historical information (see also Section 8.3 G)?: **NO**

5.2 Information on Incremental Updating
Does the metadata standard specify:

Obligation

A. __ A structure for information on incremental updates of the entire dataset?: **NO**
B. __ A structure for information on incremental updates of particular portions of the dataset?: **NO**
C. __ A structure to describe the update process for the dataset?: **NO**

SECTION 6 – Extent of the Dataset

This section examines how the standard provides for the specification of the spatial and temporal extent of the dataset.

6.1 The spatial extent of the dataset

Obligation

A. **M** Does the metadata standard allow/require for the description of horizontal extent (e.g., bounding rectangle/polygon, named location (i.e. Stockholm, etc.))?: **YES**
B. **O** Does the metadata standard allow/require for the description of vertical extent (e.g., minima/maxima heights, etc.)?: **YES**
C. **M** Does the standard require specific units for describing extent?: **YES**
D. __ Does the standard allow specification of the units for describing extent?: **NO**
E. __ Does the standard require specific spatial reference system for describing extent?: **NO**
F. **M** Does the standard allow specification of the spatial reference system for describing extent?: **YES**

6.2 The temporal extent of the dataset

Obligation

A. **O** Does the standard allow/require for the description of time period for which the dataset is relevant?: **YES**

B. __ Does the standard allow/require specification of the temporal reference system?: **NO**

SECTION 7 – Data Content of the Dataset

This section examines how the standard provides for the specification of the data content of the dataset in terms of overview elements, themes, data definitions and product specifications.

7.1 Dataset overview

Does the standard allow/require for the provision of the following information:

Obligation

A. __ Dataset overview?: **NO**

B. **M** Abstract – brief summary of the content of the dataset?: **YES**

C. **O** Purpose – summary of the intentions with which the dataset was developed?: **YES**

D. __ Usage – list of activities for which the dataset has been used? (see also Section 8.5): **NO**

E. **C** Spatial schema – (e.g., vector, raster)? (see also Section 9): **YES**

F. **M** Spatial reference system? (see also Section 12): **YES**

G. **M** Language?: **YES**

H. **M** Character set (e.g., ISO8859-1, JISX0208)?: **YES**

I. **O** List of documents providing further information about the dataset?: **YES**
(URL)

J. **O** Sample/preview dataset?: **YES**
(URL)

K. **O** Related datasets?: **YES**
(URL)

L. **O** Original size of the dataset?: **YES**

M. __ Compression? (see also Section 11.5): **NO**

N. **O** Other?: **YES**
scale from, scale to

7.2 Theme(s)

Obligation

A. **M** Does the standard allow/require the provision for the description of theme(s) (e.g., topography, land cover, transport, etc.) in the dataset?: **YES**

B. __ If A is yes, does the standard allow the referencing of an outside dictionary or thesaurus?: **NO**

C. **O** If A is yes, does the standard itself include a dictionary or thesaurus within the metadata?: **YES**
D. __ If A is yes, does the standard allow the use of unauthorized user-defined keywords?: **NO**

7.3 Data Definition

Obligation

A. **M** Does the standard allow/require the provision for the description of the data definition?: **YES**
B. If A is yes, does the standard allow/require the provision for the description of:
C 1. Features/Objects?: **YES**
C 2. Attributes?: **YES**
C 3. Relations?: **YES**
C. __ Can attributes be described separately from feature/object classes?: **NO**
D. **O** Does the standard allow/require the specification of the domain of the attributes?: **YES**
E. **O** Does the standard allow/require the specification of data type of the attributes?: **YES**
F. __ Does the standard allow/require the specification of particular classification coding schemes (e.g., FACC, OSKA, S-57 OC)?: **NO**

7.4 Product Specification

Obligation

A. __ Does the standard allow/require for the provision of information about whether the dataset meets a particular product specification?: **NO**
B. __ If A is yes, can it be internal to the Metadata?: **N/A**
C. __ If A is yes, can it be external to the Metadata?: **N/A**

SECTION 8 – Data Quality of the Dataset

This section examines how the standard deals with the question of data quality. This includes homogeneity and metrics for data quality, quality assurance, and usage.

8.1 Overall Data Quality Statement

Obligation

A. **O** Does the standard allow/require for the provision of quality to be described? If yes, please answer the questions below. (*If no, all following answers in Section* 8 *are N/A, please skip to Section* 9.): **YES**
B. __ Does the standard allow/require the provision for the quality information referencing an external quality standard, quality assurance standard, or specification?: **NO**

8.2 Homogeneity of Data Quality

Obligation

A. **O** Does the standard allow/require for the variation of the quality of the dataset to be described?: **YES**

B. What type of aggregation of quality information is allowed?

__ 1. Entire database?: **NO**

O 2. Feature/Object instance?: **YES**

__ 3. Feature/Object class?: **NO**

__ 4. Attribute instance?: **NO**

__ 5. Attribute class?: **NO**

__ 6. Layer?: **NO**

__ 7. Theme?: **NO**

__ 8. Geographic extent?: **NO**

__ 9. Other?: **NO**

8.3 Metrics for Data Quality

Does the standard allow/require the provision of values for:

Obligation

A. **O** Positional accuracy?: **YES**

1. Standard error?: **NO**
2. Maximum error?: **NO**
3. Error correlation?: **NO**
4. Confidence level (e.g., 90%, 95%)?: **NO**
5. Other?: **NO**

B. __ Attribute accuracy?: **NO**

1. Standard error?: **NO**
2. Maximum error?: **NO**
3. Error frequency?: **NO**
4. Error correlation?: **NO**
5. Confidence level (e.g., 90%, 95%)?: **NO**
6. Other?: **NO**

C. __ Temporal accuracy?: **NO**

1. Standard error?: **NO**
2. Maximum error?: **NO**
3. Error correlation?: **NO**
4. Confidence level? (e.g., 90%, 95%): **NO**
5. Other?: **NO**

D. **O** Completeness?: **YES**

1. Error of omission?: **NO**
2. Error of commission?: **NO**
3. Other?: **NO**

E. **O** Currentness?: **YES**

1. Date?: **YES**

2. Maximum age?: **NO**
3. Temporal extent?: **YES**
4. Other?: **YES**
   code of update frequency

F. **O** Logical consistency?: **YES**
   not formalized, present as a verbal description

G. **M** Lineage?: **YES**
   not formalized, present as a verbal description
   1. List of processing steps?: **YES**
   2. List of values of processing parameters?: **YES**
   3. List of printouts of statistics, etc.?: **YES**
   4. Source material?: **YES**
   5. Other?: **YES**

H. **O** Additional Quality Information?: **YES**
   semantic accuracy, spatial resolution

8.4 Quality Assurance Methods

Does the standard allow/require the provision of information about the methods used to determine:

Obligation

A. __ Positional accuracy?: **NO**
   1. Standard error in adjustments?: **NO**
   2. Repeated measurements?: **NO**
   3. Independent measurements?: **NO**
   4. Subjective evaluations?: **NO**
   5. Other?: **NO**

B. __ Attribute accuracy?: **NO**
   1. Standard error?: **NO**
   2. Error classification matrix?: **NO**
   3. Independent measurements?: **NO**
   4. Subjective evaluation?: **NO**
   5. Other?: **NO**

C. __ Temporal accuracy?: **NO**
   1. Independent measurements?: **NO**
   2. Subjective evaluation?: **NO**
   3. Other?: **NO**

D. __ Completeness?: **NO**
   1. Field checks?: **NO**
   2. Repeated measurements?: **NO**
   3. Independent measurements?: **NO**
   4. Subjective evaluation?: **NO**
   5. Error of commission?: **NO**
   6. Other?: **NO**

E. __ Currentness?: **NO**
 1. Date of source material?: **NO**
 2. Repeated measurements?: **NO**
 3. Independent measurements?: **NO**
 4. Subjective evaluation?: **NO**
 5. Other?: **NO**
F. __ Logical consistency?: **NO**
 1. Verification results of consistency checks?: **NO**
 2. Repeated processing?: **NO**
 3. Comparison with independent sources?: **NO**
 4. Subjective evaluation?: **NO**
 5. Other?: **NO**
G. __ Lineage?: **NO**
 1. Checking results of processing steps: **NO**
 2. Subjective evaluation: **NO**
 3. Other: **NO**
H. __ Additional Quality Information?: **NO**

8.5 Usage

Obligation

A. __ Does the metadata standard allow/require the provision of information about usage?: **NO**
 1. Who used it: **NO**
 2. When it was used: **NO**
 3. For what application it was used: **NO**
 4. The effectiveness of use: **NO**
 5. Other: **NO**

8.6 Alternative Quality Description: **YES**

SECTION 9 – Spatial Data Structure of the Dataset

This section examines how the standard provides for the specification of the spatial data model and spatial data types.

9.1 Spatial Data Model
Does the standard allow/require the provision for:

Obligation

A. **O** The description of the geometrical and/or topological primitives used to model the real world?: **YES**
B. __ Information about an external standard for geometrical and/or topological primitive specification?: **NO**

9.2 Spatial Data Types

Does the standard allow/require the provision for:

Obligation

A. **O** Information about the spatial data type(s) used in the dataset?: **YES**
not formalized, present as a verbal description

B. **O** If A is yes, which of the following generic spatial data types are supported? For each, please list specific data primitive levels supported by each data type.
1. Vector (e.g., spaghetti, chain-node/node-edge-face, planar graph, full topology)?: **YES**
2. Areal Raster (e.g., thematic raster – scanned and classification maps, image raster – satellite images, scanned aerial photographs)?: **YES**
3. Grid/Matrix (e.g., Digital Elevation Model)?: **YES**
4. Triangular Irregular Network (TIN)?: **YES**
5. Other?: **YES**

9.3 Other Data Types

Does the standard allow the provision of information about:

Obligation

A. **O** Video?: **YES**
B. **O** Sound?: **YES**
C. **O** Other?: **YES**

SECTION 10 – Spatial Reference of the Dataset

This section examines how the dataset provides for the specification of the spatial reference characteristics of the data set.

Does the standard allow/require the provision for the following information:

10.1 Geoidal Model

Obligation

A. __ Geoidal model name?: **NO**
B. __ Geoidal model definitions?: **NO**

10.2 Geodetic datum

Obligation

A. __ Geodetic datum name?: **NO**
B. __ Geodetic datum definitions?: **NO**

10.3 Reference Ellipsoid

Obligation

A. __ Reference ellipsoid name?: **NO**
B. __ Reference ellipsoid definitions?: **NO**

10.4 Projection

Obligation

A. **M** Projection name?: **YES**
B. __ Projection definitions?: **NO**

10.5 Coordinate Systems

Obligation

A. **M** Coordinate system type?: **YES**
B. __ Coordinate system definitions?: **NO**

10.6 Height Reference System

Obligation

A. **C** Vertical datum name?: **YES**
B. __ Types of heights?: **NO**

10.7 Ancillary Geodetic Information

Obligation

A. __ Does the standard allow/require the provision for ancillary geodetic, gravimetric, or magnetometric information?: **NO**

10.8 Transformation Parameters
Does the standard specify a method to provide information for coordinate transformation(s):

Obligation

A. __ Model name?: **NO**
B. __ Control points?: **NO**
C. __ Transformation parameters?: **NO**

10.9 Non-Coordinate Spatial Reference
Does the standard allow/require the provision for information on non-coordinate spatial reference systems (e.g., postal systems, street address)?:

Obligation

A. **C** Reference system name?: **YES**
B. __ Reference system definitions?: **NO**
C. __ Other?: **NO**

SECTION 11 – Availability and Distribution of the Dataset

This section examines how the standard provides information on the availability of the dataset?

Does the standard allow/require the provision for:

11.1 The Distributor(s) of the Dataset

Obligation

A. **M** Information about the distributor(s) of the dataset?: **YES**
B. **M** Contact information for the distributor(s)?: **YES**

11.2 Restrictions to Access and Usage

Obligation

A. __ Information about copyright on the dataset (e.g., date, period, copyright owner, etc.)?: **NO**
B. **O** Information about restrictions on the access of the dataset?: **YES**
C. **O** Information about restrictions on the usage of the dataset?: **YES**

11.3 Pricing Details

Obligation

A. **O** Information about pricing?: **YES**

11.4 Distribution of the Dataset

Obligation

A. **O** Information about distribution media?: **YES**
B. **O** Information about on-line access?: **YES**
C. **O** Information about units of distribution (e.g., tiles, layers, polygons, elements)?: **YES**

11.5 Distribution Format

Obligation

A. **O** Information about distribution formats (e.g., commercial CAD format, commercial GIS formats, National/International transfer standard)?: **YES**
B. __ Information about the encoding of the dataset (e.g., binary, ASCII)?: **NO**
C. __ Information about compression method?: **NO**
D. __ Information about compressed dataset size?: **NO**
E. **O** Information about the distribution mechanism?: **YES**
F. __ Information about the encapsulation of data (e.g., ISO 8211, ISO 8824)?: **NO**

11.6 Ordering Procedures

Obligation

A. **O** Information about ordering procedures for the dataset?: **YES**

11.7 Support Services

Obligation

A. __ Information about support services (e.g., additional data transformations, format or coordinate conversions)?: **NO**

SECTION 12 – Authorization and Verification of the Metadata

This section examines how the standard provides for the description of the authorization and verification on the metadata of a dataset (e.g., how is the metadata itself authorized and verified?).

12.1 Metadata reference

Obligation

A. **M** Does the standard allow/require the provision of information about reference dates of the metadata (e.g., last check date, last update date, review date)?: **YES**

12.2 Verification Authority

Obligation

A. __ Does the standard allow/require the provision for information about who verified/authorized the metadata?: **NO**

B. __ Does the standard specify the method for providing this information (e.g., ISO 9000): **NO**

12.3 Statistical Methods Used for Verification

Obligation

A. __ Does the standard specify the statistical methods to be used for verification of the metadata?: **NO**

B. __ Does the standard allow/require the provision for the description of the statistical methods used to verify the metadata?: **NO**

END OF CHARACTERISTICS
FIN

Denmark
Infodatabase Om Geodata
National Danish GI Metadata Service

Anders Nielsen, Peter Kjeld

I. Information about the Metadata Standard Itself

SECTION 0 – Assessment Information

This section provides the user with a general description of the scope and intended use of the standard. Information on the respondent who prepared the assessment is also provided.

0.1 Respondent

Assessment prepared by:

Anders Nielsen, Cartographer
Kort & Matrikelstyrelsen (National Survey and Cadastre, Denmark)
Rentemestervej 8
DK-2400 København NV, Denmark
Phone: +45 3587 5050
Fax: +45 3587 5053
E-mail: kms@kms.dk (personal email: asn@kms.dk)

Date assessment completed:
November 1999; updated September 2000 and May 2002, Reviewed Spring, 2005

Relationship of respondent to standard:
user
custodian

0.2 Brief Summary Statement

A. Summary of the scope and intent of the standard:
The National Danish GI Metadata Service (Infodatabase om Geodata) is a subset implementation (profile) of the CEN ENv 12657 metadata standard. In this chapter, the term "standard" is representing this profile
The National Danish GI Metadata Service is designed to describe geographic

datasets. Both geometric and non-coordinate, indirect spatially referenced systems are addressed. The service has from the very beginning been focused on the needs of regional and national government, but there are no restrictions on assigning datasets to the metadata service.

The profile provides for maintaining some degree of conformance with both formal and practical constraints in the national and international GI society. Thus meaning to be both easy-going for those providing metadata and a strict tool for those seeking.

The dataset descriptions include data about the content, representation, extent (both geometric and temporal), spatial reference, quality and administration of a geographic dataset. The primary purposes of the National Danish GI Metadata Service are to present an overview, and to provide contact information.

SECTION 1 – Administration of Metadata Standard

This section provides a general description of the administrative framework, within which the standard was developed, tested and currently resides. Significant details are also provided in terms of the developmental and managerial history as well as information on the existence of documentation, software tools and training materials. This section also provides a point of contact for further information on the standard.

1.1 Name of Standard

A. Name of standard:
   1. In original language(s)
      In Danish: "Infodatabase Om Geodata"
   2. English translation, if appropriate:
      "National Danish GI Metadata Service"
B. Version/Edition:
   (Internal Version number 1.51 dated 1999-04-04)
C. Language(s) of documentation:
   English
   Danish
D. Acronym(s):
   N/A
E. Official ID:
   N/A
F. Name(s) of recognizing standards authority(ies):
   N/A

1.2 Responsible Institutions

Contact information about those institutions that are (or were) responsible for the processes involved in the evolution of the standard:

A. Production and/or development:

Kort & Matrikelstyrelsen (National Survey and Cadastre, Denmark)
Rentemestervej 8
DK-2400 København NV, Denmark
Phone: +45 3587 5050
Fax: +45 3587 5051
E-mail: kms@kms.dk

Danmarks Tekniske Universitet (Technical University of Denmark)
Department of Planning, Surveying
DK-2800 Lyngby, Denmark

B. Testing:

N/A

C. Conformance:

N/A

D. Maintenance:

Kort & Matrikelstyrelsen (National Survey and Cadastre, Denmark)
Rentemestervej 8
DK-2400 København NV, Denmark
Phone: +45 3587 5050
Fax: +45 3587 5051
E-mail: kms@kms.dk

E. Distribution:

Kort & Matrikelstyrelsen (National Survey and Cadastre, Denmark)
Rentemestervej 8
DK-2400 København NV, Denmark
Phone: +45 3587 5050
Fax: +45 3587 5051

F. Help Desk/User Support:

Kort & Matrikelstyrelsen (National Survey and Cadastre, Denmark)
Rentemestervej 8
DK-2400 København NV, Denmark
Phone: +45 3587 5050
Fax: +45 3587 5051
E-mail: kms@kms.dk

1.3 Development History

A. Beginning of work on metadata standard:
1996-06 Previous metadatabase designed 1993 is revised with respect to the evolving standard from CEN/TC287

B. Milestones:
1997-04 new Danish service available on the Internet
1999-Q4 Service is updated for more rigid conformance with "CEN ENv 12657" and "Megrin Core Metadata"

C. Anticipated completion: **N/A**
D. Update cycle: **N/A**

1.4 Status of the Standard

A. Is the standard officially recognized: **NO**
B. Date(s) of recognition: **N/A**
C. Anticipated date of recognition: **N/A**
D. Current stage in recognition process: **N/A**
E. Steps still required to achieve recognition: **N/A**

1.5 Access to Documentation of the Standard

A. Is the standard copyrighted?: **NO**
B. Name and address of copyright owner: **N/A**
C. Do restrictions on the use of the standard apply?: **NO**
D. Is the standard documentation currently available?: **YES**
Database and other related software may be available for demonstration on request.

E. List the form/media of standard documentation: **N/A**
F. Is the standard available on the Internet?: **YES**
Services on these URLs:
http://www.geodata-info.dk

G. Price: **N/A**
H. Contact details for obtaining the standard:
Jan Hjelmager
Kort & Matrikelstyrelsen (National Survey and Cadastre, Denmark)
Rentemestervej 8
DK-2400 København NV, Denmark
Phone: +45 3587 5050
Fax: +45 3587 5051
E-mail: kms@kms.dk

I. Does the standard have ISBN/ISSN numbers?: **NO**

1.6 Available Software Tools

A. Is software available to implement the standard as a database?: **NO**
B. Is software available to test compliance of metadata with the standard?: **NO**
C. Is software available to produce metadata compliant with the standard?: **YES**
Access database holding the data for the services
Metadata update and maintenance tool
Tools for the automated generation of static html and xml documents

D. Other relevant software?: **NO**
E. Is there a test dataset available to test software tools for implementing the standard?: **NO**
F. For each product available above, please list contact information:

Jan Hjelmager
Kort & Matrikelstyrelsen (National Survey and Cadastre, Denmark)
Rentemestervej 8
DK-2400 København NV, Denmark
Phone: +45 3587 5050
Fax: +45 3587 5053
E-mail: kms@kms.dk

1.7 Available Training Materials

A. Are organized training sessions available?: **NO**
B. Contact details for training sessions: **N/A**
C. Training documentation available: **NO**

SECTION 2 – Use and Implementation of the Metadata Standard

This section provides information on the way in which the standard is used and the purpose of its use. It also provides descriptions about the level of abstraction of the standard.

2.1 Intended Use of the Standard
Can the standard be used:

A. Within a transfer data structure?: **NO**
B. To produce hardcopy catalog of datasets?: **YES**
C. To catalog data in a digital on-line catalog?: **YES**
D. Other:

Tools and concept provide for the development of metadata services on other scopes (regional, company proprietary, or international)
Tools and concept provide for multilingual services (partially implemented in the Danish service)
XML tools guarantee cross platform portability of data

2.2 Intended Purpose of the Standard
Is the intended purpose of the standard:

A. To support discovery, identification, location of data?: **YES**
B. To support how to access data?: **YES**
C. To support the determination of dataset fitness for use?: **YES**
D. To support transfer of dataset?: **NO**
E. To support an organization's data management?: **YES**

F. Other:

Service is provided to strengthen awareness about GI data exchange Service is provided to gain value from "cross-government" data usage Service is shaped to absorb new tendencies in GI management

2.3 Table of Contents

A. Provide the standard's Table of Contents

(refer to datcon11.pdf which compares this profile with CEN ENv 12657 and Megrin Core metadata)

Dataset identification

Dataset title
Alternative title
Abbreviated title
Version

Dataset overview

Dataset_abstract
Primary_organisation
Spatial_subschema_typeID
Dataset_language_Id
Dataset_language
Dataset_character_set
Purpose
Intended_application_scale
usage
Document_reference
Sample: Sample_ref;Data;Sample_description
Series_name
Related_dataset

Dataset quality elements

Source
Positional_accuracy
Temporal_accuracy
Completeness
Entry_date
Last_check_date
Last_update_date
Comment:Status_type

Spatial reference system

Indirect_reference_system_type:Name
Coordinate system: Identifier;Name;Datum;Ellipsoid;Map_projection; Height_reference_system

Extent

Extent_status
Extent_date
Planar extent: Bounding quadrangle;Extent_geographic_area
Temporal extent

Data definition

Object_type: Name;Attribute_type
Classification
Thesaurus: Name of thesaurus;Thesaurus_contact;Term;Definition;Picture

Administrative metadata

Organisation and organisation role:
Org_name;Function_of_organisation;Abbreviated_name;Organisation_address
Point of contact and point of contact role: Point of contact name;Department;Point of contact role
Distribution:
Restrictions_on_use;Copyright_owners;Unit_of_distribution;Price_information; Data_media;Formats

2.4 Level of Abstraction of the Metadata Standard Definition

A. Does the standard define metadata elements?: **YES**
B. Does the standard define a model for the relationships between the metadata elements?: **YES**
C. Provide an overview diagram of the model of the metadata, if available. **N/A**
D. Does the standard define domains and/or data types of the metadata elements?: **YES**
E. Does the standard define how the metadata should be encoded?: **NO**
F. Does the standard define a database schema for the automatic generation of a database?: **NO**

2.5 Extensibility of the Standard

A. Is the metadata description extensible (e.g., Can user add metadata domains, elements, thematic profiles, etc.)?: **NO**

2.6 Availability of Metadata
Does the standard specify:

A. How metadata are to be accessed?: **YES**
access via Internet
as html files in a stand-alone application
B. A mechanism for querying the metadata?: **YES**
via predefined categories
via search engine included in the update software
C. Particular standardized query protocols?: **NO**
D. That the metadata should be encoded for computer access?: **NO**
E. That the metadata should be encrypted for security?: **NO**

2.7 Examples

A. Does the standard provide examples of implementations?: **YES**
like the implementations listed in this chapter
B. How many examples are available?: **N/A**

SECTION 3 – Linkage and Coordination of the Metadata Standard

This section provides information on how the metadata standard is linked to other standards, particularly transfer standards, and how such linkage is implemented. If a significant linkage does exist, detailed technical information is provided on the precise nature of the linkage to the standard, and how the linkage affects implementation of the metadata standard.

3.1 Associated Transfer Standards

A. Is the metadata standard associated with a transfer standard(s)?: **NO**
B. If A is yes, provide references and contact details for each linked standard: **N/A**
C. If A is yes, are both (all) standards coordinated?: **N/A**
D. If A is yes, describe how the coordination works, both from a technical and conformance (normative) viewpoint: **N/A**

3.2 Incorporated into transfer mechanism

A. Is the metadata standard incorporated into a transfer mechanism?: **NO**
An ongoing development project is concerned with the idea of using XML (eXtensible Markup Language) for a painless cross-platform transfer of metadata. 1st generation DTDs have been derived from the standards involved.
B. If A is yes, provide references and contact details for each incorporated mechanism: **N/A**
C. If A is yes, describe how: **N/A**

3.3 Other linked, coordinated or associated standards

A. Is the metadata standard linked, coordinated, or associated with any other standard(s) not listed above?: **NO**
B. If A is yes, provide references and contact details for each standard: **N/A**
C. If A is yes, describe how for each relevant standard: **N/A**

II. Categories the Metadata Standard Uses to Describe Datasets

The following sections are categories of metadata that may be used by the metadata standard to describe data. Not all categories are relevant to different datasets and may not necessarily be present in the standard. The presence or absence of a category should not be used for judging one standard against another.

Note: For each of the questions in the following sections please provide the obligation status information. Valid obligation values are:

Obligation Level

M Mandatory This information must be entered to comply with the metadata standard.
C Conditional This information is entered based on other information in the metadata.
O Optional This information is optionally entered in the metadata.

If the answer is 'NO', the obligation status should be left blank.

SECTION 4 – Dataset Identification

This section examines how the standard provides for the specification of the various sorts of titles and identifiers for a dataset.

4.1 The title(s) of a dataset

Does the standard allow/require the provision for:

Obligation

A. __ Previous title(s)?: **NO**
B. **M** Current title(s)?: **YES**
C. __ Possible future title(s)?: **NO**
D. __ Official titles in different languages?: **NO**
E. **O** Abbreviated title(s)?: **YES**
F. **O** Alternative Title(s)?: **YES**
G. __ Does the standard support effective dates of titles?: **NO**
H. **O** Other identifiers (e.g., ISBN, National Stock Number, Uniform Product Code)?: **YES**
Dataset version #

4.2 Parties responsible for the dataset

Does the standard allow/require the provision for:

Obligation

A. **O** Author?: **YES**
B. **M** Owner?: **YES**
C. **O** Other?: **YES**

4.3 ISBD Information

Obligation

A. __ Does the metadata standard specify inclusion of mandatory ISBD information (see glossary) within the metadata?: **NO**

SECTION 5 – Status of the Dataset

This section examines how the standard provides for the specification of the current status and updating of the data set.

5.1 Status and Progress of Datasets

Does the metadata standard allow/require the provision for:

Obligation

A. **O** Information on the status of the entire dataset?: **YES**
Free text: date, coverage fraction, future development etc.

B. __ Information on the status of the particular subset (e.g., observation time, inauguration time, extraction time)?: **NO**

Does the metadata standard specify:

C. __ A structure for updating or maintenance information?: **NO**

D. __ A structure to describe historical information (see also Section 8.3 G)?: **NO**

5.2 Information on Incremental Updating

Does the metadata standard specify:

Obligation

A. __ A structure for information on incremental updates of the entire dataset?: **NO**

B. __ A structure for information on incremental updates of particular portions of the dataset?: **NO**

C. __ A structure to describe the update process for the dataset?: **NO**

SECTION 6 – Extent of the Dataset

This section examines how the standard provides for the specification of the spatial and temporal extent of the dataset.

6.1 The spatial extent of the dataset

Obligation

A. **C** Does the metadata standard allow/require for the description of horizontal extent (e.g., bounding rectangle/polygon, named location (i.e., Stockholm, etc.))?: **YES**
If the spatial extent of the dataset is defined by a named locality like a county or a municipality. This information is mandatory. If a bounding box defines the spatial extent of the dataset, reference system information and min/max values must be provided.

B. __ Does the metadata standard allow/require for the description of vertical extent (e.g., minima/maxima heights, etc.)?: **NO**

C. __ Does the standard require specific units for describing extent?: **NO**

D. **C** Does the standard allow specification of the units for describing extent?: **YES**

E. __ Does the standard require specific spatial reference system for describing extent?: **NO**

F. __ Does the standard allow specification of the spatial reference system for describing extent?: **NO**

6.2 The temporal extent of the dataset

Obligation

A. **O** Does the standard allow/require for the description of time period for which the dataset is relevant?: **YES**

B. __ Does the standard allow/require specification of the temporal reference system?: **NO**

SECTION 7 – Data Content of the Dataset

This section examines how the standard provides for the specification of the data content of the dataset in terms of overview elements, themes, data definitions and product specifications.

7.1 Dataset overview

Does the standard allow/require for the provision of the following information:

Obligation

A. **M** Dataset overview?: **YES**
B. **M** Abstract – brief summary of the content of the dataset?: **YES**
C. **M** Purpose – summary of the intentions with which the dataset was developed?: **YES**
D. **M** Usage – list of activities for which the dataset has been used? (see also Section 8.5): **YES**
E. **O** Spatial schema – (e.g., vector, raster)? (see also Section 9): **YES**
F. **M** Spatial reference system? (see also Section 12): **YES**
G. **M** Language?: **YES**
H. **O** Character set (e.g., ISO8859-1, JISX0208)?: **YES**
I. **O** List of documents providing further information about the dataset?: **YES**
J. **O** Sample/preview dataset?: **YES**
K. **O** Related datasets?: **YES**
L. __ Original size of the dataset?: **NO**
M. __ Compression? (see also Section 11.5): **NO**
N. __ Other?: **NO**

7.2 Theme(s)

Obligation

A. **O** Does the standard allow/require the provision for the description of theme(s) (e.g., topography, land cover, transport, etc.) in the dataset?: **YES**

B. __ If A is yes, does the standard allow the referencing of an outside dictionary or thesaurus?: **NO**

C. __ If A is yes, does the standard itself include a dictionary or thesaurus within the metadata?: **NO**

D. __ If A is yes, does the standard allow the use of unauthorized user-defined keywords?: **NO**

7.3 Data Definition

Obligation

A. **O** Does the standard allow/require the provision for the description of the data definition?: **YES**
B. If A is yes, does the standard allow/require the provision for the description of:
O 1. Features/Objects?: **YES**
O 2. Attributes?: **YES**
__ 3. Relations?: **NO**
C. __ Can attributes be described separately from feature/object classes?: **NO**
D. __ Does the standard allow/require the specification of the domain of the attributes?: **NO**
E. __ Does the standard allow/require the specification of data type of the attributes?: **NO**
F. __ Does the standard allow/require the specification of particular classification coding schemes (e.g., FACC, OSKA, S-57 OC)?: **NO**

7.4 Product Specification

Obligation

A. **O** Does the standard allow/require for the provision of information about whether the dataset meets a particular product specification?: **YES**
B. __ If A is yes, can it be internal to the Metadata?: **NO**
C. **C** If A is yes, can it be external to the Metadata?: **YES**

SECTION 8 – Data Quality of the Dataset

This section examines how the standard deals with the question of data quality. This includes homogeneity and metrics for data quality, quality assurance, and usage.

8.1 Overall Data Quality Statement

Obligation

A. **M** Does the standard allow/require for the provision of quality to be described? If yes, please answer the questions below. (*If no, all following answers in Section* 8 *are N/A, please skip to Section* 9.): **YES**
B. __ Does the standard allow/require the provision for the quality information referencing an external quality standard, quality assurance standard, or specification?: **NO**

8.2 Homogeneity of Data Quality

Obligation

A. **O** Does the standard allow/require for the variation of the quality of the dataset to be described?: **YES**

B. What type of aggregation of quality information is allowed?

1. Entire database?: **NO**
__ 2. Feature/Object instance?: **NO**
__ 3. Feature/Object class?: **NO**
__ 4. Attribute instance?: **NO**
__ 5. Attribute class?: **NO**
__ 6. Layer?: **NO**
__ 7. Theme?: **NO**
__ 8. Geographic extent?: **NO**
__ 9. Other?: **NO**

8.3 Metrics for Data Quality

Does the standard allow/require the provision of values for:

Obligation

A. Positional accuracy?: **NO**
1. Standard error?: **NO**
2. Maximum error?: **NO**
3. Error correlation?: **NO**
4. Confidence level (e.g., 90%, 95%)?: **NO**
5. Other?: **NO**

B. __ Attribute accuracy?: **NO**
1. Standard error?: **NO**
2. Maximum error?: **NO**
3. Error frequency?: **NO**
4. Error correlation?: **NO**
5. Confidence level (e.g., 90%, 95%)?: **NO**
6. Other?: **NO**

C. Temporal accuracy?: **NO**
1. Standard error?: **NO**
2. Maximum error?: **NO**
3. Error correlation?: **NO**
4. Confidence level? (e.g., 90%, 95%): **NO**
5. Other?: **NO**

D. Completeness?: **NO**
1. Error of omission?: **NO**
2. Error of commission?: **NO**
3. Other?: **NO**

E. Currentness?: **NO**
1. Date?: **NO**
2. Maximum age?: **NO**
3. Temporal extent?: **NO**
4. Other?: **NO**

F. __ Logical consistency?: **NO**

G.	Lineage?:	**NO**
	1. List of processing steps?:	**NO**
	2. List of values of processing parameters?:	**NO**
	3. List of printouts of statistics, etc.?:	**NO**
	4. Source material?:	**NO**
	5. Other?:	**NO**
H.	__ Additional Quality Information?:	**NO**

8.4 Quality Assurance Methods

Does the standard allow/require the provision of information about the methods used to determine:

Obligation

A.	__ Positional accuracy?:	**NO**
	1. Standard error in adjustments?:	**N/A**
	2. Repeated measurements?:	**N/A**
	3. Independent measurements?:	**N/A**
	4. Subjective evaluations?:	**N/A**
	5. Other?:	**N/A**
B.	__ Attribute accuracy?:	**NO**
	1. Standard error?:	**N/A**
	2. Error classification matrix?:	**N/A**
	3. Independent measurements?:	**N/A**
	4. Subjective evaluation?:	**N/A**
	5. Other?:	**N/A**
C.	__ Temporal accuracy?:	**NO**
	1. Independent measurements?:	**N/A**
	2. Subjective evaluation?:	**N/A**
	3. Other?:	**N/A**
D.	__ Completeness?:	**NO**
	1. Field checks?:	**N/A**
	2. Repeated measurements?:	**N/A**
	3. Independent measurements?:	**N/A**
	4. Subjective evaluation?:	**N/A**
	5. Error of commission?:	**N/A**
	6. Other?:	**N/A**
E.	__ Currentness?:	**NO**
	1. Date of source material?:	**N/A**
	2. Repeated measurements?:	**N/A**
	3. Independent measurements?:	**N/A**
	4. Subjective evaluation?:	**N/A**
	5. Other?:	**N/A**
F.	__ Logical consistency?:	**NO**
	1. Verification results of consistency checks?:	**N/A**
	2. Repeated processing?:	**N/A**

3. Comparison with independent sources?: **N/A**
4. Subjective evaluation?: **N/A**
5. Other?: **N/A**

G. __ Lineage?: **NO**
1. Checking results of processing steps: **N/A**
2. Subjective evaluation: **N/A**
3. Other: **N/A**

H. __ Additional Quality Information?: **NO**

8.5 Usage

Obligation

A. Does the metadata standard allow/require the provision of information about usage?: **NO**
1. Who used it: **N/A**
2. When it was used: **N/A**
3. For what application it was used: **N/A**
4. The effectiveness of use: **N/A**
5. Other: **N/A**

8.6 Alternative Quality Description: **NO**

SECTION 9 – Spatial Data Structure of the Dataset

This section examines how the standard provides for the specification of the spatial data model and spatial data types.

9.1 Spatial Data Model
Does the standard allow/require the provision for:

Obligation

A. __ The description of the geometrical and/or topological primitives used to model the real world?: **NO**

B. __ Information about an external standard for geometrical and/or topological primitive specification?: **NO**

9.2 Spatial Data Types
Does the standard allow/require the provision for:

Obligation

A. Information about the spatial data type(s) used in the dataset?: **NO**

B. **C** If A is yes, which of the following generic spatial data types are supported? For each, please list specific data primitive levels supported by each data type.
1. Vector (e.g., spaghetti, chain-node/node-edge-face, planar graph, full topology)?: **YES**

Only vector/non vector is specified

2. Areal Raster (e.g., thematic raster – scanned and classification maps, image raster – satellite images, scanned aerial photographs)?: **YES**

Only raster/non raster is specified

3. Grid / Matrix (e.g., Digital Elevation Model)?: **NO**
4. Triangular Irregular Network (TIN)?: **NO**
5. Other?: **YES**

Alphanumeric, indirectly referenced geodata

9.3 Other Data Types

Does the standard allow the provision of information about:

Obligation

A. __ Video?: **NO**
B. __ Sound?: **NO**
C. **O** Other?: **YES**

Alphanumeric, indirectly referenced geodata

SECTION 10 – Spatial Reference of the Dataset

This section examines how the dataset provides for the specification of the spatial reference characteristics of the data set.

Does the standard allow/require the provision for the following information:

10.1 Geoidal Model

Obligation

A. Geoidal model name?: **NO**
B. __ Geoidal model definitions?: **NO**

10.2 Geodetic datum

Obligation

A. Geodetic datum name?: **NO**
B. __ Geodetic datum definitions?: **NO**

10.3 Reference Ellipsoid

Obligation

A. Reference ellipsoid name?: **NO**
B. __ Reference ellipsoid definitions?: **NO**

10.4 Projection

Obligation

A. Projection name?: **NO**
B. __ Projection definitions?: **NO**

10.5 Coordinate Systems

Obligation

A. Coordinate system type?: **NO**
B. __ Coordinate system definitions?: **NO**

10.6 Height Reference System

Obligation

A. Vertical datum name?: **NO**
B. __ Types of heights?: **NO**

10.7 Ancillary Geodetic Information

Obligation

A. __ Does the standard allow/require the provision for ancillary geodetic, gravimetric, or magnetometric information?: **NO**

10.8 Transformation Parameters
Does the standard specify a method to provide information for coordinate transformation(s):

Obligation

A. __ Model name?: **NO**
B. __ Control points?: **NO**
C. __ Transformation parameters?: **NO**

10.9 Non-Coordinate Spatial Reference
Does the standard allow/require the provision for information on non-coordinate spatial reference systems (e.g., postal systems, street address)?:

Obligation

A. Reference system name?: **NO**
B. __ Reference system definitions?: **NO**
C. __ Other?: **NO**

SECTION 11 – Availability and Distribution of the Dataset

This section examines how the standard provides information on the availability of the dataset?
Does the standard allow/require the provision for:

11.1 The Distributor(s) of the Dataset

Obligation

A. Information about the distributor(s) of the dataset?: **NO**

B. Contact information for the distributor(s)?: **NO**

11.2 Restrictions to Access and Usage

Obligation

A. Information about copyright on the dataset (e.g., date, period, copyright owner, etc.)?: **NO**
B. Information about restrictions on the access of the dataset?: **NO**
C. Information about restrictions on the usage of the dataset?: **NO**

11.3 Pricing Details

Obligation

A. Information about pricing?: **NO**

11.4 Distribution of the Dataset

Obligation

A. Information about distribution media?: **NO**
B. Information about on-line access?: **NO**
C. Information about units of distribution (e.g., tiles, layers, polygons, elements)?: **NO**

11.5 Distribution Format

Obligation

A. Information about distribution formats (e.g., commercial CAD format, commercial GIS formats, National/International transfer standard)?: **NO**
B. Information about the encoding of the dataset (e.g., binary, ASCII)?: **NO**
C. __ Information about compression method?: **NO**
D. __ Information about compressed dataset size?: **NO**
E. __ Information about the distribution mechanism?: **NO**
F. __ Information about the encapsulation of data (e.g., ISO 8211, ISO 8824)?: **NO**

11.6 Ordering Procedures

Obligation

A. Information about ordering procedures for the dataset?: **NO**

11.7 Support Services

Obligation

A. Information about support services (e.g., additional data transformations, format or coordinate conversions)?: **NO**

SECTION 12 – Authorization and Verification of the Metadata

This section examines how the standard provides for the description of the authorization and verification on the metadata of a dataset (e.g., how is the metadata itself authorized and verified?).

12.1 Metadata reference

Obligation

A. Does the standard allow/require the provision of information about reference dates of the metadata (e.g., last check date, last update date, review date)?: **NO**

12.2 Verification Authority

Obligation

A. **M** Does the standard allow/require the provision for information about who verified/authorized the metadata?: **YES**
organisation
contact person
address
phone, fax, email

B. __ Does the standard specify the method for providing this information (e.g., ISO 9000): **NO**

12.3 Statistical Methods Used for Verification

Obligation

A. __ Does the standard specify the statistical methods to be used for verification of the metadata?: **NO**

B. __ Does the standard allow/require the provision for the description of the statistical methods used to verify the metadata?: **NO**

END OF CHARACTERISTICS
FIN

H. Moellering, H.J.G.L. Aalders & A. Crane (Editors)

Finland
JHS 137 Tietotuoteseloste; JHS 137A Tietotuoteseloste – Paikkatiedot
JHS 137 Data Product Description; JHS 137A Data Product Description – Geographic Information
JHS 137; JHS 137A

Paula Ahonen-Rainio

I. Information about the Metadata Standard Itself

SECTION 0 – Assessment Information

This section provides the user with a general description of the scope and intended use of the standard. Information on the respondent who prepared the assessment is also provided.

0.1 Respondent

Assessment prepared by:
Paula Ahonen, lecturer
Helsinki University of Technology
PO Box 1200, FIN-02015 HUT
Fax: +358 9 465 077
Phone: +358 9 451 3919
E-mail: paula.ahonen@hut.fi

Date assessment completed:
September 1999, updated January 2001 and May 2002, Reviewed Fall, 2004

Relationship of respondent to standard:
Chair of the drafting committee

0.2 Brief Summary Statement

A. Summary of the scope and intent of the standard:
The standard defines a common way to give metadata for any digital data product.

The metadata set is called a Data Product Description. The descriptions are meant to be shown and searched on www, and thus gathered in a decentralized directory service.

The standard is composed of a generic part (JHS 137) and additional parts specific to certain types of data. Therefore, the main standard is not limited to geographic data products but has an additional part (JHS 137A) specific to spatial aspects of data.

The standard defines the metadata elements at a general level allowing details to be given in referred documents on the www sites of responsible organisations. This is meant to lower the threshold of using the standard on a wide front as well as each application field to follow their standards in detailed descriptions. For example, details of the ISO 19115 for describing geographic information or relevant standards on the field of statistics can be implemented while using JHS as a national framework standard.

SECTION 1 – Administration of Metadata Standard

This section provides a general description of the administrative framework, within which the standard was developed, tested and currently resides. Significant details are also provided in terms of the developmental and managerial history as well as information on the existence of documentation, software tools and training materials. This section also provides a point of contact for further information on the standard.

1.1 Name of Standard

A. Name of standard:
   1. In original language(s):
      JHS 137 Tietotuoteseloste; JHS 137A Tietotuoteseloste – Paikkatiedot
   2. English translation, if appropriate:
      JHS 137 Data Product Description; JHS 137A Data Product Description – Geographic Information

B. Version/Edition:
   JHS 137, 1.9.1998-31.8.2001; JHS 137A, 1.11.1998-31.10.2001

C. Language(s) of documentation:
   Finnish

D. Acronym(s):
   N/A

E. Official ID:
   JHS 137; JHS 137A

F. Name(s) of recognizing standards authority(ies):
   JUHTA – Consultative Committee for Data Administration in Public Administration in Finland, c/o Ministry of the Interior PO Box 257, FIN-00171 Helsinki

Fax: +358 9 16042827
E-mail to the secretary general; for the person in charge each time, check the web site at http://www.intermin.fi/juhta/

1.2 Responsible Institutions

Contact information about those institutions that are (or were) responsible for the processes involved in the evolution of the standard:

A. Production and/or development:

National Land Survey of Finland
PO Box 84, FIN-00521 Helsinki
Fax: +358 205 415454

B. Testing:

Karttakeskus Oy (merged to Genimap Oy)
(Mrs. Anna-Maija Ainola)
PO Box 106, FIN-02601 Vantaa
Fax: +358 201 340449

Finnish Environment Institute
PO Box 140, FIN-00251 Helsinki
Fax: +358 9 40300190

Statistics Finland (Mrs. Riitta Poukka)
FIN-00022 Tilastokeskus
Fax: +358 9 17342750

National Land Survey (Mr. Ari Tella)
PO Box 84, FIN-00521 Helsinki
Fax: +358 205 414495

C. Conformance:

NO

D. Maintenance:

JUHTA – Consultative Committee for Data Administration in Public Administration in Finland, c/o Ministry of the Interior (see 1.1 F above)

E. Distribution:

JUHTA, see above

F. Help Desk/User Support:

JUHTA, see above

1.3 Development History

A. Beginning of work on metadata standard: **January 1997**

B. Milestones:

Draft standard December 1997; National commenting April 1998; Final draft divided, according to the comments, into two parts: the generic part JHS 137 and the additional part JHS 137A with elements specific to spatial aspects, both finalized in June–August 1998; Recognition of both parts October 1999.

C. Anticipated completion: **N/A**
D. Update cycle: **Three years**

1.4 Status of the Standard

A. Is the standard officially recognized: **YES**
B. Date(s) of recognition: **October 1999**
C. Anticipated date of recognition: **N/A**
D. Current stage in recognition process: **N/A**
E. Steps still required to achieve recognition: **N/A**

1.5 Access to Documentation of the Standard

A. Is the standard copyrighted?: **NO**
B. Name and address of copyright owner: **N/A**
C. Do restrictions on the use of the standard apply?: **NO**
D. Is the standard documentation currently available?: **YES**
E. List the form/media of standard documentation:

Paper, www

F. Is the standard available on the Internet?: **NO**
G. Price: **Free**
H. Contact details for obtaining the standard:

Ministry of the Interior
PO Box 257, FIN-00171 Helsinki
Fax: +358 9 1602827
(for the person in charge each time, check the web site at http://www.intermin.fi/juhta/)

I. Does the standard have ISBN/ISSN numbers?: **NO**

1.6 Available Software Tools

A. Is software available to implement the standard as a database?: **NO**
B. Is software available to test compliance of metadata with the standard?: **NO**
C. Is software available to produce metadata compliant with the standard?: **NO**
D. Other relevant software: **NO**
E. Is there a test dataset available to test software tools for implementing the standard?: **NO**
F. For each product available above, please list contact information: **N/A**

1.7 Available Training Materials

A. Are organized training sessions available?: **NO**
B. Contact details for training sessions: **N/A**
C. Training documentation available: **NO**

SECTION 2 – Use and Implementation of the Metadata Standard

This section provides information on the way in which the standard is used and the purpose of its use. It also provides descriptions about the level of abstraction of the standard.

2.1 Intended Use of the Standard
Can the standard be used:

A. Within a transfer data structure?: **NO**
B. To produce hardcopy catalog of datasets?: **YES**
However, the standard does not define layout
C. To catalog data in a digital on-line catalog?: **YES**
However, the standard does not define layout or format
D. Other: **NO**

2.2 Intended Purpose of the Standard
Is the intended purpose of the standard:

A. To support discovery, identification, location of data?: **YES**
B. To support how to access data?: **YES**
C. To support the determination of dataset fitness for use?: **YES**
D. To support transfer of dataset?: **NO**
E. To support an organization's data management?: **NO**
F. Other: **NO**

2.3 Table of Contents

A. Provide the standard's Table of Contents
(Translation from Finnish)
JHS 137 Data product description
0. Introduction
1. Scope
2. References
3. Structure of the data product description
4. Elements of the data product description
Annex 1 Remarks for the implementation
Annex 2 Examples of description
JHS 137A Data product description – Geographic information
0. Introduction
1. Scope
2. References

3. Adding spatial aspects to the data product description
4. Additional elements
Annex 1 Remarks for the implementation
Annex 2 Examples

2.4 Level of Abstraction of the Metadata Standard Definition

A. Does the standard define metadata elements?: **YES**
B. Does the standard define a model for the relationships between the metadata elements?: **YES**
C. Provide an overview diagram of the model of the metadata, if available. **YES**
D. Does the standard define domains and/or data types of the metadata elements?: **YES**
However, most of the domains are "free text"
E. Does the standard define how the metadata should be encoded?: **NO**
F. Does the standard define a database schema for the automatic generation of a database?: **NO**

2.5 Extensibility of the Standard

A. Is the metadata description extensible (e.g., Can user add metadata domains, elements, thematic profiles, etc.)?: **YES**
A user can add elements under the control of the Coordinator of the standard (i.e., JUHTA)

2.6 Availability of Metadata
Does the standard specify:

A. How metadata are to be accessed?: **NO**
B. A mechanism for querying the metadata?: **NO**
C. Particular standardized query protocols?: **NO**
D. That the metadata should be encoded for computer access?: **NO**
E. That the metadata should be encrypted for security?: **NO**

2.7 Examples

A. Does the standard provide examples of implementations?: **YES**
In annexes JHS 137 and JHS 137A provide metadata of different data products
B. How many examples are available?: **4**

SECTION 3 – Linkage and Coordination of the Metadata Standard

This section provides information on how the metadata standard is linked to other standards, particularly transfer standards, and how such linkage is implemented. If a significant linkage does exist, detailed technical information is provided on the precise nature of the linkage to the standard, and how the linkage affects implementation of the metadata standard.

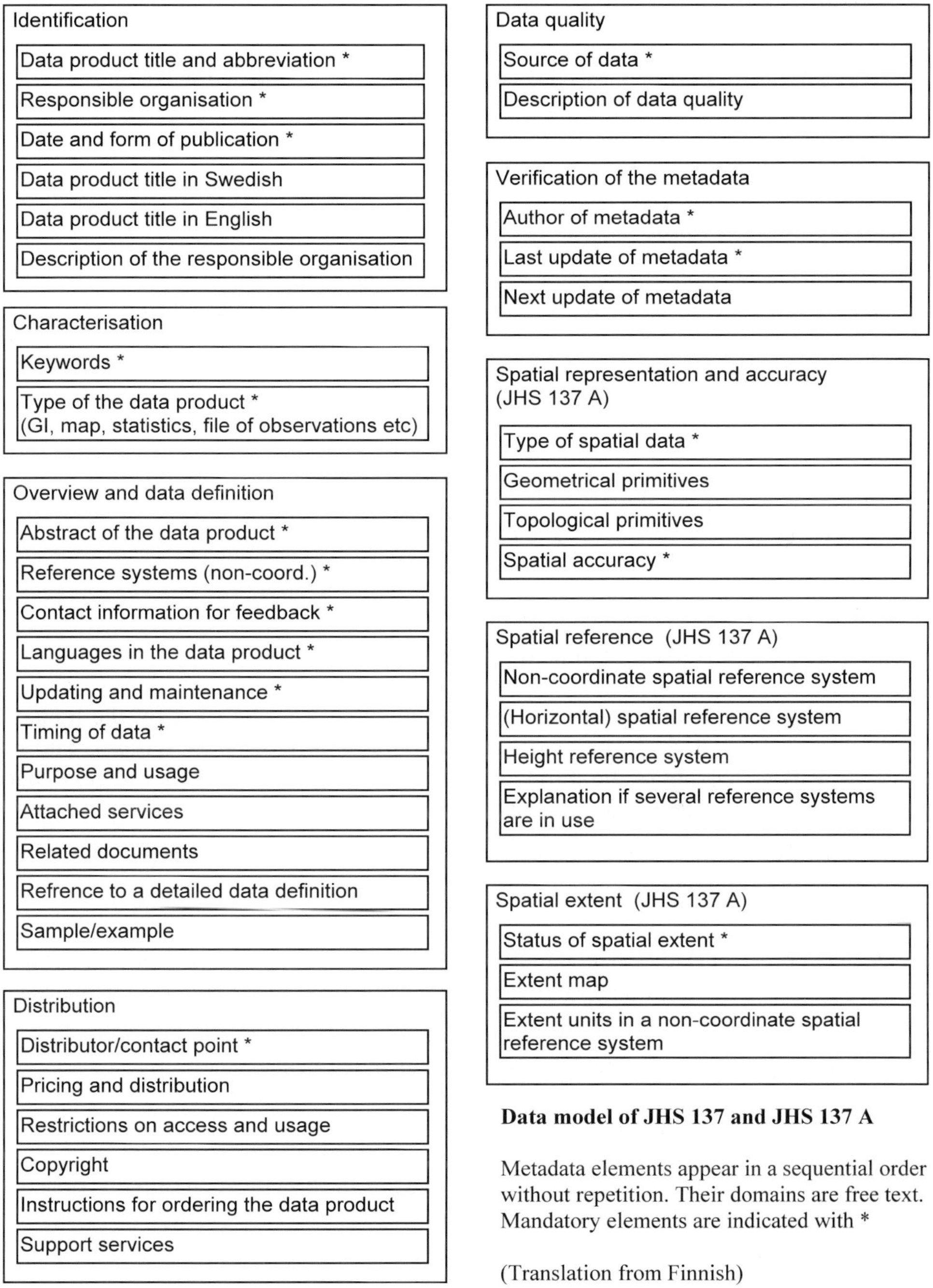

Fig. 1. Data model of JHS 137 and 137 A.

3.1 Associated Transfer Standards

A. Is the metadata standard associated with a transfer standard(s)?: **NO**
B. If A is yes, provide references and contact details for each linked standard: **N/A**
C. If A is yes, are both (all) standards coordinated?: **N/A**
D. If A is yes, describe how the coordination works, both from a technical and conformance (normative) viewpoint: **N/A**

3.2 Incorporated into transfer mechanism

A. Is the metadata standard incorporated into a transfer mechanism?: **NO**
B. If A is yes, provide references and contact details for each incorporated mechanism: **N/A**
C. If A is yes, describe how: **N/A**

3.3 Other linked, coordinated or associated standards

A. Is the metadata standard linked, coordinated, or associated with any other standard(s) not listed above?: **NO**
B. If A is yes, provide references and contact details for each standard: **N/A**
C. If A is yes, describe how for each relevant standard: **N/A**

II. Categories the Metadata Standard Uses to Describe Datasets

The following sections are categories of metadata that may be used by the metadata standard to describe data. Not all categories are relevant to different datasets and may not necessarily be present in the standard. The presence or absence of a category should not be used for judging one standard against another.

Note: For each of the questions in the following sections please provide the obligation status information. Valid obligation values are:

Obligation Level

M	Mandatory	This information must be entered to comply with the metadata standard.
C	Conditional	This information is entered based on other information in the metadata.
O	Optional	This information is optionally entered in the metadata.

If the answer is 'NO', the obligation status should be left blank.

SECTION 4 – Dataset Identification

This section examines how the standard provides for the specification of the various sorts of titles and identifiers for a dataset.

4.1 The title(s) of a dataset
Does the standard allow/require the provision for:

Obligation

A. __ Previous title(s)?: **NO**
B. **M** Current title(s)?: **YES**
C. __ Possible future title(s)?: **NO**
D. **O** Official titles in different languages?: **YES**
E. **O** Abbreviated title(s)?: **YES**
F. __ Alternative Title(s)?: **NO**
G. __ Does the standard support effective dates of titles?: **NO**
H. __ Other identifiers (e.g., ISBN, National Stock Number, Uniform Product Code)?: **NO**

4.2 Parties responsible for the dataset
Does the standard allow/require the provision for:

Obligation

A. __ Author?: **NO**
B. __ Owner?: **NO**
C. **M** Other?: **YES**
"Responsible party" without limiting to a specific role (ref. 4.2 A–B)

4.3 ISBD Information

Obligation

A. __ Does the metadata standard specify inclusion of mandatory ISBD information (see glossary) within the metadata?: **NO**

SECTION 5 – Status of the Dataset

This section examines how the standard provides for the specification of the current status and updating of the data set.

5.1 Status and Progress of Datasets
Does the metadata standard allow/require the provision for:

Obligation

A. **M** Information on the status of the entire dataset?: **YES**
Whether the dataset is under construction, finished etc. with related dates shall be given as a part of the element "Timing of data"
B. __ Information on the status of the particular subset (e.g., observation time, inauguration time, extraction time)?: **NO**
Does the metadata standard specify:
C. **M** A structure for updating or maintenance information?: **YES**
However, not a specific structure but a free text element for any aspects of updating and maintenance

D. **M** A structure to describe historical information (see also Section 8.3 G)?: **YES**
However, not a specific structure but a free text element for timing of data (ref. 5.1 A)

5.2 Information on Incremental Updating
Does the metadata standard specify:

Obligation

A. **M** A structure for information on incremental updates of the entire dataset?: **YES**
However, not a specific structure but a free text element for any aspects of updating and maintenance (ref. 5.1 C)
B. __ A structure for information on incremental updates of particular portions of the dataset?: **NO**
C. **M** A structure to describe the update process for the dataset?: **YES**
Ref. 5.1 C

SECTION 6 – Extent of the Dataset

This section examines how the standard provides for the specification of the spatial and temporal extent of the dataset.

6.1 The spatial extent of the dataset

Obligation

A. **M** Does the metadata standard allow/require for the description of horizontal extent (e.g., bounding rectangle/polygon, named location (i.e., Stockholm, etc.))?: **YES**
Textual description is mandatory, an extent map is optional as well as a named location with unit names or unit id´s in a non-coordinate spatial reference system
B. __ Does the metadata standard allow/require for the description of vertical extent (e.g., minima/maxima heights, etc.)?: **NO**
C. __ Does the standard require specific units for describing extent?: **NO**
D. __ Does the standard allow specification of the units for describing extent?: **NO**
E. __ Does the standard require specific spatial reference system for describing extent?: **NO**
However, the standard states that the Coordinator of a metadata service shall specify the reference system for extent maps.
F. **C** Does the standard allow specification of the spatial reference system for describing extent?: **YES**
Ref. 6.1 E

6.2 The temporal extent of the dataset

Obligation

A. **M** Does the standard allow/require for the description of time period for which the dataset is relevant?: **YES**
B. __ Does the standard allow/require specification of the temporal reference system?: **NO**

SECTION 7 – Data Content of the Dataset

This section examines how the standard provides for the specification of the data content of the dataset in terms of overview elements, themes, data definitions and product specifications.

7.1 Dataset overview
Does the standard allow/require for the provision of the following information:

Obligation

A. **M** Dataset overview?: **YES**
B. **M** Abstract – brief summary of the content of the dataset?: **YES**
C. **O** Purpose – summary of the intentions with which the dataset was developed?: **YES**
D. **O** Usage – list of activities for which the dataset has been used? (see also Section 8.5): **YES**
E. **M** Spatial schema – (e.g., vector, raster)? (see also Section 9): **YES**
F. **M** Spatial reference system? (see also Section 12): **YES**
G. **M** Language?: **YES**
H. __ Character set (e.g., ISO8859-1, JISX0208)?: **NO**
I. **O** List of documents providing further information about the dataset?: **YES**
J. **O** Sample/preview dataset?: **YES**
K. __ Related datasets?: **NO**
L. __ Original size of the dataset?: **NO**
M. __ Compression? (see also Section 11.5): **NO**
N. **O** Other?: **YES**
Services attached to the dataset, Reference to a detailed description of the dataset (ref. 7.3 A, 7.4 A)

7.2 Theme(s)

Obligation

A. **M** Does the standard allow/require the provision for the description of theme(s) (e.g., topography, land cover, transport, etc.) in the dataset?: **YES**
B. **O** If A is yes, does the standard allow the referencing of an outside dictionary or thesaurus?: **YES**
C. __ If A is yes, does the standard itself include a dictionary or thesaurus within the metadata?: **NO**
D. **O** If A is yes, does the standard allow the use of unauthorized user-defined keywords?: **YES**
In addition, the standard gives instructions how to formulate keywords in this kind of a situation

7.3 Data Definition

Obligation

A. **O** Does the standard allow/require the provision for the description of the data definition?: **YES**

A reference to an external description in an easily accessible document, preferably on www. The document may provide the definition at any level of detail (ref. 7.3 B-F)

B. If A is yes, does the standard allow/require the provision for the description of:

__ 1. Features/Objects?: **N/A**
__ 2. Attributes?: **N/A**
__ 3. Relations?: **N/A**

C. __ Can attributes be described separately from feature/object classes?: **N/A**

D. __ Does the standard allow/require the specification of the domain of the attributes?: **N/A**

E. __ Does the standard allow/require the specification of data type of the attributes?: **N/A**

F. __ Does the standard allow/require the specification of particular classification coding schemes (e.g., FACC, OSKA, S-57 OC)?: **N/A**

7.4 Product Specification

Obligation

A. **O** Does the standard allow/require for the provision of information about whether the dataset meets a particular product specification?: **YES**

B. __ If A is yes, can it be internal to the Metadata?: **NO**

C. **O** If A is yes, can it be external to the Metadata?: **YES**

SECTION 8 – Data Quality of the Dataset

This section examines how the standard deals with the question of data quality. This includes homogeneity and metrics for data quality, quality assurance, and usage.

8.1 Overall Data Quality Statement

Obligation

A. **M** Does the standard allow/require for the provision of quality to be described? If yes, please answer the questions below. (*If no, all following answers in Section* 8 *are N/A, please skip to Section* 9.): **YES**

See 8.6

B. __ Does the standard allow/require the provision for the quality information referencing an external quality standard, quality assurance standard, or specification?: **N/A**

8.2 Homogeneity of Data Quality

Obligation

A. __ Does the standard allow/require for the variation of the quality of the dataset to be described?: **N/A**

B. What type of aggregation of quality information is allowed? **N/A**

__ 1. Entire database?: **N/A**
__ 2. Feature/Object instance?: **N/A**
__ 3. Feature/Object class?: **N/A**
__ 4. Attribute instance?: **N/A**
__ 5. Attribute class?: **N/A**
__ 6. Layer?: **N/A**
__ 7. Theme?: **N/A**
__ 8. Geographic extent?: **N/A**
__ 9. Other?: **N/A**

8.3 Metrics for Data Quality

Does the standard allow/require the provision of values for:

Obligation

A. __ Positional accuracy?: **N/A**
1. Standard error?: **N/A**
2. Maximum error?: **N/A**
3. Error correlation?: **N/A**
4. Confidence level (e.g., 90%, 95%)?: **N/A**
5. Other?: **N/A**

B. __ Attribute accuracy?: **N/A**
1. Standard error?: **N/A**
2. Maximum error?: **N/A**
3. Error frequency?: **N/A**
4. Error correlation?: **N/A**
5. Confidence level (e.g., 90%, 95%)?: **N/A**
6. Other?: **N/A**

C. __ Temporal accuracy?: **N/A**
1. Standard error?: **N/A**
2. Maximum error?: **N/A**
3. Error correlation?: **N/A**
4. Confidence level? (e.g., 90%, 95%): **N/A**
5. Other?: **N/A**

D. __ Completeness?: **N/A**
1. Error of omission?: **N/A**
2. Error of commission?: **N/A**
3. Other?: **N/A**

E. __ Currentness?: **N/A**
1. Date?: **N/A**
2. Maximum age?: **N/A**
3. Temporal extent?: **N/A**
4. Other?: **N/A**

F. __ Logical consistency?: **N/A**
G. __ Lineage?: **N/A**
1. List of processing steps?: **N/A**
2. List of values of processing parameters?: **N/A**
3. List of printouts of statistics, etc.?: **N/A**
4. Source material?: **N/A**
5. Other?: **N/A**
H. __ Additional Quality Information?: **N/A**

8.4 Quality Assurance Methods

Does the standard allow/require the provision of information about the methods used to determine:

Obligation

A. __ Positional accuracy?: **N/A**
1. Standard error in adjustments?: **N/A**
2. Repeated measurements?: **N/A**
3. Independent measurements?: **N/A**
4. Subjective evaluations?: **N/A**
5. Other?: **N/A**
B. __ Attribute accuracy?: **N/A**
1. Standard error?: **N/A**
2. Error classification matrix?: **N/A**
3. Independent measurements?: **N/A**
4. Subjective evaluation?: **N/A**
5. Other?: **N/A**
C. __ Temporal accuracy?: **N/A**
1. Independent measurements?: **N/A**
2. Subjective evaluation?: **N/A**
3. Other?: **N/A**
D. __ Completeness?: **N/A**
1. Field checks?: **N/A**
2. Repeated measurements?: **N/A**
3. Independent measurements?: **N/A**
4. Subjective evaluation?: **N/A**
5. Error of commission?: **N/A**
6. Other?: **N/A**
E. __ Currentness?: **N/A**
1. Date of source material?: **N/A**
2. Repeated measurements?: **N/A**
3. Independent measurements?: **N/A**
4. Subjective evaluation?: **N/A**
5. Other?: **N/A**
F. __ Logical consistency?: **N/A**
1. Verification results of consistency checks?: **N/A**

2. Repeated processing?: **N/A**
3. Comparison with independent sources?: **N/A**
4. Subjective evaluation?: **N/A**
5. Other?: **N/A**

G. __ Lineage?: **N/A**
1. Checking results of processing steps: **N/A**
2. Subjective evaluation: **N/A**
3. Other: **N/A**

H. __ Additional Quality Information?: **N/A**

8.5 Usage

Obligation

A. __ Does the metadata standard allow/require the provision of information about usage?: **N/A**
1. Who used it: **N/A**
2. When it was used: **N/A**
3. For what application it was used: **N/A**
4. The effectiveness of use: **N/A**
5. Other: **N/A**

8.6 Alternative Quality Description:

Since the standard is intended to description of any kind of data no common quality model can be assumed. Therefore, the standard provides two elements for describing quality: "Source of data" is mandatory and "Description of data quality" is optional. JHS 137 A provides a mandatory element "Spatial accuracy". The description of quality may follow any quality model or metrics, at any level of detail. It may be included in the metadata or the element may provide a reference to an external description in an easily accessible document, preferably on www.

SECTION 9 – Spatial Data Structure of the Dataset

This section examines how the standard provides for the specification of the spatial data model and spatial data types.

9.1 Spatial Data Model
Does the standard allow/require the provision for:

Obligation

A. **O** The description of the geometrical and/or topological primitives used to model the real world?: **YES**

B. **O** Information about an external standard for geometrical and/or topological primitive specification?: **YES**

9.2 Spatial Data Types
Does the standard allow/require the provision for:

Obligation

A. **M** Information about the spatial data type(s) used in the dataset?: **YES**
A free text element allowing any data types to be listed (ref. 9.2 B)
B. __ If A is yes, which of the following generic spatial data types are supported? For each, please list specific data primitive levels supported by each data type.
1. Vector (e.g., spaghetti, chain-node/node-edge-face, planar graph, full topology)?: **N/A**
2. Areal Raster (e.g., thematic raster – scanned and classification maps, image raster – satellite images, scanned aerial photographs)?: **N/A**
3. Grid / Matrix (e.g., Digital Elevation Model)?: **N/A**
4. Triangular Irregular Network (TIN)?: **N/A**
5. Other?: **N/A**

9.3 Other Data Types
Does the standard allow the provision of information about:

Obligation

A. __ Video?: **NO**
B. __ Sound?: **NO**
C. **O** Other?: **YES**
In fact, any data type can be described textually but without a specified structure.

SECTION 10 – Spatial Reference of the Dataset

This section examines how the dataset provides for the specification of the spatial reference characteristics of the data set.

Does the standard allow/require the provision for the following information:

10.1 Geoidal Model

Obligation

A. **O** Geoidal model name?: **YES**
B. **O** Geoidal model definitions?: **YES**
A reference to an external www document

10.2 Geodetic datum

Obligation

A. **O** Geodetic datum name?: **YES**
B. **O** Geodetic datum definitions?: **YES**
A reference to an external www document

10.3 Reference Ellipsoid

Obligation

A. **O** Reference ellipsoid name?: **YES**
B. **O** Reference ellipsoid definitions?: **YES**
A reference to an external www document

10.4 Projection

Obligation

A. **O** Projection name?: **YES**
B. **O** Projection definitions?: **YES**
A reference to an external www document

10.5 Coordinate Systems

Obligation

A. **C** Coordinate system type?: **YES**
B. **O** Coordinate system definitions?: **YES**
A reference to an external www document

10.6 Height Reference System

Obligation

A. **C** Vertical datum name?: **YES**
B. **O** Types of heights?: **YES**
A reference to an external www document

10.7 Ancillary Geodetic Information

Obligation

A. __ Does the standard allow/require the provision for ancillary geodetic, gravimetric, or magnetometric information?: **NO**

10.8 Transformation Parameters
Does the standard specify a method to provide information for coordinate transformation(s):

Obligation

A. __ Model name?: **NO**
B. __ Control points?: **NO**
C. __ Transformation parameters?: **NO**

10.9 Non-Coordinate Spatial Reference
Does the standard allow/require the provision for information on non-coordinate spatial reference systems (e.g., postal systems, street address)?:

Obligation

A. **C** Reference system name?: **YES**
B. **O** Reference system definitions?: **YES**
C. **O** Other?: **YES**
Date/version of the reference system if the system is not fully stabile

SECTION 11 – Availability and Distribution of the Dataset

This section examines how the standard provides information on the availability of the dataset?

Does the standard allow/require the provision for:

11.1 The Distributor(s) of the Dataset

Obligation

A. __ Information about the distributor(s) of the dataset?: **NO**
B. **M** Contact information for the distributor(s)?: **YES**

11.2 Restrictions to Access and Usage

Obligation

A. **O** Information about copyright on the dataset (e.g., date, period, copyright owner, etc.)?: **YES**
B. **O** Information about restrictions on the access of the dataset?: **YES**
C. **O** Information about restrictions on the usage of the dataset?: **YES**

11.3 Pricing Details

Obligation

A. **O** Information about pricing?: **YES**

11.4 Distribution of the Dataset

Obligation

A. **O** Information about distribution media?: **YES**
B. **O** Information about on-line access?: **YES**
C. **O** Information about units of distribution (e.g., tiles, layers, polygons, elements)?: **YES**

11.5 Distribution Format

Obligation

A. **O** Information about distribution formats (e.g., commercial CAD format, commercial GIS formats, National/International transfer standard)?: **YES**
B. __ Information about the encoding of the dataset (e.g., binary, ASCII)?: **NO**

C. **O** Information about compression method?: **YES**
D. **O** Information about compressed dataset size?: **YES**
E. **O** Information about the distribution mechanism?: **YES**
F. __ Information about the encapsulation of data (e.g., ISO 8211, ISO 8824)?: **NO**

11.6 Ordering Procedures

Obligation

A. **O** Information about ordering procedures for the dataset?: **YES**

11.7 Support Services

Obligation

A. **O** Information about support services (e.g., additional data transformations, format or coordinate conversions)?: **YES**

SECTION 12 – Authorization and Verification of the Metadata

This section examines how the standard provides for the description of the authorization and verification on the metadata of a dataset (e.g., how is the metadata itself authorized and verified?).

12.1 Metadata reference

Obligation

A. **M** Does the standard allow/require the provision of information about reference dates of the metadata (e.g., last check date, last update date, review date)?: **YES**

12.2 Verification Authority

Obligation

A. **M** Does the standard allow/require the provision for information about who verified/authorized the metadata?: **YES**
Name, organization, e-mail
B. __ Does the standard specify the method for providing this information (e.g., ISO 9000): **NO**

12.3 Statistical Methods Used for Verification

Obligation

A. __ Does the standard specify the statistical methods to be used for verification of the metadata?: **NO**
B. __ Does the standard allow/require the provision for the description of the statistical methods used to verify the metadata?: **NO**

END OF CHARACTERISTICS
FIN

World Spatial Metadata Standards
H. Moellering, H.J.G.L. Aalders & A. Crane (Editors)

Hungary
KIKERES Térinformatikai Profil
KIKERES Spatial Metadata Profile

Istvan Kadar, Tamas Prajczer

I. Information about the Metadata Standard Itself

SECTION 0 – Assessment Information

This section provides the user with a general description of the scope and intended use of the standard. Information on the respondent who prepared the assessment is also provided.

0.1 Respondent

Assessment prepared by:

Tamás Prajczer
director
GeoX Ltd.
1033 Budapest, Főtér 1, Hungary
+36 1 4379458
+36 30 9608337
prajczer@geox.hu

Date assessment completed:

October 2000; updated May 2002, Reviewed Spring, 2005

Relationship of respondent to standard:

Drafting committee member

0.2 Brief Summary Statement

A. Summary of the scope and intent of the standard:
Currently the Hungarian Spatial Metadata Standard does not exists. The Spatial Metadata Profil is a "de facto" standard. It is a part (a profile) of the HGLS.
The HGLS – the metadata description of the KIKERES – provides a common set of terminology and definitions for the documentation of information resources. The HGLS does not define the technical environment of the implementation, but it was determined by the KIKERES system.

The Interministerial Committee for Information Technology (ITB) – which is the forum of the IT managers of the central public administration led by the Prime Minister's Office (MeH) – decided to create a public administration data catalogue accessible free of charge for anybody through the Internet. This service is the KIKERES from the full Hungarian name of – Hungarian Governmental Information Locator Service. The KIKERES services the most important features – in the form of metadata – of the public utility information resources found in the public administration.
The METATÉR project – the Hungarian Spatial Metadata Service – can be considered as the forerunner of the KIKERES. It was a non-hierarchical system for spatial metadata service, functioning with a single profile. After the completion of the KIKERES, the METATÉR has become one profile of the KIKERES.
The HGLS has two parts: the base profile and the extensions. The basic profile contains the common elements used in the description of information resources available in the public administration. Obviously there are subject areas, which require special metadata too to describe their information resources. For example the projection system is an integrated feature of spatial data, but not used in general GILS. The extensions can be used to define special elements which are important for specific areas

SECTION 1 – Administration of Metadata Standard

This section provides a general description of the administrative framework, within which the standard was developed, tested and currently resides. Significant details are also provided in terms of the developmental and managerial history as well as information on the existence of documentation, software tools and training materials. This section also provides a point of contact for further information on the standard.

1.1 Name of Standard

A. Name of standard:
   1. In original language(s):
      KIKERES Térinformatikai Profil
   2. English translation, if appropriate:
      KIKERES Spatial Metadata Profile
B. Version/Edition:
   version 1.0
C. Language(s) of documentation:
   Hungarian
D. Acronym(s):
   N/A
E. Official ID:
   N/A
F. Name(s) of recognizing standards authority(ies):
   N/A

1.2 Responsible Institutions

Contact information about those institutions that are (or were) responsible for the processes involved in the evolution of the standard:

A. Production and/or development:
  Office of the Government Commissioner for ICT
  1024 Budapest, Szilágyi Erzsébet fasor 11/b, Hungary
  sikolya@ikb.meh.hu

B. Testing:
  N/A

C. Conformance:
  N/A

D. Maintenance:
  N/A

E. Distribution:
  N/A

F. Help Desk/User Support:
  N/A
  Office of the Government Commissioner for ICT
  1024 Budapest, Szilágyi Erzsébet fasor 11/b, Hungary
  sikolya@ikb.meh.hu

1.3 Development History

A. Beginning of work on metadata standard: **31/01/2000**
  1st meeting of the KIKERES metadata group

B. Milestones:
  15/02/2000 – 1st draft of the KIKERES Base Profile
  29/02/2000 – 2nd draft of the KIKERES Base Profile
  30/03/2000 – Final version of the KIKERES Base Profile
  30/04/2000 – Final version of the KIKERES Spatial Metadata Profile

C. Anticipated completion: **N/A**

D. Update cycle: **Not yet decided**

1.4 Status of the Standard

A. Is the standard officially recognized: **NO**

B. Date(s) of recognition: **N/A**

C. Anticipated date of recognition: **N/A**

D. Current stage in recognition process: **N/A**
  Not planned to be recognized

E. Steps still required to achieve recognition: **N/A**

1.5 Access to Documentation of the Standard

A. Is the standard copyrighted?: **YES**

B. Name and address of copyright owner:

Miklósné Domokos
Géza Bakonyi
Péter Kardeván
Klára Koltay
Gábor Kleinheincz
Tamás Prajczer
Zsolt Sikolya
Ágnes Vadász

C. Do restrictions on the use of the standard apply?: **NO**

D. Is the standard documentation currently available?: **YES**

E. List the form/media of standard documentation:

paper
Adobe pdf / download
MS Word97 / download

F. Is the standard available on the Internet?:

http://www.kikeres.gov.hu

G. Price: **Free**

H. Contact details for obtaining the standard:

Zsolt Sikolya
Office of the Government Commissioner for ICT
1024 Budapest, Szilágyi Erzsébet fasor 11., Hungary
sikolya@ikb.meh.hu

I. Does the standard have ISBN/ISSN numbers?: **NO**

1.6 Available Software Tools

A. Is software available to implement the standard as a database?: **YES**

B. Is software available to test compliance of metadata with the standard?: **YES**

C. Is software available to produce metadata compliant with the standard?: **YES**

D. Other relevant software: **NO**

E. Is there a test dataset available to test software tools for implementing the standard?: **NO**

F. For each product available above, please list contact information:

Zsolt Sikolya
Prime Minister's Office
1024 Budapest, Szilágyi Erzsébet fasor 11., Hungary

1.7 Available Training Materials

A. Are organized training sessions available?: **NO**

B. Contact details for training sessions: **N/A**

C. Training documentation available: **N/A**

SECTION 2 – Use and Implementation of the Metadata Standard

This section provides information on the way in which the standard is used and the purpose of its use. It also provides descriptions about the level of abstraction of the standard.

2.1 Intended Use of the Standard
Can the standard be used:

A. Within a transfer data structure?: **NO**
B. To produce hardcopy catalog of datasets?: **YES**
C. To catalog data in a digital on-line catalog?: **YES**
D. Other: **NO**

2.2 Intended Purpose of the Standard
Is the intended purpose of the standard:

A. To support discovery, identification, location of data?: **YES**
B. To support how to access data?: **YES**
C. To support the determination of dataset fitness for use?: **YES**
D. To support transfer of dataset?: **NO**
E. To support an organization's data management?: **YES**
F. Other: **NO**

2.3 Table of Contents

A. Provide the standard's Table of Contents
This table of contents describes only the main chapters of the base profile and the spatial data profile.

Base profile

- **Formal information**
  - **Title**
  - **Publication date**
  - **Size**
  - **Extent**
  - **Relations**
  - **Organizations**
- **Content information**
  - **Controlled keywords**
  - **Keywords**
  - **Abstract**
- **Administrative information**
  - **Classification of metadata record**
  - **Occurrence of resource**
  - **Type**
  - **Date of creation**

- **Language**
- **Technical information for accessibility**
- **Rights**

- **Spatial data profil**
  - **Series**
  - **Status**
  - **Update**
  - **Sample**
  - **Name**
  - **Format**
  - **☐ Quality**
  - **Positional accuracy**
  - **Completeness**
  - **Objects**
  - **Lineage**
- **Spatial Reference**
  - **Indirect reference**
  - **Direct reference**
- **Identifier**
- **Unit of Distribution**

2.4 Level of Abstraction of the Metadata Standard Definition

A. Does the standard define metadata elements?: **YES**
B. Does the standard define a model for the relationships between the metadata elements?: **YES**
C. Provide an overview diagram of the model of the metadata, if available. **N/A**
D. Does the standard define domains and/or data types of the metadata elements?: **YES**
E. Does the standard define how the metadata should be encoded?: **YES**
F. Does the standard define a database schema for the automatic generation of a database?: **YES**

2.5 Extensibility of the Standard

A. Is the metadata description extensible (e.g., Can user add metadata domains, elements, thematic profiles, etc.)?: **YES**

The core set (base profile) can be completed with special metadata elements for certain subjects areas, thus further metadata profiles may be created. The profiles – worked out this way – can be added with further supplements, thus a hierarchical system – generated by origination – of the profiles can be configured. The metadata records, characterizing the individual information resources, are made according to the profile characterizing the subject area of the information resource.

- **When creating a new profile, the following general rules are to be followed:**
- **New elements of a extension can be compound elements or data elements.**

- **A new element must not change the characteristics of the elements of the parent profile.**
- **No addition can be accepted, if it contains elements which are already present in the parent profile.**
- **A new profile becomes valid when it is approved and registered by the central administration.**
- **When creating a new element – element or compound element – those characteristics and features have to be specified once used in the base profile.**

Acceptance of additions and their administration for the system are going on in a centralised way. This is important first of all to avoid the acceptance of several types of additions and the overlapping of various additions created in different branches.

2.6 Availability of Metadata

Does the standard specify:

A. How metadata are to be accessed?: **YES**
  via Internet
B. A mechanism for querying the metadata?: **YES**
  own protocol is used for querying the metadata
C. Particular standardized query protocols?: **YES**
  Z39.50
D. That the metadata should be encoded for computer access?: **YES**
E. That the metadata should be encrypted for security?: **NO**

2.7 Examples

A. Does the standard provide examples of implementations?: **NO**
B. How many examples are available?: **N/A**

SECTION 3 – Linkage and Coordination of the Metadata Standard

This section provides information on how the metadata standard is linked to other standards, particularly transfer standards, and how such linkage is implemented. If a significant linkage does exist, detailed technical information is provided on the precise nature of the linkage to the standard, and how the linkage affects implementation of the metadata standard.

3.1 Associated Transfer Standards

A. Is the metadata standard associated with a transfer standard(s)?: **NO**
B. If A is yes, provide references and contact details for each linked standard: **N/A**
C. If A is yes, are both (all) standards coordinated?: **N/A**
D. If A is yes, describe how the coordination works, both from a technical and conformance (normative) viewpoint: **N/A**

3.2 Incorporated into transfer mechanism

A. Is the metadata standard incorporated into a transfer mechanism?: **NO**

B. If A is yes, provide references and contact details for each incorporated mechanism: **N/A**
C. If A is yes, describe how: **N/A**

3.3 Other linked, coordinated or associated standards

A. Is the metadata standard linked, coordinated, or associated with any other standard(s) not listed above?: **NO**
B. If A is yes, provide references and contact details for each standard: **N/A**
C. If A is yes, describe how for each relevant standard: **N/A**

II. Categories the Metadata Standard Uses to Describe Datasets

The following sections are categories of metadata that may be used by the metadata standard to describe data. Not all categories are relevant to different datasets and may not necessarily be present in the standard. The presence or absence of a category should not be used for judging one standard against another.

Note: For each of the questions in the following sections please provide the obligation status information. Valid obligation values are:

Obligation Level

M	Mandatory	This information must be entered to comply with the metadata standard.
C	Conditional	This information is entered based on other information in the metadata.
O	Optional	This information is optionally entered in the metadata.

If the answer is 'NO', the obligation status should be left blank.

SECTION 4 – Dataset Identification

This section examines how the standard provides for the specification of the various sorts of titles and identifiers for a dataset.

4.1 The title(s) of a dataset
Does the standard allow/require the provision for:

Obligation

A. __ Previous title(s)?: **NO**
B. **M** Current title(s)?: **YES**
C. __ Possible future title(s)?: **NO**
D. __ Official titles in different languages?: **NO**
E. **O** Abbreviated title(s)?: **YES**
F. **O** Alternative Title(s)?: **YES**
G. __ Does the standard support effective dates of titles?: **NO**

H. **O** Other identifiers (e.g., ISBN, National Stock Number, Uniform Product Code)?: **YES**
URL
URN
ISBN
ISSN
other

4.2 Parties responsible for the dataset
Does the standard allow/require the provision for:

Obligation

A. **M** Author?: **YES**
B. **M** Owner?: **YES**
C. **C** Other?: **YES**

4.3 ISBD Information

Obligation

A. __ Does the metadata standard specify inclusion of mandatory ISBD information (see glossary) within the metadata?: **NO**

SECTION 5 – Status of the Dataset

This section examines how the standard provides for the specification of the current status and updating of the data set.

5.1 Status and Progress of Datasets
Does the metadata standard allow/require the provision for:

Obligation

A. **M** Information on the status of the entire dataset?: **YES**
B. __ Information on the status of the particular subset (e.g., observation time, inauguration time, extraction time)?: **NO**
Does the metadata standard specify:
C. **O** A structure for updating or maintenance information?: **YES**
D. __ A structure to describe historical information (see also Section 8.3 G)?: **NO**

5.2 Information on Incremental Updating
Does the metadata standard specify:

Obligation

A. __ A structure for information on incremental updates of the entire dataset?: **NO**
B. __ A structure for information on incremental updates of particular portions of the dataset?: **NO**

C. __ A structure to describe the update process for the dataset?: **NO**

SECTION 6 – Extent of the Dataset

This section examines how the standard provides for the specification of the spatial and temporal extent of the dataset.

6.1 The spatial extent of the dataset

Obligation

A. **M** Does the metadata standard allow/require for the description of horizontal extent (e.g., bounding rectangle/polygon, named location (i.e., Stockholm, etc.))?: **YES**

West Bounding Coordinate in EOV
East Bounding Coordinate in EOV
South Bounding Coordinate in EOV
North Bounding Coordinate in EOV
West Bounding Coordinate in WGS84
East Bounding Coordinate in WGS84
South Bounding Coordinate in WGS84
North Bounding Coordinate in WGS84
Location name YES

B. __ Does the metadata standard allow/require for the description of vertical extent (e.g., minima/maxima heights, etc.)?: **NO**

C. __ Does the standard require specific units for describing extent?: **NO**

D. __ Does the standard allow specification of the units for describing extent?: **NO**

E. **M** Does the standard require specific spatial reference system for describing extent?: **YES**

F. __ Does the standard allow specification of the spatial reference system for describing extent?: **NO**

6.2 The temporal extent of the dataset

Obligation

A. **M** Does the standard allow/require for the description of time period for which the dataset is relevant?: **YES**

B. __ Does the standard allow/require specification of the temporal reference system?: **NO**

SECTION 7 – Data Content of the Dataset

This section examines how the standard provides for the specification of the data content of the dataset in terms of overview elements, themes, data definitions and product specifications.

7.1 Dataset overview

Does the standard allow/require for the provision of the following information:

Obligation

A. **M** Dataset overview?: **YES**
B. __ Abstract – brief summary of the content of the dataset?: **NO**
C. __ Purpose – summary of the intentions with which the dataset was developed?: **NO**
D. __ Usage – list of activities for which the dataset has been used? (see also Section 8.5): **NO**
E. **M** Spatial schema – (e.g., vector, raster)? (see also Section 9): **YES**
F. **M** Spatial reference system? (see also Section 12): **YES**
G. **M** Language?: **YES**
H. __ Character set (e.g., ISO8859-1, JISX0208)?: **NO**
I. **O** List of documents providing further information about the dataset?: **YES**
J. **O** Sample/preview dataset?: **YES**
K. **O** Related datasets?: **YES**
L. **O** Original size of the dataset?: **YES**
M. __ Compression? (see also Section 11.5): **NO**
N. __ Other?: **NO**

7.2 Theme(s)

Obligation

A. **M** Does the standard allow/require the provision for the description of theme(s) (e.g., topography, land cover, transport, etc.) in the dataset?: **YES**
B. **O** If A is yes, does the standard allow the referencing of an outside dictionary or thesaurus?: **YES**
C. **M** If A is yes, does the standard itself include a dictionary or thesaurus within the metadata?: **YES**
D. **M** If A is yes, does the standard allow the use of unauthorized user-defined keywords?: **YES**

7.3 Data Definition

Obligation

A. __ Does the standard allow/require the provision for the description of the data definition?: **NO**
B. If A is yes, does the standard allow/require the provision for the description of: **N/A**
__ 1. Features/Objects?: **N/A**
__ 2. Attributes?: **N/A**
__ 3. Relations?: **N/A**
C. __ Can attributes be described separately from feature/object classes?: **NO**
D. __ Does the standard allow/require the specification of the domain of the attributes?: **NO**

E. __ Does the standard allow/require the specification of data type of the attributes?: **NO**

F. __ Does the standard allow/require the specification of particular classification coding schemes (e.g., FACC, OSKA, S-57 OC)?: **NO**

7.4 Product Specification

Obligation

A. __ Does the standard allow/require for the provision of information about whether the dataset meets a particular product specification?: **NO**

B. __ If A is yes, can it be internal to the Metadata?: **N/A**

C. __ If A is yes, can it be external to the Metadata?: **N/A**

SECTION 8 – Data Quality of the Dataset

This section examines how the standard deals with the question of data quality. This includes homogeneity and metrics for data quality, quality assurance, and usage.

8.1 Overall Data Quality Statement

Obligation

A. **O** Does the standard allow/require for the provision of quality to be described? If yes, please answer the questions below. (*If no, all following answers in Section* 8 *are N/A, please skip to Section* 9.): **YES**

B. __ Does the standard allow/require the provision for the quality information referencing an external quality standard, quality assurance standard, or specification?: **NO**

8.2 Homogeneity of Data Quality

Obligation

A. __ Does the standard allow/require for the variation of the quality of the dataset to be described?: **NO**

B. What type of aggregation of quality information is allowed?

M 1. Entire database?: **YES**

__ 2. Feature/Object instance?: **NO**

__ 3. Feature/Object class?: **NO**

__ 4. Attribute instance?: **NO**

__ 5. Attribute class?: **NO**

__ 6. Layer?: **NO**

__ 7. Theme?: **NO**

__ 8. Geographic extent?: **NO**

__ 9. Other?: **NO**

8.3 Metrics for Data Quality

Does the standard allow/require the provision of values for:

Obligation

A. **M** Positional accuracy?:	**YES**
1. Standard error?:	**YES**
2. Maximum error?:	**NO**
3. Error correlation?:	**NO**
4. Confidence level (e.g., 90%, 95%)?:	**NO**
5. Other?:	**NO**
B. __ Attribute accuracy?:	**NO**
1. Standard error?:	**N/A**
2. Maximum error?:	**N/A**
3. Error frequency?:	**N/A**
4. Error correlation?:	**N/A**
5. Confidence level (e.g., 90%, 95%)?:	**N/A**
6. Other?:	**N/A**
C. __ Temporal accuracy?:	**NO**
1. Standard error?:	**N/A**
2. Maximum error?:	**N/A**
3. Error correlation?:	**N/A**
4. Confidence level? (e.g., 90%, 95%):	**N/A**
5. Other?:	**N/A**
D. **M** Completeness?:	**YES**
1. Error of omission?:	**NO**
2. Error of commission?:	**NO**
3. Other?:	**YES**
percent of omission	
E. **O** Currentness?:	**YES**
1. Date?:	**NO**
2. Maximum age?:	**NO**
3. Temporal extent?:	**YES**
4. Other?:	**YES**
Date of last update of dataset	
F. __ Logical consistency?:	**NO**
G. **O** Lineage?:	**YES**
1. List of processing steps?:	**YES**
2. List of values of processing parameters?:	**NO**
3. List of printouts of statistics, etc.?:	**NO**
4. Source material?:	**YES**
5. Other?:	**NO**
H. __ Additional Quality Information?:	**NO**

8.4 Quality Assurance Methods

Does the standard allow/require the provision of information about the methods used to determine:

Obligation

A. __ Positional accuracy?:	**NO**
1. Standard error in adjustments?:	**N/A**
2. Repeated measurements?:	**N/A**
3. Independent measurements?:	**N/A**
4. Subjective evaluations?:	**N/A**
5. Other?:	**N/A**
B. __ Attribute accuracy?:	**NO**
1. Standard error?:	**N/A**
2. Error classification matrix?:	**N/A**
3. Independent measurements?:	**N/A**
4. Subjective evaluation?:	**N/A**
5. Other?:	**N/A**
C. __ Temporal accuracy?:	**NO**
1. Independent measurements?:	**N/A**
2. Subjective evaluation?:	**N/A**
3. Other?:	**N/A**
D. __ Completeness?:	**NO**
1. Field checks?:	**N/A**
2. Repeated measurements?:	**N/A**
3. Independent measurements?:	**N/A**
4. Subjective evaluation?:	**N/A**
5. Error of commission?:	**N/A**
6. Other?:	**N/A**
E. __ Currentness?:	**NO**
1. Date of source material?:	**N/A**
2. Repeated measurements?:	**N/A**
3. Independent measurements?:	**N/A**
4. Subjective evaluation?:	**N/A**
5. Other?:	**N/A**
F. __ Logical consistency?:	**NO**
1. Verification results of consistency checks?:	**N/A**
2. Repeated processing?:	**N/A**
3. Comparison with independent sources?:	**N/A**
4. Subjective evaluation?:	**N/A**
5. Other?:	**N/A**
G. __ Lineage?:	**NO**
1. Checking results of processing steps:	**N/A**
2. Subjective evaluation:	**N/A**
3. Other:	**N/A**
H. __ Additional Quality Information?:	**N/A**

8.5 Usage

Obligation

A. __ Does the metadata standard allow/require the provision of information about usage?: **NO**

1. Who used it: **N/A**
2. When it was used: **N/A**
3. For what application it was used: **N/A**
4. The effectiveness of use: **N/A**
5. Other: **N/A**

8.6 Alternative Quality Description: **NO**

SECTION 9 – Spatial Data Structure of the Dataset

This section examines how the standard provides for the specification of the spatial data model and spatial data types.

9.1 Spatial Data Model
Does the standard allow/require the provision for:

Obligation

A. **M** The description of the geometrical and/or topological primitives used to model the real world?: **YES**
B. __ Information about an external standard for geometrical and/or topological primitive specification?: **NO**

9.2 Spatial Data Types
Does the standard allow/require the provision for:

Obligation

A. **M** Information about the spatial data type(s) used in the dataset?: **YES**
B. __ If A is yes, which of the following generic spatial data types are supported?
It is a 'free text' field. There are no restrictions for data primitives, all types are allowed.
For each, please list specific data primitive levels supported by each data type.
1. Vector (e.g., spaghetti, chain-node/node-edge-face, planar graph, full topology)?: **YES**
2. Areal Raster (e.g., thematic raster – scanned and classification maps, image raster – satellite images, scanned aerial photographs)?: **YES**
3. Grid/Matrix (e.g., Digital Elevation Model)?: **YES**
4. Triangular Irregular Network (TIN)?: **YES**
5. Other?: **YES**

9.3 Other Data Types
Does the standard allow the provision of information about:

Obligation

A. **O** Video?: **YES**
B. **O** Sound?: **YES**
C. **O** Other?: **YES**

SECTION 10 – Spatial Reference of the Dataset

This section examines how the dataset provides for the specification of the spatial reference characteristics of the data set.

Does the standard allow/require the provision for the following information:

10.1 Geoidal Model

Obligation

A. __ Geoidal model name?: **NO**
B. __ Geoidal model definitions?: **NO**

10.2 Geodetic datum

Obligation

A. **M** Geodetic datum name?: **YES**
B. __ Geodetic datum definitions?: **NO**

10.3 Reference Ellipsoid

Obligation

A. __ Reference ellipsoid name?: **NO**
B. __ Reference ellipsoid definitions?: **NO**

10.4 Projection

Obligation

A. **M** Projection name?: **YES**
B. __ Projection definitions?: **NO**

10.5 Coordinate Systems

Obligation

A. __ Coordinate system type?: **NO**
B. __ Coordinate system definitions?: **NO**

10.6 Height Reference System

Obligation

A. **M** Vertical datum name?: **YES**
B. __ Types of heights?: **NO**

10.7 Ancillary Geodetic Information

Obligation

A. __ Does the standard allow/require the provision for ancillary geodetic, gravimetric, or magnetometric information?: **NO**

10.8 Transformation Parameters
Does the standard specify a method to provide information for coordinate transformation(s):

Obligation

A. __ Model name?: **NO**
B. __ Control points?: **NO**
C. __ Transformation parameters?: **NO**

10.9 Non-Coordinate Spatial Reference
Does the standard allow/require the provision for information on non-coordinate spatial reference systems (e.g., postal systems, street address)?:

Obligation

A. **M** Reference system name?: **YES**
B. __ Reference system definitions?: **NO**
C. __ Other?: **NO**

SECTION 11 – Availability and Distribution of the Dataset

This section examines how the standard provides information on the availability of the dataset?

Does the standard allow/require the provision for:

11.1 The Distributor(s) of the Dataset

Obligation

A. **M** Information about the distributor(s) of the dataset?: **YES**
B. **M** Contact information for the distributor(s)?: **YES**

11.2 Restrictions to Access and Usage

Obligation

A. **M** Information about copyright on the dataset (e.g., date, period, copyright owner, etc.)?: **YES**
B. **M** Information about restrictions on the access of the dataset?: **YES**
C. __ Information about restrictions on the usage of the dataset?: **NO**

11.3 Pricing Details

Obligation

A. **O** Information about pricing?: **YES**

11.4 Distribution of the Dataset

Obligation

A. __ Information about distribution media?: **NO**
B. **O** Information about on-line access?: **YES**
C. **O** Information about units of distribution (e.g., tiles, layers, polygons, elements)?: **YES**

11.5 Distribution Format

Obligation

A. **O** Information about distribution formats (e.g., commercial CAD format, commercial GIS formats, National/International transfer standard)?: **YES**
B. __ Information about the encoding of the dataset (e.g., binary, ASCII)?: **NO**
C. __ Information about compression method?: **NO**
D. __ Information about compressed dataset size?: **NO**
E. __ Information about the distribution mechanism?: **NO**
F. __ Information about the encapsulation of data (e.g., ISO 8211, ISO 8824)?: **NO**

11.6 Ordering Procedures

Obligation

A. **O** Information about ordering procedures for the dataset?: **YES**

11.7 Support Services

Obligation

A. __ Information about support services (e.g., additional data transformations, format or coordinate conversions)?: **NO**

SECTION 12 – Authorization and Verification of the Metadata

This section examines how the standard provides for the description of the authorization and verification on the metadata of a dataset (e.g., how is the metadata itself authorized and verified?).

12.1 Metadata reference

Obligation

A. **M** Does the standard allow/require the provision of information about reference dates of the metadata (e.g., last check date, last update date, review date)?: **YES**

Date of creation
Date of last update

12.2 Verification Authority

Obligation

A. __ Does the standard allow/require the provision for information about who verified/authorized the metadata?: **NO**

B. __ Does the standard specify the method for providing this information (e.g., ISO 9000): **NO**

12.3 Statistical Methods Used for Verification

Obligation

A. __ Does the standard specify the statistical methods to be used for verification of the metadata?: **NO**

B. __ Does the standard allow/require the provision for the description of the statistical methods used to verify the metadata?: **NO**

END OF CHARACTERISTICS

FIN

H. Moellering, H.J.G.L. Aalders & A. Crane (Editors)

Israel
Israel Metadata Standard 2000

Ammatzia Peled

I. Information about the Metadata Standard Itself

SECTION 0 – Assessment Information

This section provides the user with a general description of the scope and intended use of the standard. Information on the respondent who prepared the assessment is also provided.

0.1 Respondent

Assessment prepared by:

Ammatzia Peled,
Consultant for Computer Assisted Mapping and GIS, Survey of Israel
University of Haifa
Department of Geography
Mt. Carmel Campus
Haifa, 31905
ISRAEL
Tel: +(972) 4-8-343-591 (H); +(972) 4-8-240-148
Fax: +(972) 4-8-343-763 (H); +(972) 4-8-249-605
E-mail: peled@geo.haifa.ac.il

Date assessment completed:

July 2000, updated October 2000 and May 2002, Reviewed Spring, 2005

Relationship of respondent to standard:

Author, drafting committee member

0.2 Brief Summary Statement

A. Summary of the scope and intent of the standard:

The Israeli Metadata Standard (Format) IMF2000 is aimed to establish the framework and mechanism for transferring information related to spatial data. Origi-

nally, it was launched to supplement spatial data offered, by the Survey of Israel. It was developed to accommodate any spatial data regardless of its' origin, format or domain (vector, raster or textual). In addition, the standard was devised as an instrument of spatial data marketing and distribution. Thus, the standard was divided into separate chapters that each accommodate various aspects of the description and quality definition of the spatial data in terms of spatial accuracy, completeness and currentness, and information related to availability of information and its' accessibility. The major motivation in the development and the emphasis of the separate structure of individual chapters was to accommodate many optional models and a variety of data without using separate profiles but rather various parts (chapters) of the common standard.

SECTION 1 – Administration of Metadata Standard

This section provides a general description of the administrative framework, within which the standard was developed, tested and currently resides. Significant details are also provided in terms of the developmental and managerial history as well as information on the existence of documentation, software tools and training materials. This section also provides a point of contact for further information on the standard.

1.1 Name of Standard

A. Name of standard:
   IMF2000
   1. In original language(s):
      Israel Metadata Standard 2000
   2. English translation, if appropriate:
      Israel Metadata Standard 2000
B. Version/Edition:
C. Language(s) of documentation:
   Hebrew
D. Acronym(s):
   IMF2000
E. Official ID:
   No official ID (yet)
F. Name(s) of recognizing standards authority(ies):
   Survey of Israel

1.2 Responsible Institutions

Contact information about those institutions that are (or were) responsible for the processes involved in the evolution of the standard:

A. Production and/or development:

GIS Project Manager
Survey of Israel
1 Lincoln st.
Tel Aviv
Israel
Tel: +(972) 3-6231-900
Fax: +(972) 3-661-0866
E-mail: yurirg@mapi.gov.il

B. Testing:
Same

C. Conformance:
Same

D. Maintenance:
Same

E. Distribution:
Same

F. Help Desk/User Support:
Same

1.3 Development History

A. Beginning of work on metadata standard:
1991 (within transfer standard)
1998 (new metadata standard)

B. Milestones:
1991 – spatial accuracy and origin information
1995 – spatial information source index
1996 – distribution information
1997 – all country spatial information index
1998 – work started on new metadata standard
1999 – framework approved
2000 – March – first general information draft
June – chapters and 3 levels of information Approved.

C. Anticipated completion: **October 2000**

D. Update cycle: **1st review – October 2001**

1.4 Status of the Standard

A. Is the standard officially recognized: **NO**

B. Date(s) of recognition: **N/A**

C. Anticipated date of recognition: **December 2000**

D. Current stage in recognition process: **Draft was approved**

E. Steps still required to achieve recognition: **Tests by SOI special committee**

1.5 Access to Documentation of the Standard

A. Is the standard copyrighted?: **NO**
B. Name and address of copyright owner:
Survey of Israel
C. Do restrictions on the use of the standard apply?: **NO**
D. Is the standard documentation currently available?: **YES**
E. List the form/media of standard documentation:
Paper
F. Is the standard available on the Internet?: **NO**
G. Price: **100 NIS**
H. Contact details for obtaining the standard:
Survey of Israel
I. Does the standard have ISBN/ISSN numbers?: **NO**

1.6 Available Software Tools

A. Is software available to implement the standard as a database?: **NO**
B. Is software available to test compliance of metadata with the standard?: **NO**
C. Is software available to produce metadata compliant with the standard?: **NO**
D. Other relevant software: **N/A**
E. Is there a test dataset available to test software tools for implementing the standard?: **NO**
F. For each product available above, please list contact information: **N/A**

1.7 Available Training Materials

A. Are organized training sessions available?: **NO**
B. Contact details for training sessions: **N/A**
C. Training documentation available: **N/A**

SECTION 2 – Use and Implementation of the Metadata Standard

This section provides information on the way in which the standard is used and the purpose of its use. It also provides descriptions about the level of abstraction of the standard.

2.1 Intended Use of the Standard
Can the standard be used:

A. Within a transfer data structure?: **YES**
B. To produce hardcopy catalog of datasets?: **YES**
C. To catalog data in a digital on-line catalog?: **YES**
D. Other: **NO**

2.2 Intended Purpose of the Standard
Is the intended purpose of the standard:

A. To support discovery, identification, location of data?: **YES**

B. To support how to access data?: **YES**
C. To support the determination of dataset fitness for use?: **YES**
D. To support transfer of dataset?: **YES**
E. To support an organization's data management?: **YES**
F. Other: **NO**

2.3 Table of Contents

A. Provide the standard's Table of Contents

Introduction
Points of contact
Structure of Standard Levels of detail
Spatial Information
- **Vector**
- **Raster**

Non-Spatial information
Standard Chapters
- **General Information**
- **Vector 0-D**
- **Vector 1-D**
- **Vector 2-D**
- **Raster Air-Photograph**
- **Raster Orthophoto Raster spectral**
- **Raster Orthoimage**
- **Data distribution**
- **Data availability**

2.4 Level of Abstraction of the Metadata Standard Definition

A. Does the standard define metadata elements ?: **YES**
B. Does the standard define a model for the relationships between the metadata elements?: **NO**
C. Provide an overview diagram of the model of the metadata, if available. **N/A**
D. Does the standard define domains and/or data types of the metadata elements?: **NO**
E. Does the standard define how the metadata should be encoded?: **YES**
F. Does the standard define a database schema for the automatic generation of a database?: **NO**

2.5 Extensibility of the Standard

A. Is the metadata description extensible (e.g., Can user add metadata domains, elements, thematic profiles, etc.)?: **NO**

2.6 Availability of Metadata
Does the standard specify:

A. How metadata are to be accessed?: **NO**

B. A mechanism for querying the metadata?: **NO**
C. Particular standardized query protocols?: **NO**
D. That the metadata should be encoded for computer access?: **YES**
E. That the metadata should be encrypted for security?: **NO**

2.7 Examples

A. Does the standard provide examples of implementations?: **NO**
B. How many examples are available?: **0**

SECTION 3 – Linkage and Coordination of the Metadata Standard

This section provides information on how the metadata standard is linked to other standards, particularly transfer standards, and how such linkage is implemented. If a significant linkage does exist, detailed technical information is provided on the precise nature of the linkage to the standard, and how the linkage affects implementation of the metadata standard.

3.1 Associated Transfer Standards

A. Is the metadata standard associated with a transfer standard(s)?: **YES**
B. If A is yes, provide references and contact details for each linked standard:
IEF'91, Survey of Israel
C. If A is yes, are both (all) standards coordinated?: **YES**
D. If A is yes, describe how the coordination works, both from a technical and conformance (normative) viewpoint:
All information specified in the Metadata Standard may be included in the transferred data as the Transfer standard may accommodate all encoded metadata entities and any textual information (free text). The coordination is managed by the Survey of Israel GIS Forum. Free text additions to the transfer standard and other minor needed changes motivated to define an enhanced transfer standard (IEF2000) that will be recognized in December 2000 to compliment the new IMF200 metadata standard

3.2 Incorporated into transfer mechanism

A. Is the metadata standard incorporated into a transfer mechanism?: **YES**
B. If A is yes, provide references and contact details for each incorporated mechanism:
IEF'2000
C. If A is yes, describe how:
All information specified in the Metadata Standard may be included in the transferred data as the Transfer standard may accommodate all encoded metadata entities and any textual information (free text).

3.3 Other linked, coordinated or associated standards

A. Is the metadata standard linked, coordinated, or associated with any other standard(s) not listed above?: **NO**
B. If A is yes, provide references and contact details for each standard: **N/A**
C. If A is yes, describe how for each relevant standard: **N/A**

II. Categories the Metadata Standard Uses to Describe Datasets

The following sections are categories of metadata that may be used by the metadata standard to describe data. Not all categories are relevant to different datasets and may not necessarily be present in the standard. The presence or absence of a category should not be used for judging one standard against another.

Note: For each of the questions in the following sections please provide the obligation status information. Valid obligation values are:

Obligation Level

M	Mandatory	This information must be entered to comply with the metadata standard.
C	Conditional	This information is entered based on other information in the metadata.
O	Optional	This information is optionally entered in the metadata.

If the answer is 'NO', the obligation status should be left blank.

SECTION 4 – Dataset Identification

This section examines how the standard provides for the specification of the various sorts of titles and identifiers for a dataset.

4.1 The title(s) of a dataset
Does the standard allow/require the provision for:

Obligation

A. __ Previous title(s)?: **NO**
B. **M** Current title(s)?: **YES**
C. __ Possible future title(s)?: **NO**
D. __ Official titles in different languages?: **NO**
E. __ Abbreviated title(s)?: **NO**
F. __ Alternative Title(s)?: **NO**
G. __ Does the standard support effective dates of titles?: **NO**
H. **C** Other identifiers (e.g., ISBN, National Stock Number, Uniform Product Code)?: **YES**
Uniform Product Code
National or (Government) departmental Code

4.2 Parties responsible for the dataset
Does the standard allow/require the provision for:

Obligation

A. __ Author?: **NO**
B. __ Owner?: **NO**
C. __ Other?: **NO**

4.3 ISBD Information

Obligation

A. __ Does the metadata standard specify inclusion of mandatory ISBD information (see glossary) within the metadata?: **NO**

SECTION 5 – Status of the Dataset

This section examines how the standard provides for the specification of the current status and updating of the data set.

5.1 Status and Progress of Datasets
Does the metadata standard allow/require the provision for:

Obligation

A. __ Information on the status of the entire dataset?: **NO**
B. **O** Information on the status of the particular subset (e.g., observation time, inauguration time, extraction time)?: **YES**
Observation Time; Method of data acquisition

Does the metadata standard specify:

C. __ A structure for updating or maintenance information?: **NO**
D. __ A structure to describe historical information (see also Section 8.3 G)?: **NO**

5.2 Information on Incremental Updating
Does the metadata standard specify:

Obligation

A. __ A structure for information on incremental updates of the entire dataset?: **NO**
B. __ A structure for information on incremental updates of particular portions of the dataset?: **NO**
C. __ A structure to describe the update process for the dataset?: **NO**

SECTION 6 – Extent of the Dataset

This section examines how the standard provides for the specification of the spatial and temporal extent of the dataset.

6.1 The spatial extent of the dataset

Obligation

A. **O** Does the metadata standard allow/require for the description of horizontal extent (e.g., bounding rectangle/polygon, named location (i.e., Stockholm, etc.))?: **YES**
Bounding rectangle; Bounding polygon; Municipality name; Municipality code;
Administrative region (with about 15 options of issuing authorities)

B. **O** Does the metadata standard allow/require for the description of vertical extent (e.g., minima/maxima heights, etc.)?: **YES**
Minima/maxima

C. __ Does the standard require specific units for describing extent?: **NO**

D. **O** Does the standard allow specification of the units for describing extent?: **YES**

E. __ Does the standard require specific spatial reference system for describing extent?: **NO**

F. **O** Does the standard allow specification of the spatial reference system for describing extent?: **YES**

6.2 The temporal extent of the dataset

Obligation

A. **O** Does the standard allow/require for the description of time period for which the dataset is relevant?: **YES**

B. __ Does the standard allow/require specification of the temporal reference system?: **NO**

SECTION 7 – Data Content of the Dataset

This section examines how the standard provides for the specification of the data content of the dataset in terms of overview elements, themes, data definitions and product specifications.

7.1 Dataset overview
Does the standard allow/require for the provision of the following information:

Obligation

A. **O** Dataset overview?: **YES**

B. **O** Abstract – brief summary of the content of the dataset?: **YES**

C. **O** Purpose – summary of the intentions with which the dataset was developed?: **YES**

D. __ Usage – list of activities for which the dataset has been used? (see also Section 8.5): **NO**

E. **C** Spatial schema – (e.g., vector, raster)? (see also Section 9): **YES**

F. __ Spatial reference system? (see also Section 12): **NO**

G. **O** Language?: **YES**

H. __ Character set (e.g., ISO8859-1, JISX0208)?: **NO**

I. __ List of documents providing further information about the dataset?: **NO**
J. __ Sample/preview dataset?: **NO**
K. **M** Related datasets?: **YES**
L. __ Original size of the dataset?: **NO**
M. __ Compression? (see also Section 11.5): **NO**
N. **M** Other?: **YES**

Level of detail; Unique Coding schema; list and codes of issuing authorities

7.2 Theme(s)

Obligation

A. **C** Does the standard allow/require the provision for the description of theme(s) (e.g., topography, land cover, transport, etc.) in the dataset?: **YES**
B. **C** If A is yes, does the standard allow the referencing of an outside dictionary or thesaurus?: **YES**
C. **C** If A is yes, does the standard itself include a dictionary or thesaurus within the metadata?: **YES**
D. **O** If A is yes, does the standard allow the use of unauthorized user-defined keywords?: **YES**

7.3 Data Definition

Obligation

A. **C** Does the standard allow/require the provision for the description of the data definition?: **YES**
B. If A is yes, does the standard allow/require the provision for the description of:
   C 1. Features/Objects?: **YES**
   C 2. Attributes?: **YES**
   C 3. Relations?: **YES**
C. **C** Can attributes be described separately from feature/object classes?: **YES**
D. **C** Does the standard allow/require the specification of the domain of the attributes?: **YES**
E. **C** Does the standard allow/require the specification of data type of the attributes?: **YES**
F. **C** Does the standard allow/require the specification of particular classification coding schemes (e.g., FACC, OSKA, S-57 OC)?: **YES**

IEF91, 827, Infrastructure, Interior-Planning, Lamas, and more

7.4 Product Specification

Obligation

A. __ Does the standard allow/require for the provision of information about whether the dataset meets a particular product specification?: **NO**

B. __ If A is yes, can it be internal to the Metadata?: **N/A**
C. __ If A is yes, can it be external to the Metadata?: **N/A**

SECTION 8 – Data Quality of the Dataset

This section examines how the standard deals with the question of data quality. This includes homogeneity and metrics for data quality, quality assurance, and usage.

8.1 Overall Data Quality Statement

Obligation

A. **C** Does the standard allow/require for the provision of quality to be described? If yes, please answer the questions below. (*If no, all following answers in Section* 8 *are N/A, please skip to Section* 9.): **YES**
B. __ Does the standard allow/require the provision for the quality information referencing an external quality standard, quality assurance standard, or specification?: **NO**

8.2 Homogeneity of Data Quality

Obligation

A. __ Does the standard allow/require for the variation of the quality of the dataset to be described?: **NO**
B. What type of aggregation of quality information is allowed?
__ 1. Entire database?: **NO**
O 2. Feature/Object instance?: **YES**
__ 3. Feature/Object class?: **NO**
__ 4. Attribute instance?: **NO**
__ 5. Attribute class?: **NO**
O 6. Layer?: **YES**
__ 7. Theme?: **NO**
__ 8. Geographic extent?: **NO**
__ 9. Other?: **NO**

8.3 Metrics for Data Quality
Does the standard allow/require the provision of values for:

Obligation

A. **O** Positional accuracy?: **YES**
1. Standard error?: **YES**
2. Maximum error?: **NO**
3. Error correlation?: **NO**
4. Confidence level (e.g., 90%, 95%)?: **NO**
5. Other?: **NO**
B. __ Attribute accuracy?: **NO**
1. Standard error?: **N/A**

2. Maximum error?:	**N/A**
3. Error frequency?:	**N/A**
4. Error correlation?:	**N/A**
5. Confidence level (e.g., 90%, 95%)?:	**N/A**
6. Other?:	**N/A**
C. **O** Temporal accuracy?:	**YES**
1. Standard error?:	**NO**
2. Maximum error?:	**YES**
3. Error correlation?:	**NO**
4. Confidence level? (e.g., 90%, 95%):	**YES**
5. Other?:	**NO**
D. __ Completeness?:	**NO**
1. Error of omission?:	**N/A**
2. Error of commission?:	**N/A**
3. Other?:	**N/A**
E. **O** Currentness?:	**YES**
1. Date?:	**YES**
2. Maximum age?:	**NO**
3. Temporal extent?:	**YES**
4. Other?:	**NO**
F. __ Logical consistency?:	**NO**
G. __ Lineage?:	**NO**
1. List of processing steps?:	**N/A**
2. List of values of processing parameters?:	**N/A**
3. List of printouts of statistics, etc.?:	**N/A**
4. Source material?:	**N/A**
5. Other?:	**N/A**
H. __ Additional Quality Information?:	**NO**

8.4 Quality Assurance Methods

Does the standard allow/require the provision of information about the methods used to determine:

Obligation

A. __ Positional accuracy?:	**NO**
1. Standard error in adjustments?:	**N/A**
2. Repeated measurements?:	**N/A**
3. Independent measurements?:	**N/A**
4. Subjective evaluations?:	**N/A**
5. Other?:	**N/A**
B. __ Attribute accuracy?:	**NO**
1. Standard error?:	**N/A**
2. Error classification matrix?:	**N/A**
3. Independent measurements?:	**N/A**
4. Subjective evaluation?:	**N/A**

5. Other?: **N/A**

C. __ Temporal accuracy?: **NO**

1. Independent measurements?: **N/A**
2. Subjective evaluation?: **N/A**
3. Other?: **N/A**

D. __ Completeness?: **NO**

1. Field checks?: **N/A**
2. Repeated measurements?: **N/A**
3. Independent measurements?: **N/A**
4. Subjective evaluation?: **N/A**
5. Error of commission?: **N/A**
6. Other?: **N/A**

E. __ Currentness?: **NO**

1. Date of source material?: **N/A**
2. Repeated measurements?: **N/A**
3. Independent measurements?: **N/A**
4. Subjective evaluation?: **N/A**
5. Other?: **N/A**

F. __ Logical consistency?: **NO**

1. Verification results of consistency checks?: **N/A**
2. Repeated processing?: **N/A**
3. Comparison with independent sources?: **N/A**
4. Subjective evaluation?: **N/A**
5. Other?: **N/A**

G. __ Lineage?: **NO**

1. Checking results of processing steps: **N/A**
2. Subjective evaluation: **N/A**
3. Other: **N/A**

H. __ Additional Quality Information?: **NO**

8.5 Usage

Obligation

A. __ Does the metadata standard allow/require the provision of information about usage?: **NO**

1. Who used it: **N/A**
2. When it was used: **N/A**
3. For what application it was used: **N/A**
4. The effectiveness of use: **N/A**
5. Other: **N/A**

8.6 Alternative Quality Description: **YES**
Free text

SECTION 9 – Spatial Data Structure of the Dataset

This section examines how the standard provides for the specification of the spatial data model and spatial data types.

9.1 Spatial Data Model

Does the standard allow/require the provision for:

Obligation

A. __ The description of the geometrical and/or topological primitives used to model the real world?: **NO**

B. __ Information about an external standard for geometrical and/or topological primitive specification?: **NO**

9.2 Spatial Data Types

Does the standard allow/require the provision for:

Obligation

A. __ Information about the spatial data type(s) used in the dataset?: **NO**

B. __ If A is yes, which of the following generic spatial data types are supported? **N/A**

For each, please list specific data primitive levels supported by each data type.

1. Vector (e.g., spaghetti, chain-node/node-edge-face, planar graph, full topology)?: **N/A**
2. Areal Raster (e.g., thematic raster – scanned and classification maps, image raster – satellite images, scanned aerial photographs)?: **N/A**
3. Grid/Matrix (e.g., Digital Elevation Model)?: **N/A**
4. Triangular Irregular Network (TIN)?: **N/A**
5. Other?: **N/A**

9.3 Other Data Types

Does the standard allow the provision of information about:

Obligation

A. __ Video?: **NO**

B. __ Sound?: **NO**

C. __ Other?: **NO**

SECTION 10 – Spatial Reference of the Dataset

This section examines how the dataset provides for the specification of the spatial reference characteristics of the data set.

Does the standard allow/require the provision for the following information:

10.1 Geoidal Model

Obligation

A. **O** Geoidal model name?: **YES**

B. __ Geoidal model definitions?: **NO**

10.2 Geodetic datum

Obligation

A. **O** Geodetic datum name?: **YES**
B. __ Geodetic datum definitions?: **NO**

10.3 Reference Ellipsoid

Obligation

A. **O** Reference ellipsoid name?: **YES**
B. __ Reference ellipsoid definitions?: **NO**

10.4 Projection

Obligation

A. **O** Projection name?: **YES**
B. __ Projection definitions?: **NO**

10.5 Coordinate Systems

Obligation

A. **O** Coordinate system type?: **YES**
B. __ Coordinate system definitions?: **NO**

10.6 Height Reference System

Obligation

A. **O** Vertical datum name?: **YES**
B. __ Types of heights?: **NO**

10.7 Ancillary Geodetic Information

Obligation

A. __ Does the standard allow/require the provision for ancillary geodetic, gravimetric, or magnetometric information?: **NO**

10.8 Transformation Parameters
Does the standard specify a method to provide information for coordinate transformation(s):

Obligation

A. __ Model name?: **NO**
B. __ Control points?: **NO**
C. __ Transformation parameters?: **NO**

10.9 Non-Coordinate Spatial Reference
Does the standard allow/require the provision for information on non-coordinate spatial reference systems (e.g., postal systems, street address)?:

Obligation

A. __ Reference system name?: **NO**
B. __ Reference system definitions?: **NO**
C. **O** Other?: **YES**

Specific coded addresses in terms of postal system zones or street address

SECTION 11 – Availability and Distribution of the Dataset

This section examines how the standard provides information on the availability of the dataset?

Does the standard allow/require the provision for:

11.1 The Distributor(s) of the Dataset

Obligation

A. **O** Information about the distributor(s) of the dataset?: **YES**
B. **O** Contact information for the distributor(s)?: **YES**

11.2 Restrictions to Access and Usage

Obligation

A. __ Information about copyright on the dataset (e.g., date, period, copyright owner, etc.)?: **NO**
B. __ Information about restrictions on the access of the dataset?: **NO**
C. **O** Information about restrictions on the usage of the dataset?: **YES**

11.3 Pricing Details

Obligation

A. __ Information about pricing?: **NO**

11.4 Distribution of the Dataset

Obligation

A. __ Information about distribution media?: **NO**
B. __ Information about on-line access?: **NO**
C. __ Information about units of distribution (e.g., tiles, layers, polygons, elements)?: **NO**

11.5 Distribution Format

Obligation

A. **O** Information about distribution formats (e.g., commercial CAD format, commercial GIS formats, National/International transfer standard)?: **YES**

B. __ Information about the encoding of the dataset (e.g., binary, ASCII)?: **NO**
C. __ Information about compression method?: **NO**
D. __ Information about compressed dataset size?: **NO**
E. __ Information about the distribution mechanism?: **NO**
F. __ Information about the encapsulation of data (e.g., ISO 8211, ISO 8824)?: **NO**

11.6 Ordering Procedures

Obligation

A. __ Information about ordering procedures for the dataset?: **NO**

11.7 Support Services

Obligation

A. __ Information about support services (e.g., additional data transformations, format or coordinate conversions)?: **NO**

SECTION 12 – Authorization and Verification of the Metadata

This section examines how the standard provides for the description of the authorization and verification on the metadata of a dataset (e.g., how is the metadata itself authorized and verified?).

12.1 Metadata reference

Obligation

A. **O** Does the standard allow/require the provision of information about reference dates of the metadata (e.g., last check date, last update date, review date)?: **YES**
Last check date

12.2 Verification Authority

Obligation

A. **O** Does the standard allow/require the provision for information about who verified/authorized the metadata?: **YES**
Authority/Institute; Point of contact
B. __ Does the standard specify the method for providing this information (e.g., ISO 9000): **NO**

12.3 Statistical Methods Used for Verification

Obligation

A. __ Does the standard specify the statistical methods to be used for verification of the metadata?: **NO**

B. __ Does the standard allow/require the provision for the description of the statistical methods used to verify the metadata?: **NO**

END OF CHARACTERISTICS
FIN

Japan
日本メタデータプロファイル
Nihon Metadeta Purofairu
Japanese Metadata Profile

Kazuhiko Akeno, Shoji Okuyama, Shinji Takazawa

I. Information about the Metadata Standard Itself

SECTION 0 – Assessment Information

This section provides the user with a general description of the scope and intended use of the standard. Information on the respondent who prepared the assessment is also provided.

0.1 Respondent

Assessment prepared by:

Kazuhiko AKENO, Shoji OKUYAMA and Shinji TAKAZAWA
Kazuhiko AKENO,
Head of Information Access Division, Geoinformation Department
Geographical Survey Institute (GSI)
Kitasato-1, Tsukuba-shi, Ibaraki-ken, 305-0811 Japan
Fax: +81 29 864 1805
Tel: +81 29 864 1111
E-mail: gsi-ch@gsi.go.jp

Date assessment completed:
March 2004, Reviewed March, 2005

Relationship of respondent to standard:
Committee member of Japanese Standards for Geographic Information

0.2 Brief Summary Statement

A. Summary of the scope and intent of the standard:
The Japanese metadata standard defines schema required for describing spatial data. It provides information about the identification, the extent, the quality, spatial reference, and distribution of spatial data. It is applicable to spatial datasets,

dataset series, and individual features and attributes. The metadata standard defines mandatory and conditional metadata sections, metadata entities, and metadata elements. They are the core set required to serve the metadata applications for data discovery, data access, and use of digital data. Metadata sections consist of identification, data quality, reference system, distribution, and metadata reference. The Japanese metadata standard was developed as a profile of ISO 19115. It consists of core metadata and some additional elements and packages such as keywords and quality in ISO 19115. It was standardized as a part of Japanese Standards for Geographic Information (JSGI). JSGI was adopted as Technical Standard officially by the Liaison Committee of Ministries and Agencies Concerned with GIS. Metadata based on JSGI was prepared in some organizations and is utilized for clearinghouse activities.

SECTION 1 – Administration of Metadata Standard

This section provides a general description of the administrative framework, within which the standard was developed, tested and currently resides. Significant details are also provided in terms of the developmental and managerial history as well as information on the existence of documentation, software tools and training materials. This section also provides a point of contact for further information on the standard.

1.1 Name of Standard

A. Name of standard:
   1. In original language(s):
      日本メタデータプロファイル
   2. English translation, if appropriate:
      Japanese Metadata Profile

B. Version/Edition:
   Second Edition

C. Language(s) of documentation:
   Japanese

D. Acronym(s):
   JMP 2.0

E. Official ID:
   N/A

F. Name(s) of recognizing standards authority(ies):
   The Liaison Committee of Ministries and Agencies Concerned with GIS and The Committee on Japanese Standards for Geographic Information

1.2 Responsible Institutions

Contact information about those institutions that are (or were) responsible for the processes involved in the evolution of the standard:

A. Production and/or development:
GSI and private companies (joint study)
Geographical Survey Institute (GSI):
Kitasato-1, Tsukuba-shi, Ibaraki-ken, 305-0811 Japan
Fax: +81 29 864 1805
Tel: +81 29 864 1111

B. Testing:
N/A

C. Conformance:
N/A

D. Maintenance:
GSI and private companies (joint study)
Geographical Survey Institute (GSI):
Kitasato-1, Tsukuba-shi, Ibaraki-ken, 305-0811 Japan
Fax: +81 29 864 1805
Tel: +81 29 864 1111

E. Distribution:
Geographical Survey Institute (GSI)
Kitasato-1, Tsukuba-shi, Ibaraki-ken, 305-0811 Japan
Fax: +81 29 864 1805
Tel: +81 29 864 1111

F. Help Desk/User Support:
Geographical Survey Institute (GSI)
Kitasato-1, Tsukuba-shi, Ibaraki-ken, 305-0811 Japan
Fax: +81 29 864 1805
Tel: +81 29 864 1111

1.3 Development History

A. Beginning of work on metadata standard: **1996**

B. Milestones:
March 1997 First Draft
March 1999 The first edition
March 2003 The second edition

C. Anticipated completion: **N/A**

D. Update cycle:
Depend on amendment of ISO 19115.

1.4 Status of the Standard

A. Is the standard officially recognized: **YES**

B. Date(s) of recognition: **March 1999**

C. Anticipated date of recognition: **N/A**

D. Current stage in recognition process:
Authorized in government organization

E. Steps still required to achieve recognition: **N/A**

1.5 Access to Documentation of the Standard

A. Is the standard copyrighted?: **YES**

B. Name and address of copyright owner:
GSI and private companies (joint study)
Geographical Survey Institute (GSI):
Kitasato-1, Tsukuba-shi, Ibaraki-ken, 305-0811 Japan
Fax: +81 29 864 1805
Tel: +81 29 864 1111

C. Do restrictions on the use of the standard apply?: **NO**

D. Is the standard documentation currently available?: **YES**
JMP 2.0 Specification, Technical Report of Geographical Survey Institute, E1-No. 281 (in Japanese)

E. List the form/media of standard documentation:
Paper and
electronic file

F. Is the standard available on the Internet?: **YES**
Minimum information in English (including UML and XSD)
http://zgate.gsi.go.jp/ch/jmp20/jmp20_eng.html
Comprehensive information in Japanese
http://zgate.gsi.go.jp/ch/jmp20/jmp20.html

G. Price: **Free**

H. Contact details for obtaining the standard:
Geographical Survey Institute (GSI):
Kitasato-1, Tsukuba-shi, Ibaraki-ken, 305-0811 Japan
Fax: +81 29 864 1805
Tel: +81 29 864 1111

I. Does the standard have ISBN/ISSN numbers?: **NO**

1.6 Available Software Tools

A. Is software available to implement the standard as a database?: **YES**
Isite-J

B. Is software available to test compliance of metadata with the standard?: **NO**

C. Is software available to produce metadata compliant with the standard?: **YES**
JMP 2.0 Metadata Editor.

D. Other relevant software: **YES**
JMP 2.0 Metadata Converter
Clearinghouse node development kit based on JMP 2.0

E. Is there a test dataset available to test software tools for implementing the standard?: **NO**

F. For each product available above, please list contact information: **N/A**
Geographical Survey Institute (GSI):
Kitasato-1, Tsukuba-shi, Ibaraki-ken, 305-0811 Japan
Fax: +81 29 864 1805
Tel: +81 29 864 1111

1.7 Available Training Materials

A. Are organized training sessions available?: **NO**
B. Contact details for training sessions: **N/A**
C. Training documentation available: **YES**
JMP 2.0 Manual, Technical Report of Geographical Survey Institute, E1-No. 282 (in Japanese)
JMP 2.0 Metadata Editor Manual, Technical Report of Geographical Survey Institute, E1-No. 283 (in Japanese)
These manuals are downloadable from GSI web site.

SECTION 2 – Use and Implementation of the Metadata Standard

This section provides information on the way in which the standard is used and the purpose of its use. It also provides descriptions about the level of abstraction of the standard.

2.1 Intended Use of the Standard
Can the standard be used:

A. Within a transfer data structure?: **NO**
B. To produce hardcopy catalog of datasets?: **NO**
C. To catalog data in a digital on-line catalog?: **YES**
D. Other: **N/A**

2.2 Intended Purpose of the Standard
Is the intended purpose of the standard:

A. To support discovery, identification, location of data?: **YES**
B. To support how to access data?: **YES**
C. To support the determination of dataset fitness for use?: **YES**
D. To support transfer of dataset?: **YES**
E. To support an organization's data management?: **YES**
F. Other: **N/A**

2.3 Table of Contents

A. Provide the standard's Table of Contents
1. Introduction
2. Metadata packages
3. UML data model
4. Data dictionary
5. Code list
6. Metadata elements defined in external standards
Annex A Comparison between ISO 19115 and JMP 2.0
Annex B UML notation
Annex C JMP 2.0 XML Schema
Annex D List of contributing companies

2.4 Level of Abstraction of the Metadata Standard Definition

A. Does the standard define metadata elements?: **YES**
B. Does the standard define a model for the relationships between the metadata elements?: **YES**
C. Provide an overview diagram of the model of the metadata, if available. **YES**
See UML diagram at http://zgate.gsi.go.jp/ch/jmp20/jmp20UML.html
D. Does the standard define domains and/or data types of the metadata elements?: **YES**
E. Does the standard define how the metadata should be encoded?: **YES**
F. Does the standard define a database schema for the automatic generation of a database?: **NO**

2.5 Extensibility of the Standard

A. Is the metadata description extensible (e.g., Can user add metadata domains, elements, thematic profiles, etc.)?: **NO**

2.6 Availability of Metadata
Does the standard specify:

A. How metadata are to be accessed?: **NO**
Metadata access as clearinghouse is specified outside of the Metadata standards.
B. A mechanism for querying the metadata?: **NO**
Metadata access as clearinghouse is specified outside of the Metadata standards.
C. Particular standardized query protocols?: **NO**
Metadata access as clearinghouse is specified outside of the Metadata standards.

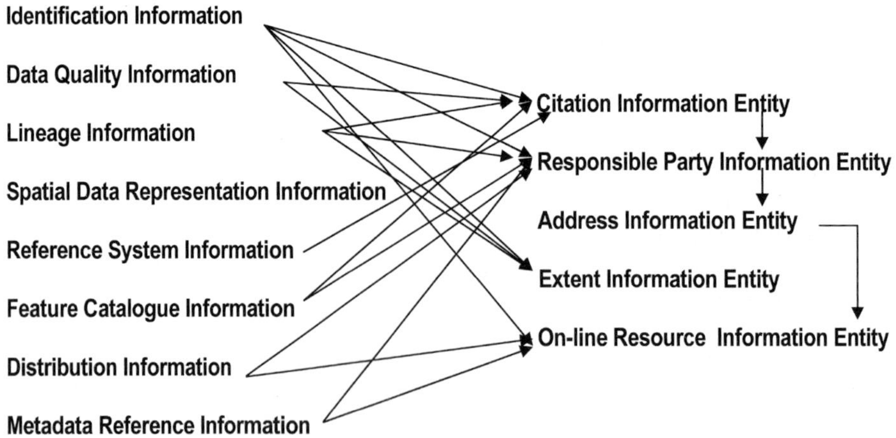

Fig. 1. Schema of metadata entities relationships.

D. That the metadata should be encoded for computer access?: **YES**
E. That the metadata should be encrypted for security?: **NO**

2.7 Examples

A. Does the standard provide examples of implementations?: **YES**

B. How many examples are available?: **1**

SECTION 3 – Linkage and Coordination of the Metadata Standard

This section provides information on how the metadata standard is linked to other standards, particularly transfer standards, and how such linkage is implemented. If a significant linkage does exist, detailed technical information is provided on the precise nature of the linkage to the standard, and how the linkage affects implementation of the metadata standard.

3.1 Associated Transfer Standards

A. Is the metadata standard associated with a transfer standard(s)?: **YES**
B. If A is yes, provide references and contact details for each linked standard:
JSGI, which is based on some of ISO 19100 Standards and includes data transfer standard.
C. If A is yes, are both (all) standards coordinated?: **YES**
D. If A is yes, describe how the coordination works, both from a technical and conformance (normative) viewpoint:
JMP 2.0 is a part of JSGI.

3.2 Incorporated into transfer mechanism

A. Is the metadata standard incorporated into a transfer mechanism?: **YES**
B. If A is yes, provide references and contact details for each incorporated mechanism:
JSGI, which is based on some of ISO 19100 Standards and includes data transfer mechanism.
C. If A is yes, describe how:
JMP 2.0 is a part of JSGI.

3.3 Other linked, coordinated or associated standards

A. Is the metadata standard linked, coordinated, or associated with any other standard(s) not listed above?: **NO**
B. If A is yes, provide references and contact details for each standard: **N/A**
C. If A is yes, describe how for each relevant standard: **N/A**

II. Categories the Metadata Standard Uses to Describe Datasets

The following sections are categories of metadata that may be used by the metadata standard to describe data. Not all categories are relevant to different datasets and may

not necessarily be present in the standard. The presence or absence of a category should not be used for judging one standard against another.

Note: For each of the questions in the following sections please provide the obligation status information. Valid obligation values are:

Obligation Level

M	Mandatory	This information must be entered to comply with the metadata standard.
C	Conditional	This information is entered based on other information in the metadata.
O	Optional	This information is optionally entered in the metadata.

If the answer is 'NO', the obligation status should be left blank.

SECTION 4 – Dataset Identification

This section examines how the standard provides for the specification of the various sorts of titles and identifiers for a dataset.

4.1 The title(s) of a dataset
Does the standard allow/require the provision for:

Obligation

A. __ Previous title(s)?: **NO**
B. **M** Current title(s)?: **YES**
C. __ Possible future title(s)?: **NO**
D. __ Official titles in different languages?: **NO**
E. __ Abbreviated title(s)?: **NO**
F. __ Alternative Title(s)?: **NO**
G. **M** Does the standard support effective dates of titles?: **YES**
H. __ Other identifiers (e.g., ISBN, National Stock Number, Uniform Product Code)?: **NO**

4.2 Parties responsible for the dataset
Does the standard allow/require the provision for:

Obligation

A. **O** Author?: **YES**
B. **O** Owner?: **YES**
C. **M** Other?: **YES**

"originator", "publisher", "custodian", "principal investigator", "resource provider", "processor", "distributor", "point of contact", "user", "author" or "owner"

4.3 ISBD Information

Obligation

A. __ Does the metadata standard specify inclusion of mandatory ISBD information (see glossary) within the metadata?: **NO**

SECTION 5 – Status of the Dataset

This section examines how the standard provides for the specification of the current status and updating of the data set.

5.1 Status and Progress of Datasets
Does the metadata standard allow/require the provision for:

Obligation

A. **M** Information on the status of the entire dataset?: **YES**
"completed", "under development", "planned", "required", "on-going", "historical archive" or "obsolete"

B. **O** Information on the status of the particular subset (e.g., observation time, inauguration time, extraction time)?: **YES**

Does the metadata standard specify:

C. __ A structure for updating or maintenance information?: **NO**

D. **O** A structure to describe historical information (see also Section 8.3 G)?: **YES**

5.2 Information on Incremental Updating
Does the metadata standard specify:

Obligation

A. **O** A structure for information on incremental updates of the entire dataset?: **YES**
It can be described in lineage information.

B. **O** A structure for information on incremental updates of particular portions of the dataset?: **YES**
It can be described in lineage information.

C. **O** A structure to describe the update process for the dataset?: **YES**
It can be described in lineage information.

SECTION 6 – Extent of the Datasct

This section examines how the standard provides for the specification of the spatial and temporal extent of the dataset.

6.1 The spatial extent of the dataset

Obligation

A. **O** Does the metadata standard allow/require for the description of horizontal extent (e.g., boundin grectengle/polygon, named location (i.e., Stockholm, etc.))?: **YES**
"bounding box", "bounding polygon", "geographic identifier"

B. **O** Does the metadata standard allow/require for the description of vertical extent (e.g., minima/maxima heights, etc.)?: **YES**

"minimum value", "maximum value"

C. **C** Does the standard require specific units for describing extent?: **YES**
D. __ Does the standard allow specification of the units for describing extent?: **NO**
E. **C** Does the standard require specific spatial reference system for describing extent?: **YES**
F. **C** Does the standard allow specification of the spatial reference system for describing extent?: **YES**

6.2 The temporal extent of the dataset

Obligation

A. **O** Does the standard allow/require for the description of time period for which the dataset is relevant?: **YES**
B. __ Does the standard allow/require specification of the temporal reference system?: **NO**

SECTION 7 – Data Content of the Dataset

This section examines how the standard provides for the specification of the data content of the dataset in terms of overview elements, themes, data definitions and product specifications.

7.1 Dataset overview
Does the standard allow/require for the provision of the following information:

Obligation

A. __ Dataset overview?: **NO**
B. **M** Abstract – brief summary of the content of the dataset?: **YES**
C. **O** Purpose – summary of the intentions with which the dataset was developed?: **YES**
D. __ Usage – list of activities for which the dataset has been used? (see also Section 8.5): **NO**
E. **O** Spatial schema – (e.g., vector, raster)? (see also Section 9): **YES**
F. **O** Spatial reference system? (see also Section 12): **YES**
G. **O** Language?: **YES**
H. **C** Character set (e.g., ISO8859-1, JISX0208)?: **YES**
I. **O** List of documents providing further information about the dataset?: **YES**
J. __ Sample/preview dataset?: **NO**
K. __ Related datasets?: **NO**
L. __ Original size of the dataset?: **NO**
M. __ Compression? (see also section 11.5): **NO**
N. **O** Other?: **YES**

"browse graphic", "keywords", "resolution" etc.

7.2 Theme(s)

Obligation

A. **O** Does the standard allow/require the provision for the description of theme(s) (e.g., topography, land cover, transport, etc.) in the dataset?: **YES**
B. __ If A is yes, does the standard allow the referencing of an outside dictionary or thesaurus?: **NO**
C. **C** If A is yes, does the standard itself include a dictionary or thesaurus within the metadata?: **YES**
D. __ If A is yes, does the standard allow the use of unauthorized user-defined keywords?: **NO**

7.3 Data Definition

Obligation

A. __ Does the standard allow/require the provision for the description of the data definition?: **NO**
B. If A is yes, does the standard allow/require the provision for the description of:
__ 1. Features/Objects?: **N/A**
__ 2. Attributes?: **N/A**
__ 3. Relations?: **N/A**
C. __ Can attributes be described separately from feature/object classes?: **N/A**
D. __ Does the standard allow/require the specification of the domain of the attributes?: **N/A**
E. __ Does the standard allow/require the specification of data type of the attributes?: **N/A**
F. __ Does the standard allow/require the specification of particular classification coding schemes (e.g., FACC, OSKA, S-57 OC)?: **N/A**

7.4 Product Specification

Obligation

A. **O** Does the standard allow/require for the provision of information about whether the dataset meets a particular product specification?: **YES**
B. If A is yes, can it be internal to the Metadata?: **NO**
C. **C** If A is yes, can it be external to the Metadata?: **YES**

SECTION 8 – Data Quality of the Dataset

This section examines how the standard deals with the question of data quality. This includes homogeneity and metrics for data quality, quality assurance, and usage.

8.1 Overall Data Quality Statement

Obligation

A. **O** Does the standard allow/require for the provision of quality to be described? If yes, please answer the questions below. (*If no, all following answers in Section* 8 *are N/A, please skip to Section* 9.): **YES**

B. **O** Does the standard allow/require the provision for the quality information referencing an external quality standard, quality assurance standard, or specification?: **YES**

8.2 Homogeneity of Data Quality

Obligation

A. **O** Does the standard allow/require for the variation of the quality of the dataset to be described?: **YES**

B. What type of aggregation of quality information is allowed?

__ 1. Entire database?: **NO**
__ 2. Feature/Object instance?: **NO**
__ 3. Feature/Object class?: **NO**
__ 4. Attribute instance?: **NO**
__ 5. Attribute class?: **NO**
__ 6. Layer?: **NO**
__ 7. Theme?: **NO**
O 8. Geographic extent?: **YES**
C 9. Other?: **YES**
"free text"

8.3 Metrics for Data Quality
Does the standard allow/require the provision of values for:

Obligation

A. **O** Positional accuracy?: **YES**
1. Standard error?: **NO**
2. Maximum error?: **NO**
3. Error correlation?: **NO**
4. Confidence level (e.g., 90%, 95%)?: **NO**
5. Other?: **YES**
"free text"

B. **O** Attribute accuracy?: **YES**
1. Standard error?: **NO**
2. Maximum error?: **NO**
3. Error frequency?: **NO**
4. Error correlation?: **NO**
5. Confidence level (e.g., 90%, 95%)?:
6. Other?: **YES**
"free text"

C. **O** Temporal accuracy?: **YES**
1. Standard error?: **NO**

2. Maximum error?:	**NO**
3. Error correlation?:	**NO**
4. Confidence level? (e.g., 90%, 95%):	**NO**
5. Other?:	**YES**
"free text"	
D. **O** Completeness?:	**YES**
1. Error of omission?:	**YES**
2. Error of commission?:	**YES**
3. Other?:	**YES**
"free text"	
E. **O** Currentness?:	**YES**
1. Date?:	**NO**
2. Maximum age?:	**NO**
3. Temporal extent?:	**NO**
4. Other?:	**YES**
"free text"	
F. **C** Logical consistency?:	**YES**
G. **O** Lineage?:	**YES**
1. List of processing steps?:	**NO**
2. List of values of processing parameters?:	**NO**
3. List of printouts of statistics, etc.?:	**NO**
4. Source material?:	**NO**
5. Other?:	**YES**
"free text"	
H. **O** Additional Quality Information?:	**YES**
N/A	

8.4 Quality Assurance Methods

Does the standard allow/require the provision of information about the methods used to determine:

Obligation

A. **O** Positional accuracy?:	**YES**
1. Standard error in adjustments?:	**NO**
2. Repeated measurements?:	**NO**
3. Independent measurements?:	**NO**
4. Subjective evaluations?:	**NO**
5. Other?:	**YES**
"free text"	
B. **O** Attribute accuracy?:	**YES**
1. Standard error?:	**NO**
2. Error classification matrix?:	**NO**
3. Independent measurements?:	**NO**

4. Subjective evaluation?: **NO**
5. Other?: **YES**
"free text"

C. **O** Temporal accuracy?: **YES**
1. Independent measurements?: **NO**
2. Subjective evaluation?: **NO**
3. Other?: **YES**
"free text"

D. **O** Completeness?: **YES**
1. Field checks?: **NO**
2. Repeated measurements?: **NO**
3. Independent measurements?: **NO**
4. Subjective evaluation?: **NO**
5. Error of commission?: **NO**
6. Other?: **YES**
"free text"

E. **O** Currentness?: **YES**
1. Date of source material?: **NO**
2. Repeated measurements?: **NO**
3. Independent measurements?: **NO**
4. Subjective evaluation?: **NO**
5. Other?: **YES**
"free text"

F. **O** Logical consistency?: **YES**
1. Verification results of consistency checks?: **NO**
2. Repeated processing?: **NO**
3. Comparison with independent sources?: **NO**
4. Subjective evaluation?: **NO**
5. Other?: **YES**
"free text"

G. __ Lineage?: **NO**
1. Checking results of processing steps: **N/A**
2. Subjective evaluation: **N/A**
3. Other: **N/A**

H. **O** Additional Quality Information?: **YES**
"free text"

8.5 Usage

Obligation

A. __ Does the metadata standard allow/require the provision of information about usage?: **NO**
1. Who used it: **N/A**

2. When it was used: **N/A**
3. For what application it was used: **N/A**
4. The effectiveness of use: **N/A**
5. Other: **N/A**

8.6 Alternative Quality Description: **NO**

SECTION 9 – Spatial Data Structure of the Dataset

This section examines how the standard provides for the specification of the spatial data model and spatial data types.

9.1 Spatial Data Model
Does the standard allow/require the provision for:

Obligation

A. __ The description of the geometrical and/or topological primitives used to model the real world?: **NO**
B. **O** Information about an external standard for geometrical and/or topological primitive specification?: **YES**

9.2 Spatial Data Types
Does the standard allow/require the provision for:

Obligation

A. **O** Information about the spatial data type(s) used in the dataset?: **YES**
B. **O** If A is yes, which of the following generic spatial data types are supported? For each, please list specific data primitive levels supported by each data type.
1. Vector (e.g., spaghetti, chain-node/node-edge-face, planar graph, full topology)?: **YES**
spaghetti, node-chain, planar graph, full topology
2. Areal Raster (e.g., thematic raster – scanned and classification maps, image raster – satellite images, scanned aerial photographs)?: **NO**
3. Grid/Matrix (e.g., Digital Elevation Model)?: **YES**
4. Triangular Irregular Network (TIN)?: **YES**
Other?: **YES**
"text", "stereo model", "video"

9.3 Other Data Types
Does the standard allow the provision of information about:

Obligation

A. **O** Video?: **YES**
B. __ Sound?: **NO**

C. __ Other?: **NO**

SECTION 10 – Spatial Reference of the Dataset

This section examines how the dataset provides for the specification of the spatial reference characteristics of the data set.

Does the standard allow/require the provision for the following information:

10.1 Geoidal Model

Obligation

A. __ Geoidal model name?: **NO**
B. __ Geoidal model definitions?: **NO**

10.2 Geodetic datum

Obligation

A. **O** Geodetic datum name?: **YES**
B. **C** Geodetic datum definitions?: **YES**
In citation

10.3 Reference Ellipsoid

Obligation

A. __ Reference ellipsoid name?: **NO**
B. __ Reference ellipsoid definitions?: **NO**

10.4 Projection

Obligation

A. __ Projection name?: **NO**
B. __ Projection definitions?: **NO**

10.5 Coordinate Systems

Obligation

A. **O** Coordinate system type?: **YES**
B. **C** Coordinate system definitions?: **YES**
In citation

10.6 Height Reference System

Obligation

A. **O** Vertical datum name?: **YES**

B. __ Types of heights?: **NO**

10.7 Ancillary Geodetic Information

Obligation

A. Does the standard allow/require the provision for ancillary geodetic, gravimetric, or magnetometric information?: **NO**

10.8 Transformation Parameters

Does the standard specify a method to provide information for coordinate transformation(s):

Obligation

A. __ Model name?: **NO**
B. __ Control points?: **NO**
C. __ Transformation parameters?: **NO**

10.9 Non-Coordinate Spatial Reference

Does the standard allow/require the provision for information on non-coordinate spatial reference systems (e.g., postal systems, street address)?:

Obligation

A. **O** Reference system name?: **YES**
B. **O** Reference system definitions?: **YES**
In citation
C. **O** Other?: **YES**
Responsible parties etc.

SECTION 11 – Availability and Distribution of the Dataset

This section examines how the standard provides information on the availability of the dataset?

Does the standard allow/require the provision for:

11.1 The Distributor(s) of the Dataset

Obligation

A. **O** Information about the distributor(s) of the dataset?: **YES**
B. **O** Contact information for the distributor(s)?: **YES**

11.2 Restrictions to Access and Usage

Obligation

A. **O** Information about copyright on the dataset (e.g., date, period, copyright owner, etc.)?: **YES**

B. **O** Information about restrictions on the access of the dataset?: **YES**
C. **O** Information about restrictions on the usage of the dataset?: **YES**

11.3 Pricing Details

Obligation

A. __ Information about pricing?: **NO**

11.4 Distribution of the Dataset

Obligation

A. **O** Information about distribution media?: **YES**
B. **O** Information about on-line access?: **YES**
C. **O** Information about units of distribution (e.g., tiles, layers, polygons, elements)?: **YES**

11.5 Distribution Format

Obligation

A. **C** Information about distribution formats (e.g., commercial CAD format, commercial GIS formats, National/International transfer standard)?: **YES**
B. **O** Information about the encoding of the dataset (e.g., binary, ASCII)?: **YES**

Regarding B-F, the information can be described in free text.

C. **O** Information about compression method?: **YES**
D. **O** Information about compressed dataset size?: **YES**
E. **O** Information about the distribution mechanism?: **YES**
F. **O** Information about the encapsulation of data (e.g., ISO 8211, ISO 8824)?: **YES**

11.6 Ordering Procedures

Obligation

A. **O** Information about ordering procedures for the dataset?: **YES**

11.7 Support Services

Obligation

A. __ Information about support services (e.g., additional data transformations, format or coordinate conversions)?: **NO**

SECTION 12 – Authorization and Verification of the Metadata

This section examines how the standard provides for the description of the authorization and verification on the metadata of a dataset (e.g., how is the metadata itself authorized and verified?).

12.1 Metadata reference

Obligation

A. **M** Does the standard allow/require the provision of information about reference dates of the metadata (e.g., last check date, last update date, review date)?: **YES**

12.2 Verification Authority

Obligation

A. __ Does the standard allow/require the provision for information about who verified/authorized the metadata?: **NO**

B. __ Does the standard specify the method for providing this information (e.g., ISO 9000): **NO**

12.3 Statistical Methods Used for Verification

Obligation

A. __ Does the standard specify the statistical methods to be used for verification of the metadata?: **NO**

B. __ Does the standard allow/require the provision for the description of the statistical methods used to verify the metadata?: **NO**

END OF CHARACTERISTICS
FIN

Korea

국가지리정보체계(NGIS)의 자료이력서(메타데이터) 잠정표준

National Geographic Information System Metadata: Interim Standard

Tae-jung Min, Heung-muk Cho

I. Information about the Metadata Standard Itself

SECTION 0 – Assessment Information

This section provides the user with a general description of the scope and intended use of the standard. Information on the respondent who prepared the assessment is also provided.

0.1 Respondent

Assessment prepared by:

Min Tae-jung, Director
Geodesy Division, National Geography Institute, 111 Wonchon-dong, Paldal-gu Suwon-shi, 442-380 Korea
Fax: +82 31 210 2644
Phone: +82 31 210 2603
E-mail: tjmin@moct.go.kr

Heung-muk Cho, Section Chief
Geographic Information Division, National Geography Institute
111 Wonchon-dong, Paldal-gu Suwon-shi, 442-380 Korea
Fax: +82 31 210 2743
Phone: +82 31 210 2730
E-mail: hmcho@moct.go.kr

Date assessment completed:

September 2000, updated February 2001 and May 2002, Reviewed Spring, 2005

Relationship of respondent to standard:

User/Drafting committee member

0.2 Brief Summary Statement

A. Summary of the scope and intent of the standard:
The temporary metadata standard (TTA.KO-10.0098: National Geographic Information System Metadata Interim-Standard) based on the draft ISO standard was prepared in 1998 to distribute and exchange the existing geospatial information until the completion of ISO 19115.
The NGIS Metadata Interim Standard is developed by the GIS Standard Subcommittee under the Steering Committee of NGIS., metadata standard was based on the ISO/TC211 15046-15 Metadata (draft) version 4.4 (then nowadays known as ISO 19115).
This NGIS metadata standard defines the schema required for describing geographic information and services. It provides information about the identification, the extent, the quality, the spatial and temporal schema, spatial reference, and distribution of digital geographic data. This standard is applicable to the cataloguing of datasets, clearinghouse activities, and the full description datasets.
Since being published in 1999, some inconsistencies and deficiencies has been identified in the NGIS Metadata Standard and the ISO/TC211 metadata standard had changes almost. Therefore, in December 1999, the National Geography Institute(NGI) initiated a review of the NGIS Metadata Standard and started to develop a new metadata standard base on the international metadata standard ISO/TC211 19115.3.
The new metadata standard of NGI will be almost compliant with the 3nd CD of ISO 19115. The NGI is responsible for the establishment of national geographic database and management digital mapping. And then NGI will be issued the new NGI metadata standard as NGIS Metadata Standard through the Geographic Information Subcommittee of NGIS. The NGIS Metadata Interim Standard will be superseded by the new NGI metadata standard.

SECTION 1 – Administration of Metadata Standard

This section provides a general description of the administrative framework, within which the standard was developed, tested and currently resides. Significant details are also provided in terms of the developmental and managerial history as well as information on the existence of documentation, software tools and training materials. This section also provides a point of contact for further information on the standard.

1.1 Name of Standard

A. Name of standard:
   1. In original language(s):
      국가지리정보체계(NGIS)의 자료이력서(메타데이터) 잠정표준
   2. English translation, if appropriate:
      National Geographic Information System Metadata Interim-Standard

B. Version/Edition:
      Version 1, 1999

C. Language(s) of documentation:
  Korean

D. Acronym(s):
  N/A

E. Official ID:
  N/A

F. Name(s) of recognizing standards authority(ies):
  The Steering Committee of National Geographic Information System, National Geography Institute

1.2 Responsible Institutions

Contact information about those institutions that are (or were) responsible for the processes involved in the evolution of the standard:

A. Production and/or development:
  National Geography Institute
  111 Wonchon-dong, Paldal-gu, Suwon-shi, 442-380 Korea
  Fax: +82 31 210 2600
  Phone: +82 31 210 2626
  E-mail: hmcho@moct.go.kr

  National Computerization Agency
  168 Jukjeon-ri, Suji-eub. Yongin-shi, 449-717 Korea
  Phone: +82 31 260 2114

B. Testing:
  National Geography Institute
  111 Wonchon-dong, Paldal-gu Suwon-shi, 442- 380 Korea
  Fax: +82 31 210 2600
  Phone: +82 31 210 2626,
  E-mail: hmcho@moct.go.kr

C. Conformance:
  National Geography Institute
  111 Wonchon-dong, Paldal-gu Suwon-shi, 442- 380 Korea
  Fax: +82 31 210 2600
  Phone: +82 31 210 2626,
  E-mail: hmcho@moct.go.kr

D. Maintenance:
  National Geography Institute
  111 Wonchon-dong, Paldal-gu Suwon-shi, 442- 380 Korea
  Fax: +82 31 210 2600
  Phone: +82 31 210 2626
  E-mail: hmcho@moct.go.kr

E. Distribution:
  National Geography Institute
  111 Wonchon-dong, Paldal-gu

Suwon-shi, 442-380 Korea
Fax: +82 31 210 2600
Phone: +82 31 210 2626
E-mail: hmcho@moct.go.kr

F. Help Desk/User Support:
National Geography Institute
111 Wonchon-dong, Paldal-gu Suwon-shi, 442- 380 Korea
Fax: +82 31 210 2600
Phone: +82 31 210 2626
E-mail: hmcho@moct.go.kr

1.3 Development History

A. Beginning of work on metadata standard: **1996**
B. Milestones:

- **Adoption of draft version for the National Digital Base Mapping in 1998**
- **Adoption of Version 1 for the NGIS Interim-Standard in June 1999**

C. Anticipated completion: **2001**
D. Update cycle:
Revised as required(developing a new metadata standard underway in 2000)

1.4 Status of the Standard

A. Is the standard officially recognized: **Expected**
B. Date(s) of recognition: **N/A**
C. Anticipated date of recognition: **2001**
D. Current stage in recognition process:
Awaiting publication of ISO Metadata Standard
E. Steps still required to achieve recognition:

- **Compile Korean profile of ISO Metadata Standard**
- **Public review, evaluate and comments from Korean geospatial community**
- **Finalize profile**
- **Submit Korean Metadata profile to the Agency of Technology and Standard of Korea for formal approval and recognition**

1.5 Access to Documentation of the Standard

A. Is the standard copyrighted?: **NO**
B. Name and address of copyright owner: **N/A**
C. Do restrictions on the use of the standard apply?: **NO**
D. Is the standard documentation currently available?: **YES**
E. List the form/media of standard documentation:
Paper
F. Is the standard available on the Internet?: **NO**

G. Price: **Free**
H. Contact details for obtaining the standard:
National Geography Institute
111 Wonchon-dong, Paldal- gu Suwon-shi, 442-380 Korea
Fax: +82 31 210 2600
Phone: +82 31 210 2626
E-mail: hmcho@moct.go.kr
I. Does the standard have ISBN/ISSN numbers?: **NO**

1.6 Available Software Tools

A. Is software available to implement the standard as a database?: **NO**
B. Is software available to test compliance of metadata with the standard?: **NO**
C. Is software available to produce metadata compliant with the standard?: **NO**
D. Other relevant software: **N/A**
E. Is there a test dataset available to test software tools for implementing the standard?: **NO**
F. For each product available above, please list contact information. **N/A**

1.7 Available Training Materials

A. Are organized training sessions available?: **NO**
B. Contact details for training sessions: **N/A**
C. Training documentation available: **N/A**

SECTION 2 – Use and Implementation of the Metadata Standard

This section provides information on the way in which the standard is used and the purpose of its use. It also provides descriptions about the level of abstraction of the standard.

2.1 Intended Use of the Standard
Can the standard be used:

A. Within a transfer data structure?: **YES**
B. To produce hardcopy catalog of datasets?: **YES**
C. To catalog data in a digital on-line catalog?: **NO**
D. Other: **N/A**

2.2 Intended Purpose of the Standard
Is the intended purpose of the standard:

A. To support discovery, identification, location of data?: **YES**
B. To support how to access data?: **YES**
C. To support the determination of dataset fitness for use?: **YES**
D. To support transfer of dataset?: **YES**
E. To support an organization's data management?: **YES**
F. Other: **NO**

2.3 Table of Contents

A. Provide the standard's Table of Contents

1. Identification Information
2. Data Quality Information
3. Lineage Information
4. Spatial Data Representation Information
5. Spatial Reference Information
6. Feature and Attribute Information
7. Distribution Information
8. Metadata Reference Information
9. Citation Information
10. Contact Information
11. Address Information

2.4 Level of Abstraction of the Metadata Standard Definition

A. Does the standard define metadata elements?: **YES**
B. Does the standard define a model for the relationships between the metadata elements?: **NO**
C. Provide an overview diagram of the model of the metadata, if available. **N/A**
D. Does the standard define domains and/or data types of the metadata elements?: **YES**
E. Does the standard define how the metadata should be encoded?: **YES**
F. Does the standard define a database schema for the automatic generation of a database?: **YES**

2.5 Extensibility of the Standard

A. Is the metadata description extensible (e.g., Can user add metadata domains, elements, thematic profiles, etc.)?: **NO**

2.6 Availability of Metadata
Does the standard specify:

A. How metadata are to be accessed?: **NO**
B. A mechanism for querying the metadata?: **NO**
C. Particular standardized query protocols?: **NO**
D. That the metadata should be encoded for computer access?: **NO**
E. That the metadata should be encrypted for security?: **NO**

2.7 Examples

A. Does the standard provide examples of implementations?: **NO**
B. How many examples are available?: **0**

SECTION 3 – Linkage and Coordination of the Metadata Standard

This section provides information on how the metadata standard is linked to other standards, particularly transfer standards, and how such linkage is implemented. If a significant linkage does exist, detailed technical information is provided on the precise nature of the linkage to the standard, and how the linkage affects implementation of the metadata standard.

3.1 Associated Transfer Standards

A. Is the metadata standard associated with a transfer standard(s)?: **NO**
B. If A is yes, provide references and contact details for each linked standard: **N/A**
C. If A is yes, are both (all) standards coordinated?: **N/A**
D. If A is yes, describe how the coordination works, both from a technical and conformance (normative) viewpoint: **N/A**

3.2 Incorporated into transfer mechanism

A. Is the metadata standard incorporated into a transfer mechanism?: **NO**
B. If A is yes, provide references and contact details for each incorporated mechanism: **N/A**
C. If A is yes, describe how: **N/A**

3.3 Other linked, coordinated or associated standards

A. Is the metadata standard linked, coordinated, or associated with any other standard(s) not listed above?: **NO**
B. If A is yes, provide references and contact details for each standard: **N/A**
C. If A is yes, describe how for each relevant standard: **N/A**

II. Categories the Metadata Standard Uses to Describe Datasets

The following sections are categories of metadata that may be used by the metadata standard to describe data. Not all categories are relevant to different datasets and may not necessarily be present in the standard. The presence or absence of a category should not be used for judging one standard against another.

Note: For each of the questions in the following sections please provide the obligation status information. Valid obligation values are:

Obligation Level

M	Mandatory	This information must be entered to comply with the metadata standard.
C	Conditional	This information is entered based on other information in the metadata.
O	Optional	This information is optionally entered in the metadata.

If the answer is ‘NO’, the obligation status should be left blank.

SECTION 4 – Dataset Identification

This section examines how the standard provides for the specification of the various sorts of titles and identifiers for a dataset.

4.1 The title(s) of a dataset
Does the standard allow/require the provision for:

Obligation

A. __ Previous title(s)?: **NO**
B. **M** Current title(s)?: **YES**
C. __ Possible future title(s)?: **NO**
D. __ Official titles in different languages?: **NO**
E. __ Abbreviated title(s)?: **NO**
F. __ Alternative Title(s)?: **NO**
G. __ Does the standard support effective dates of titles?: **NO**
H. __ Other identifiers (e.g., ISBN, National Stock Number, Uniform Product Code)?: **NO**

4.2 Parties responsible for the dataset
Does the standard allow/require the provision for:

Obligation

A. **M** Author?: **YES**
B. **M** Owner?: **YES**
C. __ Other?: **NO**

4.3 ISBD Information

Obligation

A. __ Does the metadata standard specify inclusion of mandatory ISBD information (see glossary) within the metadata?: **NO**

SECTION 5 – Status of the Dataset

This section examines how the standard provides for the specification of the current status and updating of the data set.

5.1 Status and Progress of Datasets
Does the metadata standard allow/require the provision for:

Obligation

A. __ Information on the status of the entire dataset?: **NO**
B. __ Information on the status of the particular subset (e.g., observation time, inauguration time, extraction time)?: **NO**

Does the metadata standard specify:

C. __ A structure for updating or maintenance information?: **NO**
D. __ A structure to describe historical information (see also Section 8.3 G)?: **NO**

5.2 Information on Incremental Updating
Does the metadata standard specify:

Obligation

A. __ A structure for information on incremental updates of the entire dataset?: **NO**
B. __ A structure for information on incremental updates of particular portions of the dataset?: **NO**
C. __ A structure to describe the update process for the dataset?: **NO**

SECTION 6 – Extent of the Dataset

This section examines how the standard provides for the specification of the spatial and temporal extent of the dataset.

6.1 The spatial extent of the dataset

Obligation

A. **M** Does the metadata standard allow/require for the description of horizontal extent (e.g., bounding rectangle/polygon, named location (i.e., Stockholm, etc.))?: **YES**
Bounding Rectangle Coordinates(North, South, East, West), Place name
B. **O** Does the metadata standard allow/require for the description of vertical extent (e.g., minima/maxima heights, etc.)?: **YES**
C. **M** Does the standard require specific units for describing extent?: **YES**
D. __ Does the standard allow specification of the units for describing extent?: **NO**
E. **M** Does the standard require specific spatial reference system for describing extent?: **YES**
F. __ Does the standard allow specification of the spatial reference system for describing extent?: **NO**

6.2 The temporal extent of the dataset

Obligation

A. **O** Does the standard allow/require for the description of time period for which the dataset is relevant?: **YES**
B. __ Does the standard allow/require specification of the temporal reference system?: **NO**

SECTION 7 – Data Content of the Dataset

This section examines how the standard provides for the specification of the data content of the dataset in terms of overview elements, themes, data definitions and product specifications.

7.1 Dataset overview
Does the standard allow/require for the provision of the following information:

Obligation

A. __ Dataset overview?: **NO**
B. **M** Abstract – brief summary of the content of the dataset?: **YES**
C. **M** Purpose – summary of the intentions with which the dataset was developed?: **YES**
D. __ Usage – list of activities for which the dataset has been used? (see also Section 8.5): **NO**
E. **M** Spatial schema – (e.g., vector, raster)? (see also Section 9): **YES**
F. **M** Spatial reference system? (see also Section 12): **YES**
G. **M** Language?: **YES**
H. **C** Character set (e.g., ISO8859-1, JISX0208)?: **YES**
I. **O** List of documents providing further information about the dataset?: **YES**
J. **M** Sample/preview dataset?: **YES**
K. **M** Related datasets?: **YES**
L. __ Original size of the dataset?: **NO**
M. **O** Compression? (see also Section 11.5): **YES**
N. __ Other?: **NO**

7.2 Theme(s)

Obligation

A. **M** Does the standard allow/require the provision for the description of theme(s) (e.g., topography, land cover, transport, etc.) in the dataset?: **YES**
B. __ If A is yes, does the standard allow the referencing of an outside dictionary or thesaurus?: **NO**
C. **M** If A is yes, does the standard itself include a dictionary or thesaurus within the metadata?: **YES**
D. __ If A is yes, does the standard allow the use of unauthorized user-defined keywords?: **NO**

7.3 Data Definition

Obligation

A. **C** Does the standard allow/require the provision for the description of the data definition?: **YES**
B. If A is yes, does the standard allow/require the provision for the description of: **YES**
C 1. Features/Objects?: **YES**
O 2. Attributes?: **YES**
C 3. Relations?: **YES**
C. **C** Can attributes be described separately from feature/object classes?: **YES**
D. **O** Does the standard allow/require the specification of the domain of the attributes?: **YES**
E. **M** Does the standard allow/require the specification of data type of the attributes?: **YES**
F. __ Does the standard allow/require the specification of particular classification coding schemes (e.g., FACC, OSKA, S-57 OC)?: **NO**

7.4 Product Specification

Obligation

A. __ Does the standard allow/require for the provision of information about whether the dataset meets a particular product specification?: **NO**

B. __ If A is yes, can it be internal to the Metadata?: **N/A**

C. __ If A is yes, can it be external to the Metadata?: **N/A**

SECTION 8 – Data Quality of the Dataset

This section examines how the standard deals with the question of data quality. This includes homogeneity and metrics for data quality, quality assurance, and usage.

8.1 Overall Data Quality Statement

Obligation

A. __ Does the standard allow/require for the provision of quality to be described? If yes, please answer the questions below. (*If no, all following answers in Section* 8 *are N/A, please skip to Section* 9.): **NO**

B. __ Does the standard allow/require the provision for the quality information referencing an external quality standard, quality assurance standard, or specification?: **NO**

8.2 Homogeneity of Data Quality

Obligation

A. __ Does the standard allow/require for the variation of the quality of the dataset to be described?: **N/A**

B. What type of aggregation of quality information is allowed? **N/A**

__ 1. Entire database?: **N/A**

__ 2. Feature/Object instance?: **N/A**

__ 3. Feature/Object class?: **N/A**

__ 4. Attribute instance?: **N/A**

__ 5. Attribute class?: **N/A**

__ 6. Layer?: **N/A**

__ 7. Theme?: **N/A**

__ 8. Geographic extent?: **N/A**

__ 9. Other?: **N/A**

8.3 Metrics for Data Quality

Does the standard allow/require the provision of values for:

Obligation

A. __ Positional accuracy?: **N/A**

1. Standard error?: **N/A**

2. Maximum error?: **N/A**

	3. Error correlation?:	**N/A**
	4. Confidence level (e.g., 90%, 95%)?:	**N/A**
	5. Other?:	**N/A**
B.	__ Attribute accuracy?:	**N/A**
	1. Standard error?:	**N/A**
	2. Maximum error?:	**N/A**
	3. Error frequency?:	**N/A**
	4. Error correlation?:	**N/A**
	5. Confidence level (e.g., 90%, 95%)?:	**N/A**
	6. Other?:	**N/A**
C.	__ Temporal accuracy?:	**N/A**
	1. Standard error?:	**N/A**
	2. Maximum error?:	**N/A**
	3. Error correlation?:	**N/A**
	4. Confidence level? (e.g., 90%, 95%):	**N/A**
	5. Other?:	**N/A**
D.	__ Completeness?:	**N/A**
	1. Error of omission?:	**N/A**
	2. Error of commission?:	**N/A**
	3. Other?:	**N/A**
E.	__ Currentness?:	**N/A**
	1. Date?:	**N/A**
	2. Maximum age?:	**N/A**
	3. Temporal extent?:	**N/A**
	4. Other?:	**N/A**
F.	__ Logical consistency?:	**N/A**
G.	__ Lineage?:	**N/A**
	1. List of processing steps?:	**N/A**
	2. List of values of processing parameters?:	**N/A**
	3. List of printouts of statistics, etc.?:	**N/A**
	4. Source material?:	**N/A**
	5. Other?:	**N/A**
H.	__ Additional Quality Information?:	**N/A**

8.4 Quality Assurance Methods

Does the standard allow/require the provision of information about the methods used to determine:

Obligation

A.	__ Positional accuracy?:	**N/A**
	1. Standard error in adjustments?:	**N/A**
	2. Repeated measurements?:	**N/A**
	3. Independent measurements?:	**N/A**
	4. Subjective evaluations?:	**N/A**
	5. Other?:	**N/A**

B. __ Attribute accuracy?: **N/A**
  1. Standard error?: **N/A**
  2. Error classification matrix?: **N/A**
  3. Independent measurements?: **N/A**
  4. Subjective evaluation?: **N/A**
  5. Other?: **N/A**

C. __ Temporal accuracy?: **N/A**
  1. Independent measurements?: **N/A**
  2. Subjective evaluation?: **N/A**
  3. Other?: **N/A**

D. __ Completeness?: **N/A**
  1. Field checks?: **N/A**
  2. Repeated measurements?: **N/A**
  3. Independent measurements?: **N/A**
  4. Subjective evaluation?: **N/A**
  5. Error of commission?: **N/A**
  6. Other?: **N/A**

E. __ Currentness?: **N/A**
  1. Date of source material?: **N/A**
  2. Repeated measurements?: **N/A**
  3. Independent measurements?: **N/A**
  4. Subjective evaluation?: **N/A**
  5. Other?: **N/A**

F. __ Logical consistency?: **N/A**
  1. Verification results of consistency checks?: **N/A**
  2. Repeated processing?: **N/A**
  3. Comparison with independent sources?: **N/A**
  4. Subjective evaluation?: **N/A**
  5. Other?: **N/A**

G. __ Lineage?: **N/A**
  1. Checking results of processing steps: **N/A**
  2. Subjective evaluation: **N/A**
  3. Other: **N/A**

H. __ Additional Quality Information?: **N/A**

8.5 Usage

Obligation

A. __ Does the metadata standard allow/require the provision of information about usage?: **N/A**
  1. Who used it: **N/A**
  2. When it was used: **N/A**
  3. For what application it was used: **N/A**

4. The effectiveness of use: **N/A**
5. Other: **N/A**

8.6 Alternative Quality Description:

SECTION 9 – Spatial Data Structure of the Dataset

This section examines how the standard provides for the specification of the spatial data model and spatial data types.

9.1 Spatial Data Model
Does the standard allow/require the provision for:

Obligation

A. __ The description of the geometrical and/or topological primitives used to model the real world?: **NO**
B. __ Information about an external standard for geometrical and/or topological primitive specification?: **NO**

9.2 Spatial Data Types
Does the standard allow/require the provision for:

Obligation

A. **M** Information about the spatial data type(s) used in the dataset?: **YES**
B. __ If A is yes, which of the following generic spatial data types are supported? For each, please list specific data primitive levels supported by each data type.
1. Vector (e.g., spaghetti, chain-node/node-edge-face, planar graph, full topology)?: **YES**
spaghetti, chain-node/node-edge-face, planar graph, full topology
2. Areal Raster (e.g., thematic raster – scanned and classification maps, image raster – satellite images, scanned aerial photographs)?: **YES**
thematic raster, image raster, scanned aerial photographs
3. Grid/Matrix (e.g., Digital Elevation Model)?: **NO**
4. Triangular Irregular Network (TIN)?: **NO**
5. Other?: **N/A**

9.3 Other Data Types
Does the standard allow the provision of information about:

Obligation

A. __ Video?: **NO**
B. __ Sound?: **NO**
C. __ Other?: **NO**

SECTION 10 – Spatial Reference of the Dataset

This section examines how the dataset provides for the specification of the spatial reference characteristics of the data set.

Does the standard allow/require the provision for the following information:

10.1 Geoidal Model

Obligation

A. **O** Geoidal model name?: **YES**
B. __ Geoidal model definitions?: **NO**

10.2 Geodetic datum

Obligation

A. **M** Geodetic datum name?: **YES**
B. **O** Geodetic datum definitions?: **YES**

10.3 Reference Ellipsoid

Obligation

A. **M** Reference ellipsoid name?: **YES**
B. **M** Reference ellipsoid definitions?: **YES**

10.4 Projection

Obligation

A. **M** Projection name?: **YES**
B. __ Projection definitions?: **NO**

10.5 Coordinate Systems

Obligation

A. **M** Coordinate system type?: **YES**
B. **M** Coordinate system definitions?: **YES**

10.6 Height Reference System

Obligation

A. **M** Vertical datum name?: **YES**
B. **O** Types of heights?: **YES**

10.7 Ancillary Geodetic Information

Obligation

A. __ Does the standard allow/require the provision for ancillary geodetic, gravimetric, or magnetometric information?: **NO**

10.8 Transformation Parameters

Does the standard specify a method to provide information for coordinate transformation(s):

Obligation

A. **C** Model name?: **YES**
B. __ Control points?: **NO**
C. **C** Transformation parameters?: **YES**

10.9 Non-Coordinate Spatial Reference

Does the standard allow/require the provision for information on non-coordinate spatial reference systems (e.g., postal systems, street address)?:

Obligation

A. **C** Reference system name?: **YES**
Administrative units, Postal address, Road number, etc.
B. **C** Reference system definitions?: **YES**
C. __ Other?: **NO**

SECTION 11 – Availability and Distribution of the Dataset

This section examines how the standard provides information on the availability of the dataset?

Does the standard allow/require the provision for:

11.1 The Distributor(s) of the Dataset

Obligation

A. **M** Information about the distributor(s) of the dataset?: **YES**
B. **M** Contact information for the distributor(s)?: **YES**

11.2 Restrictions to Access and Usage

Obligation

A. **O** Information about copyright on the dataset (e.g., date, period, copyright owner, etc.)?: **YES**
B. **O** Information about restrictions on the access of the dataset?: **YES**
C. **O** Information about restrictions on the usage of the dataset?: **YES**

11.3 Pricing Details

Obligation

A. **O** Information about pricing?: **YES**

11.4 Distribution of the Dataset

Obligation

A. **O** Information about distribution media?: **YES**
B. **M** Information about on-line access?: **YES**
C. **O** Information about units of distribution (e.g., tiles, layers, polygons, elements)?: **YES**

11.5 Distribution Format

Obligation

A. **M** Information about distribution formats (e.g., commercial CAD format, commercial GIS formats, National/International transfer standard)?: **YES**
B. **O** Information about the encoding of the dataset (e.g., binary, ASCII)?: **YES**
C. **O** Information about compression method?: **YES**
D. __ Information about compressed dataset size?: **NO**
E. __ Information about the distribution mechanism?: **NO**
F. __ Information about the encapsulation of data (e.g., ISO 8211, ISO 8824)?: **NO**

11.6 Ordering Procedures

Obligation

A. **O** Information about ordering procedures for the dataset?: **YES**

11.7 Support Services

Obligation

A. __ Information about support services (e.g., additional data transformations, format or coordinate conversions)?: **NO**

SECTION 12 – Authorization and Verification of the Metadata

This section examines how the standard provides for the description of the authorization and verification on the metadata of a dataset (e.g., how is the metadata itself authorized and verified?).

12.1 Metadata reference

Obligation

A. **M** Does the standard allow/require the provision of information about reference dates of the metadata (e.g., last check date, last update date, review date)?: **YES**
Created, Reviewed, Updated date

12.2 Verification Authority

Obligation

A. __ Does the standard allow/require the provision for information about who verified/authorized the metadata?: **NO**

B. __ Does the standard specify the method for providing this information (e.g., ISO 9000): **NO**

12.3 Statistical Methods Used for Verification

Obligation

A. __ Does the standard specify the statistical methods to be used for verification of the metadata?: **NO**

B. __ Does the standard allow/require the provision for the description of the statistical methods used to verify the metadata?: **NO**

END OF CHARACTERISTICS
FIN

H. Moellering, H.J.G.L. Aalders & A. Crane (Editors)

The Netherlands NCGI Metadata

Henri J.G.L. Aalders

I. Information about the Metadata Standard Itself

SECTION 0 – Assessment Information

This section provides the user with a general description of the scope and intended use of the standard. Information on the respondent who prepared the assessment is also provided.

0.1 Respondent

Assessment prepared by:

Prof. Ir. Henri J.G.L. Aalders
Section GIS Technology
OTB, Research Institute for Housing,
Urban and Mobility Studies
Jaffalaan 9
P.O. Box 5030
NL-2600 GA Delft
Phone: +31 15 278 1567
Fax: +31 14 278 2745
E-mail: h.aalders@otb.tudelft.nl

Date assessment completed:

November 1998, updated November 2000 and May 2002, Reviewed Fall, 2004

Relationship of respondent to standard:

Drafting committee member

0.2 Brief Summary Statement

A. Summary of the scope and intent of the standard:
This standard was developed for the creation of a national clearinghouse in the Netherlands after the proposal of the CEN standard ENV 12657 in 1995 and 1996. The Clearinghouse (and so the standard) became under private ownership in October 1997.

SECTION 1 – Administration of Metadata Standard

This section provides a general description of the administrative framework, within which the standard was developed, tested and currently resides. Significant details are also provided in terms of the developmental and managerial history as well as information on the existence of documentation, software tools and training materials. This section also provides a point of contact for further information on the standard.

1.1 Name of Standard

A. Name of standard:
   1. In original language(s):
   NCGI Metadata
   2. English translation, if appropriate:
B. Version/Edition:
   accepted V1.0 Sept. 1996
C. Language(s) of documentation:
   Dutch: "Aansluitdocument Nationaal Clearinghouse Geo-Informatie"
D. Acronym(s):
   NCGI Metadata
E. Official ID:
   N/A
F. Name(s) of recognizing standards authority(ies):

1.2 Responsible Institutions

Contact information about those institutions that are (or were) responsible for the processes involved in the evolution of the standard:

A. Production and/or development:
   NCGI (National Clearinghouse Geo-Information)
   POB 1442
   NL-7301 BR Apeldoorn
   Phone: +31 55 528 5869
   Fax: +31 55 528 5803
   E-mail: clhouse@euronet.nl
   Internet: www.ncgi.nl
B. Testing:
   see 1.2 A
C. Conformance:
   see 1.2 A
D. Maintenance:
   see 1.2 A
E. Distribution:
   see 1.2 A

F. Help Desk/User Support:
  see 1.2 A

1.3 Development History

A. Beginning of work on metadata standard:
  March 1995. Development by PGB Standards of National LIS-council Ravi

B. Milestones:
  - **Initiative group founded in June 1995**
  - **Development since Sept 1995**
  - **Testing during the IDEFIX period (predecessor NCGI) in January 1996 until end July 1996**

C. Anticipated completion: **N/A**

D. Update cycle: **N/A**

1.4 Status of the Standard

A. Is the standard officially recognized: **YES**

B. Date(s) of recognition: **Sept 1996**

C. Anticipated date of recognition: **N/A**

D. Current stage in recognition process:
  PGB Standards Ravi

E. Steps still required to achieve recognition: **N/A**

1.5 Access to Documentation of the Standard

A. Is the standard copyrighted?: **NO**

B. Name and address of copyright owner: **N/A**

C. Do restrictions on the use of the standard apply?: **NO**

D. Is the standard documentation currently available?: **YES**

E. List the form/media of standard documentation:
  Paper

F. Is the standard available on the Internet?: **NO**

G. Price: **Free**

H. Contact details for obtaining the standard:
  Contact NCGI website http://www.ncgi.nl

I. Does the standard have ISBN/ISSN numbers?: **NO**

1.6 Available Software Tools

A. Is software available to implement the standard as a database?: **YES**
  MS-Access application software developed by Civility and in Geokey, developed by Geodan

B. Is software available to test compliance of metadata with the standard?: **YES**
  software developed by Civility

C. Is software available to produce metadata compliant with the standard?: **YES**
 software developed by DLO Staringcentrum, available through NCGI

D. Other relevant software: **NO**

E. Is there a test dataset available to test software tools for implementing the standard?: **YES**
 developed in MS-Access database

F. For each product available above, please list contact information:
 See 1.2 A

1.7 Available Training Materials

A. Are organized training sessions available?: **YES**

B. Contact details for training sessions:
 See 1.2 A

C. Training documentation available: **NO**

SECTION 2 – Use and Implementation of the Metadata Standard

This section provides information on the way in which the standard is used and the purpose of its use. It also provides descriptions about the level of abstraction of the standard.

2.1 Intended Use of the Standard
Can the standard be used:

A. Within a transfer data structure?: **YES**

B. To produce hardcopy catalog of datasets?: **YES**

C. To catalog data in a digital on-line catalog?: **YES**

D. Other:
 Use of data within the Internet site of NCGI

2.2 Intended Purpose of the Standard
Is the intended purpose of the standard:

A. To support discovery, identification, location of data?: **YES**

B. To support how to access data?: **YES**

C. To support the determination of dataset fitness for use?: **YES**

D. To support transfer of dataset?: **YES**

E. To support an organization's data management?:
 Possible, but the standard is developed for data exploration

F. Other:
 Describes information of the structure of the datasets

2.3 Table of Contents

A. Provide the standard's Table of Contents
 Summary

1. Introduction
2. Architecture of NCGI
3. What to do when joining
4. At the end
Annexes
1. Participants NCGI
2. Dutch translation of Annex A1 of CEN Metadata standard version Febr.96
3. Example Cadastre
4. Example Ministry of Transport and Water control
5. Example Ministry of Housing, Physical planning and Environment
6. Example Province Gelderland
7. Example Central Bureau of Statistics
8. Example of DLO/Staringcentrum

2.4 Level of Abstraction of the Metadata Standard Definition

A. Does the standard define metadata elements?:
For some metadata elements
B. Does the standard define a model for the relationships between the metadata elements?: **NO**
C. Provide an overview diagram of the model of the metadata, if available. **N/A**
D. Does the standard define domains and/or data types of the metadata elements?: **YES**
E. Does the standard define how the metadata should be encoded?: **NO**
F. Does the standard define a database schema for the automatic generation of a database?: **NO**
No definition but any database schema can be included

2.5 Extensibility of the Standard

A. Is the metadata description extensible (e.g., Can user add metadata domains, elements, thematic profiles, etc.)?: **NO**

2.6 Availability of Metadata
Does the standard specify:

A. How metadata are to be accessed?: **NO**
B. A mechanism for querying the metadata?: **NO**
C. Particular standardized query protocols?: **NO**
D. That the metadata should be encoded for computer access?: **NO**
E. That the metadata should be encrypted for security?: **NO**

2.7 Examples

A. Does the standard provide examples of implementations?: **YES**
see "Aansluitdocument NCGI"
B. How many examples are available?: **8**

SECTION 3 – Linkage and Coordination of the Metadata Standard

This section provides information on how the metadata standard is linked to other standards, particularly transfer standards, and how such linkage is implemented. If a significant linkage does exist, detailed technical information is provided on the precise nature of the linkage to the standard, and how the linkage affects implementation of the metadata standard.

3.1 Associated Transfer Standards

A. Is the metadata standard associated with a transfer standard(s)?: **NO**
B. If A is yes, provide references and contact details for each linked standard: **N/A**
C. If A is yes, are both (all) standards coordinated?: **N/A**
D. If A is yes, describe how the coordination works, both from a technical and conformance (normative) viewpoint: **N/A**

3.2 Incorporated into transfer mechanism

A. Is the metadata standard incorporated into a transfer mechanism?: **NO**
B. If A is yes, provide references and contact details for each incorporated mechanism: **N/A**
C. If A is yes, describe how: **N/A**

3.3 Other linked, coordinated or associated standards

A. Is the metadata standard linked, coordinated, or associated with any other standard(s) not listed above?: **YES**
B. If A is yes, provide references and contact details for each standard:
CEN ENV 12657, Metadata standard
C. If A is yes, describe how for each relevant standard:
More metadata elements are compulsory as in the CEN Metadata standard ENV 12657, as well as the dataset-ID is compulsory

II. Categories the Metadata Standard Uses to Describe Datasets

The following sections are categories of metadata that may be used by the metadata standard to describe data. Not all categories are relevant to different datasets and may not necessarily be present in the standard. The presence or absence of a category should not be used for judging one standard against another.

Note: For each of the questions in the following sections please provide the obligation status information. Valid obligation values are:

Obligation Level

M	Mandatory	This information must be entered to comply with the metadata standard.
C	Conditional	This information is entered based on other information in the metadata.
O	Optional	This information is optionally entered in the metadata.

If the answer is 'NO', the obligation status should be left blank.

SECTION 4 – Dataset Identification

This section examines how the standard provides for the specification of the various sorts of titles and identifiers for a dataset.

4.1 The title(s) of a dataset

Does the standard allow/require the provision for:

Obligation

A. __ Previous title(s)?: **NO**
B. **M** Current title(s)?: **YES**
C. __ Possible future title(s)?: **NO**
D. __ Official titles in different languages?: **NO**
E. **O** Abbreviated title(s)?: **YES**
F. **O** Alternative Title(s)?: **YES**
G. __ Does the standard support effective dates of titles?: **NO**
H. __ Other identifiers (e.g., ISBN, National Stock Number, Uniform Product Code)?: **NO**

4.2 Parties responsible for the dataset

Does the standard allow/require the provision for:

Obligation

A. **M** Author?: **YES**
B. **O** Owner?: **YES**
C. **O** Other?: **YES**

4.3 ISBD Information

Obligation

A. __ Does the metadata standard specify inclusion of mandatory ISBD information (see glossary) within the metadata?: **NO**

SECTION 5 – Status of the Dataset

This section examines how the standard provides for the specification of the current status and updating of the data set.

5.1 Status and Progress of Datasets

Does the metadata standard allow/require the provision for:

Obligation

A. **O** Information on the status of the entire dataset?: **YES**
Yes, actual status, future status, starting/end dates of validity, percentage of readiness of the dataset

B. __ Information on the status of the particular subset (e.g., observation time, inauguration time, extraction time)?: **NO**

Does the metadata standard specify:

C. **O** A structure for updating or maintenance information?: **YES**
D. __ A structure to describe historical information (see also Section 8.3 G)?: **NO**

5.2 Information on Incremental Updating
Does the metadata standard specify:

Obligation

A. **O** A structure for information on incremental updates of the entire dataset?: **YES**
B. __ A structure for information on incremental updates of particular portions of the dataset?: **NO**
C. __ A structure to describe the update process for the dataset?: **NO**

SECTION 6 – Extent of the Dataset

This section examines how the standard provides for the specification of the spatial and temporal extent of the dataset.

6.1 The spatial extent of the dataset

Obligation

A. **M** Does the metadata standard allow/require for the description of horizontal extent (e.g., bounding rectangle/polygon, named location (i.e., Stockholm, etc.))?: **YES**
Yes, bounding box or named location
B. **M** Does the metadata standard allow/require for the description of vertical extent (e.g., minima/maxima heights, etc.)?: **YES**
min/max height
C. __ Does the standard require specific units for describing extent?: **NO**
D. **M** Does the standard allow specification of the units for describing extent?: **YES**
E. **M** Does the standard require specific spatial reference system for describing extent?: **YES**
F. **O** Does the standard allow specification of the spatial reference system for describing extent?: **YES**

6.2 The temporal extent of the dataset

Obligation

A. **M** Does the standard allow/require for the description of time period for which the dataset is relevant?: **YES**
B. __ Does the standard allow/require specification of the temporal reference system?: **NO**

SECTION 7 – Data Content of the Dataset

This section examines how the standard provides for the specification of the data content of the dataset in terms of overview elements, themes, data definitions and product specifications.

7.1 Dataset overview

Does the standard allow/require for the provision of the following information:

Obligation

A. **M** Dataset overview?: **YES**
B. **M** Abstract – brief summary of the content of the dataset?: **YES**
C. **M** Purpose – summary of the intentions with which the dataset was developed?: **YES**
D. **M** Usage – list of activities for which the dataset has been used? (see also Section 8.5): **YES**
E. **M** Spatial schema – (e.g., vector, raster)? (see also Section 9): **YES**
F. **M** Spatial reference system? (see also Section 12): **YES**
G. **M** Language?: **YES**
H. **M** Character set (e.g., ISO8859-1, JISX0208)?: **YES**
I. **O** List of documents providing further information about the dataset?: **YES**
J. __ Sample/preview dataset?: **NO**
K. __ Related datasets?: **NO**
L. __ Original size of the dataset?: **NO**
M. __ Compression? (see also Section 11.5): **NO**
N. __ Other?: **NO**

7.2 Theme(s)

Obligation

A. **O** Does the standard allow/require the provision for the description of theme(s) (e.g., topography, land cover, transport, etc.) in the dataset?: **YES**
B. **M** If A is yes, does the standard allow the referencing of an outside dictionary or thesaurus?: **YES**
C. **M** If A is yes, does the standard itself include a dictionary or thesaurus within the metadata?: **YES**
D. **O** If A is yes, does the standard allow the use of unauthorized user-defined keywords?: **YES**

7.3 Data Definition

Obligation

A. **M** Does the standard allow/require the provision for the description of the data definition?: **YES**
B. If A is yes, does the standard allow/require the provision for the description of:
M 1. Features/Objects?: **YES**

M 2. Attributes?: **YES**
M 3. Relations?: **YES**

C. **M** Can attributes be described separately from feature/object classes?: **YES**

D. **O** Does the standard allow/require the specification of the domain of the attributes?: **YES**

E. __ Does the standard allow/require the specification of data type of the attributes?: **NO**

F. **O** Does the standard allow/require the specification of particular classification coding schemes (e.g., FACC, OSKA, S-57 OC)?: **YES**

7.4 Product Specification

Obligation

A. **O** Does the standard allow/require for the provision of information about whether the dataset meets a particular product specification?: **YES**

B. __ If A is yes, can it be internal to the Metadata?: **NO**

C. **O** If A is yes, can it be external to the Metadata?: **YES**

SECTION 8 – Data Quality of the Dataset

This section examines how the standard deals with the question of data quality. This includes homogeneity and metrics for data quality, quality assurance, and usage.

8.1 Overall Data Quality Statement

Obligation

A. **M** Does the standard allow/require for the provision of quality to be described? If yes, please answer the questions below. (*If no, all following answers in Section* 8 *are N/A, please skip to Section* 9.): **YES**

B. **M** Does the standard allow/require the provision for the quality information referencing an external quality standard, quality assurance standard, or specification?: **YES**

8.2 Homogeneity of Data Quality

Obligation

A. __ Does the standard allow/require for the variation of the quality of the dataset to be described?: **NO**

B. What type of aggregation of quality information is allowed?

M 1. Entire database?: **YES**
__ 2. Feature/Object instance?: **NO**
O 3. Feature/Object class?: **YES**
__ 4. Attribute instance?: **NO**
O 5. Attribute class?: **YES**
__ 6. Layer?: **NO**
__ 7. Theme?: **NO**

__ 8. Geographic extent?: **NO**
__ 9. Other?: **N/A**

8.3 Metrics for Data Quality
Does the standard allow/require the provision of values for:

Obligation

A. Positional accuracy?:
M 1. Standard error?: **YES**
Yes, as free text
M 2. Maximum error?: **YES**
Yes, as free text
M 3. Error correlation?: **YES**
Yes, as free text
M 4. Confidence level (e.g., 90%, 95%)?: **YES**
Yes, as free text
5. Other?: **N/A**

B. Attribute accuracy?:
M 1. Standard error?: **YES**
Yes, as free text
M 2. Maximum error?: **YES**
Yes, as free text
M 3. Error frequency?: **YES**
Yes, as free text
M 4. Error correlation?: **YES**
Yes, as free text
M 5. Confidence level (e.g., 90%, 95%)?: **YES**
Yes, as free text
6. Other?: **N/A**

C. Temporal accuracy?:
M 1. Standard error?: **YES**
Yes, as free text
M 2. Maximum error?: **YES**
Yes, as free text
M 3. Error correlation?: **YES**
Yes, as free text
M 4. Confidence level? (e.g., 90%, 95%): **YES**
Yes, as free text
5. Other?: **N/A**

D. Completeness?:
M 1. Error of omission?: **YES**
Yes, as free text
M 2. Error of commission?: **YES**
Yes, as free text
3. Other?: **N/A**

E. **O** Currentness?:	**YES**
1. Date?:	**YES**
2. Maximum age?:	**YES**
3. Temporal extent?:	**YES**
4. Other?:	**NO**
F. **M** Logical consistency?:	**YES**
Yes, as free text	
G. __ Lineage?:	
M 1. List of processing steps?:	**YES**
Yes, as free text	
M 2. List of values of processing parameters?:	**YES**
Yes, as free text	
M 3. List of printouts of statistics, etc.?:	**YES**
Yes, as free text	
M 4. Source material?:	**YES**
Yes, as free text	
5. Other?:	**N/A**
H. __ Additional Quality Information?:	**N/A**

8.4 Quality Assurance Methods

Does the standard allow/require the provision of information about the methods used to determine:

Obligation

A. __ Positional accuracy?:	**NO**
1. Standard error in adjustments?:	**N/A**
2. Repeated measurements?:	**N/A**
3. Independent measurements?:	**N/A**
4. Subjective evaluations?:	**N/A**
5. Other?:	**N/A**
B. __ Attribute accuracy?:	**NO**
1. Standard error?:	**N/A**
2. Error classification matrix?:	**N/A**
3. Independent measurements?:	**N/A**
4. Subjective evaluation?:	**N/A**
5. Other?:	**N/A**
C. __ Temporal accuracy?:	**NO**
1. Independent measurements?:	**N/A**
2. Subjective evaluation?:	**N/A**
3. Other?:	**N/A**
D. **O** Completeness?:	**YES**
1. Field checks?:	**NO**
2. Repeated measurements?:	**NO**
3. Independent measurements?:	**NO**
4. Subjective evaluation?:	**YES**

5. Error of commission?: **YES**
6. Other?: **NO**

E. __ Currentness?: **NO**
1. Date of source material?: **N/A**
2. Repeated measurements?: **N/A**
3. Independent measurements?: **N/A**
4. Subjective evaluation?: **N/A**
5. Other?: **N/A**

F. __ Logical consistency?: **NO**
1. Verification results of consistency checks?: **N/A**
2. Repeated processing?: **N/A**
3. Comparison with independent sources?: **N/A**
4. Subjective evaluation?: **N/A**
5. Other?: **N/A**

G. __ Lineage?: **NO**
1. Checking results of processing steps: **N/A**
2. Subjective evaluation: **N/A**
3. Other: **N/A**

H. __ Additional Quality Information?: **NO**

8.5 Usage

Obligation

A. __ Does the metadata standard allow/require the provision of information about usage?:
M 1. Who used it: **YES**
Yes, as free text
M 2. When it was used: **YES**
Yes, as free text
M 3. For what application it was used: **YES**
Yes, as free text
M 4. The effectiveness of use: **YES**
Yes, as free text
5. Other: **N/A**

8.6 Alternative Quality Description: **NO**

SECTION 9 – Spatial Data Structure of the Dataset

This section examines how the standard provides for the specification of the spatial data model and spatial data types.

9.1 Spatial Data Model
Does the standard allow/require the provision for:

Obligation

A. **M** The description of the geometrical and/or topological primitives used to model the real world?: **YES**

B. **M** Information about an external standard for geometrical and/or topological primitive specification?: **YES**

9.2 Spatial Data Types

Does the standard allow/require the provision for:

Obligation

A. **M** Information about the spatial data type(s) used in the dataset?: **YES**

B. **M** If A is yes, which of the following generic spatial data types are supported? **YES**
For each, please list specific data primitive levels supported by each data type.

M 1. Vector (e.g., spaghetti, chain-node/node-edge-face, planar graph, full topology)?: **YES**

M 2. Areal Raster (e.g., thematic raster – scanned and classification maps, image raster – satellite images, scanned aerial photographs)?: **YES**

M 3. Grid/Matrix (e.g., Digital Elevation Model)?: **YES**

M 4. Triangular Irregular Network (TIN)?: **YES**

5. Other?: **N/A**

9.3 Other Data Types

Does the standard allow the provision of information about:

Obligation

A. __ Video?: **NO**

B. __ Sound?: **NO**

C. __ Other?: **N/A**

SECTION 10 – Spatial Reference of the Dataset

This section examines how the dataset provides for the specification of the spatial reference characteristics of the data set.

Does the standard allow/require the provision for the following information:

10.1 Geoidal Model

Obligation

A. **M** Geoidal model name?: **YES**

B. **M** Geoidal model definitions?: **YES**

10.2 Geodetic datum

Obligation

A. **M** Geodetic datum name?: **YES**

B. __ Geodetic datum definitions?: **NO**
No, is externally defined

10.3 Reference Ellipsoid

Obligation

A. **M** Reference ellipsoid name?: **YES**
B. __ Reference ellipsoid definitions?: **NO**
No, is externally defined

10.4 Projection

Obligation

A. **M** Projection name?: **YES**
B. **O** Projection definitions?: **YES**
Yes, by parameters

10.5 Coordinate Systems

Obligation

A. **M** Coordinate system type?: **YES**
B. __ Coordinate system definitions?: **NO**
No, is externally defined

10.6 Height Reference System

Obligation

A. **M** Vertical datum name?: **YES**
B. __ Types of heights?: **NO**

10.7 Ancillary Geodetic Information

Obligation

A. __ Does the standard allow/require the provision for ancillary geodetic, gravimetric, or magnetometric information?: **NO**

10.8 Transformation Parameters
Does the standard specify a method to provide information for coordinate transformation(s):

Obligation

A. __ Model name?: **NO**
B. __ Control points?: **NO**
C. __ Transformation parameters?: **NO**

10.9 Non-Coordinate Spatial Reference

Does the standard allow/require the provision for information on non-coordinate spatial reference systems (e.g., postal systems, street address)?:

Obligation

A. **M** Reference system name?: **YES**
B. __ Reference system definitions?: **NO**
C. __ Other?: **NO**

SECTION 11 – Availability and Distribution of the Dataset

This section examines how the standard provides information on the availability of the dataset?

Does the standard allow/require the provision for:

11.1 The Distributor(s) of the Dataset

Obligation

A. **M** Information about the distributor(s) of the dataset?: **YES**
B. **M** Contact information for the distributor(s)?: **YES**

11.2 Restrictions to Access and Usage

Obligation

A. **M** Information about copyright on the dataset (e.g., date, period, copyright owner, etc.)?: **YES**
B. **M** Information about restrictions on the access of the dataset?: **YES**
C. **O** Information about restrictions on the usage of the dataset?: **YES**

11.3 Pricing Details

Obligation

A. **M** Information about pricing?: **YES**

11.4 Distribution of the Dataset

Obligation

A. **M** Information about distribution media?: **YES**
B. **M** Information about on-line access?: **YES**
C. **M** Information about units of distribution (e.g., tiles, layers, polygons, elements)?: **YES**

11.5 Distribution Format

Obligation

A. **M** Information about distribution formats (e.g., commercial CAD format, commercial GIS formats, National/International transfer standard)?: **YES**
B. __ Information about the encoding of the dataset (e.g. binary, ASCII)?: **NO**
C. __ Information about compression method?: **NO**
D. __ Information about compressed dataset size?: **NO**
E. __ Information about the distribution mechanism?: **NO**
F. __ Information about the encapsulation of data (e.g., ISO 8211, ISO 8824)?: **NO**

11.6 Ordering Procedures

Obligation

A. **M** Information about ordering procedures for the dataset?: **YES**

11.7 Support Services

Obligation

A. **M** Information about support services (e.g., additional data transformations, format or coordinate conversions)?: **YES**

SECTION 12 – Authorization and Verification of the Metadata

This section examines how the standard provides for the description of the authorization and verification on the metadata of a dataset (e.g., how is the metadata itself authorized and verified?).

12.1 Metadata reference

Obligation

A. **M** Does the standard allow/require the provision of information about reference dates of the metadata (e.g., last check date, last update date, review date)?: **YES**

12.2 Verification Authority

Obligation

A. __ Does the standard allow/require the provision for information about who verified/authorized the metadata?: **NO**
B. __ Does the standard specify the method for providing this information (e.g., ISO 9000): **NO**

12.3 Statistical Methods Used for Verification

Obligation

A. __ Does the standard specify the statistical methods to be used for verification of the metadata?: **NO**

B. __ Does the standard allow/require the provision for the description of the statistical methods used to verify the metadata?: **NO**

END OF CHARACTERISTICS
FIN

World Spatial Metadata Standards
H. Moellering, H.J.G.L. Aalders & A. Crane (Editors)

Russian Federation Государственный стандарт Российской Федерации ГОСТ Р 51353-99 "Геоинформационное картографирование. Метаданные электронных карт. Состав и содержание"

State Standard of the Russian Federation GOST R 51353-99 "Geoinformatic Mapping. Metadata of the Electronic Maps. Composition and Content", November 11, 1999

Alexander I. Martynenko

I. Information about the Metadata Standard Itself

SECTION 0 – Assessment Information

This section provides the user with a general description of the scope and intended use of the standard. Information on the respondent who prepared the assessment is also provided.

0.1 Respondent

Assessment prepared by:

Alexander I. Martynenko
Institute of Informatics Problems. Russian Academy of Sciences
44, 2, Vavilova st.
119333, Moscow, Russia
E-mail: a_martynenko@mail.ru

Date assessment completed:

February 2001; updated May 2002, Reviewed Fall, 2004

Relationship of respondent to standard:

Author

0.2 Brief Summary Statement

A. Summary of the scope and intent of the standard:

This Standard provides cartographic representation of the real world and creation of Digital Earth that is based on strict geodetic base by means of using millions air- and space high-resolution images, electronic maps of various intent and scale, and textual reference information. GOST R 51353-99 allows to synthesize the digital (electronic) image of the Earth as a spatial and temporal representation of our planet. This representation is based on mathematical and semantic modeling of spatial data changing dynamically and is realized as the aggregate of distributed electronic cartographic libraries (Electronic Maps Libraries) united in whole system by means of telecommunication technologies.

GOST R 51353-99 establishes the requirements to the composition and content of common, geodetic, gravimetric, photogrammetric and cartographic metadata used for creation (updating) and application of electronic maps. Metadata of electronic maps is data that allows to describe contents, spatial extents, quality (accuracy, complexity, consistency and actuality) and other characteristics of electronic maps.

Classification of metadata and elements of metadata provides separation of metadata into subsets according to their similarity or distinction.

Common metadata corresponds to the top level in metadata hierarchy.

Common metadata must reflect the most common characteristics of spatial geodetic, gravimetric, photogrammetric and cartographic information.

Geodetic and gravimetric metadata characterize systems of geodetic parameters of the Earth, ellipsoids, coordinate and elevation systems, catalogues of geodetic stations and leveling points, formula for normal gravity acceleration, gravimetric systems, catalogues of gravimetric stations, models of gravity field, catalogues of average gravity anomalies. Photogrammetric and cartographic metadata describe materials of survey (including digital images), digital terrain elevation matrices, digital terrain models, digital orthophotos, electronic topographic and thematic maps and orthophotomaps.

The area of GOST R 51353-99 application are as following: information and analytical supply of state authorities, GIS, systems and tools for communication, transportation, navigation, ecological monitoring, order protection, defense. The particular importance is attached to the cooperation and collaboration between governments, governmental and non-governmental organizations, international organizations and institutions in creation of Digital Earth and its application for social, economic, scientific, educational and technological development.

The statements of this Standard are liable to application by all authorities and organizations located in Russian Federation, independently of their form of gov-

ernment and subjection, which are engaged in acquisition, systematization, analysis, processing and transfer of spatial data, creation and application of electronic maps, organization of bases of metadata and digital cartographic and geodetic databanks. This material taken directly from GOST R 51353-99.

SECTION 1 – Administration of Metadata Standard

This section provides a general description of the administrative framework, within which the standard was developed, tested and currently resides. Significant details are also provided in terms of the developmental and managerial history as well as information on the existence of documentation, software tools and training materials. This section also provides a point of contact for further information on the standard.

1.1 Name of Standard

A. Name of standard:
   1. In original language(s):
      Государственный стандарт Российской Федерации ГОСТ Р 51353-99 "Геоинформационное картографирование. Метаданные электронных карт. Состав и содержание"
   2. English translation, if appropriate:
      State Standard of the Russian Federation GOST R 51353-99 "Geoinformatic Mapping. Metadata of the Electronic Maps. Composition and Content"

B. Version/Edition:
   State Standard of the Russian Federation GOST R 51353-99 "Geoinformatic Mapping. Metadata of the Electronic Maps. Composition and Content", November 11, 1999

C. Language(s) of documentation:
   Russian

D. Acronym(s):
   MEM

E. Official ID:
   N/A

F. Name(s) of recognizing standards authority(ies):
   Yu. Alexandrov, V. Filatov, N. Konon,
   A. Martynenko, V. Nepoklonov, A. Serdukov, I. Zemlianov

1.2 Responsible Institutions

Contact information about those institutions that are (or were) responsible for the processes involved in the evolution of the standard:

A. Production and/or development:
   29 Research Institute of the Defense Ministry of Russian Federation
   6 Popov proezd, 107014 Moscow, Russia
   Fax: +7(095) 269-0966

B. Testing:
Moscow State University of Geodesy and Cartography
4 Gorokhovsky per., 103064 Moscow, Russia
Tel.: +7(095) 261-8659

C. Conformance:
Standardization Technical Committee 22 "Information Technologies", R&D Institute "Voskhod"
85 Udaltzova, 117607 Moscow, Russia

D. Maintenance:
Subcomittee 51 "Geoinformation technologies",
Institute of Informatics Problems
Russian Academy of Sciences
44, Korp. 2, Vavilova st., 119333 Moscow, Russia
Tel.: +7(095)137-3494

E. Distribution:
Gosstandard of Russia
9 Leninsky prosp., GSP-1, 117049 Moscow, Russia
Tel.: +7 (095) 236-6208
Fax: +7(095) 236-6231
E-mail: info@gost.ru

F. Help Desk/User Support:
Subcomittee 51 "Geoinformation technologies",
Institute of Informatics Problems
Russian Academy of Sciences
Build. 2, 44, Vavilova st., 119333 Moscow, Russia
Tel.: +7(095)137-3494
E-mail: a_martynenko@mail.ru

1.3 Development History

A. Beginning of work on metadata standard: **1995**
B. Milestones:
1997 – first version of the Standard Project
C. Anticipated completion: **1999**
D. Update cycle: **5 Years**

1.4 Status of the Standard

A. Is the standard officially recognized: **Expected**
B. Date(s) of recognition: **N/A**
C. Anticipated date of recognition: **1999**
D. Current stage in recognition process: **Gosstandard of Russia**
E. Steps still required to achieve recognition: **Conformance**

1.5 Access to Documentation of the Standard

A. Is the standard copyrighted?: **YES**

B. Name and address of copyright owner:
Standardization Technical Committee 22 "Information technologies", Subcommittee 51 "Geoinformation technologies"
C. Do restrictions on the use of the standard apply?: **NO**
D. Is the standard documentation currently available?: **YES**
E. List the form/media of standard documentation:
Paper, CD-ROM
F. Is the standard available on the Internet?: **NO**
G. Price: **N/A**
Will be established after adopting the Standard
H. Contact details for obtaining the standard:
Alexander I. Martynenko
Institute of Informatics Problems
Russian Academy of Sciences
44, 2, Vavilova st.
119333 Moscow, Russia
E-mail: a_martynenko@mail.ru
I. Does the standard have ISBN/ISSN numbers?: **NO**

1.6 Available Software Tools

A. Is software available to implement the standard as a database?: **NO**
B. Is software available to test compliance of metadata with the standard?: **NO**
C. Is software available to produce metadata compliant with the standard?: **YES**
D. Other relevant software: **NO**
E. Is there a test dataset available to test software tools for implementing the standard?: **YES**
F. For each product available above, please list contact information:
Alexander I. Martynenko
Institute of Informatics Problems
Russian Academy of Sciences
44, 2, Vavilova st.
119333 Moscow, Russia
E-mail: a_martynenko@mail.ru

1.7 Available Training Materials

A. Are organized training sessions available?: **YES**
B. Contact details for training sessions:
Alexander I. Martynenko
Institute of Informatics Problems
Russian Academy of Sciences
44, 2, Vavilova st.
119333 Moscow, Russia
E-mail: a_martynenko@mail.ru

C. Training documentation available: **YES**

The Concept of the Electronic Maps System
Alexander Martynenko
Proceedings, 16th ICA International Cartographic Conference
Cologne, 1993, vol. 2, pp. 1160–1163

The development of Concept and Methods for Creating an International System of Geographic Maps as an Universal Basis of the Earth Knowledge
Alexander Martynenko
Proceedings, 17th ICA International Cartographic Conference
Barcelona, 1995, vol. 2, pp. 33–40

Advances in electronic Mapping Systems
Alexander Martynenko
World Aerospace Technology, London
Sterling Publications Ltd., UK, 1996, pp. 81–82

GIS and Geospatial Metadata
Alexander Martynenko
Proceedings, 18th ICA International Cartographic Conference
Stockholm, 1997, vol. 1, pp. 79–82

On Theoretical Grounds of Designing National Standard for Digital and Electronic Maps in Russia
Alexander Martynenko, Sergey Glasov
Proceedings, 18th ICA International Cartographic Conference
Stockholm, 1997, vol. 1, pp. 212–218

Global Geoinformatic Mapping as the Methodology and Technology of the Future
Alexander Martynenko
Proceedings, 19th ICA International Cartographic Conference
Ottawa, 1999, vol. 1, pp. 39–42

Digital Earth based on Metadata Electronic Maps Standard
Alexander Martynenko
Proceedings, 20th ICA International Cartographic Conference
Beijing, 2001, vol. 4, pp. 2747–2752

SECTION 2 – Use and Implementation of the Metadata Standard

This section provides information on the way in which the standard is used and the purpose of its use. It also provides descriptions about the level of abstraction of the standard.

2.1 Intended Use of the Standard
Can the standard be used:

A. Within a transfer data structure?: **YES**
B. To produce hardcopy catalog of datasets?: **YES**
C. To catalog data in a digital on-line catalog?: **YES**
D. Other: **YES**

The standard is used by development and implementation of the Internet server for the electronic libraries of maps and geospatial information metadata and is considered as a new scientific and practical problem raised due to the integration of huge bunk of various information resources (including geospatial information) within Internet. Solution of this problem allows to increase the effectiveness of information supply of fundamental researches in the area of sciences about Earth and to solve the problem of the Digital Earth. It also will promote to form the united geoinformation space and develop its infrastructure. The standard provides decisions for creation of searching Internet server, structure and content of electronic maps and geospatial information metadata, its modeling, encoding, marking, methods and algorithms for metadata indexation and search, and also for implementation of the Internet server as a real working experimental example and intellectual technology.

2.2 Intended Purpose of the Standard

Is the intended purpose of the standard:

A. To support discovery, identification, location of data?: **YES**
B. To support how to access data?: **YES**
C. To support the determination of dataset fitness for use?: **YES**
D. To support transfer of dataset?: **YES**
E. To support an organization's data management?: **NO**
F. Other: **YES**

To support automatic control systems and information system

2.3 Table of Contents

A. Provide the standard's Table of Contents

Contents of the Standard

1 Sphere of application 1
2 References 1
3 Definitions 1
4 Classification of metadata 2
5 Common metadata 2
6 Requirements to the geodetic and gravimetric metadata 2
7 Requirements to the photogrammetric metadata 4
8 Requirements to the cartographic metadata 5

This materials is taken directly from GOST R 51353-9

2.4 Level of Abstraction of the Metadata Standard Definition

A. Does the standard define metadata elements ?: **NO**
B. Does the standard define a model for the relationships between the metadata elements?: **NO**
C. Provide an overview diagram of the model of the metadata, if available. **N/A**
D. Does the standard define domains and/or data types of the metadata elements?: **YES**
E. Does the standard define how the metadata should be encoded?: **YES**
F. Does the standard define a database schema for the automatic generation of a database?: **NO**

2.5 Extensibility of the Standard

A. Is the metadata description extensible (e.g., Can user add metadata domains, elements, thematic profiles, etc.)?: **YES**
   user can define custom types and domains of metadata

2.6 Availability of Metadata
Does the standard specify:

A. How metadata are to be accessed?: **YES**
   both by position (in headers) and by keywords
B. A mechanism for querying the metadata?: **YES**
C. Particular standardized query protocols?: **NO**
D. That the metadata should be encoded for computer access?: **YES**
E. That the metadata should be encrypted for security?: **NO**

2.7 Examples

A. Does the standard provide examples of implementations?: **NO**
B. How many examples are available?: **N/A**

SECTION 3 – Linkage and Coordination of the Metadata Standard

This section provides information on how the metadata standard is linked to other standards, particularly transfer standards, and how such linkage is implemented. If a significant linkage does exist, detailed technical information is provided on the precise nature of the linkage to the standard, and how the linkage affects implementation of the metadata standard.

3.1 Associated Transfer Standards

A. Is the metadata standard associated with a transfer standard(s)?: **YES**
B. If A is yes, provide references and contact details for each linked standard:
   GOST R 50828-95 95 Geoinformatic Mapping. Spatial Data, Digital and Electronic Maps. Common Requirements
   GOST R 52055 – 2003 Geoinformatic Mapping. Spatial Models of Terrain. General Requirements

C. If A is yes, are both (all) standards coordinated?: **YES**

D. If A is yes, describe how the coordination works, both from a technical and conformance (normative) viewpoint: **YES**

The Statements of Standard are liable to application by all authorities and organization located in Russian Federation, which are engaged in acquisition, analysis, processing, and transfer of spatial data, creation and application of electronic maps

3.2 Incorporated into transfer mechanism

A. Is the metadata standard incorporated into a transfer mechanism?: **YES**

B. If A is yes, provide references and contact details for each incorporated mechanism:

GOST R 50828-95 95 Geoinformatic Mapping. Spatial Data, Digital and Electronic Maps. Common Requirements

C. If A is yes, describe how:

GOST R 50828-95 95 Geoinformatic Mapping. Spatial Data, Digital and Electronic Maps. Common Requirements

3.3 Other linked, coordinated or associated standards

A. Is the metadata standard linked, coordinated, or associated with any other standard(s) not listed above?: **YES**

B. If A is yes, provide references and contact details for each standard:

Associated standards:
GOST 34.003-90 Automated Systems. Terms and Definitions.
GOST 22268-76 Geodesy. Terms and Definitions.
GOST 24284-80 Gravitational and Magnetic Exploring. Terms and Definitions.
GOST 23935-79 Aerial Photo Cameras and Photo Survey. Terms and Definitions.
GOST 21002-75 Phototopography. Terms and Definitions.
GOST 21667-76 Cartography. Terms and Definitions.
GOST 28441-90 Digital Mapping. Terms and Definitions.

C. If A is yes, describe how for each relevant standard:

GOST R 51353-99 is referenced to the listed standards

II. Categories the Metadata Standard Uses to Describe Datasets

The following sections are categories of metadata that may be used by the metadata standard to describe data. Not all categories are relevant to different datasets and may not necessarily be present in the standard. The presence or absence of a category should not be used for judging one standard against another.

Note: For each of the questions in the following sections please provide the obligation status information. Valid obligation values are:

Obligation Level

M Mandatory This information must be entered to comply with the metadata standard.
C Conditional This information is entered based on other information in the metadata.
O Optional This information is optionally entered in the metadata.

If the answer is 'NO', the obligation status should be left blank.

SECTION 4 – Dataset Identification

This section examines how the standard provides for the specification of the various sorts of titles and identifiers for a dataset.

4.1 The title(s) of a dataset
Does the standard allow/require the provision for:

Obligation

A. __ Previous title(s)?: **NO**
B. **M** Current title(s)?: **YES**
C. __ Possible future title(s)?: **NO**
D. __ Official titles in different languages?: **NO**
E. __ Abbreviated title(s)?: **NO**
F. __ Alternative Title(s)?: **NO**
G. **M** Does the standard support effective dates of titles?: **YES**
H. __ Other identifiers (e.g., ISBN, National Stock Number, Uniform Product Code)?: **NO**

4.2 Parties responsible for the dataset
Does the standard allow/require the provision for:

Obligation

A. **O** Author?: **YES**
B. **O** Owner?: **YES**
C. **O** Other?: **YES**

4.3 ISBD Information

Obligation

A. __ Does the metadata standard specify inclusion of mandatory ISBD information (see glossary) within the metadata?: **NO**

SECTION 5 – Status of the Dataset

This section examines how the standard provides for the specification of the current status and updating of the data set.

5.1 Status and Progress of Datasets
Does the metadata standard allow/require the provision for:

Obligation

A. **M** Information on the status of the entire dataset?: **YES**
B. __ Information on the status of the particular subset (e.g., observation time, inauguration time, extraction time)?: **NO**

Does the metadata standard specify:

C. **O** A structure for updating or maintenance information?: **YES**
D. __ A structure to describe historical information (see also Section 8.3 G)?: **NO**

5.2 Information on Incremental Updating
Does the metadata standard specify:

Obligation

A. **O** A structure for information on incremental updates of the entire dataset?: **YES**
B. __ A structure for information on incremental updates of particular portions of the dataset?: **NO**
C. __ A structure to describe the update process for the dataset?: **NO**

SECTION 6 – Extent of the Dataset

This section examines how the standard provides for the specification of the spatial and temporal extent of the dataset.

6.1 The spatial extent of the dataset

Obligation

A. **M** Does the metadata standard allow/require for the description of horizontal extent (e.g., bounding rectangle/polygon, named location (i.e., Stockholm, etc.))?: **YES**
B. **M** Does the metadata standard allow/require for the description of vertical extent (e.g., minima/maxima heights, etc.)?: **YES**
C. __ Does the standard require specific units for describing extent?: **NO**
D. __ Does the standard allow specification of the units for describing extent?: **NO**
E. __ Does the standard require specific spatial reference system for describing extent?: **NO**
F. __ Does the standard allow specification of the spatial reference system for describing extent?: **NO**

6.2 The temporal extent of the dataset

Obligation

A. **C** Does the standard allow/require for the description of time period for which the dataset is relevant?: **YES**

B. __ Does the standard allow/require specification of the temporal reference system?: **NO**

SECTION 7 – Data Content of the Dataset

This section examines how the standard provides for the specification of the data content of the dataset in terms of overview elements, themes, data definitions and product specifications.

7.1 Dataset overview

Does the standard allow/require for the provision of the following information:

Obligation

A. **M** Dataset overview?: **YES**
B. __ Abstract – brief summary of the content of the dataset?: **NO**
C. __ Purpose – summary of the intentions with which the dataset was developed?: **NO**
D. __ Usage – list of activities for which the dataset has been used? (see also Section 8.5): **NO**
E. **M** Spatial schema – (e.g., vector, raster)? (see also Section 9): **YES**
F. **M** Spatial reference system? (see also Section 12): **YES**
G. **O** Language?: **YES**
H. __ Character set (e.g., ISO8859-1, JISX0208)?: **NO**
I. __ List of documents providing further information about the dataset?: **NO**
J. __ Sample/preview dataset?: **NO**
K. **C** Related datasets?: **YES**
L. **M** Original size of the dataset?: **YES**
M. __ Compression? (see also Section 11.5): **NO**
N. __ Other?: **NO**

7.2 Theme(s)

Obligation

A. **O** Does the standard allow/require the provision for the description of theme(s) (e.g., topography, land cover, transport, etc.) in the dataset?: **YES**
B. **C** If A is yes, does the standard allow the referencing of an outside dictionary or thesaurus?: **YES**
C. **C** If A is yes, does the standard itself include a dictionary or thesaurus within the metadata?: **YES**
D. **O** If A is yes, does the standard allow the use of unauthorized user-defined keywords?: **YES**

7.3 Data Definition

Obligation

A. **O** Does the standard allow/require the provision for the description of the data definition?: **YES**

B. If A is yes, does the standard allow/require the provision for the description of:

O 1. Features/Objects?: **YES**

O 2. Attributes?: **YES**

O 3. Relations?: **YES**

C. **C** Can attributes be described separately from feature/object classes?: **YES**

D. **C** Does the standard allow/require the specification of the domain of the attributes?: **YES**

E. __ Does the standard allow/require the specification of data type of the attributes?: **NO**

F. __ Does the standard allow/require the specification of particular classification coding schemes (e.g., FACC, OSKA, S-57 OC)?: **NO**

7.4 Product Specification

Obligation

A. __ Does the standard allow/require for the provision of information about whether the dataset meets a particular product specification?: **NO**

B. __ If A is yes, can it be internal to the Metadata?: **NO**

C. __ If A is yes, can it be external to the Metadata?: **NO**

SECTION 8 – Data Quality of the Dataset

This section examines how the standard deals with the question of data quality. This includes homogeneity and metrics for data quality, quality assurance, and usage.

8.1 Overall Data Quality Statement

Obligation

A. **M** Does the standard allow/require for the provision of quality to be described? If yes, please answer the questions below. (*If no, all following answers in Section* 8 *are N/A, please skip to Section* 9.): **YES**

B. **O** Does the standard allow/require the provision for the quality information referencing an external quality standard, quality assurance standard, or specification?: **YES**

8.2 Homogeneity of Data Quality

Obligation

A. **O** Does the standard allow/require for the variation of the quality of the dataset to be described?: **YES**

B. What type of aggregation of quality information is allowed? **YES**

C 1. Entire database?: **YES**

O 2. Feature/Object instance?: **YES**

O 3. Feature/Object class?: **YES**

O 4. Attribute instance?:	**YES**
O 5. Attribute class?:	**YES**
O 6. Layer?:	**YES**
O 7. Theme?:	**YES**
O 8. Geographic extent?:	**YES**
__ 9. Other?:	**NO**

8.3 Metrics for Data Quality

Does the standard allow/require the provision of values for:

Obligation

A. **M** Positional accuracy?:	**YES**
1. Standard error?:	**YES**
2. Maximum error?:	**YES**
3. Error correlation?:	**YES**
4. Confidence level (e.g., 90%, 95%)?:	**YES**
5. Other?:	**NO**
B. **M** Attribute accuracy?:	**YES**
1. Standard error?:	**YES**
2. Maximum error?:	**YES**
3. Error frequency?:	**YES**
4. Error correlation?:	**YES**
5. Confidence level (e.g., 90%, 95%)?:	**YES**
6. Other?:	**NO**
C. __ Temporal accuracy?:	**NO**
1. Standard error?:	**N/A**
2. Maximum error?:	**N/A**
3. Error correlation?:	**N/A**
4. Confidence level? (e.g., 90%, 95%):	**N/A**
5. Other?:	**N/A**
D. **M** Completeness?:	**YES**
1. Error of omission?:	**YES**
2. Error of commission?:	**YES**
3. Other?:	**NO**
E. **C** Currentness?:	**YES**
1. Date?:	**YES**
2. Maximum age?:	**YES**
3. Temporal extent?:	**YES**
4. Other?:	**NO**
F. **C** Logical consistency?:	**YES**
G. **O** Lineage?:	**YES**
1. List of processing steps?:	**NO**
2. List of values of processing parameters?:	**YES**
3. List of printouts of statistics, etc.?:	**NO**
4. Source material?:	**YES**

5. Other?:	**NO**
H. **C** Additional Quality Information?:	**YES**

8.4 Quality Assurance Methods

Does the standard allow/require the provision of information about the methods used to determine:

Obligation

A. **M** Positional accuracy?:	**YES**
1. Standard error in adjustments?:	**YES**
2. Repeated measurements?:	**NO**
3. Independent measurements?:	**YES**
4. Subjective evaluations?:	**NO**
5. Other?:	**NO**
B. **M** Attribute accuracy?:	**YES**
1. Standard error?:	**YES**
2. Error classification matrix?:	**NO**
3. Independent measurements?:	**YES**
4. Subjective evaluation?:	**NO**
5. Other?:	**NO**
C. **C** Temporal accuracy?:	**YES**
1. Independent measurements?:	**YES**
2. Subjective evaluation?:	**NO**
3. Other?:	**NO**
D. **O** Completeness?:	**YES**
1. Field checks?:	**YES**
2. Repeated measurements?:	**YES**
3. Independent measurements?:	**YES**
4. Subjective evaluation?:	**YES**
5. Error of commission?:	**YES**
6. Other?:	**NO**
E. **O** Currentness?:	**YES**
1. Date of source material?:	**YES**
2. Repeated measurements?:	**YES**
3. Independent measurements?:	**YES**
4. Subjective evaluation?:	**YES**
5. Other?:	**NO**
F. **O** Logical consistency?:	**YES**
1. Verification results of consistency checks?:	**YES**
2. Repeated processing?:	**YES**
3. Comparison with independent sources?:	**YES**
4. Subjective evaluation?:	**YES**
5. Other?:	**NO**
G. **O** Lineage?:	**YES**
1. Checking results of processing steps:	**YES**

2. Subjective evaluation: **YES**
3. Other: **NO**

H. __ Additional Quality Information?: **NO**

8.5 Usage

Obligation

A. **O** Does the metadata standard allow/require the provision of information about usage?: **YES**
1. Who used it: **YES**
2. When it was used: **YES**
3. For what application it was used: **YES**
4. The effectiveness of use: **YES**
5. Other: **NO**

8.6 Alternative Quality Description: **NO**

SECTION 9 – Spatial Data Structure of the Dataset

This section examines how the standard provides for the specification of the spatial data model and spatial data types.

9.1 Spatial Data Model
Does the standard allow/require the provision for:

Obligation

A. __ The description of the geometrical and/or topological primitives used to model the real world?: **NO**

B. **C** Information about an external standard for geometrical and/or topological primitive specification?: **YES**

9.2 Spatial Data Types
Does the standard allow/require the provision for:

Obligation

A. **M** Information about the spatial data type(s) used in the dataset?: **YES**

B. **M** If A is yes, which of the following generic spatial data types are supported? **YES**
For each, please list specific data primitive levels supported by each data type.
1. Vector (e.g., spaghetti, chain-node/node-edge-face, planar graph, full topology)?: **YES**
   spaghetti, incremental vector, chain-node topology
2. Areal Raster (e.g., thematic raster – scanned and classification maps, image raster – satellite images, scanned aerial photographs)?: **YES**
   raster maps, space and aerial photo (ortho) images

3. Grid/Matrix (e.g., Digital Elevation Model)?: **YES**
a variety of matrix-based models of special intent
4. Triangular Irregular Network (TIN)?: **NO**
5. Other?: **NO**

9.3 Other Data Types
Does the standard allow the provision of information about:

Obligation

A. **O** Video?: **YES**
B. **O** Sound?: **YES**
C. __ Other?: **NO**

SECTION 10 – Spatial Reference of the Dataset

This section examines how the dataset provides for the specification of the spatial reference characteristics of the data set.

Does the standard allow/require the provision for the following information:

10.1 Geoidal Model

Obligation

A. **C** Geoidal model name?: **YES**
B. **C** Geoidal model definitions?: **YES**

10.2 Geodetic datum

Obligation

A. **M** Geodetic datum name?: **YES**
B. **C** Geodetic datum definitions?: **YES**

10.3 Reference Ellipsoid

Obligation

A. **C** Reference ellipsoid name?: **YES**
B. **C** Reference ellipsoid definitions?: **YES**

10.4 Projection

Obligation

A. **M** Projection name?: **YES**
B. **O** Projection definitions?: **YES**

10.5 Coordinate Systems

Obligation

A. **M** Coordinate system type?: **YES**
B. **O** Coordinate system definitions?: **YES**

10.6 Height Reference System

Obligation

A. **M** Vertical datum name?: **YES**
B. **O** Types of heights?: **YES**

10.7 Ancillary Geodetic Information

Obligation

A. **C** Does the standard allow/require the provision for ancillary geodetic, gravimetric, or magnetometric information?: **YES**

10.8 Transformation Parameters
Does the standard specify a method to provide information for coordinate transformation(s):

Obligation

A. __ Model name?: **NO**
B. __ Control points?: **NO**
C. __ Transformation parameters?: **NO**

10.9 Non-Coordinate Spatial Reference
Does the standard allow/require the provision for information on non-coordinate spatial reference systems (e.g., postal systems, street address)?:

Obligation

A. __ Reference system name?: **NO**
B. __ Reference system definitions?: **NO**
C. __ Other?: **NO**

SECTION 11 – Availability and Distribution of the Dataset

This section examines how the standard provides information on the availability of the dataset?

Does the standard allow/require the provision for:

11.1 The Distributor(s) of the Dataset

Obligation

A. **O** Information about the distributor(s) of the dataset?: **YES**
B. **O** Contact information for the distributor(s)?: **YES**

11.2 Restrictions to Access and Usage

Obligation

A. **O** Information about copyright on the dataset (e.g., date, period, copyright owner, etc.)?: **YES**
B. **O** Information about restrictions on the access of the dataset?: **YES**
C. **C** Information about restrictions on the usage of the dataset?: **YES**

11.3 Pricing Details

Obligation

A. **O** Information about pricing?: **YES**

11.4 Distribution of the Dataset

Obligation

A. **O** Information about distribution media?: **YES**
B. **O** Information about on-line access?: **YES**
C. __ Information about units of distribution (e.g., tiles, layers, polygons, elements)?: **NO**

11.5 Distribution Format

Obligation

A. **C** Information about distribution formats (e.g., commercial CAD format, commercial GIS formats, National/International transfer standard)?: **YES**
B. **M** Information about the encoding of the dataset (e.g., binary, ASCII)?: **YES**
C. __ Information about compression method?: **NO**
D. __ Information about compressed dataset size?: **NO**
E. **O** Information about the distribution mechanism?: **YES**
F. __ Information about the encapsulation of data (e.g., ISO 8211, ISO 8824)?: **NO**

11.6 Ordering Procedures

Obligation

A. __ Information about ordering procedures for the dataset?: **NO**

11.7 Support Services

Obligation

A. __ Information about support services (e.g., additional data transformations, format or coordinate conversions)?: **NO**

SECTION 12 – Authorization and Verification of the Metadata

This section examines how the standard provides for the description of the authorization and verification on the metadata of a dataset (e.g., how is the metadata itself authorized and verified?).

12.1 Metadata reference

Obligation

A. **C** Does the standard allow/require the provision of information about reference dates of the metadata (e.g., last check date, last update date, review date)?: **YES**

12.2 Verification Authority

Obligation

A. **O** Does the standard allow/require the provision for information about who verified/authorized the metadata?: **YES**
B. __ Does the standard specify the method for providing this information (e.g., ISO 9000): **NO**

12.3 Statistical Methods Used for Verification

Obligation

A. **C** Does the standard specify the statistical methods to be used for verification of the metadata?: **YES**
B. **C** Does the standard allow/require the provision for the description of the statistical methods used to verify the metadata?: **YES**

END OF CHARACTERISTICS
FIN

H. Moellering, H.J.G.L. Aalders & A. Crane (Editors)

South Africa
Content Standard for Digital Geospatial Data

Nic Scheepers, Antony Cooper

I. Information about the Metadata Standard Itself

SECTION 0 – Assessment Information

This section provides the user with a general description of the scope and intended use of the standard. Information on the respondent who prepared the assessment is also provided.

0.1 Respondent

Assessment prepared by:

Nic Scheepers, Head: GIS and IT Support
Municipal Demarcation Board
Private Bag X28, Hatfield, 0028, South Africa
Fax: +27 12 342 2480
Phone: +27 12 342 2481
E-mail: nic@demarcation.org.za

Antony Cooper, Divisional Fellow
Centre for Logistics and Decision Support, CSIR
PO Box 395, Pretoria, 0001, South Africa
Fax: +27 12 841 3037
Phone: +27 12 841 4121
E-mail: acooper@csir.co.za

Date assessment completed:

June 2000; updated May 2002, Reviewed Spring, 2005

Relationship of respondent to standard:

Custodian and User

0.2 Brief Summary Statement

A. Summary of the scope and intent of the standard:
Metadata or "data about data" describe the content, quality, condition and other

characteristics of spatial data. The National Spatial Information Framework currentlyadheres to the Federal Geographic Data Committee's (FGDC) Metadata Content Standard and will migrate towards the ISO standard on metadata as soon as the content standard is approved. Visit the NSIF's Spatial Data Discovery Facility (SDDF) to learn more about how metadata is used in Internet-based search engines to discover specific GIS data sets: http://www.nsif.org.za/sddf.html.

SECTION 1 – Administration of Metadata Standard

This section provides a general description of the administrative framework, within which the standard was developed, tested and currently resides. Significant details are also provided in terms of the developmental and managerial history as well as information on the existence of documentation, software tools and training materials. This section also provides a point of contact for further information on the standard.

1.1 Name of Standard

A. Name of standard:
   1. In original language(s):
      Content Standard for Digital Geospatial Data
   2. English translation, if appropriate:

B. Version/Edition:
   Version 2, 1998

C. Language(s) of documentation:
   English

D. Acronym(s):
   CSDGM

E. Official ID:
   N/A

F. Name(s) of recognizing standards authority(ies):
   National Spatial Information Framework and the South African Bureau of Standards

1.2 Responsible Institutions

Contact information about those institutions that are (or were) responsible for the processes involved in the evolution of the standard:

A. Production and/or development:
   National Spatial Information Framework, Private Bag X954
   Pretoria, 0001, South Africa
   Fax: +27 12 322 5418
   Phone: +27 12 322 5400
   E-mail: nsif@csg.pwv.gov.za

B. Testing:
National Spatial Information Framework, Private Bag X954
Pretoria, 0001, South Africa
Fax: +27 12 322 5418
Phone: +27 12 322 5400
E-mail: nsif@csg.pwv.gov.za

C. Conformance:
National Spatial Information Framework, Private Bag X954
Pretoria, 0001, South Africa
Fax: +27 12 322 5418
Phone: +27 12 322 5400
E-mail: nsif@csg.pwv.gov.za

D. Maintenance:
National Spatial Information Framework, Private Bag X954
Pretoria, 0001, South Africa
Fax: +27 12 322 5418
Phone: +27 12 322 5400
E-mail: nsif@csg.pwv.gov.za

E. Distribution:
National Spatial Information Framework, Private Bag X954
Pretoria, 0001, South Africa
Fax: +27 12 322 5418
Phone: +27 12 322 5400
E-mail: nsif@csg.pwv.gov.za

F. Help Desk/User Support:
National Spatial Information Framework, Private Bag X954
Pretoria, 0001, South Africa
Fax: +27 12 322 5418
Phone: +27 12 322 5400
E-mail: nsif@csg.pwv.gov.za

1.3 Development History

A. Beginning of work on metadata standard: **September 1997**
Milestones:
Online: June 1998
Introduction of Map Interface within Clearinghouse: February 1999
Adoption of Version 1 of the CSDGM: June 1998
Adoption of Version 2 of the CSDGM: January 1999

B. Anticipated completion: **June 1998**

C. Update cycle:
No formal update cycle – updating is an on-going process

1.4 Status of the Standard

A. Is the standard officially recognized: **Expected**

B. Date(s) of recognition: **N/A**
C. Anticipated date of recognition: **2002**
D. Current stage in recognition process:
Awaiting publication of ISO Metadata Standard
E. Steps still required to achieve recognition:
- **Compile South African profile of ISO Metadata Standard**
- **Solicit comments from South African geospatial community**
- **Finalise profile**
- **Submit South African Metadata Profile to the South African Bureau of Standards for formal approval and recognition**

1.5 Access to Documentation of the Standard

A. Is the standard copyrighted?: **NO**
B. Name and address of copyright owner: **N/A**
C. Do restrictions on the use of the standard apply?: **NO**
D. Is the standard documentation currently available?: **YES**
E. List the form/media of standard documentation:
Web Site
F. Is the standard available on the Internet?: **YES**
http://www.nsif.org.za
G. Price: **Free**
H. Contact details for obtaining the standard:
National Spatial Information Framework, Private Bag X954
Pretoria, 0001, South Africa
Fax: +27 12 322 5418
Phone: +27 12 322 5400
E-mail: nsif@csg.pwv.gov.za
I. Does the standard have ISBN/ISSN numbers?: **NO**

1.6 Available Software Tools

A. Is software available to implement the standard as a database?: **YES**
MetaLite, Webentry, Spatial Metadata Management System
B. Is software available to test compliance of metadata with the standard?: **YES**
Metadata Parser (MP)
C. Is software available to produce metadata compliant with the standard?: **YES**
MetaLite, Webentry, Spatial Metadata Management System
D. Other relevant software:
Isite Console, MapGuide
E. Is there a test dataset available to test software tools for implementing the standard?: **NO**

F. For each product available above, please list contact information:
MetaLite and _Metadata Parser_: United States Geological Survey (USGS) and the United Nations Environmental Programme (UNEP) – information available at the following
URL: http://edcnts2.cr.usgs.gov/metalite/
E-mail: webmaster@edcintl.cr.usgs.gov

Webentry: National Spatial Information Framework (NSIF) – information available at the following
URL: http://www.nsif.org.za
Tel: +27 12 322 5400
Fax: +27 12 322 5418
E-mail: nsif@csg.pwv.gov.za

Spatial Metadata Management System (SMMS): Enabling Technology – Information available at the following
URL: http://www.enabletech.com. The ET offices can be contacted
at (509) 375-7723
Fax number is (509) 375-1154
General information email can be directed to info@enabletech.com

1.7 Available Training Materials

A. Are organized training sessions available?: **YES**
B. Contact details for training sessions:
National Spatial Information Framework, Private Bag X954
Pretoria, 0001, South Africa
Fax: +27 12 322 5418
Phone: +27 12 322 5400
E-mail: nsif@csg.pwv.gov.za
C. Training documentation available:
Title: _Geospatial Metadata Cookbook_, Author: Nic Scheepers
Source: FGDC Home Page at http://www.fgdc.gov
AUSLIG Home Page: http://www.auslig.gov.au
Date: 01-02-2000
National Spatial Information Framework
Home Page: http://www.nsif.org.za – use 'Metadata' hyperlink

SECTION 2 – Use and Implementation of the Metadata Standard

This section provides information on the way in which the standard is used and the purpose of its use. It also provides descriptions about the level of abstraction of the standard.

2.1 Intended Use of the Standard
Can the standard be used:

A. Within a transfer data structure?: **NO**
B. To produce hardcopy catalog of datasets?: **YES**
C. To catalog data in a digital on-line catalog?: **YES**
D. Other: **YES**
Yes, eliminate duplication of expensive data capture

2.2 Intended Purpose of the Standard
Is the intended purpose of the standard:

A. To support discovery, identification, location of data?: **YES**
B. To support how to access data?: **YES**
C. To support the determination of dataset fitness for use?: **YES**
D. To support transfer of dataset?: **YES**
E. To support an organization's data management?: **YES**
F. Other: **YES**
To maintain an organisation's internal investment in geospatial data;
To provide information about an organisation's data holdings and catalogues;
To provide information needed to process and interpret data to be received through a transfer from an external source;
To provide access to disparate geospatial data sets, through an Internet based metadata search facility, which were previously not possible;
To eliminate duplication in the costly and time consuming capture of geospatial data sets.

2.3 Table of Contents

A. Provide the standard's Table of Contents
1. **Identification Information**
2. **Data Quality Information**
3. **Spatial Data Organisation**
4. **Spatial Reference Information**
5. **Entity and Attribute Information**
6. **Distribution Information**
7. **Metadata Reference Information**

2.4 Level of Abstraction of the Metadata Standard Definition

A. Does the standard define metadata elements ?: **YES**
B. Does the standard define a model for the relationships between the metadata elements?: **NO**
C. Provide an overview diagram of the model of the metadata, if available. **N/A**
D. Does the standard define domains and/or data types of the metadata elements?: **YES**
E. Does the standard define how the metadata should be encoded?: **YES**

F. Does the standard define a database schema for the automatic generation of a database?: **YES**

2.5 Extensibility of the Standard

A. Is the metadata description extensible (e.g., Can user add metadata domains, elements, thematic profiles, etc.)?: **NO**

2.6 Availability of Metadata
Does the standard specify:

A. How metadata are to be accessed?: **NO**
B. A mechanism for querying the metadata?: **NO**
C. Particular standardized query protocols?: **NO**
D. That the metadata should be encoded for computer access?: **NO**
E. That the metadata should be encrypted for security?: **NO**

2.7 Examples

A. Does the standard provide examples of implementations?: **NO**
B. How many examples are available?: **0**

SECTION 3 – Linkage and Coordination of the Metadata Standard

This section provides information on how the metadata standard is linked to other standards, particularly transfer standards, and how such linkage is implemented. If a significant linkage does exist, detailed technical information is provided on the precise nature of the linkage to the standard, and how the linkage affects implementation of the metadata standard.

3.1 Associated Transfer Standards

A. Is the metadata standard associated with a transfer standard(s)?: **NO**
B. If A is yes, provide references and contact details for each linked standard: **N/A**
C. If A is yes, are both (all) standards coordinated?: **N/A**
D. If A is yes, describe how the coordination works, both from a technical and conformance (normative) viewpoint: **N/A**

3.2 Incorporated into transfer mechanism

A. Is the metadata standard incorporated into a transfer mechanism?: **NO**
B. If A is yes, provide references and contact details for each incorporated mechanism: **N/A**
C. If A is yes, describe how: **N/A**

3.3 Other linked, coordinated or associated standards

A. Is the metadata standard linked, coordinated, or associated with any other standard(s) not listed above?: **NO**
B. If A is yes, provide references and contact details for each standard: **N/A**
C. If A is yes, describe how for each relevant standard: **N/A**

II. Categories the Metadata Standard Uses to Describe Datasets

The following sections are categories of metadata that may be used by the metadata standard to describe data. Not all categories are relevant to different datasets and may not necessarily be present in the standard. The presence or absence of a category should not be used for judging one standard against another.

Note: For each of the questions in the following sections please provide the obligation status information. Valid obligation values are:

	Obligation Level	
M	Mandatory	This information must be entered to comply with the metadata standard.
C	Conditional	This information is entered based on other information in the metadata.
O	Optional	This information is optionally entered in the metadata.

If the answer is 'NO', the obligation status should be left blank.

SECTION 4 – Dataset Identification

This section examines how the standard provides for the specification of the various sorts of titles and identifiers for a dataset.

4.1 The title(s) of a dataset
Does the standard allow/require the provision for:

Obligation

A. __ Previous title(s)?: **NO**
B. **M** Current title(s)?: **YES**
C. __ Possible future title(s)?: **NO**
D. __ Official titles in different languages?: **NO**
E. __ Abbreviated title(s)?: **NO**
F. **M** Alternative Title(s)?: **YES**
G. **M** Does the standard support effective dates of titles?: **YES**
H. __ Other identifiers (e.g., ISBN, National Stock Number, Uniform Product Code)?: **NO**

4.2 Parties responsible for the dataset
Does the standard allow/require the provision for:

Obligation

A. **M** Author?: **YES**
B. **M** Owner?: **YES**
C. __ Other?: **NO**

4.3 ISBD Information

Obligation

A. __ Does the metadata standard specify inclusion of mandatory ISBD information (see glossary) within the metadata?: **NO**

SECTION 5 – Status of the Dataset

This section examines how the standard provides for the specification of the current status and updating of the data set.

5.1 Status and Progress of Datasets
Does the metadata standard allow/require the provision for:

Obligation

A. **M** Information on the status of the entire dataset?: **YES**
Currency and progress
B. __ Information on the status of the particular subset (e.g., observation time, inauguration time, extraction time)?: **NO**

Does the metadata standard specify:

C. **M** A structure for updating or maintenance information?: **YES**
D. **M** A structure to describe historical information (see also Section 8.3 G)?: **YES**

5.2 Information on Incremental Updating
Does the metadata standard specify:

Obligation

A. __ A structure for information on incremental updates of the entire dataset?: **NO**
B. __ A structure for information on incremental updates of particular portions of the dataset?: **NO**
C. **M** A structure to describe the update process for the dataset?: **YES**
Maintenance and Update Frequency

SECTION 6 – Extent of the Dataset

This section examines how the standard provides for the specification of the spatial and temporal extent of the dataset.

6.1 The spatial extent of the dataset

Obligation

A. **M** Does the metadata standard allow/require for the description of horizontal extent (e.g., bounding rectangle/polygon, named location (i.e., Stockholm, etc.))?: **YES**
Bounding Coordinates (North, South, East, West), Place Keyword (place name), Place
Keyword Thesaurus (alternative place names)

B. __ Does the metadata standard allow/require for the description of vertical extent (e.g., minima/maxima heights, etc.)?: **NO**
C. **M** Does the standard require specific units for describing extent?: **YES**
D. __ Does the standard allow specification of the units for describing extent?: **NO**
E. **M** Does the standard require specific spatial reference system for describing extent?: **YES**
F. __ Does the standard allow specification of the spatial reference system for describing extent?: **NO**

6.2 The temporal extent of the dataset

Obligation

A. **M** Does the standard allow/require for the description of time period for which the dataset is relevant?: **YES**
B. __ Does the standard allow/require specification of the temporal reference system?: **NO**

SECTION 7 – Data Content of the Dataset

This section examines how the standard provides for the specification of the data content of the dataset in terms of overview elements, themes, data definitions and product specifications.

7.1 Dataset overview
Does the standard allow/require for the provision of the following information:

Obligation

A. __ Dataset overview?: **NO**
B. **M** Abstract – brief summary of the content of the dataset?: **YES**
C. **M** Purpose – summary of the intentions with which the dataset was developed?: **YES**
D. __ Usage – list of activities for which the dataset has been used? (see also Section 8.5): **NO**
E. **M** Spatial schema – (e.g., vector, raster)? (see also Section 9): **YES**
F. **O** Spatial reference system? (see also Section 12): **YES**
G. __ Language?: **NO**
H. __ Character set (e.g., ISO8859-1, JISX0208)?: **NO**
I. **O** List of documents providing further information about the dataset?: **YES**
J. **O** Sample/preview dataset?: **YES**
K. __ Related datasets?: **NO**
L. **C** Original size of the dataset?: **YES**
M. **C** Compression? (see also Section 11.5): **YES**
N. __ Other?: **NO**

7.2 Theme(s)

Obligation

A. **M** Does the standard allow/require the provision for the description of theme(s) (e.g., topography, land cover, transport, etc.) in the dataset?: **YES**

B. __ If A is yes, does the standard allow the referencing of an outside dictionary or thesaurus?: **NO**

C. __ If A is yes, does the standard itself include a dictionary or thesaurus within the metadata?: **NO**

D. **M** If A is yes, does the standard allow the use of unauthorized user-defined keywords?: **YES**

7.3 Data Definition

Obligation

A. **O** Does the standard allow/require the provision for the description of the data definition?: **YES**

B. If A is yes, does the standard allow/require the provision for the description of:
   O 1. Features/Objects?: **YES**
   O 2. Attributes?: **YES**
   __ 3. Relations?: **NO**

C. **O** Can attributes be described separately from feature/object classes?: **YES**

D. **O** Does the standard allow/require the specification of the domain of the attributes?: **YES**

E. **O** Does the standard allow/require the specification of data type of the attributes?: **YES**

F. __ Does the standard allow/require the specification of particular classification coding schemes (e.g., FACC, OSKA, S-57 OC)?: **NO**

7.4 Product Specification

Obligation

A. __ Does the standard allow/require for the provision of information about whether the dataset meets a particular product specification?: **NO**

B. __ If A is yes, can it be internal to the Metadata?: **N/A**

C. __ If A is yes, can it be external to the Metadata?: **N/A**

SECTION 8 – Data Quality of the Dataset

This section examines how the standard deals with the question of data quality. This includes homogeneity and metrics for data quality, quality assurance, and usage.

8.1 Overall Data Quality Statement

Obligation

A. **O** Does the standard allow/require for the provision of quality to be described? If yes, please answer the questions below. (*If no, all following answers in Section* 8 *are N/A, please skip to Section* 9.): **YES**

B. __ Does the standard allow/require the provision for the quality information referencing an external quality standard, quality assurance standard, or specification?: **NO**

8.2 Homogeneity of Data Quality

Obligation

A. __ Does the standard allow/require for the variation of the quality of the dataset to be described?: **NO**

B. What type of aggregation of quality information is allowed?

O 1. Entire database?: **YES**
__ 2. Feature/Object instance?: **NO**
__ 3. Feature/Object class?: **NO**
__ 4. Attribute instance?: **NO**
__ 5. Attribute class?: **NO**
__ 6. Layer?: **NO**
__ 7. Theme?: **NO**
__ 8. Geographic extent?: **NO**
__ 9. Other?: **NO**

8.3 Metrics for Data Quality
Does the standard allow/require the provision of values for:

Obligation

A. **O** Positional accuracy?: **YES**
1. Standard error?: **YES**
2. Maximum error?: **YES**
3. Error correlation?: **YES**
4. Confidence level (e.g., 90%, 95%)?: **YES**
Other?: **YES**
Accuracy Report, Quantitative Accuracy Assessment

B. **O** Attribute accuracy?: **YES**
1. Standard error?: **YES**
2. Maximum error?: **YES**
3. Error frequency?: **YES**
4. Error correlation?: **YES**
5. Confidence level (e.g., 90%, 95%)?: **YES**
6. Other?: **YES**
Attribute Accuracy Report, Quantitative Accuracy Assessment

C. __ Temporal accuracy?: **NO**
1. Standard error?: **NO**
2. Maximum error?: **NO**
3. Error correlation?: **NO**
4. Confidence level? (e.g., 90%, 95%): **NO**
5. Other?: **NO**

D. **O** Completeness?: **YES**

1. Error of omission?: **NO**
2. Error of commission?: **NO**
3. Other? **YES**
   Free text (General report)

E. **M** Currentness?: **YES**
1. Date?: **YES**
2. Maximum age?: **YES**
3. Temporal extent?: **YES**
4. Other?: **YES**
   Currentness Reference

F. **O** Logical consistency?: **YES**
   Free text

G. **O** Lineage?: **YES**
1. List of processing steps?: **YES**
2. List of values of processing parameters?: **NO**
3. List of printouts of statistics, etc.?: **NO**
4. Source material?: **YES**
5. Other?: **YES**
   Source Citations, Process Contact, Source Scale Denominator, Source Time Period of Content, Source Contribution, Process Date

H. __ Additional Quality Information?: **NO**

8.4 Quality Assurance Methods

Does the standard allow/require the provision of information about the methods used to determine:

Obligation

A. **O** Positional accuracy?: **YES**
1. Standard error in adjustments?: **YES**
2. Repeated measurements?: **YES**
3. Independent measurements?: **YES**
4. Subjective evaluations?: **YES**
5. Other?: **YES**
   Free text

B. **O** Attribute accuracy?: **YES**
1. Standard error?: **YES**
2. Error classification matrix?: **YES**
3. Independent measurements?: **YES**
4. Subjective evaluation?: **YES**
5. Other?: **YES**
   Free text

C. __ Temporal accuracy?: **NO**
1. Independent measurements?: **NO**
2. Subjective evaluation?: **NO**
3. Other?: **NO**

D. **O** Completeness?: **YES**
1. Field checks?: **YES**
2. Repeated measurements?: **YES**
3. Independent measurements?: **YES**
4. Subjective evaluation?: **YES**
5. Error of commission?: **YES**
6. Other?: **YES**
Free text

E. **O** Currentness?: **YES**
1. Date of source material?: **YES**
2. Repeated measurements?: **YES**
3. Independent measurements?: **YES**
4. Subjective evaluation?: **YES**
5. Other?: **YES**
Free text

F. **O** Logical consistency?: **YES**
1. Verification results of consistency checks?: **YES**
2. Repeated processing?: **YES**
3. Comparison with independent sources?: **YES**
4. Subjective evaluation?: **YES**
5. Other?: **YES**
Free text

G. **O** Lineage?: **YES**
1. Checking results of processing steps: **YES**
2. Subjective evaluation: **YES**
3. Other: **YES**
Free text

H. __ Additional Quality Information?: **NO**

8.5 Usage

Obligation

A. __ Does the metadata standard allow/require the provision of information about usage?: **NO**
1. Who used it: **NO**
2. When it was used: **NO**
3. For what application it was used: **NO**
4. The effectiveness of use: **NO**
5. Other: **NO**

8.6 Alternative Quality Description: **NO**

SECTION 9 – Spatial Data Structure of the Dataset

This section examines how the standard provides for the specification of the spatial data model and spatial data types.

9.1 Spatial Data Model

Does the standard allow/require the provision for:

Obligation

A. __ The description of the geometrical and/or topological primitives used to model the real world?: **NO**

B. __ Information about an external standard for geometrical and/or topological primitive specification?: **NO**

9.2 Spatial Data Types

Does the standard allow/require the provision for:

Obligation

A. **M** Information about the spatial data type(s) used in the dataset?: **YES**

B. **M** If A is yes, which of the following generic spatial data types are supported?
For each, please list specific data primitive levels supported by each data type.

1. Vector (e.g., spaghetti, chain-node/node-edge-face, planar graph, full topology)?: **YES**

undifferentiated

Areal Raster (e.g., thematic raster – scanned and classification maps, image raster – satellite images, scanned aerial photographs)?: **YES**

undifferentiated

3. Grid/Matrix (e.g., Digital Elevation Model)?: **NO**

Triangular Irregular Network (TIN)?: **NO**

Other?: **NO**

9.3 Other Data Types

Does the standard allow the provision of information about:

Obligation

A. __ Video?: **NO**

B. __ Sound?: **NO**

C. __ Other?: **NO**

SECTION 10 – Spatial Reference of the Dataset

This section examines how the dataset provides for the specification of the spatial reference characteristics of the data set.

Does the standard allow/require the provision for the following information:

10.1 Geoidal Model

Obligation

A. **O** Geoidal model name?: **YES**

B. __ Geoidal model definitions?: **NO**

10.2 Geodetic datum

Obligation

A. **O** Geodetic datum name?: **YES**
B. __ Geodetic datum definitions?: **NO**

10.3 Reference Ellipsoid

Obligation

A. **O** Reference ellipsoid name?: **YES**
B. **O** Reference ellipsoid definitions?: **YES**

10.4 Projection

Obligation

A. **C** Projection name?: **YES**
B. **C** Projection definitions?: **YES**

10.5 Coordinate Systems

Obligation

A. **C** Coordinate system type?: **YES**
B. **C** Coordinate system definitions?: **YES**

10.6 Height Reference System

Obligation

A. **C** Vertical datum name?: **YES**
B. **C** Types of heights?: **YES**

10.7 Ancillary Geodetic Information

Obligation

A. __ Does the standard allow/require the provision for ancillary geodetic, gravimetric, or magnetometric information?: **NO**

10.8 Transformation Parameters
Does the standard specify a method to provide information for coordinate transformation(s):

Obligation

A. __ Model name?: **NO**
B. __ Control points?: **NO**
C. __ Transformation parameters?: **NO**

10.9 Non-Coordinate Spatial Reference

Does the standard allow/require the provision for information on non-coordinate spatial reference systems (e.g., postal systems, street address)?:

Obligation

A. __ Reference system name?: **NO**
B. __ Reference system definitions?: **NO**
C. __ Other?: **NO**

SECTION 11 – Availability and Distribution of the Dataset

This section examines how the standard provides information on the availability of the dataset?

Does the standard allow/require the provision for:

11.1 The Distributor(s) of the Dataset

Obligation

A. **M** Information about the distributor(s) of the dataset?: **YES**
B. **M** Contact information for the distributor(s)?: **YES**

11.2 Restrictions to Access and Usage

Obligation

A. **M** Information about copyright on the dataset (e.g., date, period, copyright owner, etc.)?: **YES**
B. **M** Information about restrictions on the access of the dataset?: **YES**
C. **M** Information about restrictions on the usage of the dataset?: **YES**

11.3 Pricing Details

Obligation

A. **C** Information about pricing?: **YES**

11.4 Distribution of the Dataset

Obligation

A. **C** Information about distribution media?: **YES**
B. **C** Information about on-line access?: **YES**
C. __ Information about units of distribution (e.g., tiles, layers, polygons, elements)?: **NO**

11.5 Distribution Format

Obligation

A. **C** Information about distribution formats (e.g., commercial CAD format, commercial GIS formats, National/International transfer standard)?: **YES**
B. __ Information about the encoding of the dataset (e.g., binary, ASCII)?: **NO**
C. **C** Information about compression method?: **YES**
D. __ Information about compressed dataset size?: **NO**
E. **C** Information about the distribution mechanism?: **YES**
F. __ Information about the encapsulation of data (e.g., ISO 8211, ISO 8824)?: **NO**

11.6 Ordering Procedures

Obligation

A. **C** Information about ordering procedures for the dataset?: **YES**

11.7 Support Services

Obligation

A. __ Information about support services (e.g., additional data transformations, format or coordinate conversions)?: **NO**

SECTION 12 – Authorization and Verification of the Metadata

This section examines how the standard provides for the description of the authorization and verification on the metadata of a dataset (e.g., how is the metadata itself authorized and verified?).

12.1 Metadata reference

Obligation

A. **M** Does the standard allow/require the provision of information about reference dates of the metadata (e.g., last check date, last update date, review date)?: **YES**
Metadata Date, Metadata Review Date, Metadata Future Review Date

12.2 Verification Authority

Obligation

A. __ Does the standard allow/require the provision for information about who verified/authorized the metadata?: **NO**
B. __ Does the standard specify the method for providing this information (e.g., ISO 9000): **NO**

12.3 Statistical Methods Used for Verification

Obligation

A. __ Does the standard specify the statistical methods to be used for verification of the metadata?: **NO**

B. __ Does the standard allow/require the provision for the description of the statistical methods used to verify the metadata?: **NO**

END OF CHARACTERISTICS
FIN

World Spatial Metadata Standards
H. Moellering, H.J.G.L. Aalders & A. Crane (Editors)

Spain
MIGRA (Mecanismo de Intercambio de Información Geográfica Relacional formado por Agregación)
Aggregated Relational Geographic Information Interchange Mechanism
UNE 148001 EX

Sebastián Mas

I. Information about the Metadata Standard Itself

SECTION 0 – Assessment Information

This section provides the user with a general description of the scope and intended use of the standard. Information on the respondent who prepared the assessment is also provided.

0.1 Respondent

Assessment prepared by:

Name: Sebastián Mas
Position: Director
Organization: Centro Nacional de Información Geográfica (CNIG)
Address: General Ibáñez de Ibero, 3. 28003 MADRID. SPAIN
Fax: +34 9 1 535 17 13
Phone: + 34 91 554 14 50
E-mail: smas@cnig.ign.es

Date assessment completed:

May 1998, updated January 2001 and May 2002, Reviewed October, 2004

Relationship of respondent to standard:

President of National Technical Committee on Digital Geographic Information (AEN/CTN 148). AENOR.
AENOR's Committe AEN/CTN 148 is author and custodian of MIGRA Standard.

0.2 Brief Summary Statement

A. Summary of the scope and intent of the standard:
MIGRA standard establishes the way to transfer all kind of digital geographic information. Both for GIS applications and Computer Aided Cartography.
The standard considers:
Glossary ot Terms, Spatial Data Model, Physical Model, Metadata, Quality, Attributes, Reference System and Coding.
Main goals of this standard are:

- **To solve in a practical way problems related to digital geographic data exchange.**
- **To assist in training Spanish geographic data users and producers to manage next CEN/TC 287 and ISO/TC 211 standards.**
- **To assist in training Spanish geographic data users to manage aspects as metadata, quality, etc.**

SECTION 1 – Administration of Metadata Standard

This section provides a general description of the administrative framework, within which the standard was developed, tested and currently resides. Significant details are also provided in terms of the developmental and managerial history as well as information on the existence of documentation, software tools and training materials. This section also provides a point of contact for further information on the standard.

1.1 Name of Standard

A. Name of standard:
1. In original language(s):
MIGRA (Mecanismo de Intercambio de Información Geográfica Relacional formado por Agregación).
2. English translation, if appropriate:
Aggregated Relational Geographic Information Interchange Mechanism

B. Version/Edition:
February 1998

C. Language(s) of documentation:
Spanish

D. Acronym(s):
MIGRA (Metadata)

E. Official ID:
UNE 148001 EX

F. Name(s) of recognizing standards authority(ies):
- **AENOR (Standardisation Spanish Association)**
- **Ministry of Industry and Energy**

1.2 Responsible Institutions

Contact information about those institutions that are (or were) responsible for the processes involved in the evolution of the standard:

A. Production and/or development:
   AENOR (Standardisation Spanish Association). AEN/CTN 148 National Technical Committee for Digital Geographic Information.
   C/ Genova, 6. 28004, MADRID. SPAIN
B. Testing:
   See 1.2.A
C. Conformance:
   See 1.2.A
D. Maintenance:
   See 1.2.A
E. Distribution:
   See 1.2.A
F. Help Desk/User Support:
   See 1.2.A

1.3 Development History

A. Beginning of work on metadata standard: **1995.04.18**
B. Milestones:
   1995.12.13. First meeting AENOR, Administration, GIS Companies.
   1996.03.13. Second MIGRA general meeting.
   1996.06.26. Third MIGRA general meeting.
   1996.10.09. First MIGRA draft version.
   1997.05.22 First MIGRA official version.
   1997.06.23 MIGRA (UNE 148001 EX) is established as Spanish National Experimental Standard.
   1998.02. MIGRA (UNE 148001 EX) is published by AENOR.
C. Anticipated completion: **1997.06.23**
D. Update cycle: **3 years**

1.4 Status of the Standard

A. Is the standard officially recognized: **YES**
B. Date(s) of recognition: **1997.06.23**
C. Anticipated date of recognition: **N/A**
D. Current stage in recognition process:
   Recognized as Spanish standard and published
E. Steps still required to achieve recognition: **N/A**

1.5 Access to Documentation of the Standard

A. Is the standard copyrighted?: **YES**
B. Name and address of copyright owner:
AENOR
C/ Génova, 6. 28004 MADRID, SPAIN
Fax: + 34 91 310 40 32
Phone: +34914326000
C. Do restrictions on the use of the standard apply?: **NO**
D. Is the standard documentation currently available?: **YES**
E. List the form/media of standard documentation:
Paper, diskette
F. Is the standard available on the Internet?: **NO**
G. Price: **43,27 €**
H. Contact details for obtaining the standard:
See 1.5.B
I. Does the standard have ISBN/ISSN numbers?: **YES**
M4921:1998

1.6 Available Software Tools

A. Is software available to implement the standard as a database?: **NO**
B. Is software available to test compliance of metadata with the standard?: **NO**
C. Is software available to produce metadata compliant with the standard?: **NO**
D. Other relevant software: **N/A**
E. Is there a test dataset available to test software tools for implementing the standard?: **YES**
F. For each product available above, please list contact information:
See 1.5.B

1.7 Available Training Materials

A. Are organized training sessions available?: **YES**
Training sessions are planned yearly
B. Contact details for training sessions:
AENOR
Génova, 6
28004 MADRID, SPAIN
C. Training documentation available: **NO**

SECTION 2 – Use and Implementation of the Metadata Standard

This section provides information on the way in which the standard is used and the purpose of its use. It also provides descriptions about the level of abstraction of the standard.

2.1 Intended Use of the Standard
Can the standard be used:

A. Within a transfer data structure?: **YES**
B. To produce hardcopy catalog of datasets?: **YES**
C. To catalog data in a digital on-line catalog?: **NO**
D. Other: **N/A**

2.2 Intended Purpose of the Standard
Is the intended purpose of the standard:

A. To support discovery, identification, location of data?: **NO**
B. To support how to access data?: **NO**
C. To support the determination of dataset fitness for use?: **YES**
D. To support transfer of dataset?: **YES**
E. To support an organization's data management?: **YES**
F. Other: **N/A**

2.3 Table of Contents

A. Provide the standard's Table of Contents

7. Metadata
7.1. Introduction
7.2. Metadata list
7.3. Format
7.4. Examples

2.4 Level of Abstraction of the Metadata Standard Definition

A. Does the standard define metadata elements?: **NO**
B. Does the standard define a model for the relationships between the metadata elements?: **NO**
C. Provide an overview diagram of the model of the metadata, if available. **N/A**
D. Does the standard define domains and/or data types of the metadata elements?: **YES**
E. Does the standard define how the metadata should be encoded?: **YES**
F. Does the standard define a database schema for the automatic generation of a database?: **NO**

2.5 Extensibility of the Standard

A. Is the metadata description extensible (e.g., Can user add metadata domains, elements, thematic profiles, etc.)?: **YES**
A user can add elements

2.6 Availability of Metadata
Does the standard specify:

A. How metadata are to be accessed?: **NO**
B. A mechanism for querying the metadata?: **NO**
C. Particular standardized query protocols?: **NO**

D. That the metadata should be encoded for computer access?: **NO**
E. That the metadata should be encrypted for security?: **NO**

2.7 Examples

A. Does the standard provide examples of implementations?: **YES**
Completed example containing:
- **Standard version**
- **Data Producer**
- **Data Distributor**
- **Transfer data content**
- **Creation and updating date**
- **Transfer files descriptions**
- **Attribute files descriptions**

B. How many examples are available?: **1**

SECTION 3 – Linkage and Coordination of the Metadata Standard

This section provides information on how the metadata standard is linked to other standards, particularly transfer standards, and how such linkage is implemented. If a significant linkage does exist, detailed technical information is provided on the precise nature of the linkage to the standard, and how the linkage affects implementation of the metadata standard.

3.1 Associated Transfer Standards

A. Is the metadata standard associated with a transfer standard(s)?: **YES**
B. If A is yes, provide references and contact details for each linked standard:
Metadata Standard is part of MIGRA (UNE 148 001 EXT) (Aggregated Relational Geographic Information Interchange Mechanism)
C. If A is yes, are both (all) standards coordinated?: **YES**
D. If A is yes, describe how the coordination works, both from a technical and conformance (normative) viewpoint:
Metadata standard is part of the Standard MIGRA (UNE 148001 Ex). Metadata is used as a part of the transfer file or standalone. Both transfer and metadata are two chapters in the Standard and they are completely coordinated from a technical and conformance view point.

3.2 Incorporated into transfer mechanism

A. Is the metadata standard incorporated into a transfer mechanism?: **YES**
B. If A is yes, provide references and contact details for each incorporated mechanism:
Metadata standard is incorporated into MIGRA (UNE 148001 EX).
C. If A is yes, describe how:
See 3.1.D.

3.3 Other linked, coordinated or associated standards

A. Is the metadata standard linked, coordinated, or associated with any other standard(s) not listed above?: **YES**

B. If A is yes, provide references and contact details for each standard:

CEN/TC 287

C. If A is yes, describe how for each relevant standard:

Metadata standard MIGRA (UNE 148001 EX) is a profile done from CEN/TC 287 Metadata standard, and it is experimental (EX) to be modified according any modification in CEN/TC 287 Metadata Standard.

II. Categories the Metadata Standard Uses to Describe Datasets

The following sections are categories of metadata that may be used by the metadata standard to describe data. Not all categories are relevant to different datasets and may not necessarily be present in the standard. The presence or absence of a category should not be used for judging one standard against another.

Note: For each of the questions in the following sections please provide the obligation status information. Valid obligation values are:

Obligation Level

M	Mandatory	This information must be entered to comply with the metadata standard.
C	Conditional	This information is entered based on other information in the metadata.
O	Optional	This information is optionally entered in the metadata.

If the answer is 'NO', the obligation status should be left blank.

SECTION 4 – Dataset Identification

This section examines how the standard provides for the specification of the various sorts of titles and identifiers for a dataset.

4.1 The title(s) of a dataset

Does the standard allow/require the provision for:

Obligation

A. __ Previous title(s)?: **NO**

B. **M** Current title(s)?: **YES**

C. __ Possible future title(s)?: **NO**

D. __ Official titles in different languages?: **NO**

E. __ Abbreviated title(s)?: **NO**

F. __ Alternative Title(s)?: **NO**

G. __ Does the standard support effective dates of titles?: **NO**

H. __ Other identifiers (e.g., ISBN, National Stock Number, Uniform Product Code)?: **NO**

4.2 Parties responsible for the dataset
Does the standard allow/require the provision for:

Obligation

A. **M** Author?: **YES**
B. **M** Owner?: **YES**
C. **M** Other?: **YES**
Owner Distributor contact person

4.3 ISBD Information

Obligation

A. __ Does the metadata standard specify inclusion of mandatory ISBD information (see glossary) within the metadata?: **NO**

SECTION 5 – Status of the Dataset

This section examines how the standard provides for the specification of the current status and updating of the data set.

5.1 Status and Progress of Datasets
Does the metadata standard allow/require the provision for:

Obligation

A. **M** Information on the status of the entire dataset?: **YES**
Creation date, last updating date
B. __ Information on the status of the particular subset (e.g., observation time, inauguration time, extraction time)?: **NO**

Does the metadata standard specify:

C. __ A structure for updating or maintenance information?: **NO**
D. __ A structure to describe historical information (see also Section 8.3 G)?: **NO**

5.2 Information on Incremental Updating
Does the metadata standard specify:

Obligation

A. __ A structure for information on incremental updates of the entire dataset?: **NO**
B. __ A structure for information on incremental updates of particular portions of the dataset?: **NO**
C. __ A structure to describe the update process for the dataset?: **NO**

SECTION 6 – Extent of the Dataset

This section examines how the standard provides for the specification of the spatial and temporal extent of the dataset.

6.1 The spatial extent of the dataset

Obligation

A. **M** Does the metadata standard allow/require for the description of horizontal extent (e.g., bounding rectangle/polygon, named location (i.e., Stockholm, etc.))?: **YES**
Rectangle, named location
B. **M** Does the metadata standard allow/require for the description of vertical extent (e.g., minima/maxima heights, etc.)?: **YES**
Minima/maxima heights
C. __ Does the standard require specific units for describing extent?: **NO**
D. **M** Does the standard allow specification of the units for describing extent?: **YES**
E. __ Does the standard require specific spatial reference system for describing extent?: **NO**
F. **M** Does the standard allow specification of the spatial reference system for describing extent?: **YES**

6.2 The temporal extent of the dataset

Obligation

A. __ Does the standard allow/require for the description of time period for which the dataset is relevant?: **NO**
B. __ Does the standard allow/require specification of the temporal reference system?: **NO**

SECTION 7 – Data Content of the Dataset

This section examines how the standard provides for the specification of the data content of the dataset in terms of overview elements, themes, data definitions and product specifications.

7.1 Dataset overview
Does the standard allow/require for the provision of the following information:

Obligation

A. __ Dataset overview?: **NO**
B. **M** Abstract – brief summary of the content of the dataset?: **YES**
C. **M** Purpose – summary of the intentions with which the dataset was developed?: **YES**
D. **M** Usage – list of activities for which the dataset has been used? (see also Section 8.5): **YES**
E. **M** Spatial schema – (e.g., vector, raster)? (see also Section 9): **YES**

F. **M** Spatial reference system? (see also Section 12): **YES**
G. __ Language?: **NO**
H. **M** Character set (e.g., ISO8859-1, JISX0208)?: **YES**
I. __ List of documents providing further information about the dataset?: **NO**
J. __ Sample/preview dataset?: **NO**
K. **M** Related datasets?: **YES**
L. **M** Original size of the dataset?: **YES**
M. __ Compression? (see also Section 11.5): **NO**
N. __ Other?: **NO**

7.2 Theme(s)

Obligation

A. **M** Does the standard allow/require the provision for the description of theme(s) (e.g., topography, land cover, transport, etc.) in the dataset?: **YES**
B. **M** If A is yes, does the standard allow the referencing of an outside dictionary or thesaurus?: **YES**
C. **M** If A is yes, does the standard itself include a dictionary or thesaurus within the metadata?: **YES**
D. **M** If A is yes, does the standard allow the use of unauthorized user-defined keywords?: **YES**

7.3 Data Definition

Obligation

A. **O** Does the standard allow/require the provision for the description of the data definition?: **YES**
B. If A is yes, does the standard allow/require the provision for the description of:
O 1. Features/Objects?: **YES**
O 2. Attributes?: **YES**
__ 3. Relations?: **NO**
C. **O** Can attributes be described separately from feature/object classes?: **YES**
D. **O** Does the standard allow/require the specification of the domain of the attributes?: **YES**
E. **O** Does the standard allow/require the specification of data type of the attributes?: **YES**
F. __ Does the standard allow/require the specification of particular classification coding schemes (e.g., FACC, OSKA, S-57 OC)?: **NO**

7.4 Product Specification

Obligation

A. __ Does the standard allow/require for the provision of information about whether the dataset meets a particular product specification?: **NO**

B. __ If A is yes, can it be internal to the Metadata?: **NO**
C. __ If A is yes, can it be external to the Metadata?: **NO**

SECTION 8 – Data Quality of the Dataset

This section examines how the standard deals with the question of data quality. This includes homogeneity and metrics for data quality, quality assurance, and usage.

8.1 Overall Data Quality Statement

Obligation

A. **M** Does the standard allow/require for the provision of quality to be described? If yes, please answer the questions below. (*If no, all following answers in Section* 8 *are N/A, please skip to Section* 9.): **YES**
B. __ Does the standard allow/require the provision for the quality information referencing an external quality standard, quality assurance standard, or specification?: **NO**

8.2 Homogeneity of Data Quality

Obligation

A. **C** Does the standard allow/require for the variation of the quality of the dataset to be described?: **YES**
B. What type of aggregation of quality information is allowed?
C 1. Entire database?: **YES**
__ 2. Feature/Object instance?: **NO**
C 3. Feature/Object class?: **YES**
__ 4. Attribute instance?: **NO**
C 5. Attribute class?: **YES**
C 6. Layer?: **YES**
C 7. Theme?: **YES**
C 9. Geographic extent?: **YES**
__ 9. Other?: **NO**

8.3 Metrics for Data Quality
Does the standard allow/require the provision of values for:

Obligation

A. **C** Positional accuracy?: **YES**
1. Standard error?: **YES**
2. Maximum error?: **YES**
3. Error correlation?: **NO**
4. Confidence level (e.g., 90%, 95%)?: **YES**
5. Other?: **YES**
Relative positional accuracy
B. **C** Attribute accuracy?: **YES**

	1. Standard error?:	**YES**
	2. Maximum error?:	**YES**
	3. Error frequency?:	**YES**
	4. Error correlation?:	**NO**
	5. Confidence level (e.g., 90%, 95%)?:	**YES**
	6. Other?:	**NO**
C.	__ Temporal accuracy?:	**NO**
	1. Standard error?:	**NO**
	2. Maximum error?:	**NO**
	3. Error correlation?:	**NO**
	4. Confidence level? (e.g., 90%, 95%):	**NO**
	5. Other?:	**NO**
D.	**C** Completeness?:	**YES**
	1. Error of omission?:	**YES**
	2. Error of commission?:	**YES**
	3. Other?:	**NO**
E.	**C** Currentness?:	**YES**
	But included in lineage	
	1. Date?:	**YES**
	2. Maximum age?:	**YES**
	3. Temporal extent?:	**YES**
	4. Other?:	**NO**
F.	**C** Logical consistency?:	**YES**
	Geometric, topological and semantic procedures	
G.	**C** Lineage?:	**YES**
	1. List of processing steps?:	**YES**
	2. List of values of processing parameters?:	**NO**
	3. List of printouts of statistics, etc.?:	**NO**
	4. Source material?:	**YES**
	5. Other?:	**YES**
	Producer, usage	
H.	__ Additional Quality Information?:	**NO**

8.4 Quality Assurance Methods

Does the standard allow/require the provision of information about the methods used to determine:

Obligation

A.	**M** Positional accuracy?:	**YES**
	1. Standard error in adjustments?:	**YES**
	2. Repeated measurements?:	**YES**
	3. Independent measurements?:	**NO**
	4. Subjective evaluations?:	**YES**
	5. Other?:	**NO**
B.	**M** Attribute accuracy?:	**YES**

1. Standard error?: **YES**
2. Error classification matrix?: **NO**
3. Independent measurements?: **NO**
4. Subjective evaluation?: **YES**
5. Other?: **NO**

C. __ Temporal accuracy?: **NO**
1. Independent measurements?: **NO**
2. Subjective evaluation?: **NO**
3. Other?: **NO**

D. __ Completeness?: **NO**
1. Field checks?: **NO**
2. Repeated measurements?: **NO**
3. Independent measurements?: **NO**
4. Subjective evaluation?: **NO**
5. Error of commission?: **NO**
6. Other?: **NO**

E. **M** Currentness?: **YES**
1. Date of source material?: **YES**
2. Repeated measurements?: **NO**
3. Independent measurements?: **NO**
4. Subjective evaluation?: **YES**
5. Other?: **NO**

F. **M** Logical consistency?: **YES**
1. Verification results of consistency checks?: **YES**
2. Repeated processing?: **YES**
3. Comparison with independent sources?: **YES**
4. Subjective evaluation?: **YES**
5. Other?: **NO**

G. **M** Lineage?: **YES**
1. Checking results of processing steps: **YES**
2. Subjective evaluation: **YES**
3. Other: **NO**

H. __ Additional Quality Information?: **NO**

8.5 Usage

Obligation

A. **O** Does the metadata standard allow/require the provision of information about usage?: **YES**
1. Who used it: **YES**
2. When it was used: **YES**
3. For what application it was used: **YES**
4. The effectiveness of use: **NO**
5. Other: **NO**

8.6 Alternative Quality Description: **NO**

SECTION 9 – Spatial Data Structure of the Dataset

This section examines how the standard provides for the specification of the spatial data model and spatial data types.

9.1 Spatial Data Model
Does the standard allow/require the provision for:

Obligation

A. **M** The description of the geometrical and/or topological primitives used to model the real world?: **YES**
B. __ Information about an external standard for geometrical and/or topological primitive specification?: **NO**

9.2 Spatial Data Types
Does the standard allow/require the provision for:

Obligation

A. **M** Information about the spatial data type(s) used in the dataset?: **YES**
B. __ If A is yes, which of the following generic spatial data types are supported?
For each, please list specific data primitive levels supported by each data type.
1. Vector (e.g., spaghetti, chain-node/node-edge-face, planar graph, full topology)?: **YES**
Spaguetti, chain- node, node-edge-face, planar-graph, full topology
1. Areal Raster (e.g., thematic raster – scanned and classification maps, image raster – satellite images, scanned aerial photographs)?: **NO**
3. Grid/Matrix (e.g., Digital Elevation Model)?: **NO**
4. Triangular Irregular Network (TIN)?: **NO**
5.Other?: **NO**

9.3 Other Data Types
Does the standard allow the provision of information about:

Obligation

A. __ Video?: **NO**
B. __ Sound?: **NO**
C. __ Other?: **NO**

SECTION 10 – Spatial Reference of the Dataset

This section examines how the dataset provides for the specification of the spatial reference characteristics of the data set.

Does the standard allow/require the provision for the following information:

10.1 Geoidal Model

Obligation

A. **M** Geoidal model name?: **YES**
B. __ Geoidal model definitions?: **NO**

10.2 Geodetic datum

Obligation

A. **M** Geodetic datum name?: **YES**
B. __ Geodetic datum definitions?: **NO**

10.3 Reference Ellipsoid

Obligation

A. **M** Reference ellipsoid name?: **YES**
B. __ Reference ellipsoid definitions?: **NO**

10.4 Projection

Obligation

A. **M** Projection name?: **YES**
B. __ Projection definitions?: **NO**

10.5 Coordinate Systems

Obligation

A. **M** Coordinate system type?: **YES**
B. __ Coordinate system definitions?: **NO**

10.6 Height Reference System

Obligation

A. **M** Vertical datum name?: **YES**
B. **M** Types of heights?: **YES**

10.7 Ancillary Geodetic Information

Obligation

A. __ Does the standard allow/require the provision for ancillary geodetic, gravimetric, or magnetometric information?: **NO**

10.8 Transformation Parameters
Does the standard specify a method to provide information for coordinate transformation(s):

Obligation

A. __ Model name?: **NO**
B. __ Control points?: **NO**
C. __ Transformation parameters?: **NO**

10.9 Non-Coordinate Spatial Reference

Does the standard allow/require the provision for information on non-coordinate spatial reference systems (e.g., postal systems, street address)?:

Obligation

A. __ Reference system name?: **NO**
B. __ Reference system definitions?: **NO**
C. __ Other?: **NO**

SECTION 11 – Availability and Distribution of the Dataset

This section examines how the standard provides information on the availability of the dataset?

Does the standard allow/require the provision for:

11.1 The Distributor(s) of the Dataset

Obligation

A. **M** Information about the distributor(s) of the dataset?: **YES**
B. **M** Contact information for the distributor(s)?: **YES**

11.2 Restrictions to Access and Usage

Obligation

A. **C** Information about copyright on the dataset (e.g., date, period, copyright owner, etc.)?: **YES**
B. **C** Information about restrictions on the access of the dataset?: **YES**
C. **C** Information about restrictions on the usage of the dataset?: **YES**

11.3 Pricing Details

Obligation

A. __ Information about pricing?: **NO**

11.4 Distribution of the Dataset

Obligation

A. __ Information about distribution media?: **NO**
B. __ Information about on-line access?: **NO**
C. __ Information about units of distribution (e.g., tiles, layers, polygons, elements)?: **NO**

11.5 Distribution Format

Obligation

A. __ Information about distribution formats (e.g., commercial CAD format, commercial GIS formats, National/International transfer standard)?: **YES**
MIGRA used a fixed format
B. **C** Information about the encoding of the dataset (e.g., binary, ASCII)?: **YES**
C. __ Information about compression method?: **NO**
D. __ Information about compressed dataset size?: **NO**
E. __ Information about the distribution mechanism?: **NO**
F. __ Information about the encapsulation of data (e.g., ISO 8211, ISO 8824)?: **NO**

11.6 Ordering Procedures

Obligation

A. __ Information about ordering procedures for the dataset?: **NO**

11.7 Support Services

Obligation

A. __ Information about support services (e.g., additional data transformations, format or coordinate conversions)?: **NO**

SECTION 12 – Authorization and Verification of the Metadata

This section examines how the standard provides for the description of the authorization and verification on the metadata of a dataset (e.g., how is the metadata itself authorized and verified?).

12.1 Metadata reference

Obligation

A. **O** Does the standard allow/require the provision of information about reference dates of the metadata (e.g., last check date, last update date, review date)?: **YES**

12.2 Verification Authority

Obligation

A. __ Does the standard allow/require the provision for information about who verified/authorized the metadata?: **NO**
B. __ Does the standard specify the method for providing this information (e.g., ISO 9000): **NO**

12.3 Statistical Methods Used for Verification

Obligation

A. __ Does the standard specify the statistical methods to be used for verification of the metadata?: **NO**

B. __ Does the standard allow/require the provision for the description of the statistical methods used to verify the metadata?: **NO**

END OF CHARACTERISTICS

FIN

World Spatial Metadata Standards
H. Moellering, H.J.G.L. Aalders & A. Crane (Editors)

United States of America Content Standard for Digital Geospatial Metadata FGDC-STD-001-1998

Richard A. Pearsall, Richard Hogan

I. Information about the Metadata Standard Itself

SECTION 0 – Assessment Information

This section provides the user with a general description of the scope and intended use of the standard. Information on the respondent who prepared the assessment is also provided.

0.1 Respondent

Assessment prepared by:

Richard A. Pearsall
U.S. Geological Survey
511 National Center
12201 Sunrise Valley Drive
Reston, VA 20192, USA
Tel: + 1 703 648-4532
Fax: + 1 703 648-4722
E-mail: rpearsall@usgs.gov

Date assessment completed:

February 2001; updated May 2002, Reviewed Spring, 2005

Relationship of respondent to standard:

Maintenance Authority

0.2 Brief Summary Statement

A. Summary of the scope and intent of the standard:
This standard is intended to support the collection and processing of geospatial metadata. It is intended to be useable by all levels of government and the private sector. The standard is not intended to reflect an implementation design. An implementation design requires adapting the structure and form of the standard to meet application requirements.

The standard was developed from the perspective of defining the information required by a prospective user to determine the availability of a set of geospatial data; to determine the fitness and the set of geospatial data for an intended use; to determine the means of accessing the set of geospatial data; and to successfully transfer the set of geospatial data. As such, the standard establishes the names of data elements and compound elements to be used for these purposes, definitions of these data elements and compound elements, and information about the values that are to be provided for the data elements. The standard does not specify the means by which this information is organized in a computer system or in a data transfer, nor the means by which this information is transmitted, communicated, or presented to the user.

The objectives of the standard are to provide a common set of terminology and definitions for the documentation of digital geospatial data. The standard establishes the names of data elements and compound elements (groups of data elements) to be used for these purposes, the definitions of these compound elements and data elements, and information about the values that are to be provided for the data elements.

The standard identifies the major uses of metadata are: to maintain an organization's internal investment in geospatial data, to provide information about an organization's data holdings to data catalogues, clearinghouses, and brokerages, and to provide information needed to process and interpret data to be received through a transfer from an external source.

The information included in the standard was selected based on four roles that metadata play:

availability – data needed to determine the sets of data that exist for a geographic location.

fitness for use – data needed to determine if a set of data meets a specific need.

access – data needed to acquire an identified set of data.

transfer – data needed to process and use a set of data.

These roles form a continuum in which a user cascades through a pyramid of choices to determine what data are available, to evaluate the fitness of the data for use, to access the data, and to transfer and process the data. The exact order in which data elements are evaluated, and the relative importance of data elements, will not be the same for all users.

SECTION 1 – Administration of Metadata Standard

This section provides a general description of the administrative framework, within which the standard was developed, tested and currently resides. Significant details are also provided in terms of the developmental and managerial history as well as information on the existence of documentation, software tools and training materials. This section also provides a point of contact for further information on the standard.

1.1 Name of Standard

A. Name of standard:

1. In original language(s):
   Content Standard for Digital Geospatial Metadata
2. English translation, if appropriate:

B. Version/Edition:
   Version 2

C. Language(s) of documentation:
   English

D. Acronym(s):
   CSDGM

E. Official ID:
   FGDC-STD-001-1998

F. Name(s) of recognizing standards authority(ies):
   Federal Geographic Data Committee

1.2 Responsible Institutions

Contact information about those institutions that are (or were) responsible for the processes involved in the evolution of the standard:

A. Production and/or development:
   Federal Geographic Data Committee
   U.S. Geological Survey
   590 National Center
   12201 Sunrise Valley Drive
   Reston, VA 20192, USA
   Tel: + 1 703 648-5514
   Fax: + 1 703 648-5755
   E-mail: fgdc@www.fgdc.gov

B. Testing:
   Federal Geographic Data Committee
   U.S. Geological Survey
   590 National Center
   12201 Sunrise Valley Drive
   Reston, VA 20192, USA
   Tel: + 1 703 648-5514
   Fax: + 1 703 648-5755
   E-mail: fgdc@www.fgdc.gov

C. Conformance:
   Federal Geographic Data Committee
   U.S. Geological Survey
   590 National Center
   12201 Sunrise Valley Drive
   Reston, VA 20192, USA
   Tel: + 1 703 648-5514
   Fax: + 1 703 648-5755
   E-mail: fgdc@www.fgdc.gov

D. Maintenance:
Federal Geographic Data Committee
U.S. Geological Survey
590 National Center
12201 Sunrise Valley Drive
Reston, VA 20192, USA
Tel: + 1 703 648-5514
Fax: + 1 703 648-5755
E-mail: fgdc@www.fgdc.gov

E. Distribution:
Federal Geographic Data Committee
U.S. Geological Survey
590 National Center
12201 Sunrise Valley Drive
Reston, VA 20192, USA
Tel: + 1 703 648-5514
Fax: + 1 703 648-5755
E-mail: fgdc@www.fgdc.gov
URL: http://www.fgdc.gov/metadata/contstan.html

F. Help Desk/User Support:
Federal Geographic Data Committee
U.S. Geological Survey
590 National Center
12201 Sunrise Valley Drive
Reston, VA 20192, USA
Tel: + 1 703 648-5514
Fax: + 1 703 648-5755
E-mail: fgdc@www.fgdc.gov

1.3 Development History

A. Beginning of work on metadata standard: **1993**
B. Milestones:
Version 1 issued: 8 June 1994
Version 2 issued: 19 June 1998
C. Anticipated completion: **Completed**
D. Update cycle: **As needed**

1.4 Status of the Standard

A. Is the standard officially recognized: **YES**
B. Date(s) of recognition:
June 19, 1998
C. Anticipated date of recognition: **N/A**

D. Current stage in recognition process:
On June 19, 1998, Secretary of the Interior Babbit, Chair, FGDC Steering Committee signed the endorsement of the Content Standard for Digital Geospatial Metadata Version 2 (FGDC-STD-001-1998) as an official FGDC Standard. To date, FGDC-STD-001-1998 has been adopted by numerous Federal, State, Local, and City governments and has been adopted by several international organizations. There are also numerous developed metadata standards that have used FGDC-STD-001-1998 as a baseline.

E. Steps still required to achieve recognition: **N/A**

1.5 Access to Documentation of the Standard

A. Is the standard copyrighted?: **NO**
B. Name and address of copyright owner: **N/A**
C. Do restrictions on the use of the standard apply?: **NO**
D. Is the standard documentation currently available?: **YES**
E. List the form/media of standard documentation:
Paper, Electronic
F. Is the standard available on the Internet?: **YES**
http://www.fgdc.gov/metadata/contstan.html
G. Price: **Free**
H. Contact details for obtaining the standard:
Federal Geographic Data Committee
U.S. Geological Survey
590 National Center
12201 Sunrise Valley Drive
Reston, VA 20192, USA
Tel: + 1 703 648-5514
Fax: + 1 703 648-5755
E-mail: fgdc@www.fgdc.gov
I. Does the standard have ISBN/ISSN numbers?: **NO**

1.6 Available Software Tools

A. Is software available to implement the standard as a database?: **YES**
B. Is software available to test compliance of metadata with the standard?: **YES**
C. Is software available to produce metadata compliant with the standard?: **YES**
D. Other relevant software:
A list of software, contact information, and the software's evaluation is available at http://www.fgdc.gov/metadata/toollist/metatool.html
E. Is there a test dataset available to test software tools for implementing the standard?: **NO**

F. For each product available above, please list contact information:
A list of software, contact information, and the software's evaluation is available at http://www.fgdc.gov/metadata/toollist/metatool.html

1.7 Available Training Materials

A. Are organized training sessions available?: **YES**

B. Contact details for training sessions:
Federal Geographic Data Committee
U.S. Geological Survey
590 National Center
12201 Sunrise Valley Drive
Reston, VA 20192, USA Tel: + 1 703 648-5514
Fax: + 1 703 648-5755
E-mail: fgdc@www.fgdc.gov

A web-based training calendar showing metadata training can be accessed at http://www.fgdc.gov/cgi-bin/people/META/WebEvent/WebEvent
A web based list of metadata trainers and their training curriculum and training descriptions can be accessed at http://www.fgdc.gov/metadata/toollist/trainers.html

C. Training documentation available: **YES**
Content Standard for Digital Geospatial Metadata Workbook – Version 2, May 2000
Federal Geographic Data Committee
U.S. Geological Survey
590 National Center 12201 Sunrise Valley Drive
Reston, VA 20192, USA
Tel: + 1 703 648-5514
Fax: + 1 703 648-5755
E-mail: fgdc@www.fgdc.gov

An electronic version (Adobe PDF) copy for Version 2 is accessible at the http://www.fgdc.gov/metadata/meta_workbook.html

SECTION 2 – Use and Implementation of the Metadata Standard

This section provides information on the way in which the standard is used and the purpose of its use. It also provides descriptions about the level of abstraction of the standard.

2.1 Intended Use of the Standard

Can the standard be used:

A. Within a transfer data structure?: **YES**
B. To produce hardcopy catalog of datasets?: **YES**
C. To catalog data in a digital on-line catalog?: **YES**
D. Other: **NO**

2.2 Intended Purpose of the Standard

Is the intended purpose of the standard:

A. To support discovery, identification, location of data?: **YES**
B. To support how to access data?: **YES**
C. To support the determination of dataset fitness for use?: **YES**
D. To support transfer of dataset?: **YES**
E. To support an organization's data management?: **NO**
F. Other: **NO**

2.3 Table of Contents

A. Provide the standard's Table of Contents

Introduction iv
Organization of the Standard vii
Metadata 1
Identification Information 2
Data Quality Information 10
Spatial Data Organization Information 16
Spatial Reference Information 19
Entity and Attribute Information 37
Distribution Information 42
Metadata Reference Information 50
Citation Information 53
Time Period Information 56
Contact Information 58
Appendix A Glossary 61
Appendix B Alphabetical List of Compound Elements and Data Elements 68
Appendix C References 72
Appendix D Guidelines for Creating Extended Elements in the Content Standard for Digital Geospatial Metadata 74
Appendix E Guidelines for Creating a Profile for the Content Standard for Digital Geospatial Metadata 77

2.4 Level of Abstraction of the Metadata Standard Definition

A. Does the standard define metadata elements?: **YES**
B. Does the standard define a model for the relationships between the metadata elements?: **YES**
C. Provide an overview diagram of the model of the metadata, if available. **YES**
D. Does the standard define domains and/or data types of the metadata elements?: **YES**

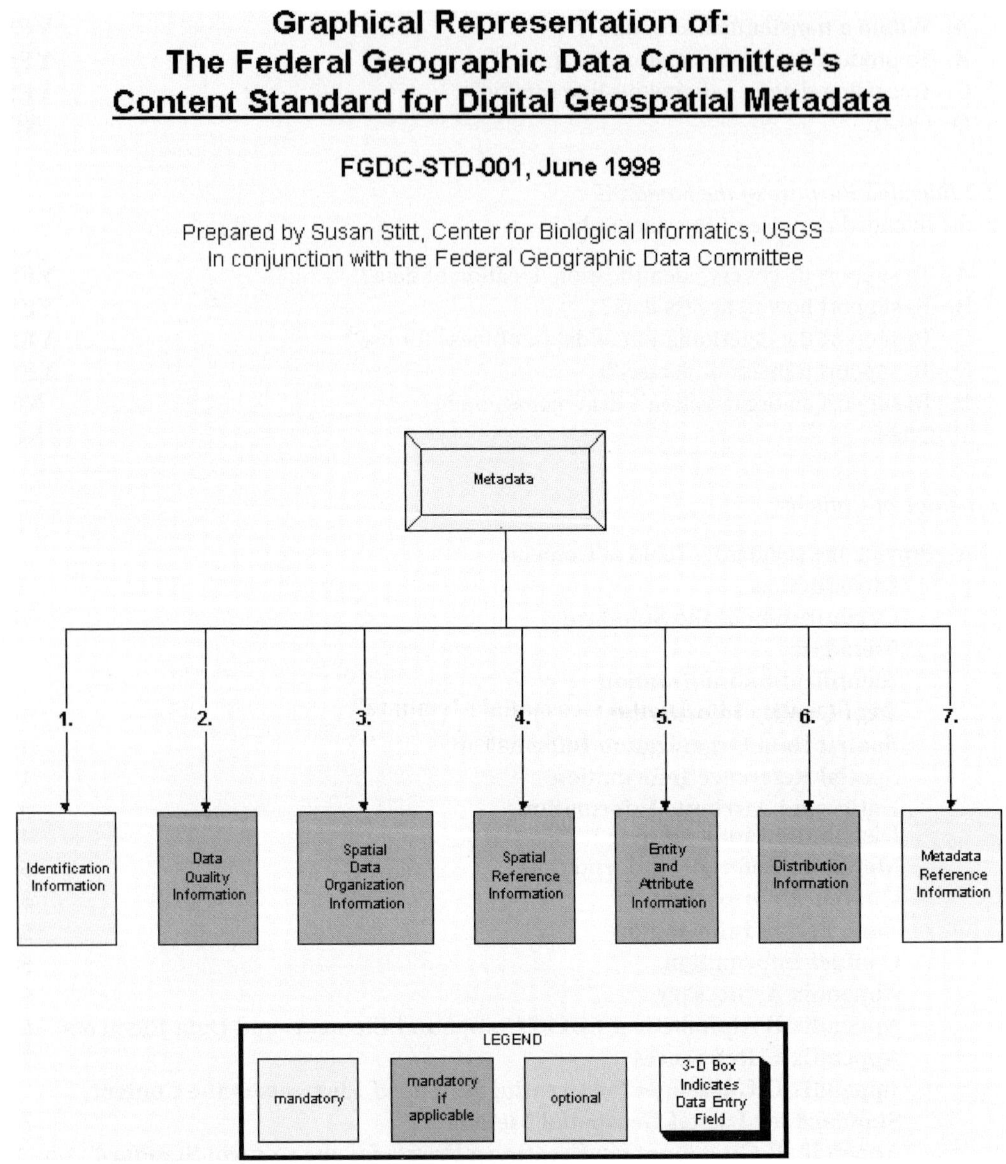

Fig. 1. Graphical representation of the FGDC Content Standard for Digital Geospatial Metadata (The graphical representation of the metadata model can be accessed at http://biology.usgs.gov/fgdc.metadata/version2).

E. Does the standard define how the metadata should be encoded?: **NO**

The FGDC has under development a standard titled "Encoding Standards for Digital Geospatial Metadata" which describes

the encoding of FGDC-STD-001-1998 for use in an online catalog/clearinghouse

F. Does the standard define a database schema for the automatic generation of a database?: **NO**

2.5 Extensibility of the Standard

A. Is the metadata description extensible (e.g., Can user add metadata domains, elements, thematic profiles, etc.)?: **YES**

Extended elements (elements outside the Standard, but needed by the data set producer) to the Standard may be defined by a data set producer or a user community. If extended elements are created, they must follow the guidelines presented in Appendix D of the standard

2.6 Availability of Metadata

Does the standard specify:

A. How metadata are to be accessed?: **NO**
B. A mechanism for querying the metadata?: **NO**
C. Particular standardized query protocols?: **NO**
D. That the metadata should be encoded for computer access?: **NO**

The FGDC has under development a standard titled "Encoding Standards for Digital Geospatial Metadata" which describes the encoding of FGDC-STD-001-1998 for use in a online catalog/clearinghouse

E. That the metadata should be encrypted for security?: **NO**

2.7 Examples

A. Does the standard provide examples of implementations?: **NO**

Examples can be found in the training workbook as well as by querying the Federal Geographic Data Committee National Spatial Data Infrastructure Clearinghouse

B. How many examples are available?: **N/A**

SECTION 3 – Linkage and Coordination of the Metadata Standard

This section provides information on how the metadata standard is linked to other standards, particularly transfer standards, and how such linkage is implemented. If a significant linkage does exist, detailed technical information is provided on the precise nature of the linkage to the standard, and how the linkage affects implementation of the metadata standard

3.1 Associated Transfer Standards

A. Is the metadata standard associated with a transfer standard(s)?: **NO**
B. If A is yes, provide references and contact details for each linked standard: **N/A**
C. If A is yes, are both (all) standards coordinated?: **N/A**

D. If A is yes, describe how the coordination works, both from a technical and conformance (normative) viewpoint: **N/A**

3.2 Incorporated into transfer mechanism

A. Is the metadata standard incorporated into a transfer mechanism?: **NO**
B. If A is yes, provide references and contact details for each incorporated mechanism: **N/A**
C. If A is yes, describe how: **N/A**

3.3 Other linked, coordinated or associated standards

A. Is the metadata standard linked, coordinated, or associated with any other standard(s) not listed above?: **NO**
B. If A is yes, provide references and contact details for each standard: **N/A**
C. If A is yes, describe how for each relevant standard: **N/A**

II. Categories the Metadata Standard Uses to Describe Datasets

The following sections are categories of metadata that may be used by the metadata standard to describe data. Not all categories are relevant to different datasets and may not necessarily be present in the standard. The presence or absence of a category should not be used for judging one standard against another.

Note: For each of the questions in the following sections please provide the obligation status information. Valid obligation values are:

Obligation Level

M	Mandatory	This information must be entered to comply with the metadata standard.
C	Conditional	This information is entered based on other information in the metadata.
O	Optional	This information is optionally entered in the metadata.

If the answer is 'NO', the obligation status should be left blank.

SECTION 4 – Dataset Identification

This section examines how the standard provides for the specification of the various sorts of titles and identifiers for a dataset.

4.1 The title(s) of a dataset
Does the standard allow/require the provision for:

Obligation

A. __ Previous title(s)?: **NO**
B. **M** Current title(s)?: **YES**
C. __ Possible future title(s)?: **NO**
D. __ Official titles in different languages?: **NO**

E. __ Abbreviated title(s)?: **NO**
F. **C** Alternative Title(s)?: **YES**
G. **M** Does the standard support effective dates of titles?: **YES**
H. **C** Other identifiers (e.g., ISBN, National Stock Number, Uniform Product Code)?: **YES**

Other identifiers include Edition, Series Information and Name, Issue Identification, Online linkage information, and information regarding relationship of this data set with a second data set (Larger Work Citation)

4.2 Parties responsible for the dataset
Does the standard allow/require the provision for:

Obligation

A. **M** Author?: **YES**
B. **C** Owner?: **YES**
C. **C** Other?: **YES**

4.3 ISBD Information

Obligation

A. __ Does the metadata standard specify inclusion of mandatory ISBD information (see glossary) within the metadata?: **NO**

SECTION 5 – Status of the Dataset

This section examines how the standard provides for the specification of the current status and updating of the data set.

5.1 Status and Progress of Datasets
Does the metadata standard allow/require the provision for:

Obligation

A. **M** Information on the status of the entire dataset?: **YES**
B. __ Information on the status of the particular subset (e.g., observation time, inauguration time, extraction time)?: **NO**

Does the metadata standard specify:

C. **M** A structure for updating or maintenance information?: **YES**
D. **C** A structure to describe historical information (see also Section 8.3 G)?: **YES**

5.2 Information on Incremental Updating
Does the metadata standard specify:

Obligation

A. **C** A structure for information on incremental updates of the entire dataset?: **YES**

B. __ A structure for information on incremental updates of particular portions of the dataset?: **NO**
C. **C** A structure to describe the update process for the dataset?: **YES**

SECTION 6 – Extent of the Dataset

This section examines how the standard provides for the specification of the spatial and temporal extent of the dataset.

6.1 The spatial extent of the dataset

Obligation

A. **M** Does the metadata standard allow/require for the description of horizontal extent (e.g., bounding rectangle/polygon, named location (i.e., Stockholm, etc.))?: **YES**
B. __ Does the metadata standard allow/require for the description of vertical extent (e.g., minima/maxima heights, etc.)?: **NO**
C. **M** Does the standard require specific units for describing extent?: **YES**
D. __ Does the standard allow specification of the units for describing extent?: **NO**
E. **M** Does the standard require specific spatial reference system for describing extent?: **YES**
F. __ Does the standard allow specification of the spatial reference system for describing extent?: **NO**

6.2 The temporal extent of the dataset

Obligation

A. **M** Does the standard allow/require for the description of time period for which the dataset is relevant?: **YES**
B. __ Does the standard allow/require specification of the temporal reference system?:**NO**

SECTION 7 – Data Content of the Dataset

This section examines how the standard provides for the specification of the data content of the dataset in terms of overview elements, themes, data definitions and product specifications.

7.1 Dataset overview

Does the standard allow/require for the provision of the following information:

Obligation

A. **O** Dataset overview?: **YES**
B. **M** Abstract – brief summary of the content of the dataset?: **YES**
C. **M** Purpose – summary of the intentions with which the dataset was developed?: **YES**
D. __ Usage – list of activities for which the dataset has been used? (see also Section 8.5): **NO**

This information may be found in some free text metadata elements

E. **C** Spatial schema – (e.g., vector, raster)? (see also Section 9): **YES**
F. **C** Spatial reference system? (see also Section 12): **YES**
G. __ Language?: **NO**
H. __ Character set (e.g., ISO8859-1, JISX0208)?: **NO**
I. **O** List of documents providing further information about the dataset?: **YES**
J. **O** Sample/preview dataset?: **YES**
K. **C** Related datasets?: **YES**
L. **O** Original size of the dataset?: **YES**
M. **C** Compression? (see also Section 11.5): **YES**
N. **C** Other?: **YES**

Other dataset overview information includes keyword (C) (place, theme, stratum, and temporal), dataset credit information (O), security information (O), and native dataset environment information (O)

7.2 Theme(s)

Obligation

A. **M** Does the standard allow/require the provision for the description of theme(s) (e.g., topography, land cover, transport, etc.) in the dataset?: **YES**
B. **C** If A is yes, does the standard allow the referencing of an outside dictionary or thesaurus?: **YES**
C. __ If A is yes, does the standard itself include a dictionary or thesaurus within the metadata?: **NO**
D. **C** If A is yes, does the standard allow the use of unauthorized user-defined keywords?: **YES**

7.3 Data Definition

Obligation

A. **C** Does the standard allow/require the provision for the description of the data definition?: **YES**
B. If A is yes, does the standard allow/require the provision for the description of:
C 1. Features/Objects?: **YES**
C 2. Attributes?: **YES**
__ 3. Relations?: **NO**
This information may be found in some free text metadata elements
C. __ Can attributes be described separately from feature/object classes?: **NO**
D. **C** Does the standard allow/require the specification of the domain of the attributes?: **YES**
E. **C** Does the standard allow/require the specification of data type of the attributes?: **YES**
F. **C** Does the standard allow/require the specification of particular classification coding schemes (e.g., FACC, OSKA, S-57 OC)?: **YES**

7.4 Product Specification

Obligation

A. **C** Does the standard allow/require for the provision of information about whether the dataset meets a particular product specification?: **YES**
B. __ If A is yes, can it be internal to the Metadata?: **NO**
C. **C** If A is yes, can it be external to the Metadata?: **YES**

SECTION 8 – Data Quality of the Dataset

This section examines how the standard deals with the question of data quality. This includes homogeneity and metrics for data quality, quality assurance, and usage.

8.1 Overall Data Quality Statement

Obligation

A. **C** Does the standard allow/require for the provision of quality to be described? If yes, please answer the questions below. (*If no, all following answers in Section* 8 *are N/A, please skip to Section* 9.): **YES**
B. **C** Does the standard allow/require the provision for the quality information referencing an external quality standard, quality assurance standard, or specification?: **YES**

8.2 Homogeneity of Data Quality

Obligation

A. __ Does the standard allow/require for the variation of the quality of the dataset to be described?: **YES**
B. What type of aggregation of quality information is allowed?
C 1. Entire database?: **YES**
C 2. Feature/Object instance?: **YES**
C 3. Feature/Object class?: **YES**
C 4. Attribute instance?: **YES**
C 5. Attribute class?: **YES**
C 6. Layer?: **YES**
C 7. Theme?: **YES**
C 8. Geographic extent?: **YES**
C 9. Other?: **YES**
Any other definable aggregation can be described

8.3 Metrics for Data Quality
Does the standard allow/require the provision of values for:

Obligation

A. **C** Positional accuracy?: **YES**
1. Standard error?: **YES**
2. Maximum error?: **YES**

3. Error correlation?: **YES**
4. Confidence level (e.g., 90%, 95%)?: **YES**
5. Other?: **NO**

B. **C** Attribute accuracy?: **YES**
1. Standard error?: **YES**
2. Maximum error?: **YES**
3. Error frequency?: **YES**
4. Error correlation?: **YES**
5. Confidence level (e.g., 90%, 95%)?: **YES**
6. Other?: **NO**

C. __ Temporal accuracy?: **NO**
1. Standard error?: **N/A**
2. Maximum error?: **N/A**
3. Error correlation?: **N/A**
4. Confidence level? (e.g., 90%, 95%): **N/A**
5. Other?: **N/A**

D. **C** Completeness?: **YES**
1. Error of omission?: **YES**
2. Error of commission?: **YES**
3. Other?: **YES**

Other completeness information includes any selection criteria, generalization, definitions used, and other rules used to derive the data set

E. **C** Currentness?: **YES**
1. Date?: **YES**
2. Maximum age?: **YES**
3. Temporal extent?: **YES**
4. Other?: **NO**

F. **C** Logical consistency?: **YES**

A free text field is provided to permit a narrative description of the logical consistency

G. **C** Lineage?: **YES**
1. List of processing steps?: **YES**
2. List of values of processing parameters?: **YES**
3. List of printouts of statistics, etc.?: **YES**
4. Source material?: **YES**
5. Other?: **YES**

Process date, process contact, source scale denominator, source media, source time period of contend, source contribution

H. **C** Additional Quality Information?: **YES**

Cloud cover

8.4 Quality Assurance Methods

Does the standard allow/require the provision of information about the methods used to determine:

Obligation

A. **C** Positional accuracy?:	**YES**
1. Standard error in adjustments?:	**YES**
2. Repeated measurements?:	**YES**
3. Independent measurements?:	**YES**
4. Subjective evaluations?:	**YES**
5. Other?:	**NO**
B. **C** Attribute accuracy?:	**YES**
1. Standard error?:	**YES**
2. Error classification matrix?:	**YES**
3. Independent measurements?:	**YES**
4. Subjective evaluation?:	**YES**
5. Other?:	**NO**
C. __ Temporal accuracy?:	**NO**
1. Independent measurements?:	**N/A**
2. Subjective evaluation?:	**N/A**
3. Other?:	**N/A**
D. **C** Completeness?:	**YES**
1. Field checks?:	**YES**
2. Repeated measurements?:	**YES**
3. Independent measurements?:	**YES**
4. Subjective evaluation?:	**YES**
5. Error of commission?:	**YES**
6. Other?:	**NO**
E. **C** Currentness?:	**YES**
1. Date of source material?:	**YES**
2. Repeated measurements?:	**YES**
3. Independent measurements?:	**YES**
4. Subjective evaluation?:	**YES**
5. Other?:	**NO**
F. **C** Logical consistency?:	**YES**
1. Verification results of consistency checks?:	**YES**
2. Repeated processing?:	**YES**
3. Comparison with independent sources?:	**YES**
4. Subjective evaluation?:	**YES**
5. Other?:	**NO**
G. **C** Lineage?:	**YES**
1. Checking results of processing steps:	**YES**
2. Subjective evaluation:	**YES**
3. Other:	**NO**
H. __ Additional Quality Information?:	**NO**

8.5 Usage

Obligation

A. __ Does the metadata standard allow/require the provision of information about usage?: **NO**

This information may be found in some free text metadata elements

1. Who used it: **N/A**
2. When it was used: **N/A**
3. For what application it was used: **N/A**
4. The effectiveness of use: **N/A**
5. Other: **N/A**

8.6 Alternative Quality Description: **NO**

SECTION 9 – Spatial Data Structure of the Dataset

This section examines how the standard provides for the specification of the spatial data model and spatial data types.

9.1 Spatial Data Model
Does the standard allow/require the provision for:

Obligation

A. **C** The description of the geometrical and/or topological primitives used to model the real world?: **YES**
B. __ Information about an external standard for geometrical and/or topological primitive specification?: **NO**

9.2 Spatial Data Types
Does the standard allow/require the provision for:

Obligation

A. **C** Information about the spatial data type(s) used in the dataset?: **YES**
B. **C** If A is yes, which of the following generic spatial data types are supported? **YES**
For each, please list specific data primitive levels supported by each data type.

1. Vector (e.g., spaghetti, chain-node/node-edge-face, planar graph, full topology)?: **YES**
1. Areal Raster (e.g., thematic raster – scanned and classification maps, image raster – satellite images, scanned aerial photographs)?: **YES**
3. Grid/Matrix (e.g., Digital Elevation Model)?: **YES**
4. Triangular Irregular Network (TIN)?: **NO**
5. Other?: **NO**

9.3 Other Data Types
Does the standard allow the provision of information about:

Obligation

A. __ Video?: **NO**

B. __ Sound?: **NO**
C. __ Other?: **NO**

SECTION 10 – Spatial Reference of the Dataset

This section examines how the dataset provides for the specification of the spatial reference characteristics of the data set.

Does the standard allow/require the provision for the following information:

10.1 Geoidal Model

Obligation

A. **C** Geoidal model name?: **YES**
B. **C** Geoidal model definitions?: **YES**

10.2 Geodetic datum

Obligation

A. **C** Geodetic datum name?: **YES**
B. **C** Geodetic datum definitions?: **YES**

10.3 Reference Ellipsoid

Obligation

A. **C** Reference ellipsoid name?: **YES**
B. **C** Reference ellipsoid definitions?: **YES**

10.4 Projection

Obligation

A. **C** Projection name?: **YES**
B. **C** Projection definitions?: **YES**

10.5 Coordinate Systems

Obligation

A. **C** Coordinate system type?: **YES**
B. **C** Coordinate system definitions?: **YES**

10.6 Height Reference System

Obligation

A. **C** Vertical datum name?: **YES**
B. **C** Types of heights?: **YES**

10.7 Ancillary Geodetic Information

Obligation

A. **C** Does the standard allow/require the provision for ancillary geodetic, gravimetric, or magnetometric information?: **YES**

10.8 Transformation Parameters

Does the standard specify a method to provide information for coordinate transformation(s):

Obligation

A. **C** Model name?: **YES**
B. **C** Control points?: **YES**
C. **C** Transformation parameters?: **YES**

10.9 Non-Coordinate Spatial Reference

Does the standard allow/require the provision for information on non-coordinate spatial reference systems (e.g., postal systems, street address)?:

Obligation

A. **C** Reference system name?: **YES**
B. **C** Reference system definitions?: **YES**
C. __ Other?: **NO**

SECTION 11 – Availability and Distribution of the Dataset

This section examines how the standard provides information on the availability of the dataset?

Does the standard allow/require the provision for:

11.1 The Distributor(s) of the Dataset

Obligation

A. **C** Information about the distributor(s) of the dataset?: **YES**
B. **C** Contact information for the distributor(s)?: **YES**

11.2 Restrictions to Access and Usage

Obligation

A. **M** Information about copyright on the dataset (e.g., date, period, copyright owner, etc.)?: **YES**
B. **M** Information about restrictions on the access of the dataset?: **YES**
C. **M** Information about restrictions on the usage of the dataset?: **YES**

11.3 Pricing Details

Obligation

A. **C** Information about pricing?: **YES**

11.4 Distribution of the Dataset

Obligation

A. **C** Information about distribution media?: **YES**
B. **C** Information about on-line access?: **YES**
C. **C** Information about units of distribution (e.g., tiles, layers, polygons, elements)?: **YES**

11.5 Distribution Format

Obligation

A. **C** Information about distribution formats (e.g., commercial CAD format, commercial GIS formats, National/International transfer standard)?: **YES**
B. **C** Information about the encoding of the dataset (e.g., binary, ASCII)?: **YES**
C. **C** Information about compression method?: **YES**
D. **C** Information about compressed dataset size?: **YES**
E. **C** Information about the distribution mechanism?: **YES**
F. **C** Information about the encapsulation of data (e.g., ISO 8211, ISO 8824)?: **YES**

11.6 Ordering Procedures

Obligation

A. **C** Information about ordering procedures for the dataset?: **YES**

11.7 Support Services

Obligation

A. **C** Information about support services (e.g., additional data transformations, format or coordinate conversions)?: **YES**

SECTION 12 – Authorization and Verification of the Metadata

This section examines how the standard provides for the description of the authorization and verification on the metadata of a dataset (e.g., how is the metadata itself authorized and verified?).

12.1 Metadata reference

Obligation

A. **M** Does the standard allow/require the provision of information about reference dates of the metadata (e.g., last check date, last update date, review date)?: **YES**

12.2 Verification Authority

Obligation

A. __ Does the standard allow/require the provision for information about who verified/authorized the metadata?: **NO**

B. __ Does the standard specify the method for providing this information (e.g., ISO 9000): **NO**

12.3 Statistical Methods Used for Verification

Obligation

A. __ Does the standard specify the statistical methods to be used for verification of the metadata?: **NO**

B. __ Does the standard allow/require the provision for the description of the statistical methods used to verify the metadata?: **NO**

END OF CHARACTERISTICS
FIN

INTERNATIONAL SPATIAL METADATA STANDARDS

World Spatial Metadata Standards
H. Moellering, H.J.G.L. Aalders & A. Crane (Editors)

The Dublin Core Metadata Element Set ISSN:1041-5653

Stuart L. Weibel

I. Information about the Metadata Standard Itself

SECTION 0 – Assessment Information

This section provides the user with a general description of the scope and intended use of the standard. Information on the respondent who prepared the assessment is also provided.

0.1 Respondent

Assessment prepared by:

Stuart L. Weibel
Executive Director
Dublin Core Metadata Initiative
OCLC Office of Research
6565 Frantz Road
Dublin, Ohio 43017, USA
Phone: +1 614 764 6081
Fax: +1 614 764 2344

Date assessment completed:

May 2002, Reviewed Spring, 2005

Relationship of respondent to standard:

Director, maintenance agency

0.2 Brief Summary Statement

A. Summary of the scope and intent of the standard:
The Dublin Core metadata element set is a standard for cross-domain information resource description. Here an information resource is defined to be anything that has identity; this is the definition used in Internet RFC 2396, "Uniform Resource Identifiers (URI): Generic Syntax", by Tim Berners-Lee et al. For Dublin Core applications a resource will typically be an electronic resource, but may be non-electronic or not accessible via the Internet as well.

This standard describes an element set only, which is generally used in the context of a specific project or application. Local or community based requirements and policies may impose additional restrictions, rules, and interpretations. It is not the purpose of this standard to define the detailed criteria by which the element set will be used within specific projects and applications.

SECTION 1 – Administration of Metadata Standard

This section provides a general description of the administrative framework, within which the standard was developed, tested and currently resides. Significant details are also provided in terms of the developmental and managerial history as well as information on the existence of documentation, software tools and training materials. This section also provides a point of contact for further information on the standard.

1.1 Name of Standard

A. Name of standard:
   1. In original language(s):
   The Dublin Core Metadata Element Set
   2. English translation, if appropriate:
   N/A

B. Version/Edition:
   ISO 15836-2003; ANSI/NISO Z39.85-2001

C. Language(s) of documentation:
   English

D. Acronym(s):
   DC; Dublin Core

E. Official ID:
   ISSN:1041-5653

F. Name(s) of recognizing standards authority(ies):
   Standardization of the Dublin Core has been pursued in three standards venues, as reflected below. The content of each rendition is substantively the same, with minor exceptions resulting from changes in other referenced standards and wording changes intended to clarify meaning. Unless otherwise indicated, statements in this publication concerning the Dublin Core refer to the ANSI/NISO version (Z39.85), as this is the most formal rendition of the standard at the time of this writing. Internet Engineering Task Force RFC 2413
   Dublin Core Metadata for Resource Discovery
   http://www.ietf.org/rfc/rfc2731.txt

CEN (CWA 13874 Dublin Core Metadata Element Set – Reference Description – Version 1.1 Download as pdf-file: http://www.cenorm.be/isss/cwa_download_area/cwa13874.pdf CEN is the European Committee for Standardization.

ANSI/NISO Z39.85 – 2001 Dublin Core Metadata Element Set Download as pdf-file: http://www.niso.org/standards/resources/Z39-85.pdf

1.2 Responsible Institutions

Contact information about those institutions that are (or were) responsible for the processes involved in the evolution of the standard:

A. Production and/or development:
Dublin Core Metadata Initiative
OCLC Office of Research
6565 Frantz Road
Dublin, Ohio 43017, USA
+1 614 764 6000
+1 614 764 2344
dcmi-feedback@dublincore.org

B. Testing:
N/A

C. Conformance:
N/A

D. Maintenance:
Dublin Core Metadata Initiative
OCLC Office of Research
6565 Frantz Road
Dublin, Ohio 43017, USA
+1 614 764 6000
+1 614 764 2344
dcmi-feedback@dublincore.org

E. Distribution:
Distribution of the standard is via the standards bodies referenced in section 1.1(F) of this chapter; copies of the standard are Web accessible as indicated in that section

F. Help Desk/User Support:
N/A

1.3 Development History

A. Beginning of work on metadata standard: **March 1995**

B. Milestones:

IETF informational RFC 2413	**September 1998**
CEN CWA 13874	**March 2000**
ANSI/NISO Z39.85-2001	**September 2001**

C. Anticipated completion: **On-going**
D. Update cycle: **5 Years**
for ANSI/NISO Z39.85-2001

1.4 Status of the Standard

A. Is the standard officially recognized: **YES**
B. Date(s) of recognition: **September 2001**
C. Anticipated date of recognition: **N/A**
D. Current stage in recognition process: **Complete**
E. Steps still required to achieve recognition: **N/A**

1.5 Access to Documentation of the Standard

A. Is the standard copyrighted?: **YES**
B. Name and address of copyright owner:
National Information Standards Organization
4733 Bethesda Avenue
Suite 300
Bethesda, MD 20814, USA
C. Do restrictions on the use of the standard apply?: **NO**
D. Is the standard documentation currently available?: **YES**
E. List the form/media of standard documentation:
Paper
Electronic
F. Is the standard available on the Internet?: **YES**
http://www.niso.org/standards/resources/Z39-85.pdf
G. Price: **Web accessible without cost**
H. Contact details for obtaining the standard:
National Information Standards Organization
4733 Bethesda Avenue
Suite 300
Bethesda, MD 20814, USA
I. Does the standard have ISBN/ISSN numbers?: **YES**
ISSN: 1041-5653

1.6 Available Software Tools

A. Is software available to implement the standard as a database?: **YES**
Tools and applications associated with implementing the Dublin Core are available from commercial organizations and as open source or free-ware from various sources. Current listings of such applications are accessible at: http://dublincore.org/tools/

B. Is software available to test compliance of metadata with the standard?: **YES**
The Dublin Core is a specification of semantics, and as such, unambiguous compliance testing is impractical. General tools for demonstration of syntactic compliance are available for various idioms in which DC metadata might be encoded, including HTML, XML, and RDF/XML

C. Is software available to produce metadata compliant with the standard?: **YES**
Tools and applications associated with implementing the Dublin Core are available from commercial organizations and as open source or free-ware from various sources. Current listings of such applications are accessible at: http://dublincore.org/tools/

D. Other relevant software: **YES**
Tools and applications associated with implementing the Dublin Core are available from commercial organizations and as open source or free-ware from various sources. Current listings of such applications are accessible at: http://dublincore.org/tools/

E. Is there a test dataset available to test software tools for implementing the standard?: **NO**

F. For each product available above, please list contact information:
See http://dublincore.org/tools/

1.7 Available Training Materials

A. Are organized training sessions available?: **YES**
OCLC Institute

B. Contact details for training sessions:
OCLC Institute
6565 Frantz Road
Dublin, Ohio 43017, USA
Tel.: +1 614 764 6000
Fax: +1 614 764 6096

C. Training documentation available:
(Note: References should include title, author, URL, source, date, number of pages)
Additional training and documentation is available at the Dublin Core Metadata Initiative Web site: http://dublincore.org

SECTION 2 – Use and Implementation of the Metadata Standard

This section provides information on the way in which the standard is used and the purpose of its use. It also provides descriptions about the level of abstraction of the standard.

2.1 Intended Use of the Standard
Can the standard be used:

A. Within a transfer data structure?: **YES**
B. To produce hardcopy catalog of datasets?: **YES**
C. To catalog data in a digital on-line catalog?: **YES**
D. Other:
The Dublin Core is intended to serve as a core element set to support cross-discipline discovery. As such, it is a general set that can be used as a foundation upon which to build richer descriptions while retaining general interoperability across other disciplines. It is expected that the Dublin Core will be implemented with discipline-specific extensions to increase the description precision

2.2 Intended Purpose of the Standard
Is the intended purpose of the standard:

A. To support discovery, identification, location of data?: **YES**
B. To support how to access data?: **YES**
C. To support the determination of dataset fitness for use?: **NO**
D. To support transfer of dataset?: **YES**
E. To support an organization's data management?: **YES**
F. Other: **N/A**

2.3 Table of Contents

A. Provide the standard's Table of Contents
Foreword
1. Scope and Purpose
2. Referenced Standards
3. Definitions
4. The Element Set
5. The Elements
Appendix A: Further Reading
Appendix B: Maintenance Agency

2.4 Level of Abstraction of the Metadata Standard Definition

A. Does the standard define metadata elements?: **YES**
B. Does the standard define a model for the relationships between the metadata elements?: **YES**
C. Provide an overview diagram of the model of the metadata, if available. **N/A**
D. Does the standard define domains and/or data types of the metadata elements?: **NO**
E. Does the standard define how the metadata should be encoded?: **NO**
Other documents at the Dublin Core website [http://dublincore.org] provide guidance as to various encoding options

F. Does the standard define a database schema for the automatic generation of a database?: **YES**

Other documents at the Dublin Core website http://dublincore.org provide guidance as to schema declaration options

2.5 Extensibility of the Standard

A. Is the metadata description extensible (e.g., Can user add metadata domains, elements, thematic profiles, etc.)?: **YES**

The Dublin Core is intended to be extensible. Implementation details of this extensibility are dependent on the encoding idiom used by a given implementation. The common encoding idioms used for Dublin Core metadata are HTML META tags, XML, and RDF/XML. Further details are available at http://dublincore.org (for the specifics of Dublin Core encoding) and at W3.org for details on XML, RDF, and supporting schema languages

2.6 Availability of Metadata

Does the standard specify:

A. How metadata are to be accessed?: **NO**
B. A mechanism for querying the metadata?: **NO**
C. Particular standardized query protocols?: **NO**
D. That the metadata should be encoded for computer access?: **YES**
E. That the metadata should be encrypted for security?: **NO**

2.7 Examples

A. Does the standard provide examples of implementations?: **YES**

Documentation available at http://dublincore.org provides examples for encoding Dublin Core metadata in HTML, XML, and RDF/XML

B. How many examples are available?:

The number changes as the documentation is updated

SECTION 3 – Linkage and Coordination of the Metadata Standard

This section provides information on how the metadata standard is linked to other standards, particularly transfer standards, and how such linkage is implemented. If a significant linkage does exist, detailed technical information is provided on the precise nature of the linkage to the standard, and how the linkage affects implementation of the metadata standard.

3.1 Associated Transfer Standards

A. Is the metadata standard associated with a transfer standard(s)?: **YES**

The Dublin Core defines semantics for cross-disciplinary resource discovery, and is implemented in a variety of transfer syntaxes. Common transfer syntaxes now in use with Dublin Core metadata include HTML encoding, XML encoding, and RDF/XML encoding

B. If A is yes, provide references and contact details for each linked standard:

Standardization of HTML, XML, and RDF/XML are all managed under the auspices of the World Wide Web Consortium (http://w3.org). Additional documentation on recommended encoding practices for Dublin Core for each of these transfer syntaxes is maintained at the Dublin Core Web Site (http:dublincore.org)

C. If A is yes, are both (all) standards coordinated?: **NO**

While these transfer syntaxes are not coordinated with Dublin Core, there is significant communication among developers in these areas resulting in updating of encoding recommendations as changes in the standards occur

D. If A is yes, describe how the coordination works, both from a technical and conformance (normative) viewpoint:

Coordination occurs informally as a result of activities of working group members involved in the various standards development

3.2 Incorporated into transfer mechanism

A. Is the metadata standard incorporated into a transfer mechanism?: **YES**

Incorporation of Dublin Core metadata into transfer mechanisms occurs as the result of project or product development. As Dublin Core was motivated by the emergence of the Web, the various encoding idioms of the Web are commonly used as transfer syntaxes as outlined in Section 3.1 above

B. If A is yes, provide references and contact details for each incorporated mechanism:

2002-04-14: Powell, Andy and Pete Johnston. Guidelines for implementing Dublin Core in XML
http://dublincore.org/documents/2002/04/14/dc-xml-guidelines/

2002-04-14: Kokkelink, Stefan and Roland Schwänzl. Expressing Qualified Dublin Core in RDF / XML
http://dublincore.org/documents/2002/04/14/dcq-rdf-xml/

2001-11-28: Beckett, Dave, Eric Miller and Dan Brickley. Expressing Simple Dublin Core in RDF / XML http://dublincore.org/documents/2001/11/28/dcmes-xml/

2000-08-15: Cox, Simon, Eric Miller & Andy Powell. Recording qualified Dublin Core metadata in HTML meta elements http://dublincore.org/documents/2000/08/15/dcq-html/

1999-12-16: Kunze, John. Encoding Dublin Core in HTML (IETF RFC 2731) http://www.ietf.org/rfc/rfc2731.txt

C. If A is yes, describe how:
The details of encoding Dublin Core are dependent on the strategy adopted by a given project (whether it is HTML-based, XML-based, or RDF/XML-based, and are elaborated in the documents cited in the previous section (3.2.B)

3.3 Other linked, coordinated or associated standards

A. Is the metadata standard linked, coordinated, or associated with any other standard(s) not listed above?: **NO**
B. If A is yes, provide references and contact details for each standard: **N/A**
C. If A is yes, describe how for each relevant standard: **N/A**

II. Categories the Metadata Standard Uses to Describe Datasets

The following sections are categories of metadata that may be used by the metadata standard to describe data. Not all categories are relevant to different datasets and may not necessarily be present in the standard. The presence or absence of a category should not be used for judging one standard against another.

Note: For each of the questions in the following sections please provide the obligation status information. Valid obligation values are:

	Obligation Level	
M	Mandatory	This information must be entered to comply with the metadata standard.
C	Conditional	This information is entered based on other information in the metadata.
O	Optional	This information is optionally entered in the metadata.

If the answer is 'NO', the obligation status should be left blank.

SECTION 4 – Dataset Identification

This section examines how the standard provides for the specification of the various sorts of titles and identifiers for a dataset.

4.1 The title(s) of a dataset
Does the standard allow/require the provision for:

Obligation

A. __ Previous title(s)?: **NO**
B. **O** Current title(s)?: **YES**
C. __ Possible future title(s)?: **NO**
D. **O** Official titles in different languages?: **YES**
E. **O** Abbreviated title(s)?: **YES**
F. **O** Alternative Title(s)?: **YES**
G. __ Does the standard support effective dates of titles?: **NO**
H. **O** Other identifiers (e.g., ISBN, National Stock Number, Uniform Product Code)?: **YES**

Any globally unique identifier scheme can be used in Dublin Core. It is the responsibility of the application to specify the scheme used (for example, URL, ISBN, UPC)

4.2 Parties responsible for the dataset
Does the standard allow/require the provision for:

Obligation

A. **O** Author?: **YES**
B. **O** Owner?: **YES**
C. **O** Other?: **YES**

4.3 ISBD Information

Obligation

A. __ Does the metadata standard specify inclusion of mandatory ISBD information (see glossary) within the metadata?: **NO**

SECTION 5 – Status of the Dataset

This section examines how the standard provides for the specification of the current status and updating of the data set.

5.1 Status and Progress of Datasets
Does the metadata standard allow/require the provision for:

Obligation

A. __ Information on the status of the entire dataset?: **NO**
B. __ Information on the status of the particular subset (e.g., observation time, inauguration time, extraction time)?: **NO**

Does the metadata standard specify:

C. __ A structure for updating or maintenance information?: **NO**
D. __ A structure to describe historical information (see also Section 8.3 G)?: **NO**

5.2 Information on Incremental Updating
Does the metadata standard specify:

Obligation

A. __ A structure for information on incremental updates of the entire dataset?: **NO**
B. __ A structure for information on incremental updates of particular portions of the dataset?: **NO**
C. __ A structure to describe the update process for the dataset?: **NO**

SECTION 6 – Extent of the Dataset

This section examines how the standard provides for the specification of the spatial and temporal extent of the dataset.

6.1 The spatial extent of the dataset

Obligation

A. **O** Does the metadata standard allow/require for the description of horizontal extent (e.g., bounding rectangle/polygon, named location (i.e., Stockholm, etc.))?: **YES**
The Coverage element provides a general mechanism for specifying both spatial and temporal extent, however the further qualification of this element is the responsibility of the application (specification of a given geographic thesaurus, coordinate scheme, bounding polygon scheme, or other controlled vocabulary or scheme)
B. **O** Does the metadata standard allow/require for the description of vertical extent (e.g., minima/maxima heights, etc.)?: **YES**
See comment in 6.1.A
C. __ Does the standard require specific units for describing extent?: **NO**
D. **O** Does the standard allow specification of the units for describing extent?: **YES**
E. __ Does the standard require specific spatial reference system for describing extent?: **NO**
F. **O** Does the standard allow specification of the spatial reference system for describing extent?: **YES**
See comment in 6.1.A

6.2 The temporal extent of the dataset

Obligation

A. **O** Does the standard allow/require for the description of time period for which the dataset is relevant?: **YES**

B. **O** Does the standard allow/require specification of the temporal reference system?: **YES**

The Coverage element provides a general mechanism for specifying both spatial and temporal extent, however the further qualification of this element is the responsibility of the application (specification of a given geographic thesaurus, coordinate scheme, bounding polygon scheme, or other controlled vocabulary or scheme)

SECTION 7 – Data Content of the Dataset

This section examines how the standard provides for the specification of the data content of the dataset in terms of overview elements, themes, data definitions and product specifications.

7.1 Dataset overview

Does the standard allow/require for the provision of the following information:

Obligation

A. **O** Dataset overview?: **YES**
B. **O** Abstract – brief summary of the content of the dataset?: **YES**
C. **O** Purpose – summary of the intentions with which the dataset was developed?: **YES**
D. __ Usage – list of activities for which the dataset has been used? (see also Section 8.5): **NO**
E. **O** Spatial schema – (e.g., vector, raster)? (see also Section 9): **YES**
F. **O** Spatial reference system? (see also Section 12): **YES**
G. **O** Language?: **YES**
H. __ Character set (e.g., ISO8859-1, JISX0208)?: **NO**
I. __ List of documents providing further information about the dataset?: **NO**
J. __ Sample/preview dataset?: **NO**
K. **O** Related datasets?: **YES**

The Relation element can be used to provide links to related data

L. __ Original size of the dataset?: **NO**
M. __ Compression? (see also Section 11.5): **NO**
N. __ Other?: **YES**

The Dublin Core architecture allows for the extension of the element set according to local application needs and the combination of Dublin Core metadata with other metadata standards through the specification of an Application Profile that is embodied in a compound schema created from existing schemas

7.2 Theme(s)

Obligation

A. __ Does the standard allow/require the provision for the description of theme(s) (e.g., topography, land cover, transport, etc.) in the dataset?: **NO**

B. __ If A is yes, does the standard allow the referencing of an outside dictionary or thesaurus?: **N/A**
C. __ If A is yes, does the standard itself include a dictionary or thesaurus within the metadata?: **N/A**
D. __ If A is yes, does the standard allow the use of unauthorized user-defined keywords?: **N/A**

7.3 Data Definition

Obligation

A. __ Does the standard allow/require the provision for the description of the data definition?: **NO**
B. If A is yes, does the standard allow/require the provision for the description of:
__ 1. Features/Objects?: **N/A**
__ 2. Attributes?: **N/A**
__ 3. Relations?: **N/A**
C. __ Can attributes be described separately from feature/object classes?: **NO**
D. __ Does the standard allow/require the specification of the domain of the attributes?: **NO**
E. __ Does the standard allow/require the specification of data type of the attributes?: **YES**
All DC metadata elements can be qualified by schemes which can restrict or specify the scope, data type, or controlled vocabulary from which a value is selected
F. __ Does the standard allow/require the specification of particular classification coding schemes (e.g., FACC, OSKA, S-57 OC)?: **YES**
Classification systems, controlled vocabulary, or encoding schemes can be specified for any Dublin Core element where appropriate

7.4 Product Specification

Obligation

A. __ Does the standard allow/require for the provision of information about whether the dataset meets a particular product specification?: **NO**
B. __ If A is yes, can it be internal to the Metadata?: **N/A**
C. __ If A is yes, can it be external to the Metadata?: **N/A**

SECTION 8 – Data Quality of the Dataset

This section examines how the standard deals with the question of data quality. This includes homogeneity and metrics for data quality, quality assurance, and usage.

8.1 Overall Data Quality Statement

Obligation

A. __ Does the standard allow/require for the provision of quality to be described? If yes, please answer the questions below. (*If no, all following answers in Section* 8 *are N/A, please skip to Section* 9.): **NO**

B. __ Does the standard allow/require the provision for the quality information referencing an external quality standard, quality assurance standard, or specification?: **N/A**

8.2 Homogeneity of Data Quality

Obligation

A. __ Does the standard allow/require for the variation of the quality of the dataset to be described?: **N/A**

B. What type of aggregation of quality information is allowed?

__ 1. Entire database?: **N/A**
__ 2. Feature/Object instance?: **N/A**
__ 3. Feature/Object class?: **N/A**
__ 4. Attribute instance?: **N/A**
__ 5. Attribute class?: **N/A**
__ 6. Layer?: **N/A**
__ 7. Theme?: **N/A**
__ 8. Geographic extent?: **N/A**
__ 9. Other?: **N/A**

8.3 Metrics for Data Quality
Does the standard allow/require the provision of values for:

Obligation

A. __ Positional accuracy?: **N/A**
1. Standard error?: **N/A**
2. Maximum error?: **N/A**
3. Error correlation?: **N/A**
4. Confidence level (e.g., 90%, 95%)?: **N/A**
5. Other?: **N/A**

B. __ Attribute accuracy?: **N/A**
1. Standard error?: **N/A**
2. Maximum error?: **N/A**
3. Error frequency?: **N/A**
4. Error correlation?: **N/A**
5. Confidence level (e.g., 90%, 95%)?: **N/A**
6. Other?: **N/A**

C. __ Temporal accuracy?: **N/A**
1. Standard error?: **N/A**
2. Maximum error?: **N/A**
3. Error correlation?: **N/A**
4. Confidence level? (e.g., 90%, 95%): **N/A**
5. Other?: **N/A**

D.	__ Completeness?:	**N/A**
	1. Error of omission?:	**N/A**
	2. Error of commission?:	**N/A**
	3. Other?:	**N/A**
E.	__ Currentness?:	**N/A**
	1. Date?:	**N/A**
	2. Maximum age?:	**N/A**
	3. Temporal extent?:	**N/A**
	4. Other?:	**N/A**
F.	__ Logical consistency?:	**N/A**
G.	__ Lineage?:	**N/A**
	1. List of processing steps?:	**N/A**
	2. List of values of processing parameters?:	**N/A**
	3. List of printouts of statistics, etc.?:	**N/A**
	4. Source material?:	**N/A**
	5. Other?:	**N/A**
H.	__ Additional Quality Information?:	**N/A**

8.4 Quality Assurance Methods

Does the standard allow/require the provision of information about the methods used to determine:

Obligation

A.	__ Positional accuracy?:	**N/A**
	1. Standard error in adjustments?:	**N/A**
	2. Repeated measurements?:	**N/A**
	3. Independent measurements?:	**N/A**
	4. Subjective evaluations?:	**N/A**
	5. Other?:	**N/A**
B.	__ Attribute accuracy?:	**N/A**
	1. Standard error?:	**N/A**
	2. Error classification matrix?:	**N/A**
	3. Independent measurements?:	**N/A**
	4. Subjective evaluation?:	**N/A**
	5. Other?:	**N/A**
C.	__ Temporal accuracy?:	**N/A**
	1. Independent measurements?:	**N/A**
	2. Subjective evaluation?:	**N/A**
	3. Other?:	**N/A**
D.	__ Completeness?:	**N/A**
	1. Field checks?:	**N/A**
	2. Repeated measurements?:	**N/A**
	3. Independent measurements?:	**N/A**
	4. Subjective evaluation?:	**N/A**
	5. Error of commission?:	**N/A**

		6. Other?:	**N/A**
E.	__	Currentness?:	**N/A**
		1. Date of source material?:	**N/A**
		2. Repeated measurements?:	**N/A**
		3. Independent measurements?:	**N/A**
		4. Subjective evaluation?:	**N/A**
		5. Other?:	**N/A**
F.	__	Logical consistency?:	**N/A**
		1. Verification results of consistency checks?:	**N/A**
		2. Repeated processing?:	**N/A**
		3. Comparison with independent sources?:	**N/A**
		4. Subjective evaluation?:	**N/A**
		5. Other?:	**N/A**
G.	__	Lineage?:	**N/A**
		1. Checking results of processing steps:	**N/A**
		2. Subjective evaluation:	**N/A**
		3. Other:	**N/A**
H.	__	Additional Quality Information?:	**N/A**

8.5 Usage

Obligation

A.	__	Does the metadata standard allow/require the provision of information about usage?:	**N/A**
		1. Who used it:	**N/A**
		2. When it was used:	**N/A**
		3. For what application it was used:	**N/A**
		4. The effectiveness of use:	**N/A**
		5. Other:	**N/A**

8.6 Alternative Quality Description: **N/A**

SECTION 9 – Spatial Data Structure of the Dataset

This section examines how the standard provides for the specification of the spatial data model and spatial data types.

9.1 Spatial Data Model
Does the standard allow/require the provision for:

Obligation

A. **O** The description of the geometrical and/or topological primitives used to model the real world?: **YES**

B. **O** Information about an external standard for geometrical and/or topological primitive specification?: **YES**

The Coverage element provides a general mechanism for specifying both spatial and temporal extent, including specification of encoding schemes of a given geographic thesaurus, coordinate scheme, bounding polygon scheme, or other controlled vocabulary or scheme

9.2 Spatial Data Types

Does the standard allow/require the provision for:

Obligation

A. **O** Information about the spatial data type(s) used in the dataset?: **YES**

B. **O** If A is yes, which of the following generic spatial data types are supported? For each, please list specific data primitive levels supported by each data type.

1. Vector (e.g., spaghetti, chain-node/node-edge-face, planar graph, full topology)?: **NO**
2. Areal Raster (e.g., thematic raster – scanned and classification maps, image raster – satellite images, scanned aerial photographs)?: **NO**
3. Grid/Matrix (e.g., Digital Elevation Model)?: **NO**

Triangular Irregular Network (TIN)?: **NO**

Other?: **YES**

Any data type can be specified within the Dublin Core, but as a general description format, it is more likely that such specific schemes would be included within extensions or application profiles that import such data types from other standards

9.3 Other Data Types

Does the standard allow the provision of information about:

Obligation

A. **O** Video?: **YES**

B. **O** Sound?: **YES**

C. **O** Other?: **YES**

Any media type can be specified within Dublin Core. Internet MIME types are the recommended typology for such specification, using the Format element

SECTION 10 – Spatial Reference of the Dataset

This section examines how the dataset provides for the specification of the spatial reference characteristics of the data set.

Does the standard allow/require the provision for the following information:

10.1 Geoidal Model

Obligation

A. __ Geoidal model name?: **NO**
B. __ Geoidal model definitions?: **NO**

10.2 Geodetic datum

Obligation

A. __ Geodetic datum name?: **NO**
B. __ Geodetic datum definitions?: **NO**

10.3 Reference Ellipsoid

Obligation

A. __ Reference ellipsoid name?: **NO**
B. __ Reference ellipsoid definitions?: **NO**

10.4 Projection

Obligation

A. __ Projection name?: **NO**
B. __ Projection definitions?: **NO**

10.5 Coordinate Systems

Obligation

A. __ Coordinate system type?: **NO**
B. __ Coordinate system definitions?: **NO**

10.6 Height Reference System

Obligation

A. __ Vertical datum name?: **NO**
B. __ Types of heights?: **NO**

10.7 Ancillary Geodetic Information

Obligation

A. __ Does the standard allow/require the provision for ancillary geodetic, gravimetric, or magnetometric information?: **NO**

10.8 Transformation Parameters
Does the standard specify a method to provide information for coordinate transformation(s):

Obligation

A. __ Model name?: **NO**
B. __ Control points?: **NO**

C. __ Transformation parameters?: **NO**

10.9 Non-Coordinate Spatial Reference

Does the standard allow/require the provision for information on non-coordinate spatial reference systems (e.g., postal systems, street address)?:

Obligation

A. __ Reference system name?: **NO**
B. __ Reference system definitions?: **NO**
C. __ Other?: **NO**

SECTION 11 – Availability and Distribution of the Dataset

This section examines how the standard provides information on the availability of the dataset?

Does the standard allow/require the provision for:

11.1 The Distributor(s) of the Dataset

Obligation

A. **O** Information about the distributor(s) of the dataset?: **YES**
B. **O** Contact information for the distributor(s)?: **YES**

11.2 Restrictions to Access and Usage

Obligation

A. x Information about copyright on the dataset (e.g., date, period, copyright owner, etc.)?: **YES**
The Rights element is intended to meet this requirement
B. x Information about restrictions on the access of the dataset?: **YES**
The Rights element can be used to meet this requirement, though it does not provide for structured management of intellectual property rights
C. x Information about restrictions on the usage of the dataset?: **YES**
The Rights element can be used to meet this requirement, though it does not provide for structured management of intellectual property rights

11.3 Pricing Details

Obligation

A. __ Information about pricing?: **NO**

11.4 Distribution of the Dataset

Obligation

A. __ Information about distribution media?: **NO**
B. **O** Information about on-line access?: **YES**
The Identifier element can include URLs linking to described resources or datasets
C. __ Information about units of distribution (e.g., tiles, layers, polygons, elements)?: **NO**

11.5 Distribution Format

Obligation

A. __ Information about distribution formats (e.g., commercial CAD format, commercial GIS formats, National/International transfer standard)?: **NO**
B. __ Information about the encoding of the dataset (e.g., binary, ASCII)?: **NO**
C. __ Information about compression method?: **NO**
D. __ Information about compressed dataset size?: **NO**
E. __ Information about the distribution mechanism?: **NO**
F. __ Information about the encapsulation of data (e.g., ISO 8211, ISO 8824)?: **NO**

Ordering Procedures

11.6 Obligation

A. __ Information about ordering procedures for the dataset?: **NO**

11.7 Support Services

Obligation

A. __ Information about support services (e.g., additional data transformations, format or coordinate conversions)?: **NO**

SECTION 12 – Authorization and Verification of the Metadata

This section examines how the standard provides for the description of the authorization and verification on the metadata of a dataset (e.g., how is the metadata itself authorized and verified?).

12.1 Metadata reference

Obligation

A. **O** Does the standard allow/require the provision of information about reference dates of the metadata (e.g., last check date, last update date, review date)?: **YES**

12.2 Verification Authority

Obligation

A. __Does the standard allow/require the provision for information about who verified/authorized the metadata?: **NO**

B. __Does the standard specify the method for providing this information (e.g., ISO 9000): **NO**

12.3 Statistical Methods Used for Verification

Obligation

A. __Does the standard specify the statistical methods to be used for verification of the metadata?: **NO**

B. __Does the standard allow/require the provision for the description of the statistical methods used to verify the metadata?: **NO**

END OF CHARACTERISTICS

FIN

World Spatial Metadata Standards
H. Moellering, H.J.G.L. Aalders & A. Crane (Editors)

CEN/TC287: Geographic Information – Data Description – Metadata prENV 12657:1998

François Salgé

I. Information about the Metadata Standard Itself

SECTION 0 – Assessment Information

This section provides the user with a general description of the scope and intended use of the standard. Information on the respondent who prepared the assessment is also provided.

0.1 Respondent

Assessment prepared by:

François Salgé
Chairman of CEN/TC 287
Institut Géographique National
136bis rue de Grenelle
75700 Paris 07SP
Fax: + 33 1 43 98 84 40
Phone: + 33 1 43 98 82 70
E-mail: francois.salge@ign.fr

Date assessment completed:

July 1998, updated September 2000 and May 2002, Reviewed Spring, 2005

Relationship of respondent to standard:

Chairman of CEN/TC 287, technical committee of Comité Européen de Normalisation in charge of standardization in the field of geographic information and having designed the standard

0.2 Brief Summary Statement

A. Summary of the scope and intent of the standard:
This European prestandard (i.e., prENV 12657:1998) establishes a conceptual schema for metadata.

Metadata can be defined at its simplest as 'data about datasets'. This European prestandard specifies those data which shall be used to describe a geographic dataset. This includes data about the content, representation, extent (both geometric and temporal), spatial reference, quality and administration of a geographic dataset.
This European prestandard also identifies those data which are mandatory for describing geographic datasets, the minimum set of metadata.
This European prestandard gives examples of how the standard may be applied but does not provide instructions or techniques for its implementation and accordingly does not concern itself with the construction of databases for holding metadata.
This European prestandard is designed primarily for use with digital geographic datasets but the principles can also be used to describe geographic datasets in other forms, such as paper maps or lists.
This European prestandard is applicable for a range of uses. It allows the geographic dataset to be described in a consistent way which makes it easy to maintain a metadata catalogue service even from remote location. This European prestandard allows metadata to be described for a number of purposes such as documentation and inclusion in a transfer file and for comparing and exploring geographic datasets.

SECTION 1 – Administration of Metadata Standard

This section provides a general description of the administrative framework, within which the standard was developed, tested and currently resides. Significant details are also provided in terms of the developmental and managerial history as well as information on the existence of documentation, software tools and training materials. This section also provides a point of contact for further information on the standard.

1.1 Name of Standard

A. Name of standard:
   1. In original language(s):
      Geographic Information – Data Description – Metadata
   2. English translation, if appropriate:
      N/A

B. Version/Edition:
   version 1998 date 1998-05-21

C. Language(s) of documentation:
   English, French and German

D. Acronym(s):
   N/A

E. Official ID:
   prENV 12657:1998

F. Name(s) of recognizing standards authority(ies):

All national standard authorities of Europe member or observer of CEN

1.2 Responsible Institutions

Contact information about those institutions that are (or were) responsible for the processes involved in the evolution of the standard:

A. Production and/or development:
CEN/TC 287 c/o AFNOR
Tour Europe
F-92xxxx Paris la Défense

B. Testing:
to be defined

C. Conformance:
to be defined

D. Maintenance:
CEN
European Committee for Standardization
Comité Européen de Normalisation
Europäisches Komitee für Normung

E. Distribution:
CEN and CEN members (national standardization bodies)

F. Help Desk/User Support:
CEN members (national standardization bodies)

1.3 Development History

A. Beginning of work on metadata standard: **1994-03**

B. Milestones:

1996-04	**First working draft**
1996-12	**six month inquiry among CEN members**
1997-09	**comment resolution meeting**
1998-05	**completion**

C. Anticipated completion:

D. Update cycle: **2 years**

1.4 Status of the Standard

A. Is the standard officially recognized: **expected**

B. Date(s) of recognition:

C. Anticipated date of recognition: **98-08-21**

D. Current stage in recognition process:
Formal voting is based on weighted vote within CEN members

E. Steps still required to achieve recognition:

1.5 Access to Documentation of the Standard

A. Is the standard copyrighted?: **YES**

B. Name and address of copyright owner: **CEN**
C. Do restrictions on the use of the standard apply?: **NO**
none other than restrictions related to copyright regime
D. Is the standard documentation currently available?: **YES**
E. List the form/media of standard documentation:
paper
F. Is the standard available on the Internet?: **YES**
http://forum.afnor.fr/afnor/WORK/AFNOR/GPN2/Z13C/indexen.htm
G. Price: **UNKNOWN**
H. Contact details for obtaining the standard:
CEN and all National standardization bodies member of CEN
I. Does the standard have ISBN/ISSN numbers?: **NO**

1.6 Available Software Tools

A. Is software available to implement the standard as a database?: **NO**
B. Is software available to test compliance of metadata with the standard?: **NO**
C. Is software available to produce metadata compliant with the standard?: **NO**
D. Other relevant software:
E. Is there a test dataset available to test software tools for implementing the standard?: **NO**
F. For each product available above, please list contact information: **N/A**

1.7 Available Training Materials

A. Are organized training sessions available?:
planed for future
B. Contact details for training sessions:
EUROGI and CEN members
C. Training documentation available:
not yet available

SECTION 2 – Use and Implementation of the Metadata Standard

This section provides information on the way in which the standard is used and the purpose of its use. It also provides descriptions about the level of abstraction of the standard.

2.1 Intended Use of the Standard
Can the standard be used:

A. Within a transfer data structure?: **YES**
B. To produce hardcopy catalog of datasets?: **YES**
C. To catalog data in a digital on-line catalog?: **YES**
D. Other: **NO**

2.2 Intended Purpose of the Standard
Is the intended purpose of the standard:

A. To support discovery, identification, location of data?: **YES**
B. To support how to access data?: **YES**
C. To support the determination of dataset fitness for use?: **YES**
D. To support transfer of dataset?: **YES**
E. To support an organization's data management?: **YES**
F. Other: **NO**

2.3 Table of Contents

A. Provide the standard's Table of Contents

Foreword
1 Scope
2 Normative references
3 Definitions
4 Metadata for describing geographic datasets
4.1 General
4.2 Geographic datasets and levels of metadata
4.3 Conformance requirements
4.4 Dates
4.5 Metadata language
4.6 Elements of metadata for describing geographic datasets
Annex A (normative) EXPRESS SCHEMA Metadata
Annex B (informative) EXPRESS-G metadata schema
B.1 Metadata schema page 1
B.2 Metadata schema page 2, dataset identification
B.3 Metadata schema page 3, dataset overview
B.4 Metadata schema page 4, dataset quality elements
B.5 Metadata schema page 5, metadata reference
B.6 Metadata schema page 6, spatial reference system
B.7 Extent metadata schema
B.8 Data definition metadata schema
B.9 Classification metadata schema
B.10 Administrative metadata schema page 1, organisation and point of contact
B.11 Administrative metadata schema page 2, distribution
Annex C (informative) Elements of metadata – Full listing
Annex D (informative) Examples of how to use the standard
D.1 Example 1 – Metadata for a utility dataset
D.2 Example 2 – Metadata for an Administrative Boundaries Dataset

2.4 Level of Abstraction of the Metadata Standard Definition

A. Does the standard define metadata elements?: **YES**

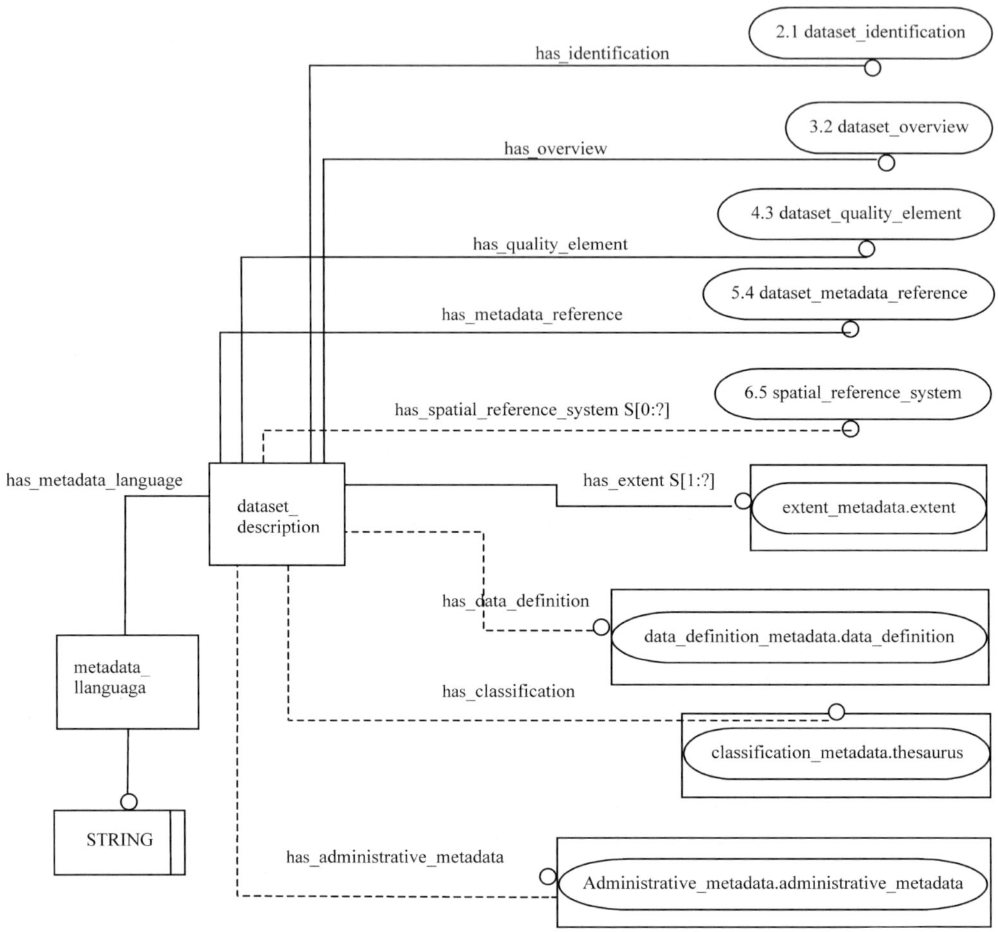

Fig. 1. EXPRESS schema CEN metadata.

B. Does the standard define a model for the relationships between the metadata elements?: **YES**
C. Provide an overview diagram of the model of the metadata, if available. **YES**
D. Does the standard define domains and/or data types of the metadata elements?: **YES**
E. Does the standard define how the metadata should be encoded?: **YES**
F. Does the standard define a database schema for the automatic generation of a database?: **NO**

2.5 Extensibility of the Standard

A. Is the metadata description extensible (e.g., Can user add metadata domains, elements, thematic profiles, etc.)?: **NO**

2.6 Availability of Metadata

Does the standard specify:

A. How metadata are to be accessed?: **NO**
B. A mechanism for querying the metadata?: **NO**
C. Particular standardized query protocols?: **NO**
D. That the metadata should be encoded for computer access?: **NO**
E. That the metadata should be encrypted for security?: **NO**

2.7 Examples

A. Does the standard provide examples of implementations?: **YES**
B. How many examples are available?: **2**

SECTION 3 – Linkage and Coordination of the Metadata Standard

This section provides information on how the metadata standard is linked to other standards, particularly transfer standards, and how such linkage is implemented. If a significant linkage does exist, detailed technical information is provided on the precise nature of the linkage to the standard, and how the linkage affects implementation of the metadata standard.

3.1 Associated Transfer Standards

A. Is the metadata standard associated with a transfer standard(s)?: **YES**
B. If A is yes, provide references and contact details for each linked standard:
    prENV 12658:1998 – Geographic information – Data description – Transfer
C. If A is yes, are both (all) standards coordinated?: **YES**
D. If A is yes, describe how the coordination works, both from a technical and conformance (normative) viewpoint:
    Coordination is done via the reference Model. CEN/TC 287 reference model describes and clarifies the field of Geographic Information, relationships and differences between Geographic Information and non-Geographic Information, identifies and defines components which can be standardized, their interfaces and relationships between components, provides safeguard against duplication of effort and avoids the creation of new standards where others already exist or are under consideration, and makes it possible for new components to be adopted as new standards and technology develop.

3.2 Incorporated into transfer mechanism

A. Is the metadata standard incorporated into a transfer mechanism?: **NO**
B. If A is yes, provide references and contact details for each incorporated mechanism: **N/A**
C. If A is yes, describe how: **N/A**

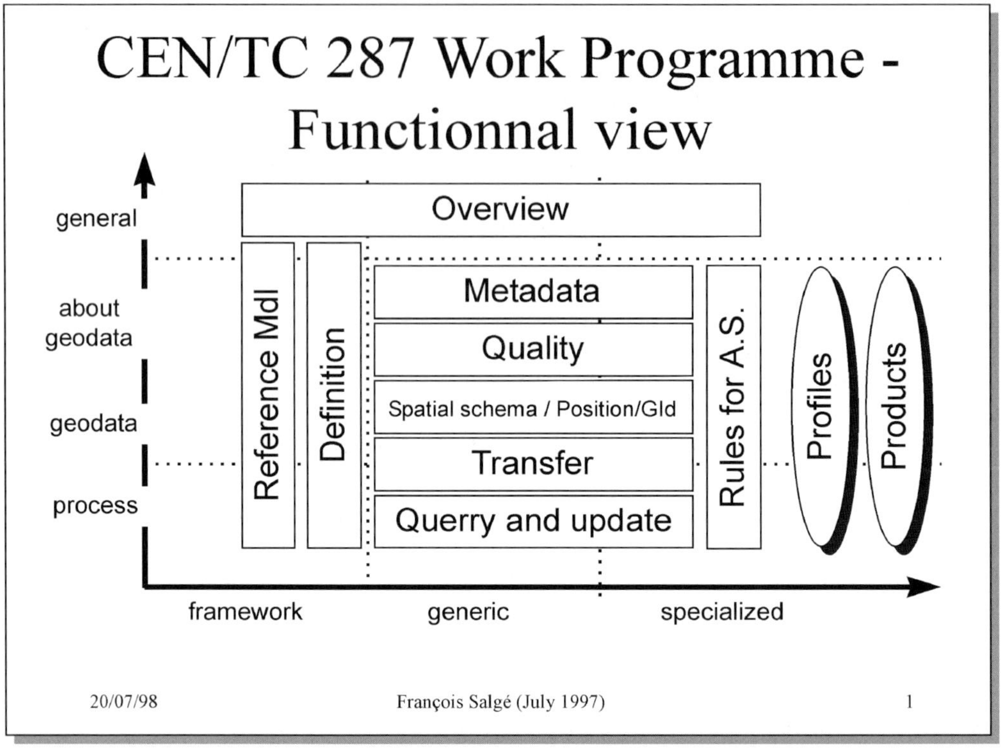

Fig. 2. CEN Work Program – Functional view.

3.3 Other linked, coordinated or associated standards

A. Is the metadata standard linked, coordinated, or associated with any other standard(s) not listed above?: **YES**
B. If A is yes, provide references and contact details for each standard:
C. If A is yes, describe how for each relevant standard:

II. Categories the Metadata Standard Uses to Describe Datasets

The following sections are categories of metadata that may be used by the metadata standard to describe data. Not all categories are relevant to different datasets and may not necessarily be present in the standard. The presence or absence of a category should not be used for judging one standard against another.

Obligation Level

M Mandatory This information must be entered to comply with the metadata standard.
C Conditional This information is entered based on other information in the metadata.
O Optional This information is optionally entered in the metadata.

If the answer is 'NO', the obligation status should be left blank.

Reference	Date	Title	Status
ENV 12009: 1997	**1997-06-23**	**Geographic information – Reference model**	**adopted**
ENV 12160: 1997	**1997-08-13**	**Geographic information – Data description – Spatial schema**	**adopted**
prENV 12656: 1998	**1998-05-21**	**Geographic information – Data description – Quality**	**formal voting**
prCR 12660: 1998	**1998-05-21**	**Geographic information – Processing – Query and update**	**formal voting**
prENV 12661: 1998	**1998-05-21**	**Geographic information – Referencing – Geographic identifiers**	**formal voting**
prENV 12762: 1998	**1998-06-25**	**Geographic information – Referencing – Position**	**formal voting**
CEN/TC 287/WG2 N45	**1996-10-26**	**Geographic information – Data description – Rules for application schema**	**WG draft**
CEN/TC 287/SWG1.4 N5.7	**1997-04-15**	**Geographic information – Fundamentals – Definition**	**SWG draft**
CEN/TC 287/N455	**1996-02-02**	**Geographic information – Fundamentals – Conceptual schema language**	**approval BT**
CEN/TC 287 N 540	**1997-02-17**	**Geographic information – Fundamentals – Overview**	**WG draft**
EN 28601:1992		**Data elements and interchange formats – Information interchange – Representation of dates and times**	
ENV ISO 10303-11:1994		**Industrial automation systems and integration – Product data representation and exchange – Part 11: Description methods: The EXPRESS language reference manual**	
ENV ISO 10303-41:1994		**Industrial automation systems and integration – Product data representation and exchange – Part 41: Integrated generic resources: Fundamentals of product description and support**	
ISO 639:1988		**Codes for the representation of names of languages**	
ISO 2788:1988		**Documentation – Guidelines for the establishment and development of monolingual thesauri**	

SECTION 4 – Dataset Identification

This section examines how the standard provides for the specification of the various sorts of titles and identifiers for a dataset.

4.1 The title(s) of a dataset
Does the standard allow/require the provision for:

Obligation

A. __ Previous title(s)?: **NO**

B. **M** Current title(s)?: **YES**
C. __ Possible future title(s)?: **NO**
D. **O** Official titles in different languages?: **YES**
E. **O** Abbreviated title(s)?: **YES**
F. **O** Alternative Title(s)?: **YES**
G. __ Does the standard support effective dates of titles?: **NO**
H. **O** Other identifiers (e.g., ISBN, National Stock Number, Uniform Product Code)?: **YES**
version

4.2 Parties responsible for the dataset
Does the standard allow/require the provision for:

Obligation

A. **M** Author?: **YES**
B. **M** Owner?: **YES**
C. **O** Other?: **YES**
Organization name and role as free text

4.3 ISBD Information

Obligation

A. __ Does the metadata standard specify inclusion of mandatory ISBD information (see glossary) within the metadata?: **NO**

SECTION 5 – Status of the Dataset

This section examines how the standard provides for the specification of the current status and updating of the data set.

5.1 Status and Progress of Datasets
Does the metadata standard allow/require the provision for:

Obligation

A. **O** Information on the status of the entire dataset?: **YES**
B. **O** Information on the status of the particular subset (e.g., observation time, inauguration time, extraction time)?: **YES**
subsets are considered as datasets, the field related datasets allows then to link subsets to datasets. Any information on the status of particular subset is thus enabled.

Does the metadata standard specify:

C. __ A structure for updating or maintenance information?: **NO**
D. __ A structure to describe historical information (see also Section 8.3 G)?: **NO**

5.2 Information on Incremental Updating

Does the metadata standard specify:

Obligation

A. __ A structure for information on incremental updates of the entire dataset?: **NO**

B. __ A structure for information on incremental updates of particular portions of the dataset?: **NO**

C. __ A structure to describe the update process for the dataset?: **NO**

SECTION 6 – Extent of the Dataset

This section examines how the standard provides for the specification of the spatial and temporal extent of the dataset.

6.1 The spatial extent of the dataset

Obligation

A. **C** Does the metadata standard allow/require for the description of horizontal extent (e.g., bounding rectangle/polygon, named location (i.e., Stockholm, etc.))?: **YES**

Bounding quadrangle, Bounding area, geographic area, name of the area covered by the geographic dataset, name of the geographic identifiers system, code of the geographic area, coverage of the geographic area

B. **C** Does the metadata standard allow/require for the description of vertical extent (e.g., minima/maxima heights, etc.)?: **YES**

minimum height value, maximum height value

C. __ Does the standard require specific units for describing extent?: **NO**

D. __ Does the standard allow specification of the units for describing extent?: **NO**

E. __ Does the standard require specific spatial reference system for describing extent?: **NO**

F. **C** Does the standard allow specification of the spatial reference system for describing extent?: **YES**

refers to ENV 12160:1998

6.2 The temporal extent of the dataset

Obligation

A. **C** Does the standard allow/require for the description of time period for which the dataset is relevant?: **YES**

refers to EN 28601:1992

B. **C** Does the standard allow/require specification of the temporal reference system?: **YES**

SECTION 7 – Data Content of the Dataset

This section examines how the standard provides for the specification of the data content of the dataset in terms of overview elements, themes, data definitions and product specifications.

7.1 Dataset overview

Does the standard allow/require for the provision of the following information:

Obligation

A. **M** Dataset overview?: **YES**
B. **M** Abstract – brief summary of the content of the dataset?: **YES**
C. **O** Purpose – summary of the intentions with which the dataset was developed?: **YES**
D. **O** Usage – list of activities for which the dataset has been used? (see also Section 8.5): **YES**
E. **M** Spatial schema – (e.g., vector, raster)? (see also Section 9): **YES**
F. **O** Spatial reference system? (see also Section 12): **YES**
G. **M** Language?: **YES**
H. **M** Character set (e.g., ISO8859-1, JISX0208)?: **YES**
I. **O** List of documents providing further information about the dataset?: **YES**
J. **O** Sample/preview dataset?: **YES**
K. **O** Related datasets?: **YES**
L. __ Original size of the dataset?: **NO**
M. __ Compression? (see also Section 11.5): **NO**
N. **O** Other?: **YES**
intended application scale

7.2 Theme(s)

Obligation

A. **C** Does the standard allow/require the provision for the description of theme(s) (e.g., topography, land cover, transport, etc.) in the dataset?: **YES**
B. **C** If A is yes, does the standard allow the referencing of an outside dictionary or thesaurus?: **YES**
C. **C** If A is yes, does the standard itself include a dictionary or thesaurus within the metadata?: **YES**
D. **C** If A is yes, does the standard allow the use of unauthorized user-defined keywords?: **YES**

7.3 Data Definition

Obligation

A. **O** Does the standard allow/require the provision for the description of the data definition?: **YES**
B. **C** If A is yes, does the standard allow/require the provision for the description of:
C 1. Features/Objects?: **YES**

C 2. Attributes?: **YES**
C 3. Relations?: **YES**

C. __ Can attributes be described separately from feature/object classes?: **NO**

D. **C** Does the standard allow/require the specification of the domain of the attributes?: **YES**

E. **C** Does the standard allow/require the specification of data type of the attributes?: **YES**

F. **C** Does the standard allow/require the specification of particular classification coding schemes (e.g., FACC, OSKA, S-57 OC)?: **YES**

7.4 Product Specification

Obligation

A. **C** Does the standard allow/require for the provision of information about whether the dataset meets a particular productspecification?: **YES**

B. **C** If A is yes, can it be internal to the Metadata?: **YES**

C. **C** If A is yes, can it be external to the Metadata?: **YES**

SECTION 8 – Data Quality of the Dataset

This section examines how the standard deals with the question of data quality. This includes homogeneity and metrics for data quality, quality assurance, and usage.

8.1 Overall Data Quality Statement

Obligation

A. **C** Does the standard allow/require for the provision of quality to be described? If yes, please answer the questions below. (*If no, all following answers in Section* 8 *are N/A, please skip to Section* 9.): **YES**

B. **O** Does the standard allow/require the provision for the quality information referencing an external quality standard, quality assurance standard, or specification?: **YES**

8.2 Homogeneity of Data Quality

Obligation

A. **C** Does the standard allow/require for the variation of the quality of the dataset to be described?: **YES**

B. **C** What type of aggregation of quality information is allowed? **YES**
C 1. Entire database?: **YES**
C 2. Feature/Object instance?: **YES**
C 3. Feature/Object class?: **YES**
C 4. Attribute instance?: **YES**
C 5. Attribute class?: **YES**
C 6. Layer?: **YES**
as datasets

C 7. Theme?:	**YES**
as datasets	
C 8. Geographic extent?:	**YES**
C 9. Other?:	**YES**
any dataset	

8.3 Metrics for Data Quality

Does the standard allow/require the provision of values for:

Obligation

A. **C** Positional accuracy?:	**YES**
1. Standard error?:	**YES**
2. Maximum error?:	**YES**
3. Error correlation?:	**YES**
4. Confidence level (e.g., 90%, 95%)?:	**YES**
5. Other?:	**YES**
free text	
B. **C** Attribute accuracy?:	**YES**
1. Standard error?:	**YES**
2. Maximum error?:	**YES**
3. Error frequency?:	**YES**
4. Error correlation?:	**YES**
5. Confidence level (e.g., 90%, 95%)?:	**YES**
6. Other?:	**YES**
free text	
C. **C** Temporal accuracy?:	**YES**
1. Standard error?:	**YES**
2. Maximum error?:	**YES**
3. Error correlation?:	**YES**
4. Confidence level? (e.g., 90%, 95%):	**YES**
5. Other?:	**YES**
free text	
D. **C** Completeness?:	**YES**
1. Error of omission?:	**YES**
2. Error of commission?:	**YES**
3. Other?:	**YES**
E. **C** Currentness?:	**YES**
1. Date?:	**YES**
2. Maximum age?:	**YES**
3. Temporal extent?:	**YES**
4. Other?:	**YES**
free text	
F. **C** Logical consistency?:	**YES**
free text	
G. **C** Lineage?:	**YES**

1. List of processing steps?: **YES**
2. List of values of processing parameters?: **YES**
3. List of printouts of statistics, etc.?: **YES**
4. Source material?: **YES**
5. Other?: **YES**

    free text

H. **O** Additional Quality Information?: **YES**

    a textual description of the quality of geographic dataset

8.4 Quality Assurance Methods

Does the standard allow/require the provision of information about the methods used to determine:

Obligation

A. __ Positional accuracy?: **NO**
1. Standard error in adjustments?:
2. Repeated measurements?:
3. Independent measurements?:
4. Subjective evaluations?:
5. Other?:

B. __ Attribute accuracy?: **NO**
1. Standard error?:
2. Error classification matrix?:
3. Independent measurements?:
4. Subjective evaluation?:
5. Other?:

C. __ Temporal accuracy?: **NO**
1. Independent measurements?:
2. Subjective evaluation?:
3. Other?:

D. __ Completeness?: **NO**
1. Field checks?:
2. Repeated measurements?:
3. Independent measurements?:
4. Subjective evaluation?:
5. Error of commission?:
6. Other?:

E. __ Currentness?: **NO**
1. Date of source material?:
2. Repeated measurements?:
3. Independent measurements?:
4. Subjective evaluation?:
5. Other?:

F. __ Logical consistency?: **NO**
1. Verification results of consistency checks?:

2. Repeated processing?:
3. Comparison with independent sources?:
4. Subjective evaluation?:
5. Other?:

G. __ Lineage?: **NO**

1. Checking results of processing steps:
2. Subjective evaluation:
3. Other:

H. __ Additional Quality Information?: **NO**

8.5 Usage

Obligation

A. **C** Does the metadata standard allow/require the provision of information about usage?: **YES**

1. Who used it: **YES**
2. When it was used: **YES**
3. For what application it was used: **YES**
4. The effectiveness of use: **YES**
5. Other: **YES**
   free text

8.6 Alternative Quality Description: **NO**

SECTION 9 – Spatial Data Structure of the Dataset

This section examines how the standard provides for the specification of the spatial data model and spatial data types.

9.1 Spatial Data Model
Does the standard allow/require the provision for:

Obligation

A. **C** The description of the geometrical and/or topological primitives used to model the real world?: **YES**
B. **C** Information about an external standard for geometrical and/or topological primitive specification?: **YES**

9.2 Spatial Data Types
Does the standard allow/require the provision for:

Obligation

A. **C** Information about the spatial data type(s) used in the dataset?: **YES**
B. **M** If A is yes, which of the following generic spatial data types are supported? **YES**
For each, please list specific data primitive levels supported by each data type.

1. Vector (e.g., spaghetti, chain-node/node-edge-face, planar graph, full topology)?: **YES**
   spaghetti, chain-node/node-edge-face, planar graph, full topology (see ENV 12160:1997 geographic information – data description – spatial schema)
2. Areal Raster (e.g., thematic raster – scanned and classification maps, image raster – satellite images, scanned aerial photographs)?: **YES**
   thematic raster – scanned and classification maps, image raster – satellite images, scanned aerial photographs (see ENV 12160:1997 geographic information – data description – spatial schema)
3. Grid/Matrix (e.g., Digital Elevation Model)?: **YES**
   (see ENV 12160:1997 geographic information – data description – spatial schema)
4. Triangular Irregular Network (TIN)?: **YES**
   (see ENV 12160:1997 geographic information – data description – spatial schema)
5. Other?: **YES**
   text

9.3 Other Data Types

Does the standard allow the provision of information about:

Obligation

A. __ Video?: **NO**
B. __ Sound?: **NO**
C. __ Other?: **NO**

SECTION 10 – Spatial Reference of the Dataset

This section examines how the dataset provides for the specification of the spatial reference characteristics of the data set.

Does the standard allow/require the provision for the following information:

10.1 Geoidal Model

Obligation

A. **O** Geoidal model name?: **YES**
B. **O** Geoidal model definitions?: **YES**

10.2 Geodetic datum

Obligation

A. **O** Geodetic datum name?: **YES**
B. **O** Geodetic datum definitions?: **YES**

10.3 Reference Ellipsoid

Obligation

A. **O** Reference ellipsoid name?: **YES**
B. **O** Reference ellipsoid definitions?: **YES**

10.4 Projection

Obligation

A. **O** Projection name?: **YES**
B. **O** Projection definitions?: **YES**

10.5 Coordinate Systems

Obligation

A. **O** Coordinate system type?: **YES**
B. **O** Coordinate system definitions?: **YES**

10.6 Height Reference System

Obligation

A. **O** Vertical datum name?: **YES**
B. **O** Types of heights?: **YES**

10.7 Ancillary Geodetic Information

Obligation

A. __ Does the standard allow/require the provision for ancillary geodetic, gravimetric, or magnetometric information?: **NO**

10.8 Transformation Parameters
Does the standard specify a method to provide information for coordinate transformation(s):

Obligation

A. __ Model name?: **NO**
B. __ Control points?: **NO**
C. __ Transformation parameters?: **NO**

10.9 Non-Coordinate Spatial Reference
Does the standard allow/require the provision for information on non-coordinate spatial reference systems (e.g., postal systems, street address)?:

Obligation

A. **C** Reference system name?: **YES**
B. **C** Reference system definitions?: **YES**
C. **C** Other?: **YES**
reference date, administrator of the indirect positioning system

SECTION 11 – Availability and Distribution of the Dataset

This section examines how the standard provides information on the availability of the dataset?

Does the standard allow/require the provision for:

11.1 The Distributor(s) of the Dataset

Obligation

A. **C** Information about the distributor(s) of the dataset?: **YES**
B. **C** Contact information for the distributor(s)?: **YES**

11.2 Restrictions to Access and Usage

Obligation

A. **C** Information about copyright on the dataset (e.g., date, period, copyright owner, etc.)?: **YES**
B. **C** Information about restrictions on the access of the dataset?: **YES**
C. **C** Information about restrictions on the usage of the dataset?: **YES**

11.3 Pricing Details

Obligation

A. **C** Information about pricing?: **YES**

11.4 Distribution of the Dataset

Obligation

A. **C** Information about distribution media?: **YES**
B. **C** Information about on-line access?: **YES**
C. **C** Information about units of distribution (e.g., tiles, layers, polygons, elements)?: **YES**

11.5 Distribution Format

Obligation

A. **C** Information about distribution formats (e.g., commercial CAD format, commercial GIS formats, National/International transfer standard)?: **YES**
B. **C** Information about the encoding of the dataset (e.g., binary, ASCII)?: **YES**
C. __ Information about compression method?: **NO**
D. __ Information about compressed dataset size?: **NO**
E. **C** Information about the distribution mechanism?: **YES**
F. **C** Information about the encapsulation of data (e.g., ISO 8211, ISO 8824)?: **YES**

11.6 Ordering Procedures

Obligation

A. **M** Information about ordering procedures for the dataset?: **YES**

11.7 Support Services

Obligation

A. __ Information about support services (e.g., additional data transformations, format or coordinate conversions)?: **NO**

SECTION 12 – Authorization and Verification of the Metadata

This section examines how the standard provides for the description of the authorization and verification on the metadata of a dataset (e.g., how is the metadata itself authorized and verified?).

12.1 Metadata reference

Obligation

A. **M** Does the standard allow/require the provision of information about reference dates of the metadata (e.g., last check date, last update date, review date)?: **YES**

12.2 Verification Authority

Obligation

A. __ Does the standard allow/require the provision for information about who verified/authorized the metadata?: **NO**

B. __ Does the standard specify the method for providing this information (e.g., ISO 9000): **NO**

12.3 Statistical Methods Used for Verification

Obligation

A. __ Does the standard specify the statistical methods to be used for verification of the metadata?: **NO**

B. __ Does the standard allow/require the provision for the description of the statistical methods used to verify the metadata?: **NO**

END OF CHARACTERISTICS
FIN

World Spatial Metadata Standards
H. Moellering, H.J.G.L. Aalders & A. Crane (Editors)

ISO/TC211: Geographic Information – Metadata ISO 19115

David M. Danko

I. Information about the Metadata Standard Itself

SECTION 0 – Assessment Information

This section provides the user with a general description of the scope and intended use of the standard. Information on the respondent who prepared the assessment is also provided.

0.1 Respondent

Assessment prepared by:

David M. Danko
Project Leader, Work Item 15, ISO TC 211
Environmental Systems Research Institute, Inc
8620 Westwood Center Drive
Vienna, VA 22182-2214
USA
Phone: +1-703-506-9515 ext 8011
Fax: +1-703-506-9514
E-mail: ddanko@esri.com

Date assessment completed:

May 1998; revised December 2000; Updated September 2004

Relationship of respondent to standard:

Project Team leader

0.2 Brief Summary Statement

A. Summary of the scope and intent of the standard:

Scope

This International Standard defines the schema required for describing geographic information and services. It provides information about the identification, the extent, the quality, the spatial and temporal schema, spatial reference, and dis-

tribution of digital geographic data.
This International Standard is applicable to:

- **The cataloguing of datasets, clearinghouse activities, and the full description of datasets;**
- **Geographic datasets, dataset series, and individual geographic features and feature properties.**

This International Standard defines:

- **mandatory and conditional metadata sections, metadata entities, and metadata elements;**
- **the minimum set of metadata required to serve the full range of metadata applications (data discovery, determining data fitness for use, data access, data transfer, and use of digital data);**
- **optional metadata elements – to allow for a more extensive standard description of geographic data, if required;**
- **a method for extending metadata to fit specialized needs.**

Though this International Standard is applicable to digital data, its principles can be extended to many other forms of geographic data such as maps, charts, and textual documents as well as non-geographic data.

NOTE: Certain mandatory metadata elements may not apply to these other forms of data.

SECTION 1 – Administration of Metadata Standard

This section provides a general description of the administrative framework, within which the standard was developed, tested and currently resides. Significant details are also provided in terms of the developmental and managerial history as well as information on the existence of documentation, software tools and training materials. This section also provides a point of contact for further information on the standard.

1.1 Name of Standard

A. Name of standard:
   1. In original language(s):
      Geographic Information – Metadata
   2. English translation, if appropriate:
      N/A

B. Version/Edition:
      International Standard, date: 2003-05-01

C. Language(s) of documentation:
      presently English, future French, possible Russian, and German

D. Acronym(s):
      N/A

E. Official ID:
ISO 19115: 2003
F. Name(s) of recognizing standards authority(ies):
International Organization for Standardization (ISO)

1.2 Responsible Institutions

Contact information about those institutions that are (or were) responsible for the processes involved in the evolution of the standard:

A. Production and/or development:
ISO Technical Committee 211 (ISO TC211), http://www.isotc211.org
B. Testing:
N/A
C. Conformance:
N/A
D. Maintenance:
ISO/TC211 **To Be Determined**
E. Distribution:
ISO TC211, http://www.isotc211.org (for draft standards) final International Standard is available from ISO Central Secretariat, Switzerland
F. Help Desk/User Support:
N/A

1.3 Development History

A. Beginning of work on metadata standard: **January 1996**
B. Milestones:

Working Drafts	
Version 1.0	**March 1996**
Version 2.0	**January 1997**
Version 3.0	**September 1997**
Version 4.0	**February 1998**
Committee Draft	**July 1998**
Committee Draft 2.0	**September 1999**
Committee Draft 3.0	**June 2000**
Draft International Standard	**August 2001**
Final Draft International Standard	**December 2002**
International Standard	**March 2003**

C. Anticipated completion: **N/A**
D. Update cycle: **TBD**

1.4 Status of the Standard

A. Is the standard officially recognized: **YES**

B. Date(s) of recognition: **May 2003**
C. Anticipated date of recognition: **N/A**
D. Current stage in recognition process: **N/A**
E. Steps still required to achieve recognition: **N/A**

1.5 Access to Documentation of the Standard

A. Is the standard copyrighted?: **YES**
B. Name and address of copyright owner:
ISO Central Secretariat, Geneva, Switzerland
C. Do restrictions on the use of the standard apply?: **NO**
D. Is the standard documentation currently available?: **YES**
E. List the form/media of standard documentation:
paper, PDF
F. Is the standard available on the Internet?: **YES**
http://www.iso.org/iso/en/CatalogueListPage.CatalogueList
G. Price: **To Be Determined**
H. Contact details for obtaining the standard:
ISO Customer Service
Post: ISO
1, rue de Varembé
CH-1211 Geneva 20
Switzerland
E-mail: sales@iso.org
Phone: + 41 22 749 01 11
Fax: + 41 22 749 09 47
I. Does the standard have ISBN/ISSN numbers?: **NO**

1.6 Available Software Tools

A. Is software available to implement the standard as a database?: **NO**
(not presently)
B. Is software available to test compliance of metadata with the standard?: **NO**
(not presently)
C. Is software available to produce metadata compliant with the standard?: **NO**
(not presently)
D. Other relevant software: **NO**
E. Is there a test dataset available to test software tools for implementing the standard?: **NO**
F. For each product available above, please list contact information: **N/A**

1.7 Available Training Materials

A. Are organized training sessions available?: **NO**
B. Contact details for training sessions: **N/A**
C. Training documentation available: **N/A**

SECTION 2 – Use and Implementation of the Metadata Standard

This section provides information on the way in which the standard is used and the purpose of its use. It also provides descriptions about the level of abstraction of the standard.

2.1 Intended Use of the Standard
Can the standard be used:

A. Within a transfer data structure?: **YES**
B. To produce hardcopy catalog of datasets?: **YES**
C. To catalog data in a digital on-line catalog?: **YES**
D. Other: **N/A**

2.2 Intended Purpose of the Standard
Is the intended purpose of the standard:

A. To support discovery, identification, location of data?: **YES**
B. To support how to access data?: **YES**
C. To support the determination of dataset fitness for use?: **YES**
D. To support transfer of dataset?: **YES**
E. To support an organization's data management?: **YES**
F. Other: **N/A**

2.3 Table of Contents

A. Provide the standard's Table of Contents
(edited for brevity)
Forward
Introduction
1 Scope
2 Conformance
2.1 Conformance requirements
2.2 Profiles
2.3 Obligation and condition
3 Normative references
4 Terms and definitions
5 Symbols and abbreviated terms
5.1 Abbreviations
5.2 UML notations
5.3 UML model relationships
5.4 UML model stereotypes
5.5 Package abbreviations
5.6 UML model/data dictionary relationships
6 Requirements
6.1 Metadata for geographic data requirement
6.2 Metadata application information

6.3 Metadata packages
6.4 Metadata datatypes
6.5 Core metadata for geographic datasets
6.6 Unified Modelling Language (UML) diagrams
6.7 Data dictionary
6.8 Metadata extensions and profiles
6.9 Abstract test suite
6.10 Comprehensive dataset metadata application profile
6.11 Metadata extension methodology
6.12 Metadata implementation
6.13 Hierarchical levels of metadata
6.14 Implementation examples
6.15 Multi-lingual support for free text fields
Annex A (normative) Metadata schemas
Annex B (normative) Data dictionary for geographic metadata
Annex C (normative) Metadata extensions and profiles
Annex D (normative) Abstract test suite
Annex E (normative) Comprehensive dataset metadata application profile
Annex F (informative) Metadata extension methodology
Annex G (informative) Metadata implementation
Annex H (informative) Hierarchical levels of metadata
Annex I (informative) Implementation examples
Annex J (informative) Multi-lingual support for free text metadata element
Bibliography

2.4 Level of Abstraction of the Metadata Standard Definition

A. Does the standard define metadata elements?: **YES**
B. Does the standard define a model for the relationships between the metadata elements?: **YES**
C. Provide an overview diagram of the model of the metadata, if available.
D. Does the standard define domains and/or data types of the metadata elements?: **YES**
E. Does the standard define how the metadata should be encoded?: **NO**
F. Does the standard define a database schema for the automatic generation of a database?: **YES**

2.5 Extensibility of the Standard

A. Is the metadata description extensible (e.g., Can user add metadata domains, elements, thematic profiles, etc.)?: **YES**

The standard defines metadata elements allowing users to define new elements along with their: name, identifier, definition, obligation, data type, and domain value. An informative annex describes an extension methodology.

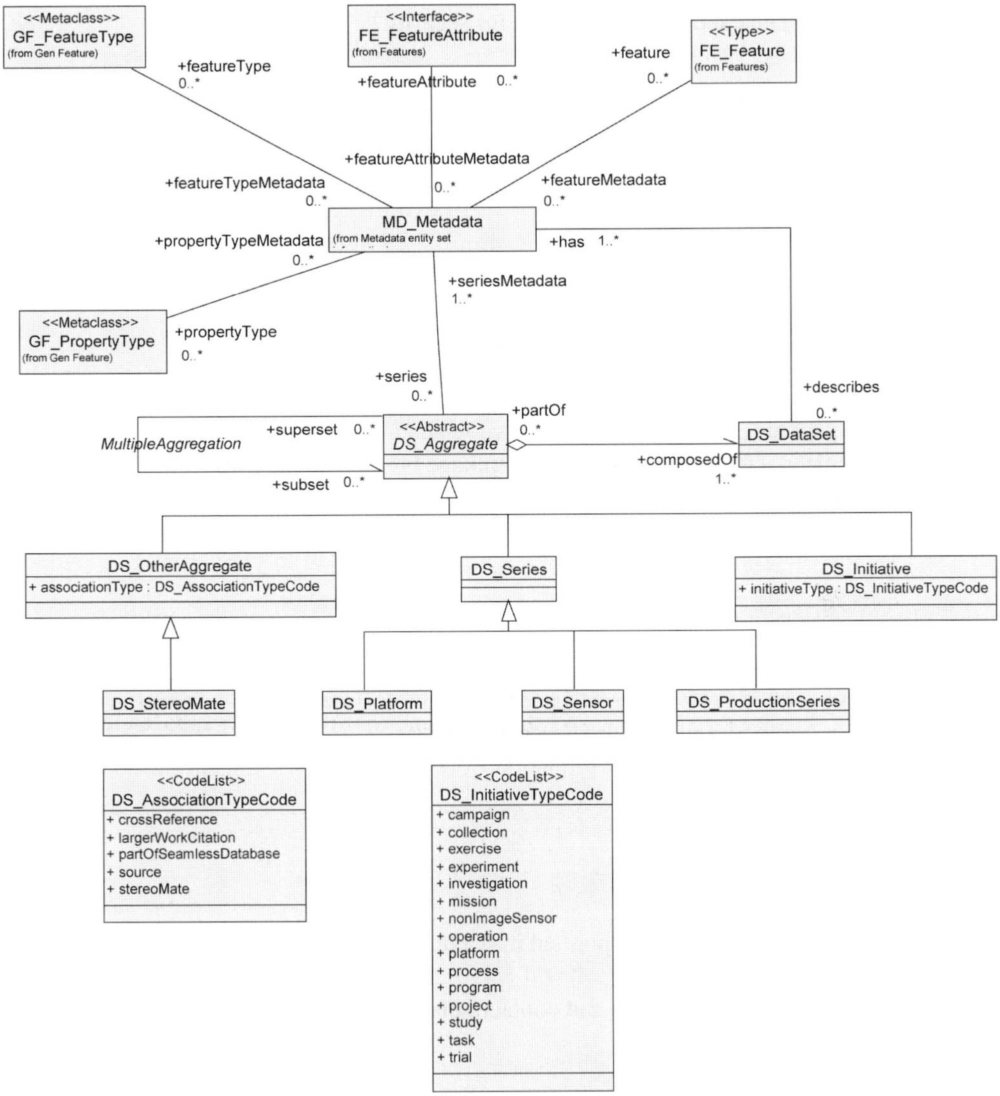

Fig. 1. UML Schema ISO DIS 19115.

2.6 Availability of Metadata

Does the standard specify:

A. How metadata are to be accessed?:	**NO**
B. A mechanism for querying the metadata?:	**NO**
C. Particular standardized query protocols?:	**NO**
D. That the metadata should be encoded for computer access?:	**YES**

E. That the metadata should be encrypted for security?: **NO**

2.7 Examples

A. Does the standard provide examples of implementations?: **YES**
using two sample metadata sets for a vector data and two examples of user extensions
B. How many examples are available?: **4**

SECTION 3 – Linkage and Coordination of the Metadata Standard

This section provides information on how the metadata standard is linked to other standards, particularly transfer standards, and how such linkage is implemented. If a significant linkage does exist, detailed technical information is provided on the precise nature of the linkage to the standard, and how the linkage affects implementation of the metadata standard.

3.1 Associated Transfer Standards

A. Is the metadata standard associated with a transfer standard(s)?: **NO**
B. If A is yes, provide references and contact details for each linked standard: **N/A**
C. If A is yes, are both (all) standards coordinated?: **N/A**
D. If A is yes, describe how the coordination works, both from a technical and conformance (normative) viewpoint: **N/A**

3.2 Incorporated into transfer mechanism

A. Is the metadata standard incorporated into a transfer mechanism?: **NO**
B. If A is yes, provide references and contact details for each incorporated mechanism: **N/A**
C. If A is yes, describe how: **N/A**

3.3 Other linked, coordinated or associated standards

A. Is the metadata standard linked, coordinated, or associated with any other standard(s) not listed above?: **YES**
B. If A is yes, provide references and contact details for each standard:
ISO 639 (all parts), Code for the representation of names of languages
ISO 3166 (all parts), Codes for the representation of names of countries and their subdivisions

ISO 4217:1995, Codes for the representation of currencies and funds
ISO 8859 (parts 1 to 15), Information technology – 8 bit single byte coded graphic character sets
ISO 8879, Information processing – Text and office systems – Standard Generalized Markup Language (SGML)
ISO/IEC 10646-1, Information technology – Universal Multiple-Octet Coded Character Set (UCS) – Part 1: Architecture and Basic Multilingual Plane
ISO/IEC 11179 (all parts), Information technology – Specification and standardization of data elements
ISO TS 19103:-), Geographic information – Conceptual schema language
ISO 19104:-1), Geographic information – Terminology
ISO 19106:-1), Geographic information – Profiles
ISO 19107:-1), Geographic information – Spatial schema
ISO 19108:-1), Geographic information – Temporal schema
ISO 19109:-1), Geographic information – Rules for application schema
ISO 19110:-1), Geographic information – Feature cataloguing methodology
ISO 19111:-1), Geographic information – Spatial referencing by coordinates
ISO 19112:-1), Geographic information – Spatial referencing by geographic identifiers
ISO 19113:-1), Geographic information – Quality principles
ISO 19114:-1), Geographic information – Quality evaluation procedures
ISO 19117:-1), Geographic information – Portrayal
ISO 19118:-1), Geographic information – Encoding

Members of IEC and ISO maintain registers of currently valid International Standards through which the above standards can be obtained.

C. If A is yes, describe how for each relevant standard:
The above normative documents contain provisions, which, through reference in the text of ISO 19115, constitute provisions of that International Standard. See http://www.statkart.no/isotc211/ for details.

II. Categories the Metadata Standard Uses to Describe Datasets

The following sections are categories of metadata that may be used by the metadata standard to describe data. Not all categories are relevant to different datasets and may

not necessarily be present in the standard. The presence or absence of a category should not be used for judging one standard against another.

Note: For each of the questions in the following sections please provide the obligation status information. Valid obligation values are:

Obligation Level

M	Mandatory	This information must be entered to comply with the metadata standard.
C	Conditional	This information is entered based on other information in the metadata.
O	Optional	This information is optionally entered in the metadata.

If the answer is 'NO', the obligation status should be left blank.

SECTION 4 – Dataset Identification

This section examines how the standard provides for the specification of the various sorts of titles and identifiers for a dataset.

4.1 The title(s) of a dataset
Does the standard allow/require the provision for:

Obligation

A. __ Previous title(s)?: **NO**
B. **M** Current title(s)?: **YES**
C. __ Possible future title(s)?: **NO**
D. __ Official titles in different languages?: **NO**
E. __ Abbreviated title(s)?: **NO**
F. **O** Alternative Title(s)?: **YES**
G. __ Does the standard support effective dates of titles?: **NO**
H. **O** Other identifiers (e.g., ISBN, National Stock Number, Uniform Product Code)?: **YES**

4.2 Parties responsible for the dataset
Does the standard allow/require the provision for:

Obligation

A. __ Author?: **NO**
B. **C** Owner?: **YES**
C. **M** Other?: **YES**

4.3 ISBD Information

Obligation

A. __ Does the metadata standard specify inclusion of mandatory ISBD information (see glossary) within the metadata?: **NO**

SECTION 5 – Status of the Dataset

This section examines how the standard provides for the specification of the current status and updating of the data set.

5.1 Status and Progress of Datasets
Does the metadata standard allow/require the provision for:

Obligation

A. **O** Information on the status of the entire dataset?: **YES**
Information as to whether dataset is: planned, completed, required, on-going, historical archive, obsolete
B. **O** Information on the status of the particular subset observation time, inauguration time, extraction time)?: **YES**
by dataset series, feature type, feature instance

Does the metadata standard specify:

C. **O** A structure for updating or maintenance information?: **YES**
see 5.1A.
D. **M** A structure to describe historical information (see also Section 8.3 G)?: **YES**
lineage

5.2 Information on Incremental Updating
Does the metadata standard specify:

Obligation

A. __ A structure for information on incremental updates of the entire dataset?: **NO**
B. __ A structure for information on incremental updates of particular portions of the dataset?: **NO**
C. __ A structure to describe the update process for the dataset?: **NO**

SECTION 6 – Extent of the Dataset

This section examines how the standard provides for the specification of the spatial and temporal extent of the dataset.

6.1 The spatial extent of the dataset

Obligation

A. **M** Does the metadata standard allow/require for the description of horizontal extent (e.g., bounding rectangle/polygon, named location (i.e., Stockholm, etc.))?: **YES**
bounding rectangle(C) or, named location (C), bounding polygon (O)
B. **O** Does the metadata standard allow/require for the description of vertical extent (e.g., minima/maxima heights, etc.)?: **YES**
Minimum value, maximum value, unit of measure

C. **M** Does the standard require specific units for describing extent?: **YES**
D. __ Does the standard allow specification of the units for describing extent?: **NO**
E. **M** Does the standard require specific spatial reference system for describing extent?: **YES**
F. **C** Does the standard allow specification of the spatial reference system for describing extent?: **YES**
when using the bounding polygon (O)

6.2 The temporal extent of the dataset

Obligation

A. **O** Does the standard allow/require for the description of time period for which the dataset is relevant?: **YES**
B. **C** Does the standard allow/require specification of the temporal reference system?: **YES**
if not Gregorian calendar (see Temporal reference system information – ISO 19108)

SECTION 7 – Data Content of the Dataset

This section examines how the standard provides for the specification of the data content of the dataset in terms of overview elements, themes, data definitions and product specifications.

7.1 Dataset overview
Does the standard allow/require for the provision of the following information:

Obligation

A. __ Dataset overview?: **NO**
B. **M** Abstract – brief summary of the content of the dataset?: **YES**
C. **O** Purpose – summary of the intentions with which the dataset was developed?: **YES**
D. **O** Usage – list of activities for which the dataset has been used? (see also Section 8.5): **YES**
E. **O** Spatial schema – (e.g., vector, raster)? (see also Section 9): **YES**
F. **O** Spatial reference system? (see also Section 12): **YES**
G. **O** Language?: **YES**
H. **C** Character set (e.g., ISO8859-1, JISX0208)?: **YES**
I. **O** List of documents providing further information about the dataset?: **YES**
J. **O** Sample/preview dataset?: **YES**
K. **O** Related datasets?: **YES**
L. __ Original size of the dataset?: **NO**
M. **O** Compression? (see also Section 11.5): **YES**
N. **O** Other?: **YES**
Resolution (equivalent scale/ground sample distance)

7.2 Theme(s)

Obligation

A. **M** Does the standard allow/require the provision for the description of theme(s) (e.g., topography, land cover, transport, etc.) in the dataset?: **YES**
B. **O** If A is yes, does the standard allow the referencing of an outside dictionary or thesaurus?: **YES**
C. **M** If A is yes, does the standard itself include a dictionary or thesaurus within the metadata?: **YES**
D. **O** If A is yes, does the standard allow the use of unauthorized user-defined keywords?: **YES**

7.3 Data Definition

Obligation

A. __ Does the standard allow/require the provision for the description of the data definition?: **NO**
B. If A is yes, does the standard allow/require the provision for the description of:
__ 1. Features/Objects?: **N/A**
__ 2. Attributes?: **N/A**
__ 3. Relations?: **N/A**
C. __ Can attributes be described separately from feature/object classes?: **NO**
D. __ Does the standard allow/require the specification of the domain of the attributes?: **NO**
E. __ Does the standard allow/require the specification of data type of the attributes?: **NO**
F. **O** Does the standard allow/require the specification of particular classification coding schemes (e.g., FACC, OSKA, S-57 OC)?: **YES**
By referencing the applicable coding scheme/specification

7.4 Product Specification

Obligation

A. **O** Does the standard allow/require for the provision of information about whether the dataset meets a particular product specification?: **YES**
By citation in data quality section
B. __ If A is yes, can it be internal to the Metadata?: **NO**
C. **O** If A is yes, can it be external to the Metadata?: **YES**

SECTION 8 – Data Quality of the Dataset

This section examines how the standard deals with the question of data quality. This includes homogeneity and metrics for data quality, quality assurance, and usage.

8.1 Overall Data Quality Statement

Obligation

A. **O** Does the standard allow/require for the provision of quality to be described? If yes, please answer the questions below. (*If no, all following answers in Section* 8 *are N/A, please skip to Section* 9.): **YES**

B. **O** Does the standard allow/require the provision for the quality information referencing an external quality standard, quality assurance standard, or specification?: **YES**

Quality information may be included in metadata, if not included metadata may point to an outside reference (product specification) and state whether the data meets that specification.

8.2 Homogeneity of Data Quality

Obligation

A. **C** Does the standard allow/require for the variation of the quality of the dataset to be described?: **YES**

B. What type of aggregation of quality information is allowed?

M 1. Entire database?: **YES**
O 2. Feature/Object instance?: **YES**
O 3. Feature/Object class?: **YES**
O 4. Attribute instance?: **YES**
O 5. Attribute class?: **YES**
O 6. Layer?: **YES**
O 7. Theme?: **YES**
O 8. Geographic extent?: **YES**
O 9. Other?: **YES**

user defined

8.3 Metrics for Data Quality
Does the standard allow/require the provision of values for:

Obligation

A. **O** Positional accuracy?: **YES**

1. Standard error?: **YES**
(user must specify)
2. Maximum error?: **YES**
(user must specify)
3. Error correlation?: **YES**
(user must specify)
4. Confidence level (e.g., 90%, 95%)?: **YES**
(user must specify)
5. Other?: **YES**
absolute/external positional accuracy, relative/internal positional accuracy, gridded data positional accuracy

B. **O** Attribute accuracy?: **YES**

(Thematic Accuracy)

1. Standard error?: **YES**
(user must specify)

2. Maximum error?: **YES**
    (user must specify)
3. Error frequency?: **YES**
    (user must specify)
4. Error correlation?: **YES**
    (user must specify)
5. Confidence level (e.g., 90%, 95%)?: **YES**
    (user must specify)
6. Other?: **YES**
    thematic classification correctness, non-quantitative attribute accuracy, quantitative attribute accuracy

C. **O** Temporal accuracy?: **YES**
1. Standard error?: **YES**
    (user must specify)
2. Maximum error?: **YES**
    (user must specify)
3. Error correlation?: **YES**
    (user must specify)
4. Confidence level? (e.g., 90%, 95%): **YES**
    (user must specify)
5. Other?: **YES**
    temporal consistency, temporal validity, accuracy of a time measurement

D. **O** Completeness?: **YES**
1. Error of omission?: **YES**
2. Error of commission?: **YES**
3. Other?: **YES**
    user defined

E. x Currentness?: **YES**
    covered under temporal extent
1. Date?: **YES**
2. Maximum age?: **NO**
3. Temporal extent?: **YES**
4. Other?: **YES**

F. **O** Logical consistency?: **YES**
    domain consistency, format consistency, topological consistency, conceptual consistency

G. **C** Lineage?: **YES**
1. List of processing steps?: **YES**
2. List of values of processing parameters?: **NO**
3. List of printouts of statistics, etc.?: **NO**
4. Source material?: **YES**
5. Other?: **NO**

H. x Additional Quality Information?: **YES**
    (user must specify)

8.4 Quality Assurance Methods

Does the standard allow/require the provision of information about the methods used to determine:

Obligation

A.	__ Positional accuracy?:	**NO**
	1. Standard error in adjustments?:	**N/A**
	2. Repeated measurements?:	**N/A**
	3. Independent measurements?:	**N/A**
	4. Subjective evaluations?:	**N/A**
	5. Other?:	**N/A**
B.	__ Attribute accuracy?:	**NO**
	1. Standard error?:	**N/A**
	2. Error classification matrix?:	**N/A**
	3. Independent measurements?:	**N/A**
	4. Subjective evaluation?:	**N/A**
	5. Other?:	**N/A**
C.	__ Temporal accuracy?:	**NO**
	1. Independent measurements?:	**N/A**
	2. Subjective evaluation?:	**N/A**
	3. Other?:	**N/A**
D.	__ Completeness?:	**NO**
	1. Field checks?:	**N/A**
	2. Repeated measurements?:	**N/A**
	3. Independent measurements?:	**N/A**
	4. Subjective evaluation?:	**N/A**
	5. Error of commission?:	**N/A**
	6. Other?:	**N/A**
E.	__ Currentness?:	**NO**
	1. Date of source material?:	**N/A**
	2. Repeated measurements?:	**N/A**
	3. Independent measurements?:	**N/A**
	4. Subjective evaluation?:	**N/A**
	5. Other?:	**N/A**
F.	__ Logical consistency?:	**NO**
	1. Verification results of consistency checks?:	**N/A**
	2. Repeated processing?:	**N/A**
	3. Comparison with independent sources?:	**N/A**
	4. Subjective evaluation?:	**N/A**
	5. Other?:	**N/A**
G.	__ Lineage?:	**NO**
	1. Checking results of processing steps:	**N/A**
	2. Subjective evaluation:	**N/A**
	3. Other:	**N/A**
H.	__ Additional Quality Information?:	**NO**

8.5 Usage

Obligation

A. **O** Does the metadata standard allow/require the provision of information about usage?: **YES**

1. Who used it: **YES**
2. When it was used: **YES**
3. For what application it was used: **YES**
4. The effectiveness of use: **NO**
5. Other: **YES**
user determined limitations

8.6 Alternative Quality Description: **YES**
Qualitative assessment

SECTION 9 – Spatial Data Structure of the Dataset

This section examines how the standard provides for the specification of the spatial data model and spatial data types.

9.1 Spatial Data Model
Does the standard allow/require the provision for:

Obligation

A. **O** The description of the geometrical and/or topological primitives used to model the real world?: **YES**
B. __ Information about an external standard for geometrical and/or topological primitive specification?: **NO**

9.2 Spatial Data Types
Does the standard allow/require the provision for:

Obligation

A. **O** Information about the spatial data type(s) used in the dataset?: **YES**
B. **O** If A is yes, which of the following generic spatial data types are supported? **YES**
For each, please list specific data primitive levels supported by each data type.

1. Vector (e.g., spaghetti, chain-node/node-edge-face, planar graph, full topology)?: **YES**
geometric object types: complexes, composites, curves, point, solid, surface; topology levels: geometry only, non-planar graph, planar graph, full planar graph, surface graph, full surface graph, full topology-3D, abstract
2. Areal Raster (e.g., thematic raster – scanned and classification maps, image raster – satellite images, scanned aerial photographs)?: **YES**
gridded data: cell geometry: point, area

3. Grid/Matrix (e.g., Digital Elevation Model)?: **YES**
gridded data: cell geometry: point, area
4. Triangular Irregular Network (TIN)?: **YES**
TIN
5. Other?: **YES**

9.3 Other Data Types

Does the standard allow the provision of information about:

Obligation

A. x Video?: **YES**
B. __ Sound?: **NO**
C. __ Other?: **NO**

SECTION 10 – Spatial Reference of the Dataset

This section examines how the dataset provides for the specification of the spatial reference characteristics of the data set.

Does the standard allow/require the provision for the following information:

10.1 Geoidal Model

Obligation

A. **C** Geoidal model name?: **YES**
B. __ Geoidal model definitions?: **NO**
see ISO 19111

10.2 Geodetic datum

Obligation

A. **C** Geodetic datum name?: **YES**
B. __ Geodetic datum definitions?: **NO**
see ISO 19111

10.3 Reference Ellipsoid

Obligation

A. **C** Reference ellipsoid name?: **YES**
B. **C** Reference ellipsoid definitions?: **YES**

10.4 Projection

Obligation

A. **C** Projection name?: **YES**
B. **C** Projection definitions?: **YES**

10.5 Coordinate Systems

Obligation

A. **C** Coordinate system type?: **YES**
B. __ Coordinate system definitions?: **NO**
see ISO 19111

10.6 Height Reference System

Obligation

A. **C** Vertical datum name?: **YES**
B. __ Types of heights?: **NO**
see ISO 19111

10.7 Ancillary Geodetic Information

Obligation

A. __ Does the standard allow/require the provision for ancillary geodetic, gravimetric, or magnetometric information?: **NO**
see ISO 19111

10.8 Transformation Parameters
Does the standard specify a method to provide information for coordinate transformation(s):

Obligation

A. __ Model name?: **NO**
B. __ Control points?: **NO**
C. __ Transformation parameters?: **NO**
see ISO 19111

10.9 Non-Coordinate Spatial Reference
Does the standard allow/require the provision for information on non-coordinate spatial reference systems (e.g., postal systems, street address)?:

Obligation

A. **C** Reference system name?: **YES**
B. __ Reference system definitions?: **NO**
see ISO 19111
C. __ Other?: **NO**

SECTION 11 – Availability and Distribution of the Dataset

This section examines how the standard provides information on the availability of the dataset?

Does the standard allow/require the provision for:

11.1 The Distributor(s) of the Dataset

Obligation

A. **O** Information about the distributor(s) of the dataset?: **YES**
B. **O** Contact information for the distributor(s)?: **YES**

11.2 Restrictions to Access and Usage

Obligation

A. **O** Information about copyright on the dataset (e.g., date, period, copyright owner, etc.)?: **YES**
B. **O** Information about restrictions on the access of the dataset?: **YES**
C. **O** Information about restrictions on the usage of the dataset?: **YES**

11.3 Pricing Details

Obligation

A. **O** Information about pricing?: **YES**

11.4 Distribution of the Dataset

Obligation

A. **O** Information about distribution media?: **YES**
B. **O** Information about on-line access?: **YES**
C. **O** Information about units of distribution (e.g., tiles, layers, polygons, elements)?: **YES**

11.5 Distribution Format

Obligation

A. **M** Information about distribution formats (e.g., commercial CAD format, commercial GIS formats, National/International transfer standard)?: **YES**
B. **O** Information about the encoding of the dataset (e.g., binary, ASCII)?: **YES**
C. **O** Information about compression method?: **YES**
D. **O** Information about compressed dataset size?: **YES**
E. **O** Information about the distribution mechanism?: **YES**
F. **O** Information about the encapsulation of data (e.g., ISO 8211, ISO 8824)?: **YES**

11.6 Ordering Procedures

Obligation

A. **O** Information about ordering procedures for the dataset?: **YES**

11.7 Support Services

Obligation

A. __ Information about support services (e.g., additional data transformations, format or coordinate conversions)?: **NO**

SECTION 12 – Authorization and Verification of the Metadata

This section examines how the standard provides for the description of the authorization and verification on the metadata of a dataset (e.g., how is the metadata itself authorized and verified?).

12.1 Metadata reference

Obligation

A. **M** Does the standard allow/require the provision of information about reference dates of the metadata (e.g., last check date, last update date, review date)?: **YES**

12.2 Verification Authority

Obligation

A. **M** Does the standard allow/require the provision for information about who verified/authorized the metadata?: **YES**
contact information for party responsible for the metadata
B. **M** Does the standard specify the method for providing this information (e.g., ISO 9000): **YES**
ISO 19115 "CI_ResponsibleParty"

12.3 Statistical Methods Used for Verification

Obligation

A. __ Does the standard specify the statistical methods to be used for verification of the metadata?: **NO**
B. __ Does the standard allow/require the provision for the description of the statistical methods used to verify the metadata?: **NO**

END OF CHARACTERISTICS
FIN

SUBJECT MATTER METADATA STANDARDS/PROFILES

World Spatial Metadata Standards
H. Moellering, H.J.G.L. Aalders & A. Crane (Editors)

EuroGeographics: LaClef Core Metadata: Topography

Marcus Wandinger, Claude Luzet

I. Information about the Metadata Standard Itself

SECTION 0 – Assessment Information

This section provides the user with a general description of the scope and intended use of the standard. Information on the respondent who prepared the assessment is also provided.

0.1 Respondent

Assessment prepared by:

Marcus Wandinger, Internet Services Manager
EuroGeographics, 6-8, Ave Blaise Pascal
Champs sur Marne, 77455 Marne la Vallée cedex 2, France
Tel.: +33 1 64153219
Fax: + 33 1 64153281
E-mail: marcus.wandinger@megrin.org

Date assessment completed:

May, 2000; Updated September, 2000, May, 2002, and May, 2005

Relationship of respondent to standard:

0.2 Brief Summary Statement

A. Summary of the scope and intent of the standard:
The standard has been developed to improve MEGRIN's previous GDDD (Geographical Data Description Directory) metadata service: It provides the European National Mapping Agencies with a standardised core metadata for use within international, national and regional metadata services. The core metadata has been designed to cater for multi-linguality; high level searching to individual map sheet level; textual and geographical searches and e-commerce implementation.

SECTION 1 – Administration of Metadata Standard

This section provides a general description of the administrative framework, within which the standard was developed, tested and currently resides. Significant details are also provided in terms of the developmental and managerial history as well as information on the existence of documentation, software tools and training materials. This section also provides a point of contact for further information on the standard.

1.1 Name of Standard

A. Name of standard:
   1. In original language(s):
      LaClef Core Metadata
   2. English translation, if appropriate:
      N/A
B. Version/Edition:
      v 2.1c May 2000
C. Language(s) of documentation:
      English
D. Acronym(s):
      N/A
E. Official ID:
      N/A
F. Name(s) of recognizing standards authority(ies):
      N/A

1.2 Responsible Institutions

Contact information about those institutions that are (or were) responsible for the processes involved in the evolution of the standard:

A. Production and/or development:
   EuroGeographics, 6-8, Ave Blaise Pascal
   Champs sur Marne, 77455 Marne la Vallée cedex 2, France
B. Testing:
   LaClef Consortium, c/o EuroGeographics
C. Conformance:
   LaClef Consortium, c/o EuroGeographics
D. Maintenance:
   EuroGeographics, 6-8, Ave Blaise Pascal, Champs sur Marne
   77455 Marne la Vallée cedex 2, France
E. Distribution:
   EuroGeographics, 6-8, Ave Blaise Pascal, Champs sur Marne
   77455 Marne la Vallée cedex 2, France
F. Help Desk/User Support:
   EuroGeographics, 6-8, Ave Blaise Pascal, Champs sur Marne
   77455 Marne la Vallée cedex 2, France

1.3 Development History

A. Beginning of work on metadata standard:
November 1994 introduction of first version of GDDD as the first implementation of CEN ENV 12657

B. Milestones:
1996 increased availability via WWW.
January 1999 start of LaClef project which aims at improving the GDDD service, e.g., with an improved core metadata.
May 1999 v. 1.0 of LaClef core metadata
May 2000 v 2.1c of LaClef core metadata
January 2003: new core profile compatible with ISO 19115
2004: prototype development of EuroMapFinder
July 2005: EuroMapFinder on-line

C. Anticipated completion: **July 2005**

D. Update cycle: **ongoing**
evolution based on INSPIRE progress, and the requirements for the implemantation of a distributed EuroMapFinder service

1.4 Status of the Standard

A. Is the standard officially recognized: **NO**

B. Date(s) of recognition: **N/A**

C. Anticipated date of recognition: **end 2000**

D. Current stage in recognition process:
When the final version of the standard is agreed at the end of 2000, it will be used by the representative body of European National Mapping Agencies

E. Steps still required to achieve recognition: **N/A**

1.5 Access to Documentation of the Standard

A. Is the standard copyrighted?: **YES**

B. Name and address of copyright owner:
LaClef Consortium, c/o EuroGeographics

C. Do restrictions on the use of the standard apply?: **YES**
Consortium agreement

D. Is the standard documentation currently available?: **YES**
on demand, after Consortium agreement

E. List the form/media of standard documentation:
paper, CD-ROM, 3.5" disc

F. Is the standard available on the Internet?: **NO**

G. Price: **Free**

H. Contact details for obtaining the standard:
EuroGeographics
6-8, Ave Blaise Pascal, Champs sur Marne
77455 Marne la Vallée cedex 2, France

I. Does the standard have ISBN/ISSN numbers?: **NO**

1.6 Available Software Tools

A. Is software available to implement the standard as a database?: **NO**
B. Is software available to test compliance of metadata with the standard?: **NO**
C. Is software available to produce metadata compliant with the standard?: **NO**
D. Other relevant software:
related software is under development
E. Is there a test dataset available to test software tools for implementing the standard?: **NO**
F. For each product available above, please list contact information: **N/A**

1.7 Available Training Materials

A. Are organized training sessions available?: **NO**
B. Contact details for training sessions: **N/A**
C. Training documentation available: **N/A**

SECTION 2 – Use and Implementation of the Metadata Standard

This section provides information on the way in which the standard is used and the purpose of its use. It also provides descriptions about the level of abstraction of the standard.

2.1 Intended Use of the Standard
Can the standard be used:

A. Within a transfer data structure?: **YES**
B. To produce hardcopy catalog of datasets?: **YES**
C. To catalog data in a digital on-line catalog?: **YES**
D. Other: **N/A**

2.2 Intended Purpose of the Standard
Is the intended purpose of the standard:

A. To support discovery, identification, location of data?: **YES**
B. To support how to access data?: **YES**
C. To support the determination of dataset fitness for use?: **YES**
D. To support transfer of dataset?: **NO**
E. To support an organization's data management?: **YES**
F. Other: **NO**

2.3 Table of Contents

A. Provide the standard's Table of Contents
Table of contents, edited, this is a core (discovery) metadata structure, consisting of 83 elements.
1. Dataset identification
2. Dataset Overview
3. Dataset Quality Elements
4. Administrative Metadata

4.1 Organisation
4.2 Point of Contact
5. Distribution
5.8. Price information for e-commerce
6. Extent
6.1. Temporal Extent
6.2. Spatial Extent
6.2.1. Spatial Reference by Coordinates
6.2.2. Spatial Reference by Geographic Identifiers
7. Metadata Reference
8. Spatial Reference System
8.1. Indirect Positioning System
8.2. Direct Positioning System
9. Classification

2.4 Level of Abstraction of the Metadata Standard Definition

A. Does the standard define metadata elements?: **YES**
B. Does the standard define a model for the relationships between the metadata elements?: **YES**
C. Provide an overview diagram of the model of the metadata, if available. **N/A**
D. Does the standard define domains and/or data types of the metadata elements?: **YES**
E. Does the standard define how the metadata should be encoded?: **YES**
F. Does the standard define a database schema for the automatic generation of a database?: **NO**

2.5 Extensibility of the Standard

A. Is the metadata description extensible (e.g., Can user add metadata domains, elements, thematic profiles, etc.)?: **YES**
The current defined standard is regarded as a 'core' model. Individual users may apply domains, elements etc., depending on particular application sectors

2.6 Availability of Metadata
Does the standard specify:

A. How metadata are to be accessed?: **NO**
B. A mechanism for querying the metadata?: **NO**
C. Particular standardized query protocols?: **NO**
D. That the metadata should be encoded for computer access?: **NO**
E. That the metadata should be encrypted for security?: **NO**

2.7 Examples

A. Does the standard provide examples of implementations?: **NO**
B. How many examples are available?: **N/A**

SECTION 3 – Linkage and Coordination of the Metadata Standard

This section provides information on how the metadata standard is linked to other standards, particularly transfer standards, and how such linkage is implemented. If a significant linkage does exist, detailed technical information is provided on the precise nature of the linkage to the standard, and how the linkage affects implementation of the metadata standard.

3.1 Associated Transfer Standards

A. Is the metadata standard associated with a transfer standard(s)?: **NO**
B. If A is yes, provide references and contact details for each linked standard: **N/A**
C. If A is yes, are both (all) standards coordinated?: **N/A**
D. If A is yes, describe how the coordination works, both from a technical and conformance (normative) viewpoint: **N/A**

3.2 Incorporated into transfer mechanism

A. Is the metadata standard incorporated into a transfer mechanism?: **NO**
B. If A is yes, provide references and contact details for each incorporated mechanism: **N/A**
C. If A is yes, describe how: **N/A**

3.3 Other linked, coordinated or associated standards

A. Is the metadata standard linked, coordinated, or associated with any other standard(s) not listed above?: **NO**
B. If A is yes, provide references and contact details for each standard: **N/A**
C. If A is yes, describe how for each relevant standard: **N/A**

II. Categories the Metadata Standard Uses to Describe Datasets

The following sections are categories of metadata that may be used by the metadata standard to describe data. Not all categories are relevant to different datasets and may not necessarily be present in the standard. The presence or absence of a category should not be used for judging one standard against another.

Note: For each of the questions in the following sections please provide the obligation status information. Valid obligation values are:

Obligation Level

M	Mandatory	This information must be entered to comply with the metadata standard.
C	Conditional	This information is entered based on other information in the metadata.
O	Optional	This information is optionally entered in the metadata.

If the answer is 'NO', the obligation status should be left blank.

SECTION 4 – Dataset Identification

This section examines how the standard provides for the specification of the various sorts of titles and identifiers for a dataset.

4.1 The title(s) of a dataset
Does the standard allow/require the provision for:

Obligation

A. __ Previous title(s)?: **NO**
B. **M** Current title(s)?: **YES**
C. __ Possible future title(s)?: **NO**
D. **M** Official titles in different languages?: **YES**
E. **O** Abbreviated title(s)?: **YES**
F. **O** Alternative Title(s)?: **YES**
G. __ Does the standard support effective dates of titles?: **NO**
H. __ Other identifiers (e.g., ISBN, National Stock Number, Uniform Product Code)?: **NO**

4.2 Parties responsible for the dataset
Does the standard allow/require the provision for:

Obligation

A. **M** Author?: **YES**
B. **M** Owner?: **YES**
C. **O** Other?: **YES**

4.3 ISBD Information

Obligation

A. __ Does the metadata standard specify inclusion of mandatory ISBD information (see glossary) within the metadata?: **NO**

SECTION 5 – Status of the Dataset

This section examines how the standard provides for the specification of the current status and updating of the data set.

5.1 Status and Progress of Datasets
Does the metadata standard allow/require the provision for:

Obligation

A. **M** Information on the status of the entire dataset?: **YES**
The element "Period Start Date" specifies the date on which the first data within the dataset was collected. A second element "Period End Date"

B. **M** Information on the status of the particular subset (e.g., observation time, inauguration time, extraction time)?: **YES**

as 5.1 A, for particular dataset units

Does the metadata standard specify:

C. **M** A structure for updating or maintenance information?: **YES**

D. **M** A structure to describe historical information (see also Section 8.3 G)?: **YES**

5.2 Information on Incremental Updating

Does the metadata standard specify:

Obligation

A. **M** A structure for information on incremental updates of the entire dataset?: **YES**

information on incremental updates and related information is included within the "Lineage" element of the metadata.

B. **M** A structure for information on incremental updates of particular portions of the dataset?: **YES**

as in 5.2 A

C. **M** A structure to describe the update process for the dataset?: **YES**

as in 5.2 A

SECTION 6 – Extent of the Dataset

This section examines how the standard provides for the specification of the spatial and temporal extent of the dataset.

6.1 The spatial extent of the dataset

Obligation

A. **C** Does the metadata standard allow/require for the description of horizontal extent (e.g., bounding rectangle/polygon, named location (i.e., Stockholm, etc.))?: **YES**

Bounding polygon, Bounding rectangle, Named locations

B. __ Does the metadata standard allow/require for the description of vertical extent (e.g., minima/maxima heights, etc.)?: **NO**

C. __ Does the standard require specific units for describing extent?: **NO**

D. **C** Does the standard allow specification of the units for describing extent?: **YES**

E. __ Does the standard require specific spatial reference system for describing extent?: **NO**

F. **C** Does the standard allow specification of the spatial reference system for describing extent?: **YES**

6.2 The temporal extent of the dataset

Obligation

A. **C** Does the standard allow/require for the description of time period for which the dataset is relevant?: **YES**

B. __ Does the standard allow/require specification of the temporal reference system?: **NO**

SECTION 7 – Data Content of the Dataset

This section examines how the standard provides for the specification of the data content of the dataset in terms of overview elements, themes, data definitions and product specifications.

7.1 Dataset overview

Does the standard allow/require for the provision of the following information:

Obligation

A. **M** Dataset overview?: **YES**
B. **M** Abstract – brief summary of the content of the dataset?: **YES**
C. **O** Purpose – summary of the intentions with which the dataset was developed?: **YES**
D. **O** Usage – list of activities for which the dataset has been used? (see also Section 8.5): **YES**
E. **M** Spatial schema – (e.g., vector, raster)? (see also Section 9): **YES**
F. **M** Spatial reference system? (see also Section 12): **YES**
G. **M** Language?: **YES**
H. **M** Character set (e.g., ISO8859-1, JISX0208)?: **YES**
I. __ List of documents providing further information about the dataset?: **NO**
J. **O** Sample/preview dataset?: **YES**
K. **O** Related datasets?: **YES**
L. __ Original size of the dataset?: **NO**
M. __ Compression? (see also Section 11.5): **NO**
N. __ Other?: **NO**

7.2 Theme(s)

Obligation

A. **O** Does the standard allow/require the provision for the description of theme(s) (e.g., topography, land cover, transport, etc.) in the dataset?: **YES**
B. __ If A is yes, does the standard allow the referencing of an outside dictionary or thesaurus?: **NO**
C. **M** If A is yes, does the standard itself include a dictionary or thesaurus within the metadata?: **YES**
D. __ If A is yes, does the standard allow the use of unauthorized user-defined keywords?: **NO**

7.3 Data Definition

Obligation

A. __ Does the standard allow/require the provision for the description of the data definition?: **NO**

B. If A is yes, does the standard allow/require the provision for the description of:

__ 1. Features/Objects?: **N/A**

__ 2. Attributes?: **N/A**

__ 3. Relations?: **N/A**

C. __ Can attributes be described separately from feature/object classes?: **NO**

D. __ Does the standard allow/require the specification of the domain of the attributes?: **NO**

E. __ Does the standard allow/require the specification of data type of the attributes?: **NO**

F. __ Does the standard allow/require the specification of particular classification coding schemes (e.g., FACC, OSKA, S-57 OC)?: **NO**

7.4 Product Specification

Obligation

A. __ Does the standard allow/require for the provision of information about whether the dataset meets a particular product specification?: **NO**

B. __ If A is yes, can it be internal to the Metadata?: **N/A**

C. __ If A is yes, can it be external to the Metadata?: **N/A**

SECTION 8 – Data Quality of the Dataset

This section examines how the standard deals with the question of data quality. This includes homogeneity and metrics for data quality, quality assurance, and usage.

8.1 Overall Data Quality Statement

Obligation

A. **C** Does the standard allow/require for the provision of quality to be described? If yes, please answer the questions below. (*If no, all following answers in Section* 8 *are N/A, please skip to Section* 9.): **YES**

B. __ Does the standard allow/require the provision for the quality information referencing an external quality standard, quality assurance standard, or specification?: **NO**

8.2 Homogeneity of Data Quality

Obligation

A. **O** Does the standard allow/require for the variation of the quality of the dataset to be described?: **YES**

B. What type of aggregation of quality information is allowed? **YES**

O 1. Entire database?: **YES**

O 2. Feature/Object instance?: **YES**

O 3. Feature/Object class?: **YES**

O 4. Attribute instance?: **YES**
O 5. Attribute class?: **YES**
O 6. Layer?: **YES**
O 7. Theme?: **YES**
O 8. Geographic extent?: **YES**
O 9. Other?: **YES**

The element "Level of Spatial Detail" provides a level of understanding of the density of spatial detail, describing the resolution of the data. Its format is dependent on the nature of the dataset.

8.3 Metrics for Data Quality

Does the standard allow/require the provision of values for:

Obligation

A. **C** Positional accuracy?: **YES**
O 1. Standard error?: **YES**
O 2. Maximum error?: **YES**
O 3. Error correlation?: **YES**
O 4. Confidence level (e.g., 90%, 95%)?: **YES**
O 5. Other?: **YES**
User defined criteria allowed, depending on nature of dataset

B. **C** Attribute accuracy?: **YES**
O 1. Standard error?: **YES**
O 2. Maximum error?: **YES**
O 3. Error frequency?: **YES**
O 4. Error correlation?: **YES**
O 5. Confidence level (e.g., 90%, 95%)?: **YES**
O 6. Other?: **YES**
as 8.3 A5

C. x Temporal accuracy?: **YES**
O 1. Standard error?: **YES**
O 2. Maximum error?: **YES**
O 3. Error correlation?: **YES**
O 4. Confidence level? (e.g., 90%, 95%): **YES**
O 5. Other?: **YES**
as 8.3 A5

D. **C** Completeness?: **YES**
O 1. Error of omission?: **YES**
O 2. Error of commission?: **YES**
O 3. Other?: **YES**
as 8.3 A5

E. **C** Currentness?: **YES**
O 1. Date?: **YES**
O 2. Maximum age?: **YES**
O 3. Temporal extent?: **YES**

O 4. Other?: **YES**

as 8.3 A5

F. **C** Logical consistency?: **YES**

G. **M** Lineage?: **YES**

O 1. List of processing steps?: **YES**

O 2. List of values of processing parameters?: **YES**

O 3. List of printouts of statistics, etc.?: **YES**

O 4. Source material?: **YES**

5. Other?: **N/A**

as 8.3 A5

H. **O** Additional Quality Information?: **YES**

The element "Level of Spatial Detail" provides a level of understanding of the density of spatial detail, describing the resolution of the data. Its format is dependent on the nature of the dataset.

8.4 Quality Assurance Methods

Does the standard allow/require the provision of information about the methods used to determine:

Obligation

A. **O** Positional accuracy?: **YES**

O 1. Standard error in adjustments?: **YES**

O 2. Repeated measurements?: **YES**

O 3. Independent measurements?: **YES**

O 4. Subjective evaluations?: **YES**

O 5. Other?: **YES**

as 8.3 A5

B. **O** Attribute accuracy?: **YES**

O 1. Standard error?: **YES**

O 2. Error classification matrix?: **YES**

O 3. Independent measurements?: **YES**

O 4. Subjective evaluation?: **YES**

O 5. Other?: **YES**

as 8.3 A5

C. **O** Temporal accuracy?: **YES**

O 1. Independent measurements?: **YES**

D. **O** 2. Subjective evaluation?: **YES**

E. **O** 3. Other?: **YES**

as 8.3 A5

O Completeness?: **YES**

O 1. Field checks?: **YES**

O 2. Repeated measurements?: **YES**

O 3. Independent measurements?: **YES**

O 4. Subjective evaluation?: **YES**

O 5. Error of commission?: **YES**

O 6. Other?: **YES**

as 8.3 A5

F. **O** Currentness?: **YES**

O 1. Date of source material?: **YES**

O 2. Repeated measurements?: **YES**

O 3. Independent measurements?: **YES**

O 4. Subjective evaluation?: **YES**

O 5. Other?: **YES**

as 8.3 A5

G. **O** Logical consistency?: **YES**

O 1. Verification results of consistency checks?: **YES**

O 2. Repeated processing?: **YES**

O 3. Comparison with independent sources?: **YES**

O 4. Subjective evaluation?: **YES**

O 5. Other?: **YES**

as 8.3 A5

H. **O** Lineage?: **YES**

O 1. Checking results of processing steps: **YES**

O 2. Subjective evaluation: **YES**

O 3. Other: **YES**

as 8.3 A5

I. **O** Additional Quality Information?: **YES**

8.5 Usage

Obligation

A. **O** Does the metadata standard allow/require the provision of information about usage?: **YES**

O 1. Who used it: **YES**

O 2. When it was used: **YES**

O 3. For what application it was used: **YES**

O 4. The effectiveness of use: **YES**

O 5. Other: **YES**

as 8.3 A5

Alternative Quality Description: **NO**

SECTION 9 – Spatial Data Structure of the Dataset

This section examines how the standard provides for the specification of the spatial data model and spatial data types.

9.1 Spatial Data Model

Does the standard allow/require the provision for:

Obligation

A. __ The description of the geometrical and/or topological primitives used to model the real world?: **NO**
B. __ Information about an external standard for geometrical and/or topological primitive specification?: **NO**

9.2 Spatial Data Types
Does the standard allow/require the provision for:

Obligation

A. **C** Information about the spatial data type(s) used in the dataset?: **YES**
B. **O** If A is yes, which of the following generic spatial data types are supported? **YES**
For each, please list specific data primitive levels supported by each data type.
1. Vector (e.g., spaghetti, chain-node/node-edge-face, planar graph, full topology)?: **YES**
all data primitive levels
2. Areal Raster (e.g., thematic raster – scanned and classification maps, image raster – satellite images, scanned aerial photographs)?: **YES**
all data primitive levels
3. Grid/Matrix (e.g., Digital Elevation Model)?: **YES**
all data primitive levels
4. Triangular Irregular Network (TIN)?: **NO**
5. textbfO Other?: **YES**
The standard allows users to provide spatial data types and primitives as appropriate to the described datasets.

9.3 Other Data Types
Does the standard allow the provision of information about:

Obligation

A. __ Video?: **NO**
B. __ Sound?: **NO**
C. __ Other?: **NO**

SECTION 10 – Spatial Reference of the Dataset

This section examines how the dataset provides for the specification of the spatial reference characteristics of the data set.

Does the standard allow/require the provision for the following information:

10.1 Geoidal Model

Obligation

A. __ Geoidal model name?: **NO**
B. __ Geoidal model definitions?: **NO**

10.2 Geodetic datum

Obligation

A. **C** Geodetic datum name?: **YES**
B. __ Geodetic datum definitions?: **NO**

10.3 Reference Ellipsoid

Obligation

A. **C** Reference ellipsoid name?: **YES**
B. __ Reference ellipsoid definitions?: **NO**

10.4 Projection

Obligation

A. **C** Projection name?: **YES**
B. __ Projection definitions?: **NO**

10.5 Coordinate Systems

Obligation

A. **C** Coordinate system type?: **YES**
B. __ Coordinate system definitions?: **NO**

10.6 Height Reference System

Obligation

A. **C** Vertical datum name?: **YES**
B. __ Types of heights?: **NO**

10.7 Ancillary Geodetic Information

Obligation

A. __ Does the standard allow/require the provision for ancillary geodetic, gravimetric, or magnetometric information?: **NO**

10.8 Transformation Parameters
Does the standard specify a method to provide information for coordinate transformation(s):

Obligation

A. __ Model name?: **NO**
B. __ Control points?: **NO**
C. __ Transformation parameters?: **NO**

10.9 Non-Coordinate Spatial Reference

Does the standard allow/require the provision for information on non-coordinate spatial reference systems (e.g., postal systems, street address)?:

Obligation

A. **C** Reference system name?: **YES**
B. __ Reference system definitions?: **NO**
C. __ Other?: **NO**

SECTION 11 – Availability and Distribution of the Dataset

This section examines how the standard provides information on the availability of the dataset?

Does the standard allow/require the provision for:

11.1 The Distributor(s) of the Dataset

Obligation

A. **M** Information about the distributor(s) of the dataset?: **YES**
B. **O** Contact information for the distributor(s)?: **YES**

11.2 Restrictions to Access and Usage

Obligation

A. **M** Information about copyright on the dataset (e.g., date, period, copyright owner, etc.)?: **YES**
B. **M** Information about restrictions on the access of the dataset?: **YES**
C. **M** Information about restrictions on the usage of the dataset?: **YES**

11.3 Pricing Details

Obligation

A. **M** Information about pricing?: **YES**

11.4 Distribution of the Dataset

Obligation

A. **M** Information about distribution media?: **YES**
B. **O** Information about on-line access?: **YES**
C. **M** Information about units of distribution (e.g., tiles, layers, polygons, elements)?: **YES**

11.5 Distribution Format

Obligation

A. Production and/or development:

SSSA Unit – Strategy and Systems for Space Applications
Space Applications Institute
EC Joint Research Centre, DG Research,
European Commission
T.P. 261
21020 Ispra (VA), Italy
Fax: +39 0332 78 5461
Phone: +39 0332 78 5425
E-mail: bernd_eckhardt@yahoo.com

B. Testing:

Bernd Eckhardt
SSSA Unit – Strategy and Systems for Space Applications
Space Applications Institute
EC Joint Research Centre, DG Research,
European Commission
T.P. 261
21020 Ispra (VA), Italy
Fax: +39 0332 78 5461
Phone: +39 0332 78 5425
E-mail: bernd_eckhardt@yahoo.com

C. Conformance:

SSSA Unit – Strategy and Systems for Space Applications
Space Applications Institute
EC Joint Research Centre, DG Research,
European Commission
T.P. 261
21020 Ispra (VA), Italy
Fax: +39 0332 78 5461
Phone: +39 0332 78 5425
E-mail: bernd_eckhardt@yahoo.com

D. Maintenance:

Bernd Eckhardt
SSSA Unit – Strategy and Systems for Space Applications
Space Applications Institute
EC Joint Research Centre, DG Research,
European Commission
T.P. 261
21020 Ispra (VA), Italy
Fax: +39 0332 78 5461
Phone: +39 0332 78 5425
E-mail: bernd_eckhardt@yahoo.com

E. Distribution:

SSSA Unit – Strategy and Systems for Space Applications

Space Applications Institute EC Joint Research Centre, DG Research, European Commission
T.P. 261
21020 Ispra (VA), Italy
Fax: +39 0332 78 5461
Phone: +39 0332 78 5425
E-mail: bernd_eckhardt@yahoo.com

F. Help Desk/User Support:

SSSA Helpdesk, Bernd Eckhardt
SSSA Unit – Strategy and Systems for Space Applications
Space Applications Institute
EC Joint Research Centre, DG Research,
European Commission
T.P. 261
21020 Ispra (VA), Italy
Fax: +39 0332 78 5461
Phone: +39 0332 78 5425
E-mail: bernd_eckhardt@yahoo.com

1.3 Development History

A. Beginning of work on metadata standard: **January 1996**
B. Milestones:
May 1997 – First public release to 700 users
February 1999 – Version 2.0 available online
C. Anticipated completion: **May 2002**
D. Update cycle: **N/A**

1.4 Status of the Standard

A. Is the standard officially recognized: **YES**
B. Date(s) of recognition:
May 1997 – First public release to 700 users
C. Anticipated date of recognition: **N/A**
D. Current stage in recognition process:
The SSSA Recommendations on Metadata document is available on INFEO-Information on Earth Observation (http://www.infeo.org)
The Metadata authoring Tool 4.1 (MAT 4.1) generated by SSSA to enable organizations to generate own metadata is available online for downloading by FTP. Access of the tool is free of charge!
E. Steps still required to achieve recognition: **N/A**

1.5 Access to Documentation of the Standard

A. Is the standard copyrighted?: **NO**

B. Name and address of copyright owner: **N/A**
C. Do restrictions on the use of the standard apply?: **NO**
D. Is the standard documentation currently available?: **YES**
E. List the form/media of standard documentation:
Online accessible per FTP, printed hardcopy on request
F. Is the standard available on the Internet?: **YES**
G. Price:
Free of charge
H. Contact details for obtaining the standard:
SSSA Unit – Strategy and Systems for Space Applications
Space Applications Institute
EC Joint Research Centre
T.P. 261
21020 Ispra (VA), Italy
Fax: +39 0332 78 5461
Phone: +39 0332 78 5425
E-mail: bernd_eckhardt@yahoo.com
I. Does the standard have ISBN/ISSN numbers?: **NO**

1.6 Available Software Tools

A. Is software available to implement the standard as a database?: **YES**
The software for our database INFEO (Information on Earth Observation)! The software tool (Metadata Authoring Tool – MAT 4.1) mentioned under C. was developed to generate XML files to ingest directly into INFEO
B. Is software available to test compliance of metadata with the standard?: **YES**
The software for our database INFEO (Information on Earth Observation) would refuse to read and ingest metadata, which is not compliant with the standard
C. Is software available to produce metadata compliant with the standard?: **YES**
Metadata Authoring Tool – MAT 4.1
D. Other relevant software:
No other software
E. Is there a test dataset available to test software tools for implementing the standard?: **YES**
F. For each product available above, please list contact information:
SSA Unit – Strategy and Systems for Space Applications
Space Applications Institute
EC Joint Research Centre
T.P. 261
21020 Ispra (VA), Italy
Fax: +39 0332 78 5461
Phone: +39 0332 78 5425
E-mail: bernd_eckhardt@yahoo.com

1.7 Available Training Materials

A. Are organized training sessions available?: **NO**
B. Contact details for training sessions: **N/A**
C. Training documentation available: **N/A**

SECTION 2 – Use and Implementation of the Metadata Standard

This section provides information on the way in which the standard is used and the purpose of its use. It also provides descriptions about the level of abstraction of the standard.

2.1 Intended Use of the Standard
Can the standard be used:

A. Within a transfer data structure?: **YES**
B. To produce hardcopy catalog of datasets?: **NO**
C. To catalog data in a digital on-line catalog?: **YES**
D. Other: **N/A**

2.2 Intended Purpose of the Standard
Is the intended purpose of the standard:

A. To support discovery, identification, location of data?: **NO**
B. To support how to access data?: **NO**
C. To support the determination of dataset fitness for use?: **YES**
D. To support transfer of dataset?: **YES**
E. To support an organization's data management?: **YES**
F. Other: **N/A**

2.3 Table of Contents

A. Provide the standard's Table of Contents

1 Introduction
1.1 Context
1.2 Aims
1.3 Rationale
1.4 Contents

2 Metadata
2.1 Introduction
2.2 Definition and purpose of metadata
2.3 CEO approach to metadata

3 Describing your resources
3.1 "Resource Types"
3.2 Describing a resource
3.3 Key
3.4 Element structure and format

3.5 Element tables of the identified "Resource Types"
3.5.1 Organisation
3.5.2 Person
3.5.3 Dataset
3.5.4 Document
3.5.5 Service
3.5.6 Project
3.5.7 Event
3.5.8 Software/Model
3.5.9 Discussion Forum
3.5.10 Promotion
3.5.11 Library

4 Controlled keywords and formats
4.1 Controlled keywords
4.1.1 Language
4.1.2 Discipline
4.1.3 Organisation type
4.1.4 Dataset type
4.1.5 Parameter
4.1.6 Document type
4.1.7 Service type
4.1.8 Project Type
4.1.9 Event type
4.1.10 Software/model type
4.1.11 Discussion Forum type
4.1.12 Promotion type
4.1.13 Library type
4.1.14 Location
4.1.15 Scale
4.1.16 Resolution
4.2 Field Formats
4.2.1 Date/time
4.2.2 Latitude/longitude

5 The CEO metadata authoring tool
6 CEO information services
7 References

Appendix A: Controlled Lists

2.4 Level of Abstraction of the Metadata Standard Definition

A. Does the standard define metadata elements?: **YES**
B. Does the standard define a model for the relationships between the metadata elements?: **YES**

Resource Types	Organisation	Person	Dataset	Document	Service	Project	Event	Software/Model	Discussion Forum	Promotion	Library
Descriptive Elements											
Title	✔	✔	✔	✔	✔	✔	✔	✔	✔	✔	✔
Organisation		✔	✔	✔	✔	✔	✔	✔	✔	✔	✔
Abstract	✔	✔	✔	✔	✔	✔	✔	✔	✔	✔	✔
Point of contact	✔	✔	✔	✔	✔	✔	✔	✔	✔	✔	✔
Language of resource	✔	✔	✔	✔	✔	✔	✔	✔	✔	✔	✔
Author			✔	✔		✔		✔			
Controlled subject index	✔	✔	✔	✔	✔	✔	✔	✔	✔	✔	✔
Spatial Coverage	✔	✔	✔	✔	✔	✔	✔	✔	✔	✔	✔
Temporal Coverage			✔	✔	✔	✔	✔	✔	✔	✔	✔
Cross Reference	✔	✔	✔	✔	✔	✔	✔	✔	✔	✔	✔
Original Resource Identifier			✔	✔	✔	✔	✔	✔		✔	
Scale			✔								
Resolution			✔								
Method/Quality			✔								
Supply details			✔	✔				✔			✔
Constraints			✔	✔			✔	✔	✔		✔
Publisher				✔							
Version				✔							
Publication Date				✔							
Administrative Elements											
Record Identifier	✔	✔	✔	✔	✔	✔	✔	✔	✔	✔	✔
ResourceType	✔	✔	✔	✔	✔	✔	✔	✔	✔	✔	✔
Language of Record	✔	✔	✔	✔	✔	✔	✔	✔	✔	✔	✔
Record Provider	✔	✔	✔	✔	✔	✔	✔	✔	✔	✔	✔
Record Creation Date	✔	✔	✔	✔	✔	✔	✔	✔	✔	✔	✔
Record Last Modification date	✔	✔	✔	✔	✔	✔	✔	✔	✔	✔	✔
Record Expiry date	✔	✔	✔	✔	✔	✔	✔	✔	✔	✔	✔

Fig. 1. Record Identification.

C. Provide an overview diagram of the model of the metadata, if available. **YES**

Element applicability

The elements, which are applicable to each "Resource Type", are summarised in the following table. It is recommended that the completion of certain elements is made mandatory for each

particular "Resource Type" to ensure that useful descriptions are obtained. These are described in detail in Sections 3.5.1 to 3.5.11

D. Does the standard define domains and/or data types of the metadata elements?: **YES**
E. Does the standard define how the metadata should be encoded?: **NO**
F. Does the standard define a database schema for the automatic generation of a database?: **NO**

2.5 Extensibility of the Standard

A. Is the metadata description extensible (e.g., can user add metadata domains, elements, thematic profiles, etc.)?: **NO**

2.6 Availability of Metadata
Does the standard specify:

A. How metadata are to be accessed?: **YES**
Given through the 11 EO related "Resource Types" defined by us: Users of our standard should try to ingest their data into this scheme where it is possible
B. A mechanism for querying the metadata?: **NO**
C. Particular standardized query protocols?: **NO**
D. That the metadata should be encoded for computer access?: **NO**
E. That the metadata should be encrypted for security?: **NO**

2.7 Examples

A. Does the standard provide examples of implementations?: **YES**
For each "Resource Type" defined in our standard an example is given to explain which information can fit under which of these "Resource Types"
B. How many examples are available?: **11**

SECTION 3 – Linkage and Coordination of the Metadata Standard

This section provides information on how the metadata standard is linked to other standards, particularly transfer standards, and how such linkage is implemented. If a significant linkage does exist, detailed technical information is provided on the precise nature of the linkage to the standard, and how the linkage affects implementation of the metadata standard.

3.1 Associated Transfer Standards

A. Is the metadata standard associated with a transfer standard(s)?: **NO**
B. If A is yes, provide references and contact details for each linked standard: **N/A**
C. If A is yes, are both (all) standards coordinated?: **N/A**

D. If A is yes, describe how the coordination works, both from a technical and conformance (normative) viewpoint: **N/A**

3.2 Incorporated into transfer mechanism

A. Is the metadata standard incorporated into a transfer mechanism?: **NO**
B. If A is yes, provide references and contact details for each incorporated mechanism: **N/A**
C. If A is yes, describe how: **N/A**

3.3 Other linked, coordinated or associated standards

A. Is the metadata standard linked, coordinated, or associated with any other standard(s) not listed above?: **NO**
B. If A is yes, provide references and contact details for each standard: **N/A**
C. If A is yes, describe how for each relevant standard: **N/A**

II. Categories the Metadata Standard Uses to Describe Datasets

The following sections are categories of metadata that may be used by the metadata standard to describe data. Not all categories are relevant to different datasets and may not necessarily be present in the standard. The presence or absence of a category should not be used for judging one standard against another.

Note: For each of the questions in the following sections please provide the obligation status information. Valid obligation values are:

Obligation Level

M	Mandatory	This information must be entered to comply with the metadata standard.
C	Conditional	This information is entered based on other information in the metadata.
O	Optional	This information is optionally entered in the metadata.

If the answer is 'NO', the obligation status should be left blank.

SECTION 4 – Dataset Identification

This section examines how the standard provides for the specification of the various sorts of titles and identifiers for a dataset.

4.1 The title(s) of a dataset
Does the standard allow/require the provision for:

Obligation

A. __ Previous title(s)?: **NO**
B. **M** Current title(s)?: **YES**
C. __ Possible future title(s)?: **NO**
D. __ Official titles in different languages?: **NO**

E. __ Abbreviated title(s)?: **NO**
F. __ Alternative Title(s)?: **NO**
G. __ Does the standard support effective dates of titles?: **NO**
H. **O** Other identifiers (e.g., ISBN, National Stock Number, Uniform Product Code)?: **YES**

Original Resource Identifier O
Record Identifier M

4.2 Parties responsible for the dataset
Does the standard allow/require the provision for:

Obligation

A. **O** Author?: **YES**
B. **M** Owner?: **YES**
C. **O** Other?: **YES**

4.3 ISBD Information

Obligation

A. __ Does the metadata standard specify inclusion of mandatory ISBD information (see glossary) within the metadata?: **NO**

SECTION 5 – Status of the Dataset

This section examines how the standard provides for the specification of the current status and updating of the data set.

5.1 Status and Progress of Datasets
Does the metadata standard allow/require the provision for:

Obligation

A. **M/O** Information on the status of the entire dataset?: **YES**

Difficult to say in our case: Datasets can be described by a certain amount of elements (some mandatory some optional) in our standard, for some datasets it will cover the entire status for others not. E.g., the status of a satellite image is fixed in some way after it has been recorded. Elements available are:

B. **M/O** Information on the status of the particular subset (e.g., observation time, inauguration time, extraction time)?: **YES**

Does the metadata standard specify:

C. __ A structure for updating or maintenance information?: **NO**
D. __ A structure to describe historical information (see also Section 8.3 G)?: **NO**

5.2 Information on Incremental Updating
Does the metadata standard specify:

Obligation

A. __ A structure for information on incremental updates of the entire dataset?: **NO**
B. __ A structure for information on incremental updates of particular portions of the dataset?: **NO**
C. __ A structure to describe the update process for the dataset?: **NO**

SECTION 6 – Extent of the Dataset

This section examines how the standard provides for the specification of the spatial and temporal extent of the dataset.

6.1 The spatial extent of the dataset

Obligation

A. **M** Does the metadata standard allow/require for the description of horizontal extent (e.g., bounding rectangle/polygon, named location (i.e., Stockholm, etc.))?: **YES**
For spatial data a point, or bounding shape for the contents of the dataset has to be specified. It is suggested that a location keyword (from a controlled list) should be used if this is more appropriate
B. __ Does the metadata standard allow/require for the description of vertical extent (e.g., minima/maxima heights, etc.)?: **NO**
C. **O** Does the standard require specific units for describing extent?: **YES**
D. __ Does the standard allow specification of the units for describing extent?: **NO**
E. __ Does the standard require specific spatial reference system for describing extent?: **NO**
F. __ Does the standard allow specification of the spatial reference system for describing extent?: **NO**

6.2 The temporal extent of the dataset

Obligation

A. **M** Does the standard allow/require for the description of time period for which the dataset is relevant?: **YES**
B. __ Does the standard allow/require specification of the temporal reference system?: **NO**

SECTION 7 – Data Content of the Dataset

This section examines how the standard provides for the specification of the data content of the dataset in terms of overview elements, themes, data definitions and product specifications.

7.1 Dataset overview

Does the standard allow/require for the provision of the following information:

Obligation

A. **M** Dataset overview?: **YES**
B. **O** Abstract – brief summary of the content of the dataset?: **YES**
C. __ Purpose – summary of the intentions with which the dataset was developed?: **NO**
D. __ Usage – list of activities for which the dataset has been used? (see also Section 8.5): **NO**
E. __ Spatial schema – (e.g., vector, raster)? (see also Section 9): **NO**
F. __ Spatial reference system? (see also Section 12): **NO**
G. **M** Language?: **YES**
H. **O** Character set (e.g., ISO8859-1, JISX0208)?: **YES**
I. **O** List of documents providing further information about the dataset?: **YES**
J. __ Sample/preview dataset?: **NO**
K. **O** Related datasets?: **YES**
L. __ Original size of the dataset?: **NO**
M. __ Compression? (see also Section 11.5): **NO**
N. **M** Other?: **YES**
M **Discipline Keywords/Valids related to that dataset,**
M **Dataset type Keywords/Valids related to that dataset,**
M **Parameter Keywords/Valids related to that dataset (7.2 Theme)**

7.2 Theme(s)

Obligation

A. **M** Does the standard allow/require the provision for the description of theme(s) (e.g., topography, land cover, transport, etc.) in the dataset?: **YES**
B. __ If A is yes, does the standard allow the referencing of an outside dictionary or thesaurus?: **NO**
C. x If A is yes, does the standard itself include a dictionary or thesaurus within the metadata?: **YES**
A short description is given in the standard to each keyword/valid that can be used describing data.
D. __ If A is yes, does the standard allow the use of unauthorized user-defined keywords?: **NO**

7.3 Data Definition

Obligation

A. **M** Does the standard allow/require the provision for the description of the data definition?: **YES**
B. If A is yes, does the standard allow/require the provision for the description of:
__ 1. Features/Objects?: **NO**
__ 2. Attributes?: **NO**
__ 3. Relations?: **NO**
C. __ Can attributes be described separately from feature/object classes?: **NO**

D. __ Does the standard allow/require the specification of the domain of the attributes?: **NO**

E. __ Does the standard allow/require the specification of data type of the attributes?: **NO**

F. __ Does the standard allow/require the specification of particular classification coding schemes (e.g., FACC, OSKA, S-57 OC)?: **NO**

7.4 Product Specification

Obligation

A. __ Does the standard allow/require for the provision of information about whether the dataset meets a particular product specification?: **NO**

B. __ If A is yes, can it be internal to the Metadata?: **NO**

C. __ If A is yes, can it be external to the Metadata?: **NO**

SECTION 8 – Data Quality of the Dataset

This section examines how the standard deals with the question of data quality. This includes homogeneity and metrics for data quality, quality assurance, and usage.

8.1 Overall Data Quality Statement

Obligation

A. **O** Does the standard allow/require for the provision of quality to be described? If yes, please answer the questions below. (*If no, all following answers in Section* 8 *are N/A, please skip to Section* 9.): **YES**

It's a free text area in our standard and organizations can add attributes describing quality related to their quality standard. See 8.6 about the description we give to the users of our metadata standard to explain them which information can be ingested under the "Method/Quality" area!

B. __ Does the standard allow/require the provision for the quality information referencing an external quality standard, quality assurance standard, or specification?: **NO**

8.2 Homogeneity of Data Quality

Obligation

A. __ Does the standard allow/require for the variation of the quality of the dataset to be described?: **NO**

B. What type of aggregation of quality information is allowed? **NO**

__ 1. Entire database?: **NO**

__ 2. Feature/Object instance?: **NO**

__ 3. Feature/Object class?: **NO**

__ 4. Attribute instance?: **NO**

__ 5. Attribute class?: **NO**

__ 6. Layer?: **NO**
__ 7. Theme?: **NO**
__ 8. Geographic extent?: **NO**
__ 9. Other?: **NO**

8.3 Metrics for Data Quality
Does the standard allow/require the provision of values for:

Obligation

A. __ Positional accuracy?: **NO**
 1. Standard error?: **NO**
 2. Maximum error?: **NO**
 3. Error correlation?: **NO**
 4. Confidence level (e.g., 90%, 95%)?: **NO**
 5. Other?: **NO**
B. __ Attribute accuracy?: **NO**
 1. Standard error?: **NO**
 2. Maximum error?: **NO**
 3. Error frequency?: **NO**
 4. Error correlation?: **NO**
 5. Confidence level (e.g., 90%, 95%)?: **NO**
 6. Other?: **NO**
C. __ Temporal accuracy?: **NO**
 1. Standard error?: **NO**
 2. Maximum error?: **NO**
 3. Error correlation?: **NO**
 4. Confidence level? (e.g., 90%, 95%): **NO**
 5. Other?: **NO**
D. __ Completeness?: **NO**
 1. Error of omission?: **NO**
 2. Error of commission?: **NO**
 3. Other?: **NO**
E. __ Currentness?: **NO**
 1. Date?: **NO**
 2. Maximum age?: **NO**
 3. Temporal extent?: **NO**
 4. Other?: **NO**
F. __ Logical consistency?: **NO**
G. __ Lineage?: **NO**
 1. List of processing steps?: **NO**
 2. List of values of processing parameters?: **NO**
 3. List of printouts of statistics, etc.?: **NO**
 4. Source material?: **NO**
 5. Other?: **NO**
H. __ Additional Quality Information?: **NO**

8.4 Quality Assurance Methods

Does the standard allow/require the provision of information about the methods used to determine:

Obligation

A.	__ Positional accuracy?:	**NO**
	1. Standard error in adjustments?:	**NO**
	2. Repeated measurements?:	**NO**
	3. Independent measurements?:	**NO**
	4. Subjective evaluations?:	**NO**
	5. Other?:	**NO**
B.	__ Attribute accuracy?:	**NO**
	1. Standard error?:	**NO**
	2. Error classification matrix?:	**NO**
	3. Independent measurements?:	**NO**
	4. Subjective evaluation?:	**NO**
	5. Other?:	**NO**
C.	__ Temporal accuracy?:	**NO**
	1. Independent measurements?:	**NO**
	2. Subjective evaluation?:	**NO**
	3. Other?:	**NO**
D.	__ Completeness?:	**NO**
	1. Field checks?:	**NO**
	2. Repeated measurements?:	**NO**
	3. Independent measurements?:	**NO**
	4. Subjective evaluation?:	**NO**
	5. Error of commission?:	**NO**
	6. Other?:	**NO**
E.	__ Currentness?:	**NO**
	1. Date of source material?:	**NO**
	2. Repeated measurements?:	**NO**
	3. Independent measurements?:	**NO**
	4. Subjective evaluation?:	**NO**
	5. Other?:	**NO**
F.	__ Logical consistency?:	**NO**
	1. Verification results of consistency checks?:	**NO**
	2. Repeated processing?:	**NO**
	3. Comparison with independent sources?:	**NO**
	4. Subjective evaluation?:	**NO**
	5. Other?:	**NO**
G.	__ Lineage?:	**NO**
	1. Checking results of processing steps:	**NO**
	2. Subjective evaluation:	**NO**
	3. Other:	**NO**

H. __ Additional Quality Information?: **NO**

8.5 Usage

Obligation

A. __ Does the metadata standard allow/require the provision of information about usage?: **NO**
1. Who used it: **NO**
2. When it was used: **NO**
3. For what application it was used: **NO**
4. The effectiveness of use: **NO**
5. Other: **NO**

8.6 Alternative Quality Description: **YES**

The Method/Quality element of our standard allows describing the following 5 terms in free text fields: Data acquisition process, Processing history, Quality, Accuracy, Completeness!
The following description explains our metadata users what information can be ingested into this 'Method/Quality' free text area:
"Provide a short description of the method by which the base data was collected, e.g., in-situ, airborne photography or by satellite.
List the primary steps involved in creation of the product and the names of any algorithms used.
Describe any quality procedures applied to the dataset. Include indicators of data quality, standard methods used and recognised or potential problems with quality. Established quality control mechanisms should be listed and references to calibration/validation datasets or models should be made.
Provide a brief outlines of the accuracy of the dataset contents.
Describe the completeness of the dataset, e.g., 'complete', 'partial', etc."

SECTION 9 – Spatial Data Structure of the Dataset

This section examines how the standard provides for the specification of the spatial data model and spatial data types.

9.1 Spatial Data Model
Does the standard allow/require the provision for:

Obligation

A. __ The description of the geometrical and/or topological primitives used to model the real world?: **NO**
B. __ Information about an external standard for geometrical and/or topological primitive specification?: **NO**

9.2 Spatial Data Types

Does the standard allow/require the provision for:

Obligation

A. **M** Information about the spatial data type(s) used in the dataset?: **YES**

B. __ If A is yes, which of the following generic spatial data types are supported? For each, please list specific data primitive levels supported by each data type.

1. Vector (e.g., spaghetti, chain-node/node-edge-face, planar graph, full topology)?: **NO**
2. Areal Raster (e.g., thematic raster – scanned and classification maps, image raster – satellite images, scanned aerial photographs)?: **NO**
3. Grid/Matrix (e.g., Digital Elevation Model)?: **NO**
4. Triangular Irregular Network (TIN)?: **NO**
5. Other?: **YES**

1-Level-Keywords	2-Level-Keywords	Description
Airborne data		any kind of data recorded with a sensor carried by an aircraft
	Bi-dimensional data	dataset with a bi-dimensional extent (generally raster imagery)
	Mono-dimensional data	dataset with a mono-dimensional extent (generally profile measurements)
	Point data	dataset corresponding to a particular point location
	Video	dataset acquired with a camera
	Aerial photos	photos taken from a plane
Ground based information		any kind of data recorded on ground
	Bi-dimensional data	dataset with a bi-dimensional extent (generally raster imagery)
	Mono-dimensional data	dataset with a mono-dimensional extent (generally profile measurements)
	Point data	dataset corresponding to a particular point location
	Maps	geocoded dataset corresponding to a given area
Non-spatial data		any kind of data which cannot be tied to a particular geographic location

	Statistics	dataset resulting from the statistical processing of a former source of information
	Databases	collection of measurements related to a particular activity or subject
Satellite data		any kind of data recorded by a sensor carried by a satellite
	Bi-dimensional data	dataset with a bi-dimensional extent (generally raster imagery)
	Mono-dimensional data	dataset with a mono-dimensional extent (generally profile measurements)
	Point data	dataset corresponding to a particular point location
Satellite/airborne derived information		any kind of data resulting from the processing of satellite or airborne data
	Bi-dimensional information	dataset with a bi-dimensional extent (generally raster imagery)
	Mono-dimensional information	dataset with a mono-dimensional extent (generally profile measurements)
	Point information	dataset corresponding to a particular point location
	Maps	geocoded dataset corresponding to a given area

9.3 Other Data Types
Does the standard allow the provision of information about:

Obligation

A. **O** Video?: **YES**
B. __ Sound?: **NO**
C. __ Other?: **NO**

SECTION 10 – Spatial Reference of the Dataset

This section examines how the dataset provides for the specification of the spatial reference characteristics of the data set.

Does the standard allow/require the provision for the following information:

10.1 Geoidal Model

Obligation

A. __ Geoidal model name?: **NO**

B. __ Geoidal model definitions?: **NO**

10.2 Geodetic datum

Obligation

A. __ Geodetic datum name?: **NO**
B. __ Geodetic datum definitions?: **NO**

10.3 Reference Ellipsoid

Obligation

A. __ Reference ellipsoid name?: **NO**
B. __ Reference ellipsoid definitions?: **NO**

10.4 Projection

Obligation

A. __ Projection name?: **NO**
B. __ Projection definitions?: **NO**

10.5 Coordinate Systems

Obligation

A. __ Coordinate system type?: **NO**
B. __ Coordinate system definitions?: **NO**

10.6 Height Reference System

Obligation

A. __ Vertical datum name?: **NO**
B. __ Types of heights?: **NO**

10.7 Ancillary Geodetic Information

Obligation

A. __ Does the standard allow/require the provision for ancillary geodetic, gravimetric, or magnetometric information?: **NO**

10.8 Transformation Parameters
Does the standard specify a method to provide information for coordinate transformation(s):

Obligation

A. __ Model name?: **NO**
B. __ Control points?: **NO**
C. __ Transformation parameters?: **NO**

10.9 Non-Coordinate Spatial Reference
Does the standard allow/require the provision for information on non-coordinate spatial reference systems (e.g., postal systems, street address)?:

Obligation

A. __ Reference system name?: **NO**
B. __ Reference system definitions?: **NO**
C. **M** Other?: **YES**
Address field; Contact name and E-mail address are mandatory

SECTION 11 – Availability and Distribution of the Dataset

This section examines how the standard provides information on the availability of the dataset?

Does the standard allow/require the provision for:

11.1 The Distributor(s) of the Dataset

Obligation

A. **O** Information about the distributor(s) of the dataset?: **YES**
B. **M** Contact information for the distributor(s)?: **YES**

11.2 Restrictions to Access and Usage

Obligation

A. __ Information about copyright on the dataset (e.g., date, period, copyright owner, etc.)?: **NO**
B. **O** Information about restrictions on the access of the dataset?: **YES**
C. **O** Information about restrictions on the usage of the dataset?: **YES**

11.3 Pricing Details

Obligation

A. **O** Information about pricing?: **YES**

11.4 Distribution of the Dataset

Obligation

A. **O** Information about distribution media?: **YES**
B. **O** Information about on-line access?: **YES**
C. **O** Information about units of distribution (e.g., tiles, layers, polygons, elements)?: **YES**

11.5 Distribution Format

Obligation

A. __ Information about distribution formats (e.g., commercial CAD format, commercial GIS formats, National/International transfer standard)?: **NO**
B. __ Information about the encoding of the dataset (e.g., binary, ASCII)?: **NO**
C. __ Information about compression method?: **NO**
D. __ Information about compressed dataset size?: **NO**
E. __ Information about the distribution mechanism?: **NO**
F. __ Information about the encapsulation of data (e.g., ISO 8211, ISO 8824)?: **NO**

11.6 Ordering Procedures

Obligation

A. **O** Information about ordering procedures for the dataset?: **YES**

11.7 Support Services

Obligation

A. __ Information about support services (e.g., additional data transformations, format or coordinate conversions)?: **NO**

SECTION 12 – Authorization and Verification of the Metadata

This section examines how the standard provides for the description of the authorization and verification on the metadata of a dataset (e.g., how is the metadata itself authorized and verified?).

12.1 Metadata reference

Obligation

A. **M** Does the standard allow/require the provision of information about reference dates of the metadata (e.g., last check date, last update date, review date)?: **YES**

12.2 Verification Authority

Obligation

A. __ Does the standard allow/require the provision for information about who verified/authorized the metadata?: **NO**
B. __ Does the standard specify the method for providing this information (e.g., ISO 9000): **NO**

12.3 Statistical Methods Used for Verification

Obligation

A. __ Does the standard specify the statistical methods to be used for verification of the metadata?: **NO**

B. __ Does the standard allow/require the provision for the description of the statistical methods used to verify the metadata?: **NO**

END OF CHARACTERISTICS
FIN

H. Moellering, H.J.G.L. Aalders & A. Crane (Editors)

EuroGeoSurveys: GEIXS: Geological Exchange Information System – Description – Metadata

Denis Bonnefoy

I. Information about the Metadata Standard Itself

SECTION 0 – Assessment Information

This section provides the user with a general description of the scope and intended use of the standard. Information on the respondent who prepared the assessment is also provided.

0.1 Respondent

Assessment prepared by:

Denis Bonnefoy
EuroGeoSurveys / BRGM
Avenue Claude Guillemin
F-45060 Orléans Cedex
Fax: +33 2 38 64 33 33
Phone: +33 2 38 64 37 79
E-mail: d.bonnefoy@brgm.fr

Date assessment completed:

June 2000; updated January 2001 and May 2002, Reviewed Spring, 2005

Relationship of respondent to standard:

User of CEN/TC 287

0.2 Brief Summary Statement

A. Summary of the scope and intent of the standard:
GEIXS system uses the European prestandard (i.e., prENV 12657:1998, adapted to specific needs of geological data.

SECTION 1 – Administration of Metadata Standard

This section provides a general description of the administrative framework, within which the standard was developed, tested and currently resides. Significant details are also provided in terms of the developmental and managerial history as well as information on the existence of documentation, software tools and training materials. This section also provides a point of contact for further information on the standard.

1.1 Name of Standard

A. Name of standard:
 1. In original language(s):
 GEIXS: Geological Exchange Information System – Description – Metadata
 2. English translation, if appropriate:
B. Version/Edition:
 version 1998 date 1998-12-15
C. Language(s) of documentation:
 English
D. Acronym(s)
 N/A
E. Official ID:
 N/A
F. Name(s) of recognizing standards authority(ies):
 All national geological surveys of Europe member of EuroGeo-Surveys plus Lithuania, Russia, Latvia, Hungaria, Estonia, Armenia

1.2 Responsible Institutions

Contact information about those institutions that are (or were) responsible for the processes involved in the evolution of the standard:

A. Production and/or development:
 EuroGeoSurveys
 Rue du Luxembourg, 3
 1000 Bruxelles, Belgium
B. Testing:
 To be defined
C. Conformance:
 To be defined
D. Maintenance:
 BRGM (Geological surveys of France) for technical maintenance and NITG-TNO (Geological Surveys of Netherlands) for thematic administration
E. Distribution:
 Geological Surveys of Europe
F. Help Desk/User Support:
 Jean Yves Breton
 BRGM
 Avenue Claude Guillemin
 F-45060 Orléans Cedex

Fax: +33 2 38 64 33 33
Phone: +33 2 38 64 34 48
E-mail: jy.breton@brgm.fr

1.3 Development History

A. Beginning of work on metadata standard: **1997–05**
B. Milestones:
1997 Implementation of data model
1998 Development of data entry application and web site
1999 Completion – end of project
C. Anticipated completion: **N/A**
D. Update cycle: **none**

1.4 Status of the Standard

A. Is the standard officially recognized: **Expected**
B. Date(s) of recognition: **N/A**
C. Anticipated date of recognition: **1998**
D. Current stage in recognition process:
Recognition by European Geological Surveys
E. Steps still required to achieve recognition: **N/A**

1.5 Access to Documentation of the Standard

A. Is the standard copyrighted?: **NO**
B. Name and address of copyright owner: **N/A**
C. Do restrictions on the use of the standard apply?: **NO**
D. Is the standard documentation currently available?: **YES**
E. List the form/media of standard documentation:
SQL script, Export Oracle
F. Is the standard available on the Internet?: **NO**
G. Price: **Unknown**
H. Contact details for obtaining the standard:
EuroGeoSurveys
I. Does the standard have ISBN/ISSN numbers?: **NO**

1.6 Available Software Tools

A. Is software available to implement the standard as a database?: **YES**
B. Is software available to test compliance of metadata with the standard?: **NO**
C. Is software available to produce metadata compliant with the standard?: **NO**
D. Other relevant software:
E. Is there a test dataset available to test software tools for implementing the standard?: **NO**
F. For each product available above, please list contact information: **N/A**

1.7 Available Training Materials

A. Are organized training sessions available?: **NO**
B. Contact details for training sessions: **N/A**
C. Training documentation available:
available on GEIXS web site (http://www.eurogeosurveys.org)

SECTION 2 – Use and Implementation of the Metadata Standard

This section provides information on the way in which the standard is used and the purpose of its use. It also provides descriptions about the level of abstraction of the standard.

2.1 Intended Use of the Standard
Can the standard be used:

A. Within a transfer data structure?: **NO**
B. To produce hardcopy catalog of datasets?: **YES**
C. To catalog data in a digital on-line catalog?: **YES**
D. Other:
To set up others catalog at national, regional or local level

2.2 Intended Purpose of the Standard
Is the intended purpose of the standard:

A. To support discovery, identification, location of data?: **YES**
B. To support how to access data?: **YES**
C. To support the determination of dataset fitness for use?: **YES**
D. To support transfer of dataset?: **YES**
E. To support an organization's data management?: **YES**
F. Other:
None

2.3 Table of Contents

A. Provide the standard's Table of Contents
Foreword
1 Scope
2 Normative references
3 Definitions
4 Metadata for describing geographic datasets
4.1 General
4.2 Geographic datasets and levels of metadata
4.3 Conformance requirements
4.4 Dates
4.5 Metadata language
4.6 Elements of metadata for describing geographic datasets

Annex A (normative) EXPRESS SCHEMA Metadata
Annex B (informative) EXPRESS-G metadata schema
B.1 Metadata schema page 1
B.2 Metadata schema page 2, dataset identification
B.3 Metadata schema page 3, dataset overview
B.4 Metadata schema page 4, dataset quality elements
B.5 Metadata schema page 5, metadata reference
B.6 Metadata schema page 6, spatial reference system
B.7 Extent metadata schema
B.8 Data definition metadata schema
B.9 Classification metadata schema
B.10 Administrative metadata schema page 1, organisation and point of contact
B.11 Administrative metadata schema page 2, distribution
Annex C (informative) Elements of metadata – Full listing
Annex D (informative) Examples of how to use the standard
D.1 Example 1 – Metadata for a utility dataset
D.2 Example 2 – Metadata for an Administrative Boundaries Dataset

2.4 Level of Abstraction of the Metadata Standard Definition

A. Does the standard define metadata elements?: **YES**
B. Does the standard define a model for the relationships between the metadata elements?: **YES**
C. Provide an overview diagram of the model of the metadata, if available. **NO**
D. Does the standard define domains and/or data types of the metadata elements?: **NO**
E. Does the standard define how the metadata should be encoded?: **YES**
F. Does the standard define a database schema for the automatic generation of a database?: **NO**

2.5 Extensibility of the Standard

A. Is the metadata description extensible (e.g., Can user add metadata domains, elements, thematic profiles, etc.)?: **NO**

2.6 Availability of Metadata
Does the standard specify:

A. How metadata are to be accessed?: **NO**
B. A mechanism for querying the metadata?: **NO**
C. Particular standardized query protocols?: **NO**
D. That the metadata should be encoded for computer access?: **NO**
E. That the metadata should be encrypted for security?: **NO**

2.7 Examples

A. Does the standard provide examples of implementations?: **YES**
B. How many examples are available?:
Around 500

SECTION 3 – Linkage and Coordination of the Metadata Standard

This section provides information on how the metadata standard is linked to other standards, particularly transfer standards, and how such linkage is implemented. If a significant linkage does exist, detailed technical information is provided on the precise nature of the linkage to the standard, and how the linkage affects implementation of the metadata standard.

3.1 Associated Transfer Standards

A. Is the metadata standard associated with a transfer standard(s)?: **NO**
B. If A is yes, provide references and contact details for each linked standard: **N/A**
C. If A is yes, are both (all) standards coordinated?: **N/A**
D. If A is yes, describe how the coordination works, both from a technical and conformance (normative) viewpoint: **N/A**

3.2 Incorporated into transfer mechanism

A. Is the metadata standard incorporated into a transfer mechanism?: **NO**
B. If A is yes, provide references and contact details for each incorporated mechanism: **N/A**
C. If A is yes, describe how: **N/A**

3.3 Other linked, coordinated or associated standards

A. Is the metadata standard linked, coordinated, or associated with any other standard(s) not listed above?: **NO**
B. If A is yes, provide references and contact details for each standard: **N/A**
C. If A is yes, describe how for each relevant standard: **N/A**

II. Categories the Metadata Standard Uses to Describe Datasets

The following sections are categories of metadata that may be used by the metadata standard to describe data. Not all categories are relevant to different datasets and may not necessarily be present in the standard. The presence or absence of a category should not be used for judging one standard against another.

Note: For each of the questions in the following sections please provide the obligation status information. Valid obligation values are:

Obligation Level

M	Mandatory	This information must be entered to comply with the metadata standard.
C	Conditional	This information is entered based on other information in the metadata.
O	Optional	This information is optionally entered in the metadata.

If the answer is 'NO', the obligation status should be left blank.

SECTION 4 – Dataset Identification

This section examines how the standard provides for the specification of the various sorts of titles and identifiers for a dataset.

4.1 The title(s) of a dataset
Does the standard allow/require the provision for:

Obligation

A. __ Previous title(s)?: **NO**
B. **M** Current title(s)?: **YES**
C. __ Possible future title(s)?: **NO**
D. **O** Official titles in different languages?: **YES**
E. **O** Abbreviated title(s)?: **YES**
F. **O** Alternative Title(s)?: **YES**
G. __ Does the standard support effective dates of titles?: **NO**
H. **O** Other identifiers (e.g., ISBN, National Stock Number, Uniform Product Code)?: **YES**
Version

4.2 Parties responsible for the dataset
Does the standard allow/require the provision for:

Obligation

A. **M** Author?: **YES**
B. **M** Owner?: **YES**
C. **O** Other?: **YES**
Organization name and role as free text

4.3 ISBD Information

Obligation

A. __ Does the metadata standard specify inclusion of mandatory ISBD information (see glossary) within the metadata?: **NO**

SECTION 5 – Status of the Dataset

This section examines how the standard provides for the specification of the current status and updating of the data set.

5.1 Status and Progress of Datasets
Does the metadata standard allow/require the provision for:

Obligation

A. **O** Information on the status of the entire dataset?: **YES**
B. **O** Information on the status of the particular subset (e.g., observation time, inauguration time, extraction time)?: **NO**

Does the metadata standard specify:

C. __ A structure for updating or maintenance information?: **NO**
D. __ A structure to describe historical information (see also Section 8.3 G)?: **NO**

5.2 Information on Incremental Updating
Does the metadata standard specify:

Obligation

A. __ A structure for information on incremental updates of the entire dataset?: **NO**
B. __ A structure for information on incremental updates of particular portions of the dataset?: **NO**
C. __ A structure to describe the update process for the dataset?: **NO**

SECTION 6 – Extent of the Dataset

This section examines how the standard provides for the specification of the spatial and temporal extent of the dataset.

6.1 The spatial extent of the dataset

Obligation

A. **C** Does the metadata standard allow/require for the description of horizontal extent (e.g., bounding rectangle/polygon, named location (i.e., Stockholm, etc.))?: **YES**
Bounding quadrangle, Bounding area, geographic area, name of the area covered by the geographic dataset, name of the geographic identifiers system, code of the geographic area, coverage of the geographic area
B. Does the metadata standard allow/require for the description of vertical extent (e.g., minima/maxima heights, etc.)?: **NO**
C. __ Does the standard require specific units for describing extent?: **NO**
D. __ Does the standard allow specification of the units for describing extent?: **NO**
E. __ Does the standard require specific spatial reference system for describing extent?: **Yes**
F. __ Does the standard allow specification of the spatial reference system for describing extent?: **NO**

6.2 The temporal extent of the dataset

Obligation

A. **C** Does the standard allow/require for the description of time period for which the dataset is relevant?: **YES**
B. __ Does the standard allow/require specification of the temporal reference system?: **NO**

SECTION 7 – Data Content of the Dataset

This section examines how the standard provides for the specification of the data content of the dataset in terms of overview elements, themes, data definitions and product specifications.

7.1 Dataset overview

Does the standard allow/require for the provision of the following information:

Obligation

A. **M** Dataset overview?: **YES**
B. **M** Abstract – brief summary of the content of the dataset?: **YES**
C. **O** Purpose – summary of the intentions with which the dataset was developed?: **YES**
D. Usage – list of activities for which the dataset has been used? (see also Section 8.5): **NO**
E. Spatial schema – (e.g., vector, raster)? (see also Section 9): **NO**
not available due to the high heterogeneity of geological data
F. **O** Spatial reference system? (see also Section 12): **YES**
G. **M** Language?: **YES**
H. Character set (e.g., ISO8859-1, JISX0208)?: **NO**
no such practice exist in geology
I. **O** List of documents providing further information about the dataset?: **YES**
J. **O** Sample/preview dataset?: **YES**
K. **O** Related datasets?: **YES**
L. __ Original size of the dataset?: **NO**
M. __ Compression? (see also Section 11.5): **NO**
N. __ Other?: **NO**

7.2 Theme(s)

Obligation

A. **C** Does the standard allow/require the provision for the description of theme(s) (e.g., topography, land cover, transport, etc.) in the dataset?: **YES**
B. **C** If A is yes, does the standard allow the referencing of an outside dictionary or thesaurus?: **YES**
C. **C** If A is yes, does the standard itself include a dictionary or thesaurus within the metadata?: **YES**
D. If A is yes, does the standard allow the use of unauthorized user-defined keywords?: **NO**

7.3 Data Definition

Obligation

A. **O** Does the standard allow/require the provision for the description of the data definition?: **YES**

B. **C** If A is yes, does the standard allow/require the provision for the description of:
C 1. Features/Objects?: **YES**
C 2. Attributes?: **YES**
3. Relations?: **NO**
see 7.1.E

C. __ Can attributes be described separately from feature/object classes?: **YES**
by using normalized tool

D. Does the standard allow/require the specification of the domain of the attributes?: **NO**
unknown notion in geology

E. **C** Does the standard allow/require the specification of data type of the attributes?: **YES**

F. Does the standard allow/require the specification of particular classification coding schemes (e.g., FACC, OSKA, S-57 OC)?: **NO**
unknown practice in geology

7.4 Product Specification

Obligation

A. Does the standard allow/require for the provision of information about whether the dataset meets a particular product specification?: **NO**

B. If A is yes, can it be internal to the Metadata?:

C. If A is yes, can it be external to the Metadata?:

SECTION 8 – Data Quality of the Dataset

This section examines how the standard deals with the question of data quality. This includes homogeneity and metrics for data quality, quality assurance, and usage.

8.1 Overall Data Quality Statement

Obligation

A. **C** Does the standard allow/require for the provision of quality to be described? If yes, please answer the questions below. (*If no, all following answers in Section* 8 *are N/A, please skip to Section* 9.): **YES**

B. **O** Does the standard allow/require the provision for the quality information referencing an external quality standard, quality assurance standard, or specification?: **YES**
free text

8.2 Homogeneity of Data Quality

Obligation

A. **C** Does the standard allow/require for the variation of the quality of the dataset to be described?: **NO**

very difficult to appreciate in geology

B. **C** What type of aggregation of quality information is allowed?

C 1. Entire database?: **YES**

2. Feature/Object instance?: **NO**
3. Feature/Object class?: **NO**
4. Attribute instance?: **NO**
5. Attribute class?: **NO**
6. Layer?: **NO**
7. Theme?: **NO**
8. Geographic extent?: **NO**
9. Other?:

All those information are too much detailed for geology. Exhaustive list of geological objects doesn't exist

8.3 Metrics for Data Quality

Does the standard allow/require the provision of values for:

Metrics for Data quality is provided as quality parameters describing performance of the geographic dataset. It is implemented as free text one of them being mandatory, whichever it is, the others being optional

Obligation

A. **C** Positional accuracy?: **YES**

1. Standard error?: **NO**
2. Maximum error?: **NO**
3. Error correlation?: **NO**
4. Confidence level (e.g., 90%, 95%)?: **NO**
5. Other?: **YES**

free text

B. **C** Attribute accuracy?: **YES**

1. Standard error?: **NO**
2. Maximum error?: **NO**
3. Error frequency?: **NO**
4. Error correlation?: **NO**
5. Confidence level (e.g., 90%, 95%)?: **NO**
6. Other?: **YES**

free text

C. **C** Temporal accuracy?: **YES**

1. Standard error?: **NO**
2. Maximum error?: **NO**
3. Error correlation?: **NO**
4. Confidence level? (e.g., 90%, 95%): **NO**

5. Other?: **YES**
free text

D. Completeness?: **NO**
1. Error of omission?:
2. Error of commission?:
3. Other?:

E. **C** Currentness?: **YES**
1. Date?: **YES**
2. Maximum age?: **YES**
3. Temporal extent?: **YES**
4. Other?: **NO**

F. Logical consistency?: **NO**
notion not used in geology

G. Lineage?: **NO**
1. List of processing steps?:
2. List of values of processing parameters?:
3. List of printouts of statistics, etc.?:
4. Source material?:
5. Other?:

H. Additional Quality Information?: **NO**

8.4 Quality Assurance Methods

Does the standard allow/require the provision of information about the methods used to determine:

Obligation

A. __ Positional accuracy?: **NO**
1. Standard error in adjustments?:
2. Repeated measurements?:
3. Independent measurements?:
4. Subjective evaluations?:
5. Other?:

B. __ Attribute accuracy?: **NO**
1. Standard error?:
2. Error classification matrix?:
3. Independent measurements?:
4. Subjective evaluation?:
5. Other?:

C. __ Temporal accuracy?: **NO**
1. Independent measurements?:
2. Subjective evaluation?:
3. Other?:

D. __ Completeness?: **NO**
1. Field checks?:
2. Repeated measurements?:

3. Independent measurements?:
4. Subjective evaluation?:
5. Error of commission?:
6. Other?:

E. __ Currentness?: **NO**
1. Date of source material?:
2. Repeated measurements?:
3. Independent measurements?:
4. Subjective evaluation?:
5. Other?:

F. __ Logical consistency?: **NO**
1. Verification results of consistency checks?:
2. Repeated processing?:
3. Comparison with independent sources?:
4. Subjective evaluation?:
5. Other?:

G. __ Lineage?: **NO**
1. Checking results of processing steps:
2. Subjective evaluation:
3. Other:

H. __ Additional Quality Information?: **NO**

8.5 Usage

Obligation

A. **C** Does the metadata standard allow/require the provision of information about usage?: **NO**
1. Who used it:
2. When it was used:
3. For what application it was used:
4. The effectiveness of use:
5. Other:

8.6 Alternative Quality Description: **NO**

SECTION 9 – Spatial Data Structure of the Dataset

This section examines how the standard provides for the specification of the spatial data model and spatial data types.

9.1 Spatial Data Model
Does the standard allow/require the provision for:

Obligation

A. The description of the geometrical and/or topological primitives used to model the real world?: **NO**

the geological modeling (3D, 4D ...) are very complex processes

B. Information about an external standard for geometrical and/or topological primitive specification?: **NO**

9.2 Spatial Data Types

Does the standard allow/require the provision for:

Obligation

A. Information about the spatial data type(s) used in the dataset?: **NO**

B. If A is yes, which of the following generic spatial data types are supported? For each, please list specific data primitive levels supported by each data type.

1. Vector (e.g., spaghetti, chain-node/node-edge-face, planar graph, full topology)?: **N/A**
2. Areal Raster (e.g., thematic raster – scanned and classification maps, image raster – satellite images, scanned aerial photographs)?: **N/A**
3. Grid/Matrix (e.g., Digital Elevation Model)?: **N/A**
4. Triangular Irregular Network (TIN)?: **N/A**
5. Other?: **N/A**

9.3 Other Data Types

Does the standard allow the provision of information about:

Obligation

A. __ Video?: **NO**

B. __ Sound?: **NO**

C. __ Other?: **NO**

SECTION 10 – Spatial Reference of the Dataset

This section examines how the dataset provides for the specification of the spatial reference characteristics of the data set.

Does the standard allow/require the provision for the following information:

10.1 Geoidal Model

Obligation

A. **O** Geoidal model name?: **YES**

B. **O** Geoidal model definitions?: **YES**

10.2 Geodetic datum

Obligation

A. **O** Geodetic datum name?: **YES**

B. **O** Geodetic datum definitions?: **YES**

10.3 Reference Ellipsoid

Obligation

A. **O** Reference ellipsoid name?: **YES**
B. **O** Reference ellipsoid definitions?: **YES**

10.4 Projection

Obligation

A. **O** Projection name?: **YES**
B. **O** Projection definitions?: **YES**

10.5 Coordinate Systems

Obligation

A. **O** Coordinate system type?: **YES**
B. **O** Coordinate system definitions?: **YES**

10.6 Height Reference System

Obligation

A. Vertical datum name?: **NO**
B. Types of heights?: **NO**

10.7 Ancillary Geodetic Information

Obligation

A. __ Does the standard allow/require the provision for ancillary geodetic, gravimetric, or magnetometric information?: **NO**

10.8 Transformation Parameters
Does the standard specify a method to provide information for coordinate transformation(s):

Obligation

A. __ Model name?: **NO**
B. __ Control points?: **NO**
C. __ Transformation parameters?: **NO**

10.9 Non-Coordinate Spatial Reference
Does the standard allow/require the provision for information on non-coordinate spatial reference systems (e.g., postal systems, street address)?:

Obligation

A. **C** Reference system name?: **YES**
B. **C** Reference system definitions?: **YES**
C. Other?: **NO**

SECTION 11 – Availability and Distribution of the Dataset

This section examines how the standard provides information on the availability of the dataset?

Does the standard allow/require the provision for:

11.1 The Distributor(s) of the Dataset

Obligation

A. **C** Information about the distributor(s) of the dataset?: **YES**
B. **C** Contact information for the distributor(s)?: **YES**

11.2 Restrictions to Access and Usage

Obligation

A. **C** Information about copyright on the dataset (e.g., date, period, copyright owner, etc.)?: **YES**
B. **C** Information about restrictions on the access of the dataset?: **YES**
C. **C** Information about restrictions on the usage of the dataset?: **YES**

11.3 Pricing Details

Obligation

A. **C** Information about pricing?: **YES**

11.4 Distribution of the Dataset

Obligation

A. **C** Information about distribution media?: **YES**
B. **C** Information about on-line access?: **YES**
C. **C** Information about units of distribution (e.g., tiles, layers, polygons, elements)?: **YES**

11.5 Distribution Format

Obligation

A. **C** Information about distribution formats (e.g., commercial CAD format, commercial GIS formats, National/International transfer standard)?: **YES**
B. **C** Information about the encoding of the dataset (e.g., binary, ASCII)?: **YES**
C. __ Information about compression method?: **NO**
D. __ Information about compressed dataset size?: **NO**
E. **C** Information about the distribution mechanism?: **YES**
F. Information about the encapsulation of data (e.g., ISO 8211, ISO 8824)?: **NO**

11.6 Ordering Procedures

Obligation

A. __ Information about ordering procedures for the dataset?: **NO**

11.7 Support Services

Obligation

A. __ Information about support services (e.g., additional data transformations, format or coordinate conversions)?: **NO**

SECTION 12 – Authorization and Verification of the Metadata

This section examines how the standard provides for the description of the authorization and verification on the metadata of a dataset (e.g., how is the metadata itself authorized and verified?).

12.1 Metadata reference

Obligation

A. **M** Does the standard allow/require the provision of information about reference dates of the metadata (e.g., last check date, last update date, review date)?: **YES**

12.2 Verification Authority

Obligation

A. __ Does the standard allow/require the provision for information about who verified/authorized the metadata?: **NO**

B. __ Does the standard specify the method for providing this information (e.g., ISO 9000): **NO**

12.3 Statistical Methods Used for Verification

Obligation

A. __ Does the standard specify the statistical methods to be used for verification of the metadata?: **NO**

B. __ Does the standard allow/require the provision for the description of the statistical methods used to verify the metadata?: **NO**

END OF CHARACTERISTICS
FIN

H. Moellering, H.J.G.L. Aalders & A. Crane (Editors)

USA/NBII: Biological Data Profile of the Content Standard for Digital Geospatial Metadata, FGDC-STD-001.1-1999

Susan Stitt

I. Information about the Metadata Standard Itself

SECTION 0 – Assessment Information

This section provides the user with a general description of the scope and intended use of the standard. Information on the respondent who prepared the assessment is also provided.

0.1 Respondent

Assessment prepared by:

Susan Stitt
Biological Scientist (Remote Sensing)
U.S. Geological Survey
Biological Resources Division
Center for Biological Informatics
P.O. Box 25046
Denver, CO 80225-0046, USA
Phone: + 1 303 202 4351
Fax: + 1 303 202 4354
E-mail: susan_stitt@usgs.gov

Date assessment completed:

July 1999; reviewed February 2001 and May 2002, Reviewed July, 2005

Relationship of respondent to standard:

Maintenance Authority, Author, and User

0.2 Brief Summary Statement

A. Summary of the scope and intent of the standard:

Objective. It has been determined that modifications to the FGDC Content Standard for Digital Geospatial Metadata are needed to create meaningful metadata for biological data sets, thus the objective of the profile is to provide a common set of terminology and definitions for the documentation of biological data through the creation of extended elements and a profile of the FGDC Content Standard for Digital Geospatial Metadata.

Scope. This profile is intended to support the collection and processing of biological data. It is intended to be useable by all levels of government and the private sector.

The profile was developed by defining information required by a prospective user to determine the availability of a set of biological data; to determine the fitness of a set of biological data for an intended use; to determine the means of accessing the set of biological data; and to successfully transfer the set of biological data. As such, the profile establishes the names of extended data elements and compound elements (groups of data elements) to be used for documenting biological data, the definitions of these extended compound elements and data elements, and information about the values to be provided for the data elements. The profile also describes any modifications to the optionality or repeatability of non-mandatory elements and any modifications to the domains of standard elements in the FGDC's Content Standard for Digital Geospatial Metadata.

The standard is not intended to reflect an implementation design. An implementation design requires adapting the structure and form of the profile to meet application requirements. The profile does not specify the means by which this information is organized in a computer system or in a data transfer, nor the means by which this information is transmitted, communicated, or presented to the user.

SECTION 1 – Administration of Metadata Standard

This section provides a general description of the administrative framework, within which the standard was developed, tested and currently resides. Significant details are also provided in terms of the developmental and managerial history as well as information on the existence of documentation, software tools and training materials. This section also provides a point of contact for further information on the standard.

1.1 Name of Standard

A. Name of standard:
   1. In original language(s):
      USA/NBII: Biological Data Profile of the Content Standard for Digital Geospatial Metadata, FGDC-STD-001.1-1999
   2. English translation, if appropriate:
      N/A

B. Version/Edition:
   FGDC-STD-001.1-1999

C. Language(s) of documentation:
   English
D. Acronym(s):
   Biological Data Profile, Biological Profile, NBII Metadata Standard
E. Official ID:
   FGDC-STD-001.1-1999
F. Name(s) of recognizing standards authority(ies):
   Federal Geographic Data Committee (FGDC), National Biological Information Infrastructure (NBII)

1.2 Responsible Institutions

Contact information about those institutions that are (or were) responsible for the processes involved in the evolution of the standard:

A. Production and/or development:

   U.S. Geological Survey
   Center for Biological Informatics
   P.O. Box 25046, MS 302
   Denver, CO 80225, USA
   Tel.: + 1 303.202.4220
   Fax: + 1 303.202.4219

   American Institute of Biological Sciences (AIBS)
   Scientific Peer Advisory and Review Services
   107 Carpenter Drive, Ste. 100
   Sterling, VA 20164, USA
   Tel.: +1 703.834.0812
   Fax: +1 703.834.1160
   E-mail: neldridge@aibs.org

   FGDC Biological Data Working Group
   FGDC Secretariat (attn: Jennifer Fox)
   U.S. Geological Survey
   590 National Center
   12201 Sunrise Valley Drive
   Reston, VA 20192, USA
   Tel.: +1 703.648.5514
   Fax: +1 703.648.5755
   E-mail: fgdc@www.fgdc.gov
   URL: http://biology.usgs.gov/fgdc.bio/

B. Testing:

   National Biological Information Infrastructure (NBII)
   National Program Office
   U.S. Geological Survey

302 National Center
12201 Sunrise Valley Drive (MS 302)
Reston, VA 20192, USA
Tel.: + 1 703 648-4216
Fax: + 1 703 648-4224
E-mail: biologywebteam@usgs.gov

And

U.S. Geological Survey
Center for Biological Informatics
P.O. Box 25046, MS 302
Denver, CO 80225, USA
Tel.: +1 303.202.4220
Fax: +1 303.202.4219

C. Conformance:

Executive Order 12906, voluntary, and
National Biological Information Infrastructure (NBII)
National Program Office
U.S. Geological Survey
302 National Center
12201 Sunrise Valley Drive (MS 302)
Reston, VA 20192, USA
Tel.: + 1 703 648-4216
Fax: + 1 703 648-4224
E-mail: biologywebteam@usgs.gov

D. Maintenance:

U.S. Geological Survey
Biological Data Profile Questions
C/O USGS Center for Biological Informatics
P.O. Box 25046
Denver Federal Center MS 302
Denver CO 80225-0046, USA
Tel.: + 1 303.202.4220
Fax: + 1 303.202.4219

E. Distribution:

Federal Geographic Data Committee
U.S. Geological Survey
590 National Center
12201 Sunrise Valley Drive
Reston, VA 20192, USA
Tel.: + 1 703 648-5514 Fax: + 1 703 648-5755
E-mail: fgdc@www.fgdc.gov
URL: http://www.fgdc.gov/standards/status/sub5_2.html

F. Help Desk/User Support:

U.S. Geological Survey
Biological Data Profile Questions
C/O USGS Center for Biological Informatics
P.O. Box 25046
Denver Federal Center MS 302
Denver, CO 80225-0046, USA
Tel.: + 1 303 202.4220
Fax: + 1 303 202.4219

1.3 Development History

A. Beginning of work on metadata standard: **1994**

B. Milestones:

11/1994 Ad hoc working group met to develop a "strawman" metadata standard

05/1995 American Institute of Biological Sciences workshop to peer review "strawman" metadata standard

02/1996 FGDC Standards Working Group supported interim implementation of the "Content Standard for NBII Metadata"

02/1997 FGDC Standards Working Group approved the NBII Metadata Standard as being in Draft Stage

07/1998 FGDC Biological Data Working Group released the Biological Data Profile of the Content Standard for Digital Geospatial Metadata for public review

07/1999 FGDC Biological Data Working Group submitted the Biological Data Profile of the Content Standard for Digital Geospatial Metadata to the FGDC Standards Working Group for approval

10/1999 FGDC Steering Committee officially endorsed the Content Standard for Digital Geospatial Metadata, Part 1: Biological Data Profile

C. Anticipated completion: **Completed**

D. Update cycle: **As needed**

1.4 Status of the Standard

A. Is the standard officially recognized: **YES**

B. Date(s) of recognition:

October 26, 1999

C. Anticipated date of recognition: **N/A**

D. Current stage in recognition process:
On October 26, 1999 Mark Schaefer, Assistant Secretary for Water and Science, acting for Secretary Babbitt in his role as FGDC Chair, signed to endorse the standard at the FGDC Steering Committee meeting. The Content Standard for Digital Geospatial Metadata, Part 1: Biological Data Profile, was assigned document number FGDC-STD-001.1-1999. To date, FGDC-STD-001.1-1999 has been adopted by the National Biological Information Infrastructure (NBII) and by several Federal government agencies

E. Steps still required to achieve recognition: **N/A**

1.5 Access to Documentation of the Standard

A. Is the standard copyrighted?: **NO**
B. Name and address of copyright owner: **N/A**
C. Do restrictions on the use of the standard apply?: **NO**
D. Is the standard documentation currently available?: **YES**
E. List the form/media of standard documentation:
Paper, Electronic (PDF, HTML, ASCII, WordPerfect)
F. Is the standard available on the Internet?: **YES**
http://www.fgdc.gov/standards/status/sub5_2.html
G. Price: **Free**
H. Contact details for obtaining the standard:
Federal Geographic Data Committee
U.S. Geological Survey
590 National Center
12201 Sunrise Valley Drive
Reston, VA 20192, USA
Tel.: + 1 703 648-5514
Fax: + 1 703 648-5755
E-mail: fgdc@www.fgdc.gov
Internet: http://www.fgdc.gov/standards/status/sub5_2.html
I. Does the standard have ISBN/ISSN numbers?: **NO**

1.6 Available Software Tools

A. Is software available to implement the standard as a database?: **YES**
M^{3}CAT
SMMS has implemented the Profile in a database
B. Is software available to test compliance of metadata with the standard?: **YES**
"mp" with the Biological Data Profile extensions file
C. Is software available to produce metadata compliant with the standard?: **YES**
SMMS, TKME, XTME, and the National Park Service (NPS) Metadata Extension for ArcGIS.

D. Other relevant software:
   Check at http://wisclinc.state.wi.us/metadata/metods.html
E. Is there a test dataset available to test software tools for implementing the standard?: **NO**
F. For each product available above, please list contact information:
   http://www.nbii.gov/datainfo/metadata/tools/index.html
   SMMS:
   http://imgs.intergraph.com/smms/
   TKME, XTME "cns" and "mp":
   http://geology.us.gov/tools/metadata/
   NPS Metadata Extension of ArcGIS:
   http://www.nature.nps.gov/im/units/mwr/gis/metadata/metadata_tools.htm
   M³CAT
   http://www.geoconnections.org/projects/geoinnovations/1999/Mcat3/m3cat-promo-en.html#BM1_1

1.7 Available Training Materials

A. Are organized training sessions available?: **YES**
B. Contact details for training sessions:
   Federal Geographic Data Committee
   U.S. Geological Survey
   590 National Center
   12201 Sunrise Valley Drive
   Reston, VA 20192, USA
   Tel.: + 1 703 648-5514
   Fax: + 1 703 648-5755
   E-mail: fgdc@www.fgdc.gov

   A web-based training calendar showing metadata training can be accessed at http://www.fgdc.gov/calendar.html
   A web based list of metadata trainers and their training curriculum and training descriptions can be accessed at http://www.fgdc.gov/metadata/toollist/trainers.html.

   See also:
   http://www.nbii.gov/datainfo/metadata/training/index.html
C. Training documentation available:
   Content Standard for Digital Geospatial Metadata Workbook – Version 2, May 2000
   Federal Geographic Data Committee
   U.S. Geological Survey
   590 National Center
   12201 Sunrise Valley Drive
   Reston, VA 20192, USA
   Tel.: + 1 703 648-5514

Fax: + 1 703 648-5755
E-mail: fgdc@www.fgdc.gov

An electronic version (Adobe PDF) copy for CSDGM Version 2 is accessible at: http://www.fgdc.gov/metadata/meta_workbook.html

See also: http://www.nbii.gov/datainfo/metadata/training/index.html

SECTION 2 – Use and Implementation of the Metadata Standard

This section provides information on the way in which the standard is used and the purpose of its use. It also provides descriptions about the level of abstraction of the standard.

2.1 Intended Use of the Standard
Can the standard be used:

A. Within a transfer data structure?:	**YES**
B. To produce hardcopy catalog of datasets?:	**YES**
C. To catalog data in a digital on-line catalog?:	**YES**
D. Other:	**NO**

2.2 Intended Purpose of the Standard
Is the intended purpose of the standard:

A. To support discovery, identification, location of data?:	**YES**
B. To support how to access data?:	**YES**
C. To support the determination of dataset fitness for use?:	**YES**
D. To support transfer of dataset?:	**YES**
E. To support an organization's data management?:	**NO**
F. Other:	**NO**

2.3 Table of Contents

A. Provide the standard's Table of Contents

1.1 Introductory Material **1-1**
1.1.1 Objective **1-1**
1.1.2. Scope **1-1**
1.1.3. Applicability **1-1**
1.1.4. Related Standards **1-2**
1.1.5. Standards Development Procedures **1-3**
1.1.6. Maintenance Authority **1-4**
1.2 Elements of the Content Standard for Digital Geospatial Metadata **1-5**
1.3 Changes to the Content Standard for Digital Geospatial Metadata **1-7**
1.3.1 Conditionality Changes **1-7**
1.3.2 Domain Changes **1-9**

1.4 Extended Elements **1-13**
1.5 References **1-49**
1.6 Index to Extended and Modified Elements **1-51**
1.7 Appendix 1.A (Informative) Glossary **1-53**

2.4 Level of Abstraction of the Metadata Standard Definition

A. Does the standard define metadata elements?: **YES**
B. Does the standard define a model for the relationships between the metadata elements?: **YES**
C. Provide an overview diagram of the model of the metadata, if available: **YES**
D. Does the standard define domains and/or data types of the metadata elements?: **YES**
E. Does the standard define how the metadata should be encoded?: **NO**
The FGDC has under development a standard titled "Encoding standards for Digital Geospatial Metadata" which describes the encoding of FGDC-STD-001-1998 for use in an online catalog/clearinghouse
F. Does the standard define a database schema for the automatic generation of a database?: **NO**

2.5 Extensibility of the Standard

A. Is the metadata description extensible (e.g., Can user add metadata domains, elements, thematic profiles, etc.)?: **YES**
Extended elements (elements outside the Standard, but needed by the data set producer) to the Standard may be defined by a data set producer or a user community. If extended elements are created, they must follow the guidelines presented in Appendix D of the FGDC Content Standard for Digital Geospatial Metadata FGDC-STD-001-1998

2.6 Availability of Metadata
Does the standard specify:

A. How metadata are to be accessed?: **NO**
B. A mechanism for querying the metadata?: **NO**
C. Particular standardized query protocols?: **NO**
D. That the metadata should be encoded for computer access?: **NO**
The FGDC has under development a standard titled "Encoding Standards for Digital Geospatial Metadata" which describes the encoding of FGDC-STD-001-1998 for use in an online catalog/clearinghouse
E. That the metadata should be encrypted for security?: **NO**

2.7 Examples

Graphical Representation of: The Federal Geographic Data Committee's Content Standard for Digital Geospatial Metadata

FGDC-STD-001, June 1998

Prepared by Susan Stitt, Center for Biological Informatics, USGS
In conjunction with the Federal Geographic Data Committee

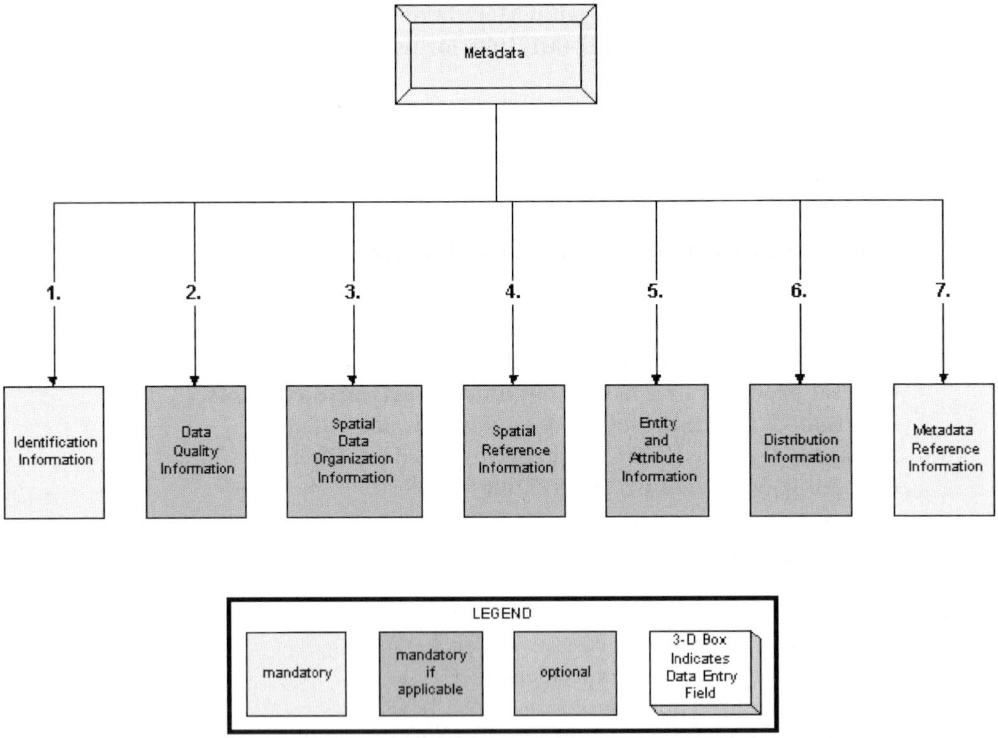

Fig. 1. Graphical representation of the FGDC Content Standard for Digital Geospatial Metadata.

The above graphic and information, although directly describing the FGDC Content Standard for Digital Geospatial Metadata, also pertains to the Biological Data Profile. Graphics that illustrate the extensions to the base standard defined in the Profile are available at http://www.nbii.gov/datainfo/metadata/standards/bdp.html.

A. Does the standard provide examples of implementations?: **NO**
Examples can be found by querying the National Biological Information Infrastructure Metadata Clearinghouse
B. How many examples are available?: **N/A**

SECTION 3 – Linkage and Coordination of the Metadata Standard

This section provides information on how the metadata standard is linked to other standards, particularly transfer standards, and how such linkage is implemented. If a significant linkage does exist, detailed technical information is provided on the precise nature of the linkage to the standard, and how the linkage affects implementation of the metadata standard.

3.1 Associated Transfer Standards

A. Is the metadata standard associated with a transfer standard(s)?: **NO**
B. If A is yes, provide references and contact details for each linked standard: **N/A**
C. If A is yes, are both (all) standards coordinated?: **N/A**
D. If A is yes, describe how the coordination works, both from a technical and conformance (normative) viewpoint: **N/A**

3.2 Incorporated into transfer mechanism

A. Is the metadata standard incorporated into a transfer mechanism?: **NO**
B. If A is yes, provide references and contact details for each incorporated mechanism: **N/A**
C. If A is yes, describe how: **N/A**

3.3 Other linked, coordinated or associated standards

A. Is the metadata standard linked, coordinated, or associated with any other standard(s) not listed above?: **YES**
The FGDC Content Standard for Digital Geospatial Metadata FGDC-STD-001-1998
B. If A is yes, provide references and contact details for each standard:
Federal Geographic Data Committee. FGDC-STD-001-1998. Content Standard for Digital Geospatial Metadata (revised June 1998). Federal Geographic Data Committee. Washington, DC

Federal Geographic Data Committee
U.S. Geological Survey
590 National Center
12201 Sunrise Valley Drive
Reston, VA 20192, USA
Tel.: + 1 703 648-5514
Fax: + 1 703 648-5755
E-mail: fgdc@www.fgdc.gov

C. If A is yes, describe how for each relevant standard:

As a profile of the Content Standard for Digital Geospatial Metadata (CSDGM), the Biological Data Profile includes additions to and slight modifications of the CSDGM. It defines domain changes and extended elements to the CSDGM that are needed to create meaningful metadata for biological data sets

II. Categories the Metadata Standard Uses to Describe Datasets

The following sections are categories of metadata that may be used by the metadata standard to describe data. Not all categories are relevant to different datasets and may not necessarily be present in the standard. The presence or absence of a category should not be used for judging one standard against another.

Note: For each of the questions in the following sections please provide the obligation status information. Valid obligation values are:

	Obligation Level	
M	Mandatory	This information must be entered to comply with the metadata standard.
C	Conditional	This information is entered based on other information in the metadata.
O	Optional	This information is optionally entered in the metadata.

If the answer is 'NO', the obligation status should be left blank.

SECTION 4 – Dataset Identification

This section examines how the standard provides for the specification of the various sorts of titles and identifiers for a dataset.

4.1 The title(s) of a dataset
Does the standard allow/require the provision for:

Obligation

A. __ Previous title(s)?: **NO**
B. **M** Current title(s)?: **YES**
C. __ Possible future title(s)?: **NO**
D. __ Official titles in different languages?: **NO**
E. __ Abbreviated title(s)?: **NO**
F. **C** Alternative Title(s)?: **YES**
G. **M** Does the standard support effective dates of titles?: **YES**
H. **C** Other identifiers (e.g., ISBN, National Stock Number, Uniform Product Code)?: **YES**

Other identifiers include Edition, Series Information and Name, Issue Identification, Online linkage information, and information regarding relationship of this data set with a second data set (Larger Work Citation)

4.2 Parties responsible for the dataset
Does the standard allow/require the provision for:

Obligation

A. **M** Author?: **YES**
B. **C** Owner?: **YES**
C. **C** Other?: **YES**

4.3 ISBD Information

Obligation

A. __ Does the metadata standard specify inclusion of mandatory ISBD information (see glossary) within the metadata?: **NO**

SECTION 5 – Status of the Dataset

This section examines how the standard provides for the specification of the current status and updating of the data set.

5.1 Status and Progress of Datasets
Does the metadata standard allow/require the provision for:

Obligation

A. **M** Information on the status of the entire dataset?: **YES**
B. __ Information on the status of the particular subset (e.g., observation time, inauguration time, extraction time)?: **NO**

Does the metadata standard specify:

C. **M** A structure for updating or maintenance information?: **YES**
D. **C** A structure to describe historical information (see also Section 8.3 G)?: **YES**

5.2 Information on Incremental Updating
Does the metadata standard specify:

Obligation

A. **C** A structure for information on incremental updates of the entire dataset?: **YES**
B. __ A structure for information on incremental updates of particular portions of the dataset?: **NO**
C. **C** A structure to describe the update process for the dataset?: **YES**

SECTION 6 – Extent of the Dataset

This section examines how the standard provides for the specification of the spatial and temporal extent of the dataset.

6.1 The spatial extent of the dataset

Obligation

A. **M** Does the metadata standard allow/require for the description of horizontal extent (e.g., bounding rectangle/polygon, named location (i.e., Stockholm, etc.))?: **YES**
Bounding rectangle, bounding polygon, named location, and description of extent are all allowable. If the dataset is geospatial, a bounding rectangle is required, a description or named location is required, place keywords and bounding polygon are conditional. If the dataset is not geospatial, bounding rectangle, description of geographic extent, place keywords, and bounding polygon are all optional

B. **O** Does the metadata standard allow/require for the description of vertical extent (e.g., minima/maxima heights, etc.)?: **YES**
Optional minimum and maximum elevations can be specified

C. **M** Does the standard require specific units for describing extent?: **YES**

D. __ Does the standard allow specification of the units for describing extent?: **NO**

E. **M** Does the standard require specific spatial reference system for describing extent?: **YES**

F. __ Does the standard allow specification of the spatial reference system for describing extent?: **NO**

6.2 The temporal extent of the dataset

Obligation

A. **M** Does the standard allow/require for the description of time period for which the dataset is relevant?: **YES**

B. **C** Does the standard allow/require specification of the temporal reference system?: **YES**
If a geologic age is specified, then provide "geologic time scale" plus "geologic age estimate" plus conditionally "geologic age uncertainty" and conditionally "geologic age explanation" and optionally "geologic citation"

SECTION 7 – Data Content of the Dataset

This section examines how the standard provides for the specification of the data content of the dataset in terms of overview elements, themes, data definitions and product specifications.

7.1 Dataset overview
Does the standard allow/require for the provision of the following information:

Obligation

A. **O** Dataset overview?: **YES**

B. **M** Abstract – brief summary of the content of the dataset?: **YES**
C. **M** Purpose – summary of the intentions with which the dataset was developed?: **YES**
D. __ Usage – list of activities for which the dataset has been used? (see also Section 8.5): **NO**
E. **C** Spatial schema – (e.g., vector, raster)? (see also Section 9): **YES**
F. **C** Spatial reference system? (see also Section 12): **YES**
G. __ Language?: **NO**
H. __ Character set (e.g., ISO8859-1, JISX0208)?: **NO**
I. **O** List of documents providing further information about the dataset?: **YES**
J. **O** Sample/preview dataset?: **YES**
K. **C** Related datasets?: **YES**
L. **O** Original size of the dataset?: **YES**
M. **C** Compression? (see also Section 11.5): **YES**
N. **C** Other?: **YES**

Other dataset overview information includes taxonomy, analytical tools, keyword (place theme, stratum, and temporal) dataset credit information, security information, and native dataset environment information. Taxonomy consists of information on the taxa (1 or more) included in the dataset, taxonomic keywords, taxonomic system, taxonomic coverage, and taxonomic classification system information. Analytical tool is defined as tools, models, or statistical procedures that were used to generate the dataset, and thus the dataset is intrinsically bound to them. Examples include reconstructions of phylogenies, population viability analyses, community ordinations, atmospheric and hydrological transport analyses, and inferences on the effects of climate change on forest composition and productivity

7.2 Theme(s)

Obligation

A. **M** Does the standard allow/require the provision for the description of theme(s) (e.g., topography, land cover, transport, etc.) in the dataset?: **YES**
B. **C** If A is yes, does the standard allow the referencing of an outside dictionary or thesaurus?: **YES**
C. __ If A is yes, does the standard itself include a dictionary or thesaurus within the metadata?: **NO**
D. **C** If A is yes, does the standard allow the use of unauthorized user-defined keywords?: **YES**

7.3 Data Definition

Obligation

A. **C** Does the standard allow/require the provision for the description of the data definition?: **YES**

B. If A is yes, does the standard allow/require the provision for the description of:
C 1. Features/Objects?: **YES**
C 2. Attributes?: **YES**
__ 3. Relations?: **NO**
This information may be found in some free text metadata elements

C. __ Can attributes be described separately from feature/object classes?: **NO**

D. **C** Does the standard allow/require the specification of the domain of the attributes?: **YES**

E. **C** Does the standard allow/require the specification of data type of the attributes?: **YES**

F. **C** Does the standard allow/require the specification of particular classification coding schemes (e.g., FACC, OSKA, S-57 OC)?: **YES**

7.4 Product Specification

Obligation

A. **C** Does the standard allow/require for the provision of information about whether the dataset meets a particular product specification?: **YES**

B. __ If A is yes, can it be internal to the Metadata?: **NO**

C. **C** If A is yes, can it be external to the Metadata?: **YES**

SECTION 8 – Data Quality of the Dataset

This section examines how the standard deals with the question of data quality. This includes homogeneity and metrics for data quality, quality assurance, and usage.

8.1 Overall Data Quality Statement

Obligation

A. **C** Does the standard allow/require for the provision of quality to be described? If yes, please answer the questions below. (*If no, all following answers in Section* 8 *are N/A, please skip to Section* 9.): **YES**

B. **C** Does the standard allow/require the provision for the quality information referencing an external quality standard, quality assurance standard, or specification?: **YES**

8.2 Homogeneity of Data Quality

Obligation

A. **C** Does the standard allow/require for the variation of the quality of the dataset to be described?: **YES**

B. What type of aggregation of quality information is allowed?
C 1. Entire database?: **YES**
C 2. Feature/Object instance?: **YES**

C 3. Feature/Object class?: **YES**
C 4. Attribute instance?: **YES**
C 5. Attribute class?: **YES**
C 6. Layer?: **YES**
C 7. Theme?: **YES**
C 8. Geographic extent?: **YES**
C 9. Other?: **YES**
Any other definable aggregation can be described

8.3 Metrics for Data Quality
Does the standard allow/require the provision of values for:

Obligation

A. **C** Positional accuracy?: **YES**
1. Standard error?: **YES**
2. Maximum error?: **YES**
3. Error correlation?: **YES**
4. Confidence level (e.g., 90%, 95%)?: **YES**
5. Other?: **NO**

B. **C** Attribute accuracy?: **YES**
1. Standard error?: **YES**
2. Maximum error?: **YES**
3. Error frequency?: **YES**
4. Error correlation?: **YES**
5. Confidence level (e.g., 90%, 95%)?: **YES**
6. Other?: **NO**

C. __ Temporal accuracy?: **NO**
1. Standard error?: **N/A**
2. Maximum error?: **N/A**
3. Error correlation?: **N/A**
4. Confidence level? (e.g., 90%, 95%): **N/A**
5. Other?: **N/A**

D. **C** Completeness?: **YES**
1. Error of omission?: **YES**
2. Error of commission?: **YES**
3. Other?: **YES**
Other completeness information includes any selection criteria, generalization, definitions, used, and other rules used to derive the data set

E. **C** Currentness?: **YES**
1. Date?: **YES**
2. Maximum age?: **YES**
3. Temporal extent?: **YES**
4. Other?: **NO**

F. **C** Logical consistency?: **YES**

A free text field is provided to permit a narrative description of the logical consistency

G. **C** Lineage?: **YES**

1. List of processing steps?: **YES**
2. List of values of processing parameters?: **YES**
3. List of printouts of statistics, etc.?: **YES**
4. Source material?: **YES**
5. Other?: **YES**

Process date, process contact, source scale denominator, source media, source time period of contend, source contribution

H. **C** Additional Quality Information?: **YES**

Cloud cover. Methodology, which is information about a single step of field and/or laboratory work

8.4 Quality Assurance Methods

Does the standard allow/require the provision of information about the methods used to determine:

Obligation

A. **C** Positional accuracy?: **YES**

1. Standard error in adjustments?: **YES**
2. Repeated measurements?: **YES**
3. Independent measurements?: **YES**
4. Subjective evaluations?: **YES**
5. Other?: **NO**

B. **C** Attribute accuracy?: **YES**

1. Standard error?: **YES**
2. Error classification matrix?: **YES**
3. Independent measurements?: **YES**
4. Subjective evaluation?: **YES**
5. Other?: **NO**

C. __ Temporal accuracy?: **NO**

1. Independent measurements?: **N/A**
2. Subjective evaluation?: **N/A**
3. Other?: **N/A**

D. **C** Completeness?: **YES**

1. Field checks?: **YES**
2. Repeated measurements?: **YES**
3. Independent measurements?: **YES**
4. Subjective evaluation?: **YES**
5. Error of commission?: **YES**
6. Other?: **NO**

E. **C** Currentness?: **YES**

1. Date of source material?: **YES**

2. Repeated measurements?: **YES**
3. Independent measurements?: **YES**
4. Subjective evaluation?: **YES**
5. Other?: **NO**

F. **C** Logical consistency?: **YES**
1. Verification results of consistency checks?: **YES**
2. Repeated processing?: **YES**
3. Comparison with independent sources?: **YES**
4. Subjective evaluation?: **YES**
5. Other?: **NO**

G. **C** Lineage?: **YES**
1. Checking results of processing steps: **YES**
2. Subjective evaluation: **YES**
3. Other: **NO**

H. __ Additional Quality Information?: **NO**

8.5 Usage

Obligation

A. __ Does the metadata standard allow/require the provision of information about usage?: **NO**

This information may be found in some free text metadata elements

1. Who used it: **N/A**
2. When it was used: **N/A**
3. For what application it was used: **N/A**
4. The effectiveness of use: **N/A**
5. Other: **N/A**

8.6 Alternative Quality Description: **NO**

SECTION 9 – Spatial Data Structure of the Dataset

This section examines how the standard provides for the specification of the spatial data model and spatial data types.

9.1 Spatial Data Model
Does the standard allow/require the provision for:

Obligation

A. **C** The description of the geometrical and/or topological primitives used to model the real world?: **YES**

B. __ Information about an external standard for geometrical and/or topological primitive specification?: **NO**

9.2 Spatial Data Types

Does the standard allow/require the provision for:

Obligation

A. **C** Information about the spatial data type(s) used in the dataset?: **YES**
B. **C** If A is yes, which of the following generic spatial data types are supported? **YES**
For each, please list specific data primitive levels supported by each data type.
1. Vector (e.g., spaghetti, chain-node/node-edge-face, planar graph, full topology)?:
Any and all **YES**
2. Areal Raster (e.g., thematic raster – scanned and classification maps, image raster – satellite images, scanned aerial photographs)?:
Any and all **YES**
3. Grid/Matrix (e.g., Digital Elevation Model)?:
Any and all **YES**
4. Triangular Irregular Network (TIN)?: **NO**
5. Other?: **NO**

9.3 Other Data Types

Does the standard allow the provision of information about:

Obligation

A. __ Video?: **NO**
B. __ Sound?: **NO**
C. __ Other?: **NO**

SECTION 10 – Spatial Reference of the Dataset

This section examines how the dataset provides for the specification of the spatial reference characteristics of the data set.

Does the standard allow/require the provision for the following information:

10.1 Geoidal Model

Obligation

A. **C** Geoidal model name?: **YES**
B. **C** Geoidal model definitions?: **YES**

10.2 Geodetic datum

Obligation

A. **C** Geodetic datum name?: **YES**
B. **C** Geodetic datum definitions?: **YES**

10.3 Reference Ellipsoid

Obligation

A. **C** Reference ellipsoid name?: **YES**
B. **C** Reference ellipsoid definitions?: **YES**

10.4 Projection

Obligation

A. **C** Projection name?: **YES**
B. **C** Projection definitions?: **YES**

10.5 Coordinate Systems

Obligation

A. **C** Coordinate system type?: **YES**
B. **C** Coordinate system definitions?: **YES**

10.6 Height Reference System

Obligation

A. **C** Vertical datum name?: **YES**
B. **C** Types of heights?: **YES**

10.7 Ancillary Geodetic Information

Obligation

A. **C** Does the standard allow/require the provision for ancillary geodetic, gravimetric, or magnetometric information?: **YES**

10.8 Transformation Parameters
Does the standard specify a method to provide information for coordinate transformation(s):

Obligation

A. **C** Model name?: **YES**
B. **C** Control points?: **YES**
C. **C** Transformation parameters?: **YES**

10.9 Non-Coordinate Spatial Reference
Does the standard allow/require the provision for information on non-coordinate spatial reference systems (e.g., postal systems, street address)?:

Obligation

A. **C** Reference system name?: **YES**
B. **C** Reference system definitions?: **YES**
C. __ Other?: **NO**

SECTION 11 – Availability and Distribution of the Dataset

This section examines how the standard provides information on the availability of the dataset?

Does the standard allow/require the provision for:

11.1 The Distributor(s) of the Dataset

Obligation

A. **C** Information about the distributor(s) of the dataset?: **YES**
B. **C** Contact information for the distributor(s)?: **YES**

11.2 Restrictions to Access and Usage

Obligation

A. **M** Information about copyright on the dataset (e.g., date, period, copyright owner, etc.)?: **YES**
B. **M** Information about restrictions on the access of the dataset?: **YES**
C. **M** Information about restrictions on the usage of the dataset?: **YES**

11.3 Pricing Details

Obligation

A. **C** Information about pricing?: **YES**

11.4 Distribution of the Dataset

Obligation

A. **C** Information about distribution media?: **YES**
B. **C** Information about on-line access?: **YES**
C. **C** Information about units of distribution (e.g., tiles, layers, polygons, elements)?: **YES**

11.5 Distribution Format

Obligation

A. **C** Information about distribution formats (e.g., commercial CAD format, commercial GIS formats, National/International transfer standard)?: **YES**
B. **C** Information about the encoding of the dataset (e.g., binary, ASCII)?: **YES**
C. **C** Information about compression method?: **YES**
D. **C** Information about compressed dataset size?: **YES**
E. **C** Information about the distribution mechanism?: **YES**
F. **C** Information about the encapsulation of data (e.g., ISO 8211, ISO 8824)?: **YES**

11.6 Ordering Procedures

Obligation

A. **C** Information about ordering procedures for the dataset?: **YES**

11.7 Support Services

Obligation

A. **C** Information about support services (e.g., additional data transformations, format or coordinate conversions)?: **YES**

SECTION 12 – Authorization and Verification of the Metadata

This section examines how the standard provides for the description of the authorization and verification on the metadata of a dataset (e.g., how is the metadata itself authorized and verified?).

12.1 Metadata reference

Obligation

A. **M** Does the standard allow/require the provision of information about reference dates of the metadata (e.g., last check date, last update date, review date)?: **YES**

12.2 Verification Authority

Obligation

A. __ Does the standard allow/require the provision for information about who verified/authorized the metadata?: **NO**

B. __ Does the standard specify the method for providing this information (e.g., ISO 9000): **NO**

12.3 Statistical Methods Used for Verification

Obligation

A. __ Does the standard specify the statistical methods to be used for verification of the metadata?: **NO**

B. __ Does the standard allow/require the provision for the description of the statistical methods used to verify the metadata?: **NO**

END OF CHARACTERISTICS
FIN

PART 5

CROSSTABLE OF NATIONAL AND INTERNATIONAL SPATIAL METADATA STANDARDS AND ASSOCIATED CHARACTERISTICS

Compiled and edited by

Harold Moellering

H. Moellering, H.J.G.L. Aalders & A. Crane (Editors)

Introduction to the Crosstable of Spatial Metadata Standards and Their Scientific and Technical Characteristics

Harold Moellering

Department of Geography Ohio State University Columbus, Ohio, 43210 United States of America
e-mail: geohal+@osu.edu

1. Introduction

Background

The Crosstable discussed here is included as a module of Part 5, and is intended to be a quick overall reference to the complex set of scientific and technical characteristics contained in each Part 4 chapter on each of the National and International spatial metadata standards, and a few application profiles. Every National spatial metadata standard in the world is represented, plus the International standards from the Dublin Core project (ANSI, 2001), the European CEN/TC287 standard (CEN/TC 287, 1998), and the ISO/TC211 19115 world metadata standard(ISO/TC211, 19115, 2002). Here the TC211 standard is the broadest and most important because it serves as the covering set of metadata elements and definitions with which the other National and Regional metadata standards will be harmonized.

The four profiles included in this work are examples of specific application subsets of other National metadata standards that have been defined for use in a specific subject matter area of use. For example, the USA/National Biological Information Infrastructure (NBII) Profile (FGDC, 1999) is a subset of the more general United States metadata profile, which was developed by the National Biological Survey, now the Biological Division of the U.S. Geological Survey. It uses the USA main FGDC metadata standard as a basing point, but then includes specific variables in the area of biology.

Likewise, the three European profiles included in this work focus on a specific subject matter: EuroGeographics/LeClef (Eurogeographics, 2000) is oriented towards topography, while the EU Community/SSSA/ES focuses on space Imagery in the European area, and finally the EuroGeoSurveys/GEIXS profile (EuroGeoSurveys, 1997) is focused on geology in Europe.

The Crosstable included as part of this book is intended to be a companion piece to the Commission's Crosstable on spatial database transfer standards as part of the ICA book titled "Spatial Database Transfer Standards 2" (Moellering and Hogan, 1997). There, as here, are a wealth of scientific and technical characteristics which have been winnowed down to a selected set of characteristics that express the essence of the assessments and evaluation of those standards. This new Crosstable on spatial metadata standards is another more sophisticated and authoritative step because it contains obligation codes expressing whether a particular metadata item is Obligatory, Conditional, or Optional. This added sophistication provides the reader with more insight and understanding on what conditions are associated with that particular spatial metadata element being encoded and included in the record for that dataset.

Development of Metadata Characteristics and the Crosstable

The initial meeting of the ICA Spatial Data Commission (the current name) to begin developing the spatial metadata characteristics was held in Costa Brava, Spain in 1995 just before the 1995 International Cartographic Conference in Barcelona. From that beginning the Commission worked to precisely define the scientific and technical characteristics (see Introduction to Part 3) so they were applicable across more than the dozen different languages in which the original metadata standards are written. Extensive field testing and peer review at Commission meetings played an integral part of that development and refinement process.

By 1998, at the Commission meeting at Aix-en-Provence, France, the members began to analyze and sift through the 250 tertiary characteristics with a goal of extracting the true essence of spatial metadata standards characteristics. By the end of that four day meeting an initial set of about 70 essential characteristics were identified, refined and included in the proposed master Crosstable variable list. This proposed list of metadata characteristics was then circulated widely within the Commission, field tested by some, and reviewed by all member directly involved in this process. At the following 1999 Commission at Strathmere, Ontario, Canada, the Commission adjusted the list of proposed variables somewhat, and then formally voted to approve the final list of 70 Crosstable variables. This final list is used herein.

2. Use of this metadata standards crosstable

This Spatial Metadata Standards Crosstable is intended to distill and summarize a very large and complex amount of scientific information into a comprehendible form. It is intended to be a quick reference resource for the seasoned spatial standards professional, and also serve as a beginning point for those desiring to learn about spatial metadata standards. The huge advantage of this Crosstable is that it uses a uniform set of characteristics developed by the ICA Standards Commission members. It is then applied to each National and International Standard, and subject matter profile in the *native language* of that standard by a Commission member who was an active member of the standards body that originally developed the standard in the first place. This has been a huge and complex undertaking, but the results provide this scientific and technical information in a most usable form for all in a common ICA scientific language, which in this case happens to be English. Overall, this Crosstable is a wonderful summary resource that can be wall mounted and easily used.

The crosstable can be used very effectively to learn the essential and relevant characteristics of a particular standard by carefully scanning down a column of interest. The answers in the

Crosstable have been made purposely simple with YES/NO or N/A (Not Available, or not known) answers. In many cases there are further qualifying terms, but they are too voluminous to include in the Crosstable, and hence the reader must refer back to the Part 4 chapter of interest to ascertain this information.

If it turns out that the reader is particularly interested in one or two particular scientific and technical characteristics, then one can just scan along the row(s) of that variable(s) to the comparative responses. Here again, qualifying terms may also be associated with the simple answers, so in those cases the reader would then refer back to the Part 4 chapter of interest for those details.

3. List of the 70 crosstable variables and explanations by group

The discussion here of the variables included in the Part 5 Crosstable will be conducted as an overview by the 12 Sections of the Scientific and Technical characteristics themselves. The detailed definitions can be found in Part 3 Scientific and Technical characteristics. This is intended so the reader new to these concepts can more readily learn about the contents of the characteristics in the crosstable.

Header for each standard or profile in the crosstable

This begins with Part I – Information about the Metadata Standard Itself. This first set of variables references information about the standard/profile, while the second set references information about data contained within the dataset itself.

The vertical header for each standard contains the following variables associated with the correct identification of the standard itself.

Country/organization of origin – Large/Bold Name
Full name the standard (1.1.A.1)
Acronym (1.1.D)
Version number (1.1.B)
Official ID (1.1.E)
Standards authority (1.1.F)
Maintenance Institution (1.2.D)

Crosstable rows of scientific and technical characteristics

Section 0 – Assessment Information: this identifies who actually encoded these characteristics for this standard. It also includes a brief summary statement, preferable taken out of the standard itself. This information is not included in the Crosstable.

Section 1 – Administration of Metadata Standard: this section includes many variables telling the Name references, Institutions Responsible for the standard, Development History, Status of the Standard, and access to documentation software tools and training about the standard. Many of the Names variables are included in the Name Header above, while four others are included as the first four rows in the Crosstable.

1. 1.4.A Official Recognition
2. 1.5.C Restrictions on Use of the Standard
3. 1.6.A Database Software Available
4. 1.7 Training Available

Section 2 – Use and Implementation of the Metadata Standard: this section includes variables concerning Intended Use and Purpose of the Standard, Table of Contents, Level of Abstraction and Extensibility of the Standard, Availability of Metadata and Examples. Included in the Crosstable are:

5. 2.1.A Intended Use Within Transfer Data Structure
6. 2.1.C Use with On-line Catalog
7. 2.2.A Support Discovery, etc. of Data
8. 2.2.B Support How to Access Data
9. 2.2.C Support Determination of Data Fitness
10. 2.2.D Support Transfer of Dataset
11. 2.4.B Define Metadata Model Relationships
12. 2.4.D Define Metadata Domains/Types
13. 2.5.A Metadata Description Extensible
14. 2.7.B Number of Examples Available

Section 3 – Linkage and Coordination of the Metadata Standard: this section tells whether the metadata standard is integrally linked to any other transfer standard, Incorporated into transfer mechanism, or linked somehow with another standard. The variables included here are:

15. 3.1.A Associated with Transfer Standard
16. 3.2.A Incorporated into Transfer Mechanism
17. 3.3.A Associated with Other Standard

Part II. – This part begins the Categories the Metadata Standard Uses to Describe Datasets. This includes the actual kinds of data they could be contained within an actual spatial dataset. These variables now include an Obligation Code associated with the metadata characteristic listed, signifying whether the characteristic is Mandatory, Conditional, or Optional.

These **M**, C, and o codes are defined as follows. Please note that the variables in Sections 1.1–3.3 have no associated Obligation codes. However, in Section 4.1 and following, each YES answer *must* have an obligation code associated with it. If that obligation code is missing, an x is inserted into the place for the obligation code for that variable. If the answer is NO or N/A, then no obligation code is expected, and that slot is left blank, the same as the chapters.

The obligation codes for the crosstable are taken directly from the Part 4 chapters. Here they have been enhanced to make them more readable in the crosstable. But things like Conditional Obligation descriptions do not appear in the crosstable because of their length and complexity, and must be read in the Part 4 chapters themselves.

The Obligation codes for the crosstable are defined as follows:

M – Mandatory Obligation ⟨CAPS, Bold⟩;

C – Conditional Obligation ⟨CAPS only⟩, the condition specifications cannot be provided in the crosstable. If the reader wants to know the conditions for the C code, then he/she must look at the Part 4 chapter for those kinds of details and associated descriptions.

o – Optional Obligation ⟨lower case⟩, the option specifications cannot be provided in the crosstable. If the reader wants to know the specifications for the o code, then he/she must look at the Part 4 chapter for those kinds of details and associated descriptions.
x – When an Obligation code is expected because the answer for the variable is YES, but no Obligation code is provided, then an x code is assigned to the Obligation code slot.

Please note that Variables 10.1/6, 10.2/3, and 10.4/5 have double answers with double associated Obligation codes.

Section 4 – Dataset Identification: this section identifies the correct title for the spatial dataset, parties responsible to the dataset, ISBN (International Standard Book Number) and ISBD (International Standard Bibliographic Description for Electronic Resources), if any. The critical characteristics for the Crosstable are:

18. 4.1.B Dataset – Current Titles
19. 4.2 Parties Responsible for Dataset ⟨YES if any of the 3 are YES⟩

Section 5 – Status of the Dataset: this references the status and Progress of the Dataset, as well as Information on Incremental Updating. It reflects the current development of a spatial dataset in progress and what kind of incremental updating arrangements have been made. The salient variables included in the Crosstable here are:

20. 5.1.A Status of Entire Dataset
21. 5.1.C Structure for Update Information
22. 5.2 Provide Incremental Updating Information ⟨YES if any are YES⟩

Section 6 – Extend of the Dataset: this includes the spatial and temporal extents of the dataset, and includes:

23. 6.1.A Description of Horizontal Extent
24. 6.1.B Description of Vertical Extent
25. 6.1.E Require Specific Spatial Reference System
26. 6.2.A Allow Description of Relevant Time Period

Section 7 – Data Content of the Dataset: this section includes subsections such as Dataset Overview, Themes, Data Definition and Product Specification. This gives the potential user of the dataset a reasonable idea of the kinds of spatial information contained within the dataset. Specific characteristics included in this section of the Crosstable here are:

27. 7.1 Number Dataset Overview Items Selected ⟨Count number YES⟩
28. 7.2.A Allow Description of Themes
29. 7.3.A Allow Description of Data Definition
30. 7.4.A Allow Information on Product Specification

Section 8 – Data Quality of the Dataset: this section indicates whether there is an Overall Data Quality Statement, Homogeneity of the Quality Data, Metrics utilized of measuring Data Quality, Quality Assurance Methods, and possible information about Usage. The key characteristics included in this section of the Crosstable here are:

31. 8.1.A Allow Quality Description
32. 8.1.B Allow Referencing External Quality Standard

33. 8.2.A Describe Data Quality Variation
34. 8.3.A Describe Positional Accuracy
35. 8.3.B Describe Attribute Accuracy
36. 8.3.C Describe Temporal Accuracy
37. 8.3.D Describe Completeness
38. 8.3.E Describe Currentness
39. 8.3.F Describe Logical Consistency
40. 8.3.G Describe Lineage
41. 8.3.H Provide Additional Quality Metrics Information
42. 8.4.A Positional Accuracy Quality Assurance Methods
43. 8.4.B Attribute Accuracy Quality Assurance Methods
44. 8.4.C Temporal Accuracy Quality Assurance Methods
45. 8.4.D Completeness Quality Assurance Methods
46. 8.4.E Currentness Quality Assurance Methods
47. 8.4.F Logical Consistency Quality Assurance Methods
48. 8.4.G Lineage Quality Assurance Methods
49. 8.4.H Provide Additional Quality Assurance Information
50. 8.5.A Provide Information About Usage

Section 9 – Spatial Data Structure of the Dataset: this section indicates the kind of spatial data structure used to spatially organize data itself. This includes subsections such as Spatial Data Model, Spatial Data Types, and Other data types. The critical characteristics included in this section of the Crosstable here are:

51. 9.1.A Describe Data Model Primitives
52. 9.2.A Describe Spatial Data Types
53. 9.3.A Allow Other Data Types

Section 10 – Spatial Reference of the Dataset: included in this section are a host of groups of information that relate to the Geoidal Model and Datum, Projection, horizontal and vertical Coordinate Systems, Transformation Parameters, and other systems of spatial referencing, e.g., street address, etc. Characteristics included here are:

54. 10.2/3 Describe Datum/Ellipsoid ⟨Double answers⟩
55. 10.1/6 Describe Geoidal Model/Height Reference System ⟨Double answers⟩
56. 10.4/5 Describe Projection/Coordinate Systems ⟨Double Answers⟩
57. 10.9 Describe Non-Coordinate Spatial Reference

Section 11 – Availability and Distribution of the Dataset: this section indicates how the dataset can be obtained as well as such things as price, restrictions for use, ordering procedures, etc. The Items contained in this section of the Crosstable are:

58. 11.1.B Distributor Contact Information
59. 11.2 Restrictions to Access and Usage
60. 11.3.A Information about Pricing
61. 11.4.A Information about Distribution Media
62. 11.4.B Information about On-line Access
63. 11.5.A Information about Distribution Formats

64. 11.5.B Information about Dataset Encoding
65. 11.5.F Information about Data Encapsulation
66. 11.6.A Information about Ordering Procedures
67. 11.7.A Information about Support Services

Section 12 – Authorization and Verification of the Metadata: here the characteristics in this section assess how the standard provides for the description of the metadata elements contained in the dataset in reference to how the metadata is authorized and verified. The Characteristic items included in the Crosstable here are:

68. 12.1.A Allow Metadata Reference Dates
69. 12.2.A Allow Information on Verification Authority
70. 12.3.A Specify Statistical Verification Methods

4. Summary and Conclusions

As can be seen, great efforts have been spent by the ICA Spatial Data Standards Commission over a number of years to develop an authoritative set of scientific and technical characteristics of these National and International spatial metadata standards, and several application profiles. The Commission spend three years developing, peer reviewing, and field testing their effectiveness and efficiency. From there, 70 of the most revealing characteristics have been selected by the Commission that represent the real essence of this set of spatial metadata standards being considered here.

With the discussions and clarifications presented above, it is hoped that beginners in spatial metadata standards as well as seasoned veterans will benefit from this comprehensive and concise Crosstable. It should again be pointed out that this Crosstable is designed to be a companion reference to the Commission's earlier 1996 "Crosstable of National and International Spatial Database Transfer Characteristics", (Moellering, Ed., 1996). They are designed to work in tandem to the scientific benefit of everyone involved in the world spatial data community.

References

[ANSI 2001]
The Dublin Core Metadata Element Set, ANSI/NISO Z3985-2001, ISSN 1041-5635, URL: http://www.niso.org/standards/resources/Z39-85.pfd.

[CENTC287 1998]
Geographic Information-data description-metadata, prENV 12657:1998, URL:http://forum.afnor.fr/servlet/Serv; etFprum?form_name+cFOrumPage&file_name=Z13C%2FPUBLIC %2FWEB%2FENGLISH%2Fpren.htm&login=invite&password=invite

[EuroGeographics 2000]
LaCLef Core Metadata, LaCLef Consortium, MEGRIN GIE, Marne La Vallée, France.

[EuroGeoSurveys 1997]
GEIXS, Project SABE, MEGRIN, URL: http//geixs.brgm.fr.

[FGDC 1999]
Content Standard for Digital Geospatial Metadata – Biological Data Profile, FGDC-STD-001-1998; Part 1 (1999).

[ISO/TC211 2002]
ISO19115, 2004; Geographic Information – Metadata, URL: www.statkart.no/isotc211.

[H.Moellering (ed.) 1996]
Crosstable of National and International Spatial Database Transfer Characteristics, in: H. Moellering, R. Hogan (eds), (1997), *Spatial Database Transfer Standards 2: Characteristics for Assessing Standards and Full Descriptions of the National and International Standards of the World,* London: Pergamon/Elsevier Science, 373 pp. plus Crosstable, ISBN 0080424325.

[H.Moellering, R.Hogan (eds) 1997]
Spatial Database Transfer Standards 2: Characteristics for Assessing Standards and Full Descriptions of the National and International Standards of the World, London: Pergamon/Elsevier Science, 373 pp. plus Crosstable, ISBN 0080424325.

APPENDICES

ANNEX A – ACRONYMS USED

ANNEX B – ICA STANDARDS COMMISSION

ANNEX C – BOOK CONTRIBUTORS

INDEX

Edited by

Henri J.G.L. Aalders

H. Moellering, H.J.G.L. Aalders & A. Crane (Editors)

Annex A: Acronyms Used

Acronym	**Full description**
3-D	3-Dimensional
3i	Information Interoperability Institute (Canada)
A	
ACZISC	Atlantic Coastal Zone Information Steering Committee (Canada)
ADOS	African Doppler Survey
AENOR	Asociación Española de Normalización y Certificación (Spanish organisation for standardisation)
AFIGÉO	Association Française Informatique Géographique (French Association for Geographic Information)
AFNOR	Association Française de NORmalisation (French Standardisation Institute)
AGI	Association for Geographic Information (UK)
AGRHYMET	African Geographical centre on HYdrography and METeorology
AISI	African Information Society Initiative
AMWG	ANZLIC Metadata Working Group (Australia/New Zealand)
AM/FM	Automated Mapping and Facility Management
ANS	American National Standard
ANSI	American National Standards Institute
ANSI NCITS L1	American National Standards Institute National Committee on Information Technology Standards Committee L1 on Geographic Information (U.S.A.)
ANZLIC	Australia and New Zealand Land Information Council
AOCRS	African Organisation for Cartography and Remote Sensing
APA	Association of Precise Survey and Applied technology (Japan)
APSDI	Asia-Pacific Spatial Data Infrastructure
ASDD	Australian Spatial Data Directory
ASDI	Australian Spatial Data Infrastructure
ASDI DN	ASDI Distribution Network (Australia)
AUSLIG	Australian Surveying and Land Information Group
AVID	Added Value Information Dissemination (for hydrographic datasets in Europe)
B	
BENELUX	Belgium, The Netherlands and Luxemburg

BKG	Bundesamt für Kartographie und Geodäsie (Federal Institute of Cartography and Geodesy in Germany)
BLM	Bureau of Land Management (U.S.A.)
BSS	Business Support Systems

C

CAC-TC211	Canadian Advisory Committee for ISO TC 211
CAGI	Czech Association for Geographic Information
CAP	Co-operative Agreements Program (U.S.A.)
CASD	Co-ordinating Committee on African Statistical Development
CCOG	Canadian Council on Geomatics
CCRS	Canada Centre for Remote Sensing, Compound Co-ordinate Reference System
CD	Committee Draft (ISO)
CD-ROM	Compact Disc – Read Only Memory
CEDAMA	Committee on Environmental Data Management (Lesotho)
CEDARE	Centre for Environment and Sustainable Development in the Arab Region and Europe
CEN	Comité Européen de Normalisation (European Standardisation Committee)
CEONet	Canadian Earth Observation Network
CEOS IDN	Committee on Earth Observation Satellites International Direction
CERCO	Comité européen de Responsable de la Cartographie Officielle (European Committee responsible for Official Cartography)
CERTU	Centre d'Études sur les Réseaux, les Transports, l'Urbanisme et les constructions publiques (Centre for road transport and urban development and governmental constructions in France)
CFS	Canadian Forest Service
CGDI	Canadian Geo-spatial Data Infrastructure
CGDI-TAP	Canadian Geo-spatial Data Infrastructure – Technology Advisory Panel
CGIS-DIGEST	Canadian Geomatics Interchange Standard – DIGEST, *see also DIGEST*
CGIS-SAIF	Canadian Geomatics Interchange Standard – Spatial Archive and Interchange Format
CGSB	Canadian General Standards Board
CGSB-CoG	CGSB – Committee on Geomatics
CHS	Canadian Hydrographic Service
CIAT	Centro Internacional de Agricultura Tropical (International Centre on Tropical Agriculture)
CIGEO	Centro de Información Geográfica CIGEO (Geographical Information Centre in Honduras)
CLEAR	spatial data CLEARing house (Europe)
CLEF	Cross-Language Evaluation Forum (Europe)
CNDG	Clearinghouse Nacional de Datos Geográficos (National Clearinghouse for Geographic Data – Uruguay)

CNIDR	Center for Networked Information Discovery and Retrieval (U.S.A.)
CNIG	Conseil National de l'Information Géographique (National Council for Geographic Information in France)
CNIG	Centro National de Información Geográfica (National Council for Geographic Information in Spain)
CNR	National Registry Centre (El Salvador)
CODI	Committee on Development Information (UN-ECA)
COCGC	County Offices of Cadastre, Geodesy and Cartography (in Romania)
CODI GEO	CODI Subcommittee on Geo-information
CoG	Committee on Geomatics (Canada)
COM/OLE	Common Object Model/Object Linking and Embedding
CP IDEA	Comité Permanente para la Infraestructura de Data Geo-espaciales de las Américas, *see also PC IDEA*
CPRM	Serviço Geológico do Brasil (Geological Survey of Brazil)
CR	Commission Report (CEN)
CSA	Canadian Standards Association
CSDGM	Content Standard Digital Geospatial Metadata
CRS	Co-ordinate Reference System
CRTS	Centre Royal de Télédétection Spatiale (Royal Centre for Remote Sensing, Morocco)
CSBTS/TC 230	National technical commission of geographic information standardisation (China)
CSL	Conceptual Schema Language
CTC	Comité Técnico Consultivo para Información Geográfica (Technical Committee for Consultancy on Geographic Information, Panama)
CVL	Central Veterinary Laboratory (Kenya)
CZISC	Atlantic Coastal Zone Information Steering Committee (Canada)

D

DC	Dublin Core
DCM	Dublin Core Metadataset
DCMI	Dublin Core Metadata Initiative
DEM	Digital Elevation Model
DG	Directorate General (EU)
DGI	Direction Générale des Impôts (National Tax Office, France)
DGIWG	Digital Geographic Information Working Group (NATO)
DIGEST	DIgital Geographic Information Exchange Standard (NATO)
DIGSA	Directores de Institutos Geográficos de Sudamericanos, España y Portugal (Directory of Geographic Institutes from South America, Spain and Portugal)
DIS	Draft International Standard
DND	Department of National Defence (Canada)
DOL	Department of Lands (Thailand)

DRC	Democratic Republic of Congo
DRSRS	Department of Resource Survey and Remote Sensing (Kenya)
DSM	Department of Surveys and Mapping (Botswana)
DSMM	Department of Survey and Mapping Malaysia
DTD	Document Type Definition
DTM	Digital Terrain Model

E

EC	European Commission
EDIS	Earth Data Information Systems
EEA	European Environmental Agency
EIS	Environmental Information System Programme for Sub-Saharan Africa
EMA	Ethiopian Mapping Authority
ENv	European (provisional) experimental standards
EO	Executive Order (U.S.A.)
EOS	Earth Observation System (NASA, U.S.A.)
EPA	Environmental Protection Agency in Ireland
EROS	Earth Resources Observation System
ESDI	European Spatial Data Infrastructure
ESMI	European Spatial Metadata Infrastructure
ESP	Environmental Support Project (Africa)
ETeMII	European Territorial Management Information Infrastructure
EU	European Union
EUREF	European Reference System
EuroGeographics	Association of European National Mapping Agencies
EuroGeoSurveys	Association of European National Geological Surveys
EUROGI	European Umbrella Organisation for Geographic Information
EuroSDR	European Spatial Data Research

F

FACC	Feature and Attribute Coding Catalogue
FAO	Food and Agriculture Organisation
FEI	Finnish Environment Institute
FEWS	Famine Early Warning System
FGDC	Federal Geographic Data Committee (U.S.A.)
FGDC STD	FGDC Standard
FIG	Fédération Internationale des Géomètres (International Federation of Surveyors)
FLIC	Fiji Land Information Council
FLIS	Fiji Land Information System
FP-4	fourth Framework Programmes for R&D (EU)
FP-5	fifth Framework Programmes for R&D (EU)
FP-6	sixth Framework Programmes for R&D (EU)
FTP	File Transfer Protocol

G

GBKN	Grootschalige Basiskaart van Nederland (Large Scale Base Map of the Netherlands)
GCM	General Command of Mapping (Turkey)
GCMD	Global Change Master Directory (Canada)
GDDD/LaClef	Geographical Data Description Directory (Europe)
GDI	Government Data Infrastructure (Thailand)
GEF/UNDP	Operational Strategy of Global Environmental Facility/United Nations Development Program
GEIXS	Geo-Scientific Electronic Information Exchange System (Europe)
GEOD-IDN	Committee on Earth Observation Satellites – International Direction
GEONet	Canadian Geospatial Network
GERMP	Government Environmental Resources Management Project (Ghana)
GGM	Global Geographic information Mapping
GI	Geographic Information
GII	Geographic Information Infrastructure
GIAC	Geomatics Industry Association of Canada
GIE/MEGRIN	Groupe d'Intéresse Economique (Economic Interest Group MEGRIN, Europe)
GILS	Global Information Locator Service
GINIE	Geographic Information Network In Europe
GIS	Geographical Information System
GISCO	GIS for the (European) Commission
GISEDI	Electronic Trade for Geographic Information (Europe)
GISTDA	Geo-Informatics and Space Technology Development Agency (Thailand)
GML	Geography Mark-up Language
GOST	State Standard of Russian Federation
GPS	Global Positioning System
GSDI	Global Spatial Data Infrastructure
GSI	Geological Survey Ireland, Geographical Survey Institute (Japan)

H

HTML	Hyper Text Mark-up Language

I

IAG	International Association for Geodesy
IATFGI	Inter-Agency Task Force on Geographic Information (Philippines)
IBGE	Instituto Brasileiro de Geografia e Estatística (Bureau of Census in Brazil)
ICA	International Cartographic Association
ICDE	Infraestructura Colombiana de Datos Espaciales (Colombian Spatial Data Infrastructure)
ICMA	International City/County Management Association
ICSM	Intergovernmental Committee on Surveying and Mapping (Australia)
ICT	Information and Communication Technology

ID	Identification
IDERC	Infraestructura de Datos Espaciales de la República de Cuba (Spatial Data Infrastructure of the Cuban Republic)
IDN	International Direction
IEC	International Electro-technical Commission
IEEE	Institute for Electrical and Electronic Engineers
IGAC	Instituto Geográfico Agustin Codazzi (Agustin Codazzi Geographic Institute of Colombia)
IGAD	Intergovernmental Authority on Development (East Africa)
IGDN	Inter-American Geo-spatial Data Network
IGM	Military Geographic Institute
IGN	Institut Géographique National (National Geographic Institute)
IGPCC	Institute of Geodesy, Photogrammetry, Cartography and Cadastre (Romania)
IGVSB	Instituto Geográfico de Venezuela Simón Bolívar (Geographic Institute of Venezuela Simón Bolívar)
IHO	International Hydrographic Organisation
IMF2000	Israel Metadata Format 2000
IMO	International Maritime Organisation
INCITS	InterNational Committee for Information Technology Standards
INEC	Instituto Nacional de Estadística y Censos (Statistics and Census National Institute in Nicaragua)
INEGI	Instituto Nacional de Estadística, Geografía e Informática (National Institute of Statistics, Geography and Informatics in Mexico)
INETER	Instituto Nicaragüense de Estudios Territoriales (Nicaraguan Institute for Territorial Studies)
INPE	Instituto Nacional de Pesquisas Espaciais (National Institute for Space Research in Brazil)
INS	Inertial Navigation System
INSEE	Institut national de la statistique et des études économiques (National Statistical Institute, France)
INSPIRE	Infrastructure for Spatial Information in Europe
IOC/UNESCO	Intergovernmental Oceanographic Commission of UNESCO
IS	International Standard
IS 191xx	International Standard on Geographic Information/Geomatics
ISBD	International Standard Bibliographic Description
ISBN	International Standard Book Number
ISO	International Organisation for Standardisation
ISO/TC 211	SO Technical Committee 211 for Geographic Information/Geomatics
ISO/TR	ISO Technical Report
ISPRS	International Society for Photogrammetry and Remote Sensing
ISSN	International Standard Serials Number
IT	Information Technology
I-Team	Geospatial Information Initiative (I-Team Initiative, U.S.A.)

ITI	Information Technology Industry Council (U.S.A.)
ITU	International Telecommunication Union
J	
JMP	Japanese Metadata Profile
JNSGI	Japanese National Standards for Geographic Information
JRC	Directorate General Joint Research Council of the EC
K	
KATS	Korean Agency for Technology and Standards
KII	Korean Information Infrastructure
KIKERES	Hungary – Spatial Metadata Profile
KMFRI	Kenya Marine Fisheries Research Institute
KMS-DK	Kort & Matrikelstyrelsens (National Survey and Cadastre in Denmark)
KWS	Kenya Wildlife Service
L	
LBMS	Location-Based Mobile Services
LBS	Location-Based Services
LINZ	Land Information New Zealand
LIS	Land Information System
LOM	Learning Object Metadata Working Group
LSV	Landelijk SamenwerkingsVerband (Netherlands Co-operative Foundation for the GBKN)
M	
MADAME	Methods for Access to Data and Metadata in Europe
MAGFOR	Ministerio de Agricultura, Ganadería y Forestaría (Ministry of Agriculture, Livestock-Farming and Forestry, Nicaragua)
MARC	Machine-Readable Cataloging
MARENA	Ministerio de Ambiente y Recursos Naturales (Ministry of Environment and Natural Resources, Nicaragua)
MARN	Ministerio del Medio Ambiente y Recursos Naturales (Ministry of Environment and Natural Resources in El Salvador)
MEGRIN	Multi-purpose European Ground Related Information Network
MEM	Государственный стандарт Российской Федерации ГОСТ Р 51353-99 "Геоинформационное картографирование. Метаданные электронных карт. Состав и содержание" State Standard of the Russian Federation GOST R 51353-99 "Geoinformatic Mapping – Metadata of the Electronic Maps: Composition and Content" (Russian Federation)
MET	ANZLIC Metadata Entry Tool, *see also ANZLIC*
MIGRA	Mecanismo de Intercambio de Información Geográfica Relacional formado por Agregación (Aggregated Relational Geographic Information Interchange Mechanism in Spain)
MLUH	Ministry of Land Use and Habitat (The Seychelles)
MMS	Minerals Management Service (U.S.A.)

MoU	Memorandum of Understanding (North America)
MUIENR	Makerere University Institute of Environment and Natural Resources (Uganda)

N

NACI	National Agency for Communication and Information (in Romania)
NACo	National Association of Counties (U.S.A.)
NAFGIM	National Framework for Geo-spatial Information Management (Ghana)
NaLIS	National Infrastructure for Land Information System (Malaysia)
NAMRIA	National Mapping and Resource Information Authority (Philippines)
NAS GCMD	Global Change Master Directory (Canada)
NASA	National Aeronautical and Space Administration (U.S.A.)
NASA EOS	NASA Earth Observation System (U.S.A.)
NATO	North Atlantic Treaty Organisation
NBII	National Biological Information Infrastructure (U.S.A.)
NCA	National Computerisation Agency (Korea)
NCGI	National Clearinghouse Geo-Informatie (National Clearinghouse for Geographic Information in the Netherlands)
NEMA	National Environment Management Authority (Uganda)
NEN	Nederlands Normalisatie Instituut (Netherlands Standardisation Institute) Nederlandse Norm, Netherlands Standard
NES	National Environment Secretariat (Kenya)
NFGIS	National Fundamental Geographic Information System (China)
NFIS	National Forest Information System
NGCC	National Geomatics Centre of China
NGI	National Geographic Institute (Korea)
NGIC	National Geographic Information Council
NGII	National Geographic Information Infrastructure
NGIS	National Geographic Information System (Korea), National Geographic Information Centre (Norway)
NGO	Non-Governmental Organisation
NIMA	National Imaging and Mapping Agency (USA)
NIN	Northern Information Network (Canada)
NIS	National Information Strategy (Romania)
NISCC	Nairobi Informal Settlements Co-ordinating Committee
NITG-TNO	Nederlands Instituut voor toegepaste Geo-wetenschappen – Nederlandse Organisatie voor toegepastnatuurwetenschappelijk onderzoek (Netherlands Institute for Applied Geo-sciences – Netherlands Organisation for applied physical Research)
NLC	National League of Cities (U.S.A.)
NLS	National Land Survey (Finland)
NLTB	Native Land Trust Board (Fiji)
NMA	National Mapping Agency
NMK	National Museums of Kenya

NOAA	National Oceanic and Atmospheric Administration (U.S.A.)
NOCGC	National Office of Cadastre, Geodesy and Cartography (Romania)
NRCan	National Resources Canada
NSDI	National Spatial Data Infrastructure
NSGIC	National States Geographic Information Council (U.S.A.)
NSGII	National System for GII (Kingdom of Saudi Arabia)
NSIF	National Spatial Information Framework (South Africa)
NTC	Norma Técnico Colombia (Colombian Technical Norm)
NTF	National Transfer Format (U.K.)
NZGLS	New Zealand Government Locator Service
O	
OAI	Open Archives Initiative
OCGI	New Zealand Officials Committee for Geo-spatial Information
OCLC	(formerly Online Cooperative Library Center); associated with the Dublin Core standard
OEEPE	Organisation d'Etude Européenne Photogrammétrie Expérimentale (European organisation for experimental photogrammetric studies)
OGC	OpenGIS® Consortium
OGDI	OpenGIS® Data Interchange
OMB	Office of Management and the Budget (U.S.A.)
OMG	Object Management Group
OSI	Ordnance Survey Ireland
OSKA	Objekt Schlüssel Katalog
P	
PAGSI	Plan d'action géographique de Société Informatique (Plan for the Information Society, France)
PAIGH	Pan American Institute of Geography and History
PC GIAP	Permanent Committee on GIS Infrastructure for Asia and the Pacific
PCI	Plan Cadastral Informatique (Digital Cadastral Plan, France)
PC IDEA	Permanent Committee on Spatial Data Infrastructure for the Americas
PP	Personal Productivity
PROCIG	Proyecto Centroamericano de Información Geográfica (Central American Development Project for GIS)
PSMA	Public Sector Mapping Agencies (Australia)
R	
Ravi	Stichting Overlegorgaan voor Vastgoedinformatie (Netherlands Council for Geographic Information)
RDB	Relational Database
RGE	Référentiel Géographique à grande Échelle (Large Scale Geographic Reference Dataset, France)
RIABIN	Red Interamericana de Información sobre la Biodiversidad (Inter-Americas Biodiversity Information Network, Nicaragua)
RIIS	Regional Integrated Information Systems (South America)

RRSU	Regional Remote Sensing Unit (Harare, Zimbabwe)
RS	Remote Sensing
RSAU	Remote Sensing Application Unit (University of Ghana)
RTSD	Royal Thai Survey Department
RTTCP	Regional Tsetse and Trypanosomosis Control Programme (Southern Africa)

S

S-57 OC	Object Catalogue of the International Hydrographic Organisation (IHO) Transfer Standard for Digital hydrographic Data
SABE	Seamless Administrative Boundaries in Europe
SADC	Southern African Development Community
SAIF	Spatial Archive and Interchange Format (Canada)
SAR	Side-looking Airborne Radar
SBSM	State Bureau of Surveying and Mapping (China)
SCC	Standards Council of Canada
SDDF	Spatial Data Discovery Facility (South Africa)
SDI	Spatial Data Infrastructure
SDTS	Spatial Data Transfer Standard (U.S.A.)
SENACYT	Secretaría Nacional de Ciencia y Tecnología (National Secretariat for Science, Technology and Innovation, Panama)
SFA	Simple Feature Access
SHOM	Service Hydrographique et Océanographique de la Marine (Military and Hydrographical Service, France)
SINIA	Sistema Nacional de Información Ambiental (Environmental Information National System, Honduras)
SoE	State of Environment (UN)
SOER	State of Environment Reporting Programme (Namibia)
SPOT	Système Probatoire d'Observation de la Terre (Experimental Earth Observation satellite)
SQL	Structured Query Language
SQL/MM	SQL/Multi Media
SSSA	EU/JRC/Strategy and Systems for Space Research

T

TAG	Technical Advisory Group (U.S.A.)
TAP	Technical Advisory Panel (Canada)
TC	Technical Committee
TC211	ISO/Technical Committee 211 for Geographic Information/Geomatics
TCP/IP	Transmission Control Protocol/Internet Protocol
TDN	Topografische Dienst Nederland (Netherlands Topographic Service)
TISI/TC 904	Thai Industrial Standard Institute/Technical Committee 904: Geographic Information – Thailand
TR	Technical Report (ISO)
TS	Technical Specification (ISO)

U

UCGIS	University Consortium for Geographic Information Science (U.S.A.)
UGIS	Urban Geographic Information System
UML	Unified Modelling Language
UN	United Nations
UN ECA	United Nations Economic Commission for Africa
UNEP	United Nations Environmental Programme
UNEP GRID	UNEP – Global Resource Information Database
UNESCO	United Nations Education, Science and Cultural Organisation
UNHCR	United Nations High Commissioner for Refugees
UNRCC-AP	United Nations Regional Cartographic Conference for Asia and the Pacific
UoD	Universe of Discourse
URISA	Urban and Regional Information Systems Association (in U.S.A.)
URL	Unified Resource Locator
US	United States (of America)
U.S.A. (USA)	United States of America
USAID	United States Agency for International Development
USGS/EROS	United States Geological Survey/Earth Resources Observation System

V

V-GIS	Virtual GIS

W

WFP	World Food Programme
WTO	World Trade Organisation
WWF	World-Wide Fund for Nature
WWW	World Wide Web

X

XML	eXtensible Markup Language

Annex B: ICA Standards Commission Membership List

ICA Commission on spatial data standards/
ACI Commission sur les normes relatives aux des données localisées
Commission WWW Home Page: http://ncl.sbs.ohio-state.edu/ica

Commission leadership

Chair

Prof. Harold Moellering, Chair
Geography Department
Ohio State University
Columbus, Ohio, U.S.A.

Executive Secretary

Mr. Aaron Crane
Navigation Technologies
Rosemont, Illinois, U.S.A.

Vice Chairs

European Co-ordinator
Mr. François Salgé
CNIG, Paris, France

Asia-Pacific Co-ordinator
Mr. Graham Baker
ANZLIC, Belconnen, ACT, Australia

Africa/Middle East Co-ordinator
Mr. Tony Cooper
CSIR, Pretoria, South Africa

Latin America Co-ordinator
Dr. Tatiana Delgado,
GEOCUBA Computer Division
La Habana 11300, Cuba

ICA Executive Committee

Commission Contact and SDI co-ordinator
Prof. Milan Konecny
Masaryk University
Brno, Czech Republic

ICA Standards Commission national member representatives

Austria	Dipl.-Ing. Peter Haunold, Technical University of Vienna
Australia	Mr. Craig Macauley, NATREM Program, Environmental and Geographic Information Department for Environment and Heritage, South Australia, Adelaide

Canada	Ms. Valerie Hume, Indian & Northern Affairs, Hull, Quebec
China	Prof. Du Dao-sheng, Wuhan Technical University of Surveying and Mapping, Wuhan
Cuba	Dr. Tatiana Delgado, Director GEOCUBA Computer Division, La Habana
Czech Republic	Dr. Petr Rapant, Technical University of Ostrava, Ostrava-Poruba
Denmark	Mr. Anders Nielsen, National Survey and Cadastre, Copenhagen
Finland	Mrs. Paula Ahonen, Helsinki University of Technology
France	Mr. François Salgé, GNIG, Paris
Germany	Mr. Ulrich Düren, Landesvermessungsamt Nordrhein-Westfalen, Bonn
Hungary	Mr. Istvan Kadar, Mapping Agency, Hungarian Defence Forces, Budapest
Israel	Dr. Ammatzia Peled, University of Haifa
Japan	Mr. Kazuhiko Akeno, Geographical Survey Institute, Tsukuba
Korea	Dr. T-j Min, National Geography Institute, Seoul, South Korea
Netherlands	Prof. Henri J.G.L. Aalders, Delft University of Technology, Research Institute for Housing, Urban and Mobility Studies
Norway	Mr. Olaf Østensen, The Norwegian Mapping Authority, Hønefoss
Russia	Prof. Alexander I. Martynenko, Moscow State University for Geodesy and Cartography, Moscow, Russian Federation
South Africa	Mr. Antony Cooper, CSIR, Pretoria
Spain	Mr. Sebastián Más Mayoral, Centro National de Información Geografica, Madrid
Sweden	Mr. Bengt Rystedt, National Land Survey, Gävle
Switzerland	Dr. Steffen Frischknecht, MGIS Gesellschaft für Consulting und Innovative Software GMBH, München, Germany
U.S.A.	Mr. Richard Hogan, US Geological Survey, Reston, Virginia, United States of America

ICA Standards Commission corresponding members

Chile	Mr. Patricio Zavala, Instituto Geográfico Militar, Santiago
Columbia	Mr. Jaime Maya, Instituto Geográfico Agustín Codazzi, Santafé de Bogotá
Croatia	Dr. Miljenko Lapaine, University of Zagreb
Czech Republic	Prof. Ing. Bohuslav Veverka, Prague
Guinea	Dr. Mohamed Hassimou Fofana, Institut de Topographie et de Cartographie de Guinée, Conakry, République de Guinée
Greece	Dr. Marinos Kavouras, National Technical University of Athens, Zografos
Iceland	Mr. Thorbjörg Kjartansdóttir, Director, LÍSA, Reykjavik
New Zealand	Mr. Patrick van Berkel, LINZ, Heaphy House, Wellington
Slovakia	Mr. Milan Hájek, Cartographic Society of Slovak Republic, Bratislava, Slovak Republic
U.K.	Mr. Sam Sowton, Chair AGI Standards Committee, Twyford, Winchester, Hants, United Kingdom
Yugoslavia	Dr. Ivan Nestorev, Univerziteta U Beogradu, Belgrade

ICA Standards Commission international observers

Comité Européen de Normalisation, CEN/TC278/WG7
Dr. Wolfgang Zechnall, Robert Bosch GMBH, Hildesheim, Germany

Comité Européen de Normalisation, CEN/TC 287
Comité Européen des Responsables de la Cartographie Officielle
Mr. François Salgé, GNIG, Paris, France

Digital Geographic Information Working Group
Mr. Stuart Haynes, Tolworth, Surrey, United Kingdom
Mr. Herman Dohmann, NIMA/SOSEI, Reston, Virginia, U.S.A.

Dublin Core Metadata Group
Dr. Stuart Weibel, Dublin Core Metadata Initiative, OCLC Office of Research, Dublin, Ohio, U.S.A.

International Federation of Library Associations and Institutions
Mr. Jan Smits, Nederlandse Vereniging voor Kartografie, p/a Koninklijke Bibliotheek, The Hague, The Netherlands

International Geographical Union
Prof. Dr. Wolfgang Kainz, University of Vienna, Austria

International Hydrographic Organisation
Mr. Michel Huet, International Hydrographic Bureau, Monaco

International Society of Photogrammetry and Remote Sensing
Prof. Dr. Manfred Ehlers, ISPA, University of Vechta, Germany

International Standards Organisation, ISO/TC211/WG3/T15/Metadata
Mr. David M. Danko, Environmental Systems Research Institute, Reston, Virginia, U.S.A.

United Nations Headquarters
Mr. Miklos Pinther, Chief Cartographer, Ms. Alice L.K. Chow, United Nations, New York, NY, U.S.A.

World Bank Headquarters
Mr. Ulrich Boegli, Chief Cartography Section, The World Bank, Washington, D.C., U.S.A.

Ex offico members and observers

Mr. Bengt Rystedt, President, ICA, National Land Survey, Gävle, Sweden
Prof. Dr. Ferjan Ormeling, Secretary General, ICA, Utrecht University, The Netherlands

Annex C: Book Contributors

Authors

Henri J.G.L. Aalders (Netherlands)	Associate Editor, Parts 1, 2 and 4, Appendices
Paula Ahonen-Raino (Finland)	Part 4
Kazuhiko Akeno (Japan)	Parts 2 and 4
Julie Binder Maitra (U.S.A.)	Part 2
Luc Bonnefoy (France)	Part 4
Martha Ivette Chaparro-Dominguez (Co)	Part 2
Heung-muk Cho (Korea)	Part 4
Antony Cooper (South Africa)	Parts 2 and 4
Aaron Crane (U.S.A.)	Associate Editor, Parts 3 and 4
David M. Danko (U.S.A.)	Parts 2 and 4
Du Dao-sheng (China)	Part 2
Tatiana Delgado-Fernandez (Cuba)	Part 2
Bernd Eckhardt (Germany)	Part 4
Kian Fadaie (Canada)	Parts 2 and 4
Anne Frondorf (U.S.A.)	Part 4
Elizabeth J.O. Gavin (South Africa)	Part 2
Richard Hogan (U.S.A.)	Part 4
Bronislava Horacova	Part 4
Valerie Hume (Canada)	Parts 2 and 4
Jiang Jing-tong (China)	Part 4
Istvan Kadar (Hungary)	Part 4
Peter Kjeld (Denmark)	Part 4
Liu Ruo-mei (China)	Part 4
Claude Luzet (France)	Part 4
Craig Macauley (Australia)	Parts 2 and 4
Alexander I. Martynenko (Russia)	Parts 2 and 4
Sebastian Mas (Spain)	Part 4
Tae-Jung Min (Korea)	Part 4
Harold Moellering (U.S.A.)	Editor-in-Chief, Foreword, Parts 2, 3, 4 and 5
Anders Nielsen (Denmark)	Part 4
Shoji Okuyama (Japan)	Part 4
Olaf Østensen (Norway)	Part 2
Richard Pearsall (U.S.A.)	Part 4

Ammatzia Peled (Israel)	Part 4
Tamas Prajczer (Hungary)	Part 4
Petr Rapant (Czech Republic)	Part 4
Dora Inez Rey-Martinez (PCIDEA)	Part 2
Ruhl, Charles (U.S.A.)	Part 4, editorial assistant
Jan Ruzicka (Czech Republic)	Part 4
François Salgé (France)	Parts 2 and 4
Nic Scheepers (South Africa)	Part 4
Susan Stitt (U.S.A.)	Part 4
Shinji Takazawa (Japan)	Part 4
Marcus Wandinger (France)	Part 4
Stuart L. Weibel (U.S.A.)	Part 4

National and Other Part 2 Book Contributors

Henri J.G.L. Aalders	Delft University of Technology, The Netherlands
Maria Loreto Advis Neira	Military Geographic Institute, Chile
Tanya Aguayo	National Co-ordinator of Metadata, Guatemala
Paula Ahonen-Raino	Helsinki University of Technology, Finland
Kazuhiko Akeno	Geographical Survey Institute of Japan
Lilia Patricia Arias	Geographic Institute of Colombia
Urzúa Rafael Arrioja	INEGI, Mexico
Julie Binder Maitra	FGDC Standards Co-ordinator, U.S.A.
Dirk Boon	NSIF Directorate
Martha Ivette Chaparro-Dominguez	Colombian Standardisation Committee, Colombia
Heung-muk Cho	National Geography Institute, Korea
Antony Cooper	CSIR, Pretoria, South Africa
Vijay Datadin	Iwokrama International Centre for Rain Forest, Conservation and Development, Guyana
Du Dao-sheng	Wuhan University, China
Francisca N Dayrit	GIS Consultant, The Philippines
Tatiana Delgado-Fernandez	GEOCUBA, Cuba
Ulrich Düren	Landvermessungsamt Nordrhein-Westfalen, Germany
Kian Fadaie	Canada Center for Remote Sensing Natural Resources
Marlys S.García	Institution Relationships and International Co-operation, Venezuela
Jacob Gyamfi-Aidoo	formerly EIS Africa
Valerie Hume	Indian and Northern Affairs Canada Northern Program
Angela Ionita	ICI, Bucharest, Romania
Hasan Jamil	Geodetic Division, Malaysia
Jiang Jing-tong	National Geomatics Centre of China

Glenn Johnstone	National Mapping Division, Geoscience, Australia
Thorbjörg Kjartansdóttir	LÍSA, Reykjavik, Iceland
Milan Konecny	Masaryk University, Brno, Czech Republic
Kate Lance	USGS/EROS Data Center, U.S.A.
Miljenko Lapaine	University of Zagreb, Zagreb, Croatia
Craig Macauley	DEH, Government of South Australia
Pedro Martínez Fernández	GEOCUBA, Cuba
Sebastián Más Mayoral	Centro National de Información Geografica, Spain
Alvaro Medina Aravena	Ministry of National Property, Chile
Roberto López Meyer	National Geographic Institute, El Salvador
Tae-jung Min	National Geographic Institute, Korea
Carlos López-Vázquez	National Clearinghouse of Geographic Data, Uruguay
Alexander I. Martynenko	Moscow State University of Geodesy and Cartography, Russian Federation
Pedro Martínez Fernández	GEOCUBA, Cuba
Alvaro Medina Aravena	Ministry of National Property, Chile
Ricardo César Millet	IGM Director, Argentina
Harold Moellering	Ohio State University, U.S.A.
Ilaitia Navunisaravi	Department of Lands and Surveys, Fiji
Douglas Nebert	Co-ordinator for the US FGDC, U.S.A.
Anders Nielsen	National Survey and Cadastre, Denmark
Orlando Nino-Fluck	UN ECA
C. Douglas O'Brien	Idon Technologies, Ottawa, Canada
José Luis Ornelas de Anda	INEGI, Mexico
Olaf Østensen	Norwegian Mapping Authority, Norway
Wojciech Pachelski	Space Research Centre, Warsaw, Poland
Linda San Diego Papa	National Mapping and Resource Information Authority, The Philippines
Ammatzia Peled	Univeristy of Haifa, Israel
Amashani Reddy	NSIF Directorate
Dora Inez Rey-Martinez	PCIDEA, Colombia
Bengt Rystedt	National Land Survey, Gävle, Sweden
François Salgé	CNIG, France
Bryan Teahan	Consultant, Terralink International Ltd, New Zealand
Mack Thompson	Office Surveyor-General Land Information, New Zealand
Chanin Tinnachote	Engineering Chulalongkorn University, Thailand
Ricardo Antonio Torchetti	IGM Technical Area Head, Argentina
Rafael Arrioja Urzúa	INEGI, Mexico
Cameron Wilson	Natural Resources Canada

Subject Index

Instruction for Index use

This index consists of a set of main terms and sub-terms. In the sub-terms, the main term is to be read by the "–" sign, that may appear in front or behind the sub-term. On top the main term may appear in the middle of the sub-term, where the term is repeated in text.

Example:

In Index		*Read as*	
Africa	24, 123, 431	Africa	24, 123, 431
– Environmental Support Project, ESP	125, 128	Africa Environmental Support Project, ESP	125, 128
East –	128	East Africa	128

A

Added Value Information Dissemination, AVID 19, 40
AFNOR 50
Africa 24, 123ff, 431ff
– Environmental Support Project, ESP 125, 128
East – 128
North – 131
South – 24, 123, 125, 132, 133, 159, 431
Southern – 123, 126, 132, 133
West – 24, 126, 127
Environmental Information System Programme for Sub-Saharan Africa, EIS 123, 124, 126
African
– Information Society Initiative, AISI 124
– Organisation for Cartography and Remote Sensing, AOCRS 126
Co-ordinating Committee on African Statistical Development, CASD 124
Southern African Development Community, SADC 24, 126
AGRHYMET Centre 126
Algeria 123, 131
National Centre for Spatial Techniques 131
National Council for Geographic Information, CNIG 131
National Institute for Cartography and Remote Sensing 131
National Centre for Spatial Techniques 131
AM/FM European Division, Automated Mapping and Facility Management 31
America
Central – 24, 103ff, 120, 124
North – 21, 63ff, 66, 67, 76
South – 24, 63, 103ff, 104
Permanent Committee on SDI for the Americas, PC IDEA 24, 63, 103, 104, 107, 109, 110, 113, 117, 119
American
– Executive Order, EO 18, 64, 69, 577, 585
– National Standard, ANS 21
– National Standards Institute, ANSI 79
Inter-American Geo-spatial Data Network, IGDN 18, 105
National Imagery and Mapping Agency, NIMA 64, 114
Pan American Institute of Geography and History, PAIGH 21, 24, 63, 105, 110
U.S. Technical Advisory Group, TAG 67
ANS, *see* American National Standard
ANSI, *see* American National Standards Institute
– Z39.50-1995, ISO 10163-1995 53, 76, 105, 116
ANZLIC, *see* Australian and New Zealand's Land Information Council 201ff
– metadata core elements 23, 91
– metadata guidelines 23, 99, 201, 204

– metadata working group 91, 201, 203
Archive
Canadian Geomatics Interchange Standard - Spatial Archive and Interchange Format, CGIS-SAIF 66
Dublin Core Metadata Initiative 4, 258, 493
Argentina 106, 110
– Cadastral and Territorial Information Direction 106
Asia 83ff
– -Pacific 22, 83ff
– metadata initiatives, PC GIAP 4, 18, 22, 23, 83, 84, 85, 95, 99, 100, 104
– region 23
– Spatial Data Infrastructure, APSDI 18, 22, 23, 99
Assessment characteristic 169ff, 179ff
Assessment of standards Part 4, 197ff
Association
– for Geographic Information, AGI (UK) 152
– of EU National Geological Surveys 37, 602
– of European National Mapping Agencies 559, 561
– of Precise Survey and Applied technology, APA (Japan) 88
Canadian Standards Association, CSA 65
Geomatics Industry Association of Canada, GIAC 75
Czech Association for Geographic Information, CAGI 47, 257, 258, 259, 260, 261, 262
French Association for Geographic Information, AFIGÉO 50
International Association for Geodesy, IAG 125
International City/County Management Association, ICMA 71
National Association of Counties NACo (U.S.A.) 71, 72
Attribute 173
Australia 85, 91, 92, 96, 201ff
– New Zealand's Land Information Council, ANZLIC 15, 23, 25, 84, 91, 96, 99, 111, 117, 128, 157, 201ff
Geo-science Australia – National Mapping Division (formerly AUSLIG) 85, 435
Australian
– Land Information Council 23, 157
– Office of Spatial Data Management 85
– Spatial Data Directory, ASDD 91, 96, 202, 206
– Spatial Data Infrastructure, ASDI 84, 91, 97
– Surveying and Land Information Group, AUSLIG 85, 96, 435
ASDI Distribution Network, ASDI DN 92, 97
Public Sector Mapping Agencies, PSMA Australia Ltd. 84, 85

B

Benin 127
ASECNA 127
CENATEL 127
Business Support Systems, BSS 142, 143
Botswana 132, 134, 135
Brazil 106, 110
– Geological Survey 107

C

CAGI (Czech Republic) 47, 257ff
Canada 22, 63ff, 65, 72, 75, 78, 79, 81, 219ff
Atlantic Coastal Zone Information Steering Committee, ACZISC 79
Environment Canada 74
GeoConnections 18, 22, 63, 64, 75, 78, 79, 81
Geomatics Industry Association of Canada, GIAC 75
Northern Information Network, NIN 79
Canadian
– Center for Remote Sensing, CCRS 74
– Council on Geomatics, CCOG 67
– Committee on Geomatics, CoG 66, 72, 74, 80
– Department of Energy, Mines and Resources, DEMR 65
– Department of National Defence, DND 74, 75
– Department of Natural Resources, CDNR 65, 74, 223
– Federal Department of Public Works and Government Services 65
– General Standards Board, CGSB 65, 72, 73, 74, 75, 79, 80, 219, 220
– General Standards Board - Committee on Geomatics, CGSB-CoG 66, 72, 74, 80
– GEONet, Canadian Geospatial Network 18
– Geospatial Data Infrastructure, CGDI (Canada) 22, 66, 67, 72, 73, 78
– Geospatial Data Infrastructure - Technology Advisory Panel, CGDI-TAP (Canada) 73
– Geomatics Dataset Cataloguing Rules 66
– Geomatics Interchange Standard - Spatial Archive and Interchange Format, CGIS-SAIF 66
– Hydrographic Service, CHS 74
– Land Inventory Database 65
– Metadata Standard 78, 219
– Natural Resources Canada, NRCan 74
– Standards Association, CSA 65
– Standards Council of Canada, SCC 21, 65, 66, 73, 79, 80
– Land Inventory Database 65
– Technology Advisory Panel, TAP 73

Cardinality 8, 16, 162
Caribbean 18, 20, 22, 63, 103ff, 104
Cartography
  African Organisation for Cartography and Remote Sensing, AOCRS 126
  European Committee responsible for Official Cartography, CERCO 32, 35, 36, 42
  Federal Institute of Cartography and Geodesy in Germany, BKG 51
  International Cartographic Association, ICA xvii ff, 5, 123, 152, 169, 170, 172, 173, 174, 175, 197, 198, 416, 645, 646, 651
Catalog, Catalogue 18, 39, 42, 43, 50, 53, 54, 58, 92, 93, 148, 150, 157, 161, 164, 175
  – Services Specification (OGC) 72
  feature –, FACC 173, 186, 568
Cataloging, Cataloguing 85, 111, 157
  data – 145, 173
  feature – 148, 173
  – datasets 18, 95, 150, 152, 155, 220, 238, 376
  – rules 66
  – tool 50
  ISO 19110 Feature cataloguing methodology 543
  MAchine-Readable Cataloging, MARC 76
CEN (European Standardisation Committee) 6, 19, 31, 32, 44, 46, 48, 50, 51, 52, 158, 515ff
  – /TC 287, Technical Committee on Geographic Information 19, 20, 25, 31, 32, 36, 42, 45, 46, 54, 111, 170, 515ff, 601ff, 645
  ENv 12XXX (provisional European Standard, experimental European Standards) 20, 36, 561
  ENv 12657 Metadata 15, 18, 19, 20, 48, 52, 53, 54, 58, 60, 173, 174, 601
Census 24, 71, 104, 106, 108, 113, 136
Central America 24, 103ff, 104, 120
  – GIS project for development, PROCIG 24, 104, 108, 115, 116, 120
Centre, Center
  – for Biological Informatics, NBII (U.S.A.) 619
  – for Environment and Sustainable Development in the Arab Region and Europe, CEDARE 126
  – for road transport and urban development and governmental constructions, CERTU (France) 50
  Central Veterinary Laboratory, CVL (Kenya) 129
  Geographical Information Centre in Honduras, CIGEO 115
  International Centre on Tropical Agriculture, CIAT 104
  National Centre for Spatial Techniques 131
  National Registry Centre (El Salvador) 108
  Royal Centre for Remote Sensing, CRTS (Morocco) 131
Characteristics of metadata 92, 169ff, 179ff
Chile 107, 110
China 85, 86, 87, 92, 93, 97, 107, 159, 172, 237ff
Clearinghouse 22, 24, 91, 95, 97, 99, 106, 110, 112, 128, 134, 627, 629
  National Clearinghouse for Geographic Data, CNGD (Uruguay) 109
  National Geospatial Data Clearinghouse (U.S.A.) 18, 64
  Netherlands Clearinghouse for Geographic Information, NCGI 53, 393
  spatial data CLEARing house, CLEAR (Europe) 40
CNIG
  Centro Nacional de Información Geográfica 451
  Comité National de l'Information Géographique, France, Morocco 50, 58, 131
Colombia 107, 111
  Spatial Data Infrastructure, ICDE 107, 112, 117
Comité
  – européen de Normalisation, *see* CEN
  – européen des Responsables de la Cartographie Officielle, *see* CERCO
Commission
  Ethiopian Science and Technology Commission 128
  European Commission, EC, *see* EC
  Geo-spatial Information co-ordination Commission of China 86
  ICA Spatial Data Standards Commission xvii ff
  International Electrotechnical Commission, IEC 142, 543
  National Technical Commission of GI Standardisation, CSBTS/TC 230 (China) 87
  Polish Standardisation Commission 54
  United Nations Economic Commission for Africa 124
Committee
  – Draft, CD (ISO) 95, 155, 240, 376
  – on Development Information 92, 124
  – on Earth Observation Satellites - International Direction, GEOS-IDN 81
  – on Environmental Data Management, CEDAMA 134
  Canadian Committee on Geomatics 74
  Co-ordinating Committee on African Statistical Development, CASD 124
  InterNational Committee for Information Technology Standards, INCITS 21, 69, 70, 76, 79, 80, 81

Nairobi Informal Settlements Co-ordinating Committee, NISCC (Kenya) 129
Subcommittee on Geo-information, CODI 124, 125
Technical Commission, TC, *see* CEN/TC or ISO/TC
Technical Committee for Consultancy on Geographic Information, CTC (Panama) 109
Compact Disc - Read Only Memory, CD-ROM 42, 134, 135, 415
Conceptual schema language, CSL 32, 145, 151, 154, 523
Co-ordination
– of Information on the Environment, CORINE 33
Co-ordinating Committee on African Statistical Development, CASD 124
Co-ordinate
– Reference System, CRS 149
Costa Rica 112
National Geographic Institute, IGN 107
National Geo-Spatial Data Infrastructure 107, 112
Council
Canadian Council on Geomatics, CCOG 67
Icelandic Council for Standardisation 52
Information Technology Industry Council, ITI (U.S.A.) 69
Fiji Land Information Council, FLIC 87, 95
National Council for Geographic Information, CNIG 41, 50, 51, 131, 451
National States Geographic Information Council, NSGIC (U.S.A.) 71, 72
Netherlands Council for Geographic Information, Ravi 53, 395
Nairobi City Council 129
Spatial Information Council for Australia and New Zealand, ANZLIC, *see* Australian and New Zealand's Land Information Council
Standards Council of Canada, *see* Canada
Ususimaa Regional Council, Finland 43
Croatia, Republic of – 46, 47
Civil Engineering and Environment Protection Administration 46
Geodetic Administration 46, 47
Hydrographic Institute 46, 47
Institute for Standardisation and Measuring 46, 47
Crosstable, critical characteristics 645ff
Crosstable of national and international spatial metadata standards 645ff
Cross-Language Evaluation Forum, CLEF 20
CSDGM
South Africa 431ff
U.S.A. 469ff
U.S.A. – NBII 619ff
Cuba 107, 113
GEOCUBA 107
Hydrography and Geodesy Service 107
National Office of Hydrography and Geodesy 107
Spatial Data Infrastructure, IDERC 113
Czech (– Republic) 47, 257ff
– Association for Geographic Information, CAGI 47, 257, 258, 259
– Forest Management Institute 47
– Geological Survey 47
– Land Survey Office 47
– Military Topographic Institute 47

D

Data
– classification 93, 244
– definition 7, 11, 14, 58, 186, 649
– dictionary standards 84, 146, 152, 160, 162, 173, 539
– homogeneity 12, 13, 186
– lineage 12, 13, 174, 650
– model 5, 16, 37, 59, 60, 169, 174, 175, 189, 650
– origin 13, 15
– source 13, 69
– standard 96
– structure 60, 94, 150, 176, 181, 648
– transfer 18, 21, 37, 39, 43, 51, 92, 94, 100, 175, 198
– transfer standard 76, 136, 145, 158
Canadian Geospatial data Infrastructure, CGDI 22, 66, 67, 72, 73, 78
GeoData Alliance (U.S.A.) 22, 64, 72
metadata model 18, 162, 164, 648
Dataset
Canadian Datasets Cataloguing Rules, *see* Canadian
Canadian Directory Information describing Digital Geo-Referenced Datasets (Canadian Metadata Standard) 15, 18
metadataset 6, 7, 8, 11, 14, 15, 24, 161
subset 11, 148, 161, 174, 184, 645
De facto standards 54, 130, 134, 317
De jure standards 14, 54
Denmark 46, 48, 277ff
National Survey and Cadastre, KMS-DK 48, 277
Department
– of Lands, DOL (Thailand) 90
– of Resource Survey and Remote Sensing, DRSRS (Kenya) 129
– of Survey and Mapping Malaysia, DSMM 89
– of Surveys and Mapping, DSM (Botswana) 134, 135

Canadian Department of Energy Mines and Resources, *see* Canadian
Canadian Department of National Defence, DND, *see* Canadian
Canadian Department of Natural Resources, *see* Canadian
Canadian Federal Department of Public Works and Government Services, *see* Canadian
Forestry Department of Kenya 129
Digital
– data 35, 51, 106, 112, 128, 129, 130, 131, 132
– elevation model, DEM 47, 48, 173, 252
– Geographic Information Exchange Standard, DIGEST 15, 31, 47, 66, 75, 117, 152, 158
– Geographic Information Working Group, DGIWG 31, 74, 75, 117, 152
– metadata 81
Directorate General of the EU 32, 33
Directory
– metadata 11, 12, 559
– of Geographic Institutes from South America, Spain and Portugal, DIGSA 24, 106
Document Type Definition, DTD 91, 206, 207, 263
Draft International Standard, DIS 32, 155, 541
Dublin
– Core Metadata Initiative, DCMI 4, 26, 493ff, 645
– Core Metadataset, DCM 15, 493ff, 645

E

Earth Data Information Systems, EDIS 125
East Africa, *see* Africa
EC, European Commission 20, 33, 34, 40, 42, 44, 45, 46
El Salvador 108, 114
– National Geographic Institute 108
– National Registry Centre, NCR 108
EN, *see* European Norm
ENv, *see* European Norm
Environmental
– Agency 52
– Civil Engineering and Environment Protection Administration (Croatia) 46
– Information System Programme for Sub-Saharan Africa, EIS, *see* Africa
– Protection Agency (Ghana, Ireland) 52, 127
– Protection Authority (Ethiopia) 128
– Support Project, ESP (Africa) 128, 135
Environment Canada 74
Finnish Environment Institute 48, 49, 299
Government Environmental Resources Management Project, GERMP (Ghana) 127
National Environment Management Authority, NEMA (Uganda) 130
National Environment Secretariat, NES (Kenya) 129
National Environmental Information System, SINIA (Honduras) 115
Operational Strategy of Global Environmental Facility/United Nations Development Program, GEF/UNDP 113
State of Environment, SoE 126
State of Environment Reporting Programme, SOER (Namibia) 125
Ethiopia 128
– Environmental Protection Authority 128
– Institute for Geological Surveys 128
– Mapping Authority 128
– Science and Technology Commission 128
EuroGeographics 36, 559ff
EuroGeoSurveys 39, 601ff
Europe 31ff
fifth Framework Programmes for R&D in the EU, FP-5 33, 34
fourth Framework Programmes for R&D in the EU, FP-4 33
sixth Framework Programmes for R&D in the EU, FP-6 20, 34, 35, 44
Geo-Scientific Electronic Information Exchange System, GEIXS 19, 37, 39, 40, 50, 52, 601, 645
European
– Commission, EC 20, 34, 44, 45, 46, 577
– Committee on Standardisation, CEN, *see* Comité européen de Normalisation
– Economic Interest Group MEGRIN 18, 32, 33, 35, 36, 37, 38, 41, 47, 50, 52, 280, 559
– experimental ENv 15, 18, 19, 20, 25, 31, 48, 50, 52, 53, 58
– Joint Research Centre, JRC 44, 577ff
– Norm, EN 12, 20, 32
– organisation for experimental photogrammetric studies, OEEPE 59
– provisional EN, *see* ENv
– Reference System, EUREF 61
– Spatial Data Infrastructure, ESDI 33, 45
– Spatial Metadata Infrastructure, ESMI 18, 20, 41, 52
– Territorial Management Information Infrastructure, ETeMII 34, 35, 60
– Umbrella Organisation for Geographic Information, EUROGI 32, 33, 43, 46, 104, 518
– Union, EU 25, 33, 34, 45, 46, 577, 645
Association of European National Mapping Agencies, *see* Association

experimental/provisional European Norm, *see* ENv
Executive Order, EO 18, 64, 69, 577, 578, 585
eXtensible Markup Language, XML 4, 37, 39, 48, 54, 72, 151, 206, 280, 281, 284, 359, 479, 500, 501, 581

F

Faeroe Islands 48
Feature
- attribute 161, 173
- Attribute Coding Catalogue, FACC 146, 152, 173, 186
- tables 152
- type 240, 250, 545
geographic – 14, 153, 157, 249
general feature model 148
simple feature access 20, 146, 152, 154
web feature service 72
z39.50 features 76
Federal
- Geographic Data Committee, FGDC 15, 21, 24, 25, 67, 71, 78, 81, 97, 99, 105, 115, 119, 127, 132, 158, 170, 645
- Institute of Cartography and Geodesy in Germany, BKG, *see* Cartography
Canadian Federal department of Public Works and Government Services, *see* Canadian
FGDC, *see* Federal Geographic Data Committee
- Content Standard for Digital Geospatial Metadata, US FGDC STD-001-1998, Version 2.0 22, 64, 69, 76, 77, 105, 113, 114, 116, 118, 469
- Cooperative Agreements Program, CAP 71
Fiji 87, 95
- Land Information Council, FLIC 87, 95
- Land Information System, FLIS 87, 95, 99
File Transfer Protocol, FTP 119, 580
Finland 48, 279ff
Finnish Environment Institute 48, 299
Ususimaa Regional Council 43
Framework 33, 67, 69, 83, 84, 98, 113, 125, 131, 133, 142, 143, 147, 148, 150, 155, 164, 179
- for transfer 17
geodetic – 85
geo-information – 22
infrastructure – 20
interoperability – 89
legal – 128
National Data – 118
National Spatial Information – 123, 125, 132, 432
France 49
French
- Association for Geographic Information, AFIGÉO 50
- Digital Cadastral Plan, PCI 49
- Standardisation Institute, AFNOR 50, 517
- Military Hydrographical service, SHOM 49, 74
- large-scale dataset, RGE 49, 50
- National Council for Geographic Information, CNIG 41, 50, 51, 131
- National Geographic Institute, IGN 49, 50
- National Statistical Institute, INSEE 49
- National Tax Office, DGI 49
- Plan for the Information Society, PAGSI 50
Food and Agriculture Organisation, FAO 132
Fundamentals of metadata 5, 169ff

G

GEIXS 601ff
GeoConnections (Canada) 18, 22, 63, 64, 75, 78, 79, 81, 625
GeoData Alliance (U.S.A.) 22, 64, 72
Geodesy 3, 54, 55, 56, 59, 74, 84, 88, 108, 109, 119, 151, 375, 414, 419
Federal Institute of Cartography and Geodesy in Germany, BKG 51
National Office of Hydrography and Geodesy (Cuba) 107
Geographic Information
- Centre, CIGEO (Honduras) 115
- for the Commission, GISCO 33
Electronic Trade for Geographic Information, GISEDI (Europe) 20, 42
Geographic Information Council 64
Inter-Agency Task Force on Geographic Information, IATFGI (Phillipines) 89, 96
National Geographic Information Centre (Norway) 54
National Geographic Information Infrastructure 53, 83, 89, 136
National Geographic Information System (Korea) 88, 376
National Geographic Information Infrastructure Framework Plan (Phillipines) 89
National States Geographic Information Council, NSGIC (U.S.A.) 71, 72
Japanese National Standards for Geographic Information, JNSGI 95
Geographic
- Data Description Directory, GDDD/LaClef 19, 35, 36, 37, 52
- Information Centre in Honduras 108, 115
ISO 19115 Geographic Information – Metadata 18, 54, 146

Geographical Survey Institute, GSI (Japan) 18, 52, 88, 95, 97, 355
Japanese National Standards for Geographic Information, JNSGI 95
Military Institute, IGM (Argentina) 106, 110
Geography Mark-up Language, GML 54
Geomatics
– Data Sets Cataloguing Rules (Canada) 66
Canadian Council on Geomatics, CCOG 67
Canadian General Standards Board - Committee on Geomatics, CGSB-CoG 66, 74, 80
Canadian Geomatics Interchange Standard - Spatial Archive and Interchange Format, CGIS-SAIF, *see* Canadian
Geomatics Industry Association of Canada 66, 75
Geo-Informatics and Space Technology Development Agency, GISTDA (Thailand) 90
ISO Technical Committee 211 for Geographic Information/Geomatics, ISO/TC 211 141ff, 535ff
Geological
– Czech Geological Survey 47
– Survey Brazil, CPRM 107, 110
– Survey Ireland, GSI 18, 52, 87, 88, 95, 97, 355
– Geo-Scientific Electronic Information Exchange System, GEIXS 19, 37, 39, 40, 50, 52, 601ff, 645
GEOMATICA 113
GEONet, *see* Canada
Geo-referenced
Geo-spatial
– community 3, 378, 434
– data 5, 18, 20, 21, 64, 67, 69, 71, 72, 73, 75, 78, 103, 105, 110, 126, 150, 157, 160
– data infrastructure 22, 67, 73, 78, 103, 107, 112, 126
– datasets 79, 119
– information 21, 66, 86, 99, 103, 117, 126, 154
– Information co-ordination Commission 86
– Information One-Stop Project, *see* One-stop
– metadata 15, 23, 76, 76, 78, 99, 108, 117, 158
– One-Stop Board of Directors, *see* One-stop
– standards 23, 68, 70, 73
Canadian Geospatial Data Infrastructure, CGDI 22, 66, 72, 78
FGDC Content Standard for Digital Geospatial Metadata, *see* Federal
Inter-American Geo-spatial Data Network, IGDN 18, 105
I-Team Geospatial Information Initiative, I-Team Initiative (U.S.A.), *see* I-Team
National Framework for Geo-spatial Information Management, NAFGIM 127
National Geospatial Data Clearinghouse 18, 64, 78
Germany 51
Federal Institute of Cartography and Geodesy, BKG, *see* Cartography
Ghana 127
– Environmental Resources Management Project, GERMP 127
– GIS, Geographic Information System 127
– National Framework for Geo-spatial Information Management, NAFGIM, *see* Geo-spatial
Global
– Global Geographic information Mapping, GCM 55, 59
– Information Locator Service, GILS 76, 318
– interoperability 156, 162
– Map 124, 128
– Positioning System, GPS 32, 59
– Resource Information Database, GRID 126, 133
– Spatial Data Infrastructure, GSDI 33, 35, 63, 110, 119, 125
– standard 4
GML, *see* Geography Markup Language 54
GOST 53, 411ff
Greenland 48
Grid, Gridded 174, 175, 189
GSDI, *see* Global
Guatemala 108, 114, 116, 119
– National Geographic Institute "Ingeniero Alfredo Obiols Gómez", IGN 108, 114
Guide to Standardisation for Urban Geographic Information System, UGIS (China) 94
Guyana 108, 115, 119
– International Centre for Rain Forest Conservation and Development, Iwokrama 108, 115

H

Honduras 108, 115, 119
– Geographical Information Centre, CIGEO 115
– National Environmental Information System, SINIA 115
Hungary
KIKERES (Hungary) 317ff
Hydrography
Added Value Information Dissemination for hydrographic Datasets, AVID 19, 40
Canadian Hydrographic Service, CHS 74
French Military Hydrographical service, SHOM 49, 74
International Hydrographic Organisation, IHO 15, 47, 74, 117, 172
International Maritime Organisation, IMO 47
Hyper Text Mark-up Language, HTML 39, 48, 64, 69, 84, 105, 107, 112, 133

I

Iceland 51
- Council for Standardisation 51
- LÍSA, GIS co-ordination body 51
Imagery 19, 37, 49, 52, 69, 74, 85, 116, 128, 146, 151, 152, 154, 155, 164
National Imagery and Mapping Agency, NIMA (U.S.A.) 64, 114
IMF 2000 (Israel) 337ff
Inertial Navigation System, INS 59
Information
- and Communication Technology, ICT 48, 55, 124, 319
- Infrastructure, KII (Korea) 88
- Society 50, 55, 110, 113, 124
- Technology, IT 8, 42, 45, 46, 52, 88, 317, 431
- Technology Industry Council, ITI (U.S.A.) 6
InterNational Committee for Information Technology Standards, INCITS 21, 69, 70, 76, 79, 80, 81
I-Team Geospatial Information Initiative, I-Team Initiative (U.S.A.), *see* I-Team Initiative
Land Information Council, FLIC (Fiji), *see* Fiji
Land Information System, FLIS (Fiji), *see* Fiji
National Environmental Information System, SINIA (Honduras), *see* Honduras
National States Geographic Information Council, NSGIC 71, 72
Plan for the Information Society, PAGSI (France) 50
Infrastructure
Asia-Pacific Spatial Data Infrastructure, APSDI 18, 22, 23, 99
Australian Spatial Data Infrastructure, ASDI 84, 85, 91, 92, 97, 201
Canadian Geo-spatial Data Infrastructure, CGDI 22, 66, 67, 72, 73, 78
Colombian Spatial Data Infrastructure, ICDE 107, 112, 117
European Spatial Data Infrastructure, ESDI 33, 45
European Spatial Metadata Infrastructure, ESMI 18, 20, 37, 41, 52
European Territorial Management Information Infrastructure, ETeMII 34, 35, 60
Geo-spatial Data Infrastructure 22, 67, 73, 78, 103, 107, 112, 126
Government Data Infrastructure, GDI (Thailand) 99
Global Spatial Data Infrastructure, GSDI 33, 35, 63, 110, 119, 125
Korean Information Infrastructure, KII 8, 42, 45, 46, 52, 88, 317, 431
National Geographic Information Infrastructure, NGII 46, 96
National Geographic Information Infrastructure Framework Plan (Philippines) 89
National Infrastructure for Land Information System, NaLIS, *see* Malaysia
National Spatial Data Infrastructure, NSDI 21, 64, 65, 67, 69, 71, 72, 78, 87, 105, 110, 112, 116
Permanent Committee on GIS Infrastructure for Asia and the Pacific, PC GIAP 18, 22, 23, 83, 84, 99, 100, 104
Permanent Committee on Spatial Data Infrastructure for the Americas, PC IDEA 24, 63, 103, 104, 107, 109, 110, 113, 117, 119
Spatial Data Infrastructure, SDI 17, 23, 24, 31, 33, 35, 46, 63, 64, 84, 103, 124, 125, 132
Spatial Data Infrastructure of the Cuban Republic, IDERC 113
Initiative
I-Team –, *see* I-Team Initiative
INSPIRE 20, 33, 44, 55, 61
Institutional framework 113, 131
Interchange 3, 39, 43, 56, 66, 74, 147, 150, 451, 523
Spatial Archive and Interchange Format, SAIF, *see* Canadian
Intergovernmental Authority on Development, IGAD 128
International
- Association for Geodesy, IAG 125
- Cartographic Association, ICA xvii, 123
- Centre on Tropical Agriculture, CIAT 104, 105, 112, 115, 116
- Centre for Rain Forest Conservation and Development, Iwokrama 108, 115
- City/Council Management Association, ICMA 71
- Committee for Information Technology Standards, INCITS (U.S.A.) 21, 69, 70, 76, 79, 80, 81
- Committee on Earth Observation Satellites - International Direction, GEOS-IDN 81
- Electrotechnical Commission, IEC 142, 543
- Hydrographic Organisation, IHO, *see* Hydrographic
- Maritime Organisation, IMO, *see* Hydrographic
- Organisation for Standardisation, ISO 141
- Society for Photogrammetry and Remote Sensing, ISPRS 152
- Standard, IS 155
- Telecommunications Union, ITU 142
Draft International Standard, DIS 155
Internationalisation 8

Interoperability 4, 10, 20, 34, 39, 42, 45, 72, 75, 117, 143, 153, 156, 163, 498
Ireland, Republic of Ireland 52
Environmental Protection Agency, EPA 52
Geological Survey Ireland, GSI 18, 52, 87, 95, 97, 355
Ordnance Survey Ireland, OSI 52
IS, *see* ISO IS
ISO, International Organisation for Standardisation
– 19115 Metadata 2003 141ff, 150, 155, 535ff
– 19115 Geographic information – Metadata, *see* ISO 19115 Metadata 2003
– 10163-1995, *see also* ANSI Z39.50-1995
– /TC 211, Technical Committee on Geographic Information/Geomatics 141ff
– /TR, ISO Technical Report 9, 20, 146, 172
– UML world metadata diagram 541
– co-operation/liaisons 152, 160
– DIS, Draft International Standard 155
– future work 155
– metadata standard IS 19115, *see* ISO 19115 Metadata
– work programme 145, 146
Israel 136, 337ff
– Metadata Standard, IMF2000 136, 337, 338
– transfer standard, IEF'91 136, 342
I-Team Geospatial Information Initiative, I-Team Initiative (U.S.A.) 21, 64, 72
Ivory Coast 127

J

Japan 87, 95, 97, 355ff
– Association of Precise Survey and Applied Technology, APA 88
– Geographical Survey Institute, GSI 18, 52, 87, 95, 97, 355
– Metadata Profile, JMP 95, 356ff
– National Standards for Geographic Information, JNSGI 95
JHS 137, JHS 137A (Finland) 297ff
JMP 2.0 (Japan) 355ff

K

Kenya 129
– Central Veterinary Laboratory, CVL 129
– Department of Natural Resources 129
– Department of Resource Survey and Remote Sensing, DRSRS 129
– Forestry Department 129
– Marine Fisheries Research Institute, KMFRI 129
– National Environment Secretariat, NES 129
– Wildlife Service, KWS 129
National Museums of Kenya, NMK 129
Nairobi Informal Settlements Co-ordinating Committee, NISCC 129
Surveys of Kenya 129
KIKERES (Hungary) 317ff
KMS-DK (Denmark) 48, 277ff
Korea 88, 95, 98, 375ff
– Information Infrastructure, KII 8, 42, 45, 46, 52, 88, 317, 431
– Agency for Technology and Standards, KATS 88
NGIS masterplan 98
NGIS metadata standard 95, 96, 376, 378
National Computerisation Agency, NCA 88
National Geography Institute, NGI 95, 98

L

LaClef 33, 38, 39, 52, 559ff
LANDSAT 107, 127
Lesotho 134
– Department of Environment 134
– Department of Lands and Survey 134
– Department of Statistics 134
– Highlands Development Authority 134
– National Environment Secretariat 134
– SADC Environment and Land Management Sector 134
– SADC Water Sector 134
Localisation 8, 19, 65, 112
Location-based services, LBS 25, 95, 146, 154, 155

M

MAchine-Readable Cataloging, MARC 6, 19, 31, 32, 44, 46, 48, 50, 51, 52, 158
Malaysia 88, 96, 98, 159
– Department of Survey and Mapping Malaysia, DSMM 89
– National Infrastructure for Land Information System, NaLIS 88, 96, 98
Map
– information 5
– production 59, 129
– projection 15, 149, 228, 282
– objects 5
– viewer 78
base (basic) – 52, 53, 54
cadastral – 49
digital – 57, 85
electronic – 55
global – 124, 128
hard-copy – 5
topographic – 90
web – 53, 72, 78, 116, 146, 153
Large Scale Base Map of the Netherlands 52

Mapping agencies
Association of European National Mapping Agencies, EuroGeoGraphics 32, 33, 36, 559ff, 645
Automated Mapping and Facility Management, AM/FM 31
Department of Surveys and Mapping (Botswana) 134
Department of Survey and Mapping Malaysia, DSMM 89
Ethiopian Mapping Authority 128
General Command of Mapping, GCM (Turkey) 59
Geo-science Australia – National Mapping Division (formerly AUSLIG) 85, 435
see also Australia
Intergovernmental Committee on Surveying and Mapping, ICSM (Australia) 84
National Imagery and Mapping Agency, NIMA (U.S.A.), *see* Imagery
National Mapping Agency, NMA 15, 32, 35, 36, 39, 52
Norwegian Mapping Agency 'Statens Kartverk' 54
Public Sector Mapping Agencies, PSMA (Australia) 84, 85
State Bureau of Surveying and Mapping, SBSM (China) 86, 87
Marginalia 5
MEM (Russia) 411ff
Memorandum of Understanding, MOU (North America) 21, 67, 81
MERCATOR 58
Metadata 174
– distribution services 16, 96, 117
– elements 4, 10, 11, 15, 74, 76, 91, 92, 97, 100, 117, 147, 150, 156, 161, 174, 181
– extensibility 7, 76, 182, 648
– information, *see* Australia 5, 36, 51, 81, 111, 137, 169
– inventory 16, 17, 40, 51, 65, 99, 115
– level 92, 93, 94, 97, 99, 111
– model 8, 162, 164, 467, 648
– modularity 7, 8, 10
– of metadata 11
– profile 45, 76, 79, 117, 174, 317, 355, 378, 434, 645
– query 16, 96, 100, 117
– set 4, 6, 7, 8, 11, 14, 15, 24, 60, 150, 156, 161, 202, 220, 298, 536, 645
– standard 4, 21, 25, 48, 66, 67, 78, 92, 95, 96, 115, 136, 137, 156, 157, 171, 176, 179
administrative – 12, 283, 519, 562, 605
conditional – 6, 12, 161, 356, 536
core – 4, 8, 15, 25, 52, 91, 95, 97, 157, 161
dictionary – 11, 14
directory – 11, 12, 559
discovery – 15, 16, 562
geo-spatial – 15, 23, 76, 78, 99, 108, 109, 117, 158
mandatory – 202, 220, 536
optional – 4, 6, 163, 220, 536
Canadian Directory Information describing Digital Geo-Referenced Datasets (Canadian Metadata Standard) 15, 158
minimal metadataset 6, 15
METEOSAT 127
Mexico 23, 64, 66, 81, 105, 106, 115, 119
– National Institute of Statistics, Geography and Informatics, INEGI 105, 108, 115
Middle East 24, 123ff
MIGRA (Spain) 451ff
Morocco 131
– Royal Centre for Remote Sensing, CRTS 131
Multi-purpose European Ground Related Information Network, MEGRIN 18, 31, 33, 35, 36, 37, 38, 41, 48, 50, 52, 280, 282

N

Nairobi Informal Settlements Co-ordinating Committee, NISCC (Kenya) 129
Namibia 135
– National Remote Sensing Centre 135
– Surveyor General 135
National
– Agency for Communications and Information, NACI (Romania) 55
– Association of Counties NACo (U.S.A.), *see* Association
– Centre for Spatial Techniques (Algeria) 131
– Computerisation Agency, NCA (Korea) 88
– Environment Management Authority, NEMA (Uganda), *see* Environment
– Environment Secretariat, NES (Kenya), *see* Environment
– Environmental Information System, SINIA (Honduras), *see* Environment
– Fundamental Geographic Information System, NFGIS (China) 86, 97
– Framework for Geo-spatial Information Management, NAFGIM (Ghana) 127
– Geographic Information Council, NGIC 99
– Geographic Information Centre, NGIC (Norway) 99
– Geographic Information Infrastructure, NGII 46, 96
– Geographic Information Infrastructure Framework Plan (Philippines) 89
– Geographic Information System, NGIS 54, 88, 95, 96, 98, 376
– Geographic Institute, NGI 88, 95, 376

– Geomatics Centre of China, NGCC 86, 87, 99, 237
– Geospatial Data Clearinghouse (U.S.A.) 18, 64, 78
– Imagery and Mapping Agency, NIMA (U.S.A.), *see* Imagery
– Information Strategy, NIS (Romania) 55, 340
– Infrastructure for Land Information System, NaLIS, *see* Malaysia
– Institute for Cartography and Remote Sensing, *see* Algeria
– Inventory for Seychelles 129
– Japanese National Standards for Geographic Information (JNSGI), *see* Japan
– Land Survey, NLS (Finland) 48
– League of Cities, NLC (U.S.A.) 71
– Mapping Agency, NMA, *see* Mapping
– Mapping and Resource Information Authority, NAMRIA (Philippines) 98, 99
– Museums of Kenya, NMK 129
– Oceanic and Atmospheric Administration, NOAA (U.S.A.) 69, 71, 127
– Office of Hydrography and Geodesy, *see* Cuba
– Office of Cadastre, Geodesy and Cartography, NOCGC (Romania) 55
– Registry Centre, NCR, *see* El Salvador
– Resources Canada, NRCan 74
– Secretariat for Science, Technology and Innovation, SENACYT, *see* Panama
– Spatial Data Infrastructure, NSDI, *see* Infrastructure
– Spatial Information Framework, NSIF (South Africa) 122, 132, 133, 134, 138, 432
– States Geographic Information Council, NSGIC (U.S.A.), *see* Information
– Statistical Institute, INSEE (France) 49
– Tax Office, DGI (France) 49
– Transfer Format, NTF (UK) 31
American National Standard (ANS) 76, 81
NATO, *see also* Digital Geographic Information Working Group DGIWG
NBII (U.S.A.) 619ff
NCGI (Netherlands) 393ff
Netherlands 31, 32, 41, 46, 52, 132, 169, 172, 393ff
Alterra 52
– Clearinghouse for Geographic Information, NCGI 53, 131, 393
– Council for Geographic Information, Ravi 53, 395
– Institute for Applied Geo-sciences, NITG 602
– Large Scale Base Map 52
– Normalisation Institute, NEN 32, 53
– Survey Department of the Ministry of Traffic and Water Control 52
– Topographic Service, TDN 52
New Zealand 89, 99, 201ff
– Government Locator Service, NZGLS 99
Land Information New Zealand, LINZ 89
Official Committee for Geospatial Information, OCGI 99
Terralink 89
Nicaragua 108, 116
– Institute for Territorial Studies, INETER 108, 109
Statistics and Census National Institute, INEC 108
Nihon (Japan) 355ff
NIMA, *see* National Imaging and Mapping Agency
NOAA, *see* National Oceanic and Atmospheric Administration
North Africa, *see* Africa
North America, *see also* America 63
– standards infrastructure 68, 70, 73
Norway 54
– Mapping Agency 'Statens Kartverk', *see* Mapping
Geovekst 54
National Geographic Information Centre, NGIS 54
NSDI, *see* National Spatial Data Infrastructure
– Geospatial One Stop Project, *see* Geo-spatial
NSIF, *see* National Spatial Information Framework
NTF, *see* National Transfer Format

O

Object-orientation
Object-based 8
Object-oriented 10
OCLC (Dublin Core) 493ff
OEEPE, *see* European organisation for experimental photogrammetric studies
Office of Management and Budget, OMB (U.S.A.) 65, 69, 72
OGC, *see* OpenGIS® Consortium
OMB, *see* Office
Ontology 7, 8
One-stop 37, 65, 69, 71
GDDD one-stop shop 37
Geospatial Information One-Stop Project 65, 71
Geospatial One-Stop Board of Directors 65, 71
Open Archive
– Forum, OAF 20
– Initiative, OAI 20
– metadata protocol 20
OpenGIS®
– Consortium, OGC 21, 37, 42, 47, 71, 75, 110, 117, 153, 160
– Data Interchange, OGDI 39
– GML, *see also* GML 54

– Web Map Service 72, 78
– Web Feature Service 72
Oracle 51, 58, 603
Ortho
– imagery 69, 426
– photo 49, 111
– photomap 3, 59

P

Pacific 22, 23, 83ff
– ANZLIC, *see* ANZLIC
– PC GIAP, *see* PC GIAP
PAIGH 21, 24, 63, 105
Panama 109, 116
National Secretariat for Science, Technology and Innovation, SENACYT 116
Technical Committee for Consultancy on Geographic Information, CTC 109
Tommy Guardia National Geographic Institute 109, 116
PC
– GIAP, *see* Infrastructure
– IDEA, *see* Infrastructure
Permanent Committee, *see* PC
Philippines 89, 96, 99
Inter-Agency Task Force on Geographic Information, IATFGI 89, 96
National Geographic Information Infrastructure Framework Plan, *see* Infrastructure
National Mapping and Resource Information Authority, NAMRIA, *see* National
Poland 54
Committee of Geodesy of the Polish Academy of Sciences, Section of Geo-informatics 54
Head Office for Geodesy and Cartography 54
Polish Society for Spatial Data 54
Polish Standardisation Committee 54
Portrayal 146, 150, 155, 252
ISO 19117 Portrayal 155, 543
Position
– accuracy 13, 25, 136, 149, 174, 187, 650
– reference system 12
direct-indirect positioning 532, 563
Global Positioning System, GPS, *see* Global
ISO 19116 Positioning services 20, 85, 186
Profile
application – 10, 163, 504, 540
ANZLIC – 91, 92
community – 163
GEIXS – 601ff
GEO – 18, 76, 97, 117
Geo-spatial metadata – 117
GILS – 76
NBII – 619ff
Provisional European Norm, *see* ENv
PSMA, *see* Australian Public Sector Mapping Agencies Ltd.

Q

Quality 174
– conceptual schema 11
– definition 16, 137, 174, 338
– description 189, 649
– information 11, 149, 174, 186, 189
– model 311
– of metadata 174, 452
– parameters 13, 16, 37, 611
– standard 94, 186, 649
CEN ENv 12656 – 31
data – 5, 17, 37, 93, 94, 111, 130, 146, 149, 161, 172, 649
meta – 16, 174
ISO 19113 Quality principles 120, 146, 149, 154, 543
ISO 19114 Quality evaluation procedures 20, 146, 150, 154, 543
Query 32, 43, 47, 61, 117, 152, 153, 164, 175, 182
– forms 112
– metadata 16, 41, 96, 100, 117, 119
– node 118
– protocol, ISO 23950 117
– system 116, 118,119
complex – 78
remote – 164

R

Ravi (Netherlands) 53, 395
Reference model
ENv 12009 Reference Model 31, 523
ISO 19101 Reference Model 20, 145, 146, 154
reference system
geodetic reference system 15, 24, 85, 106, 125
positional reference system 12
Regional Integrated Information Systems, RIIS (Africa) 128, 130
Regional summaries 1
Relation
entity-relation model, ER-model 159
relational database 48
relationship 10, 14, 162, 164, 173
Romania 55
Academy of Sciences 55
National Agency for Communications and Information, NACI 55
County Offices of Cadastre, Geodesy and Cartography, COCGC 55

Forum for Information Society 55
Institute of Geodesy, Photogrammetry, Cartography and Cadastre, IGPCC 55
National Information Strategy, NIS 55
National Office of Cadastre, Geodesy and Cartography, NOCGC 55
Remote Sensing 3, 7, 21, 24, 74, 77, 89, 109, 125, 126, 129, 130, 131, 132, 135, 151, 152
– ground station 107
– satellites 59, 81, 85, 90, 106, 107, 133, 189
African Organisation for Cartography and Remote Sensing, AOCRS 126
Canadian Center for Remote Sensing, CCRS 74
FGDC Standard for Digital Geospatial Metadata: Extensions for Remote Sensing Metadata 19, 77
Regional Remote Sensing unit (Harare, Zimbabwe) 125, 126, 132
Rules for application schema 31, 53, 145
ENv 13376 – 31
ISO 19109 – 53, 146, 148, 151, 155, 543
Russia, Russian Federation 12, 55, 56, 57, 411, 602
– Information Society 55
Standard of Russian Federation GOST R 1.0-92 55
Metadata of electronic maps: composition and content, GOST R 51353-99 57, 58, 411ff

S

SAFARI 2000 133
Saudi Arabia 135
Schema
application – 14, 131, 146, 147, 148, 150, 155, 242, 523
conceptual – 10, 11, 32, 145, 151, 173, 515, 523, 543
database – 173, 182
metadata – 149, 159, 519, 605
spatial – 13, 20, 31, 146, 147, 154, 175, 185
SQL – 152
temporal – 20, 95, 146, 147, 154, 156, 220, 376, 535, 543
SDDF, *see* Spatial Data Discovery Facility
SDI, *see* Spatial Data Infrastructure
GSDI, Global – 33, 35, 63, 64, 110, 125
NSDI, National – 21, 64, 65, 67, 69, 71, 72, 78, 87, 105, 110, 112, 116
SDTS, Spatial Data Transfer Standard 21, 76, 158
Semantic 7, 10, 11, 12, 17, 59, 60, 61, 76, 173, 270, 462, 497, 500
– definition 15
– model(ling) 58, 412
Service
– provider 26, 163
– user 164
ISO 19119 – 146, 151, 155, 164
metadata for – 163
Web Map Service specification (OGC) 72
Seychelles 129, 132
National Inventory for the Seychelles 129
South Africa 24, 133, 431ff
– National Spatial Information Framework, NSIF 123, 125, 138, 432
South America 24, 103ff
Directory of Geographic Institutes from South America, Spain and Portugal, DIGSA, *see* Spain
Southern Africa 132
Environmental Information System Programme for Sub-Saharan Africa, EIS, *see* Africa
SAFARI 2000 133
Southern African Development Community, SADC, *see* African
Spain 58, 451ff
Directory of Geographic Institutes from South America, Spain and Portugal, DIGSA 24, 106
General Directorate for Cadastre 58
Geographic Centre of the Ministry of Defence 58
Grafcan (Canary Islands) 58
MERCATOR project 58
Municipality of Madrid 58
National Centre for Geographic Information 58
National Geographic Institute 58
Spanish organisation for standardisation, AENOR 58, 451
Tracasa (Municipality Foral de Navarra) 58
Spatial
– Archive and Interchange Format, SAIF, *see* Canadian
– attribute 15, 147, 173
– computer analysis 8
– data 4, 5, 16, 33, 41, 46, 54, 58, 83, 89, 103, 105, 110, 123, 189, 198
– data CLEARing house, CLEAR 18, 22, 41, 64, 71, 76, 97, 105, 110, 134, 220, 393, 433, 470, 536
– Data Discovery Facility, SDDF 134, 432
– Data Infrastructure, SDI 17
– Data Infrastructure of the Cuban Republic, IDERC 113
– data organisation 47, 52, 58, 83, 85, 103, 106, 124, 436
– dataset 20, 25, 48, 76, 79, 99, 117, 119, 123, 126, 157, 355, 649
– Data Transfer Standard, SDTS 21, 76, 158
– extent 59, 169, 174

– feature 72, 173
– information 21, 33, 44, 48, 83, 91, 110, 124, 133, 142, 649
– metadata 24, 47, 123, 125, 137, 174, 176, 646
– object 147, 174
– process 6, 154
– quality 16, 174
– reference system 15, 71, 89, 152, 157, 185, 649
– relationship 147
– schema 13, 20, 31, 146, 154, 175, 185
Global Spatial Data Infrastructure, GSDI, *see* SDI
National Spatial Data Infrastructure, NSDI, *see* SDI
SPOT 107, 127
SSSA (EU) 557ff
Stakeholders 21, 34, 69, 71, 74, 75, 129, 135
Standard, *see also* assessment Metadata standard, Characteristics of metadata
– application 6
– crosstable 172, 198, 645
– domain 6, 173
de facto – 54, 317
de jure – 14
ENv 12657 Metadata –, *see* ENv
global – 4, 25
international – (ISO, International Standard), *see* International
metadata – 10, 20, 51, 60, 64, 66, 155
national – 7, 21, 23, 53, 77, 94, 97, 134
regional – 6, 170
state – 5, 55, 57, 411
American National Standard, ANS, *see* Standardisation
Canadian Directory Information describing Digital Geo-Referenced Datasets (Canadian Metadata Standard), *see* Metadata
Canadian Geomatics Interchange Standard, CGIS, *see* Canadian
Canadian General Standards Board, *see* Canadian
Canadian General Standards Board - Committee on Geomatics, *see* Canadian
Canadian Standards Association, *see* Canadian
DIS (ISO Draft International Standard), *see* Draft International Standard
FGDC Content Standard for Digital Geospatial Metadata, *see* Federal
InterNational Committee for Information Technology Standards, INCITS, *see* Committee
Japanese National Standards for Geographic Information, JNSGI, *see* Japan
Korean Agency for Technology and Standards, KATS, *see* Korea
NGIS metadata standard 95, 376
Spatial Data Transfer Standard, SDTS (U.S.A.) 21, 76, 158
Standards Council of Canada, *see* Canada
Standardisation
– committee xvii ff, 31–33, 54, 103, 141ff
– of geographic information 66, 144, 145
– of geo-spatial information 67
– of metadata 24, 75, 92, 137
– of spatial data 145
European Standardisation Committee, *see* CEN
Guide to Standardisation for Urban Geographic Information System, UGIS (China) 94
International Organisation for Standardisation, ISO, *see* International
ISO/TC 211 for Geographic Information/Geomatics, *see* International
semantic – 59, 60
State of Environment, SoE 126
Sudan 130
– National Survey Corporation 130
Sweden 58, 169
Lanmäteriet (NMA) 58
Syntax 10, 16, 17, 117, 150, 493

T

TC211 141ff, 535ff
Terminology 18, 37, 51, 94, 146, 147, 150, 155, 160, 171, 172, 173
CEN/CR 13435 Vocabulary 32
ISO 19104 – terminology 146, 147, 155, 172, 543
Thailand 90, 96, 99
Department of Lands, DOL 90
Geo-Informatics and Space Technology Development Agency, GISTDA (Thailand), *see* Geomatics
Government Data Infrastructure, GDI, *see* Infrastructure
Royal Thai Survey Department, RTSD 90
Thai Industrial Standards Institute/Technical Committee 904: Geographic Information, TISI/TC 904 90
Topology, Topologic(al) 52, 147, 154, 173, 175, 189
Transfer 5, 7, 17, 21, 31, 37, 48, 64, 84, 92, 118, 158, 175, 192
– metafile 3, 169
Turkey 59
General Command of Mapping, GCM 55, 59
General Directorate of Land Registry and Cadastre 59
General Directorate of Highways 59
General Directorate of Forestry 59

U

Uganda 130
Makerere University Institute of Environment and Natural Resources, MUIENR 130
National Biomass Study 130
National Environment Management Authority, NEMA 130
UML, *see* Universal Modelling Language
UN, United Nations
– Economic Commission for Africa, UN EC 124
– Environment Program, UNEP 105
– Development Program, UNDP 113
– Operational Strategy of Global Environmental Facility GEF/UNDP 113
– United Nations High Commissioner for Refugees (UNHCR) 130
Universal Modelling Language, UML 18, 151, 159, 160, 162, 358, 539
– metadata schema 541
Universe of Discourse, UoD 8
Uruguay 109, 116
U.S.A. (USA) 63ff, 469ff
American National Standard, ANS, *see* American National Standard
American National Standards Institute, ANSI, *see* American National Standard Institute
Bureau of Land Management, BLM 69
Bureau of Transportation Statistics 71
Center for Biological Informatics 619ff
Content Standard for Digital Geospatial Metadata, *see* Spatial Data Transfer Standard
Department of Commerce 69, 71
Department of the Interior 5, 69
Department of Transportation 71
Federal Geographic Data Committee, FGDC, *see* Federal Geographic Data Committee
FGDC Geo-Spatial Profile for Metadata, *see* Spatial Data Transfer Standard
Information Technology Industry Council, ITI 69
Intertribal GIS Council 71
I-Team Geospatial Information Initiative, *see* I-Team Initiative
Minerals Management Service, MMS 69
National Association of Counties, NACo, *see* Association
NBII 619ff
National League of Cities, NLC, *see* National
National Oceanic and Atmospheric Administration, NOAA, *see* National Oceanic and Atmospheric Administration
National Geodetic Survey 71
National Spatial Reference System 71, 89
National States Geographic Information Council, NSGIC, *see* Council
Spatial Data Transfer Standard, SDTS, *see* Spatial Data Transfer Standard
University Consortium for Geographic Information Science, UCGIS 152
U.S. Army Corps of Engineers 69
U.S. Geological Survey, USGS 24, 77, 105, 115, 435, 622
U.S. Technical Advisory Group, TAG, *see* American
User requirements 10, 40, 50, 118, 145, 150

V

Value space restriction 10
Venezuela 109, 116
Geographic Institute Simón Bolívar, IGVSB 109, 116
Virtual 6, 25, 125
– catalogue 125
– GIS, V-GIS 6, 25
Vocabulary 10, 15, 503, 505, 509
CEN/CR 13435 – Vocabulary 32

W

Web
– Feature Service specification (OGC), *see* OpenGIS
– Map Service specification (OGC), *see* OpenGIS
West Africa, *see* Africa
World 141ff
– metadata schema 541
– Meteorological Organisation, WMO 74
– Trade Organisation, WTO 142

X

XML, *see* eXtensible Markup Language

Z

Z39.50-1995, ISO 10163-1995, *see* ANSI
Zambia 135